T0310270

SMALL CELL NETWORKS

SMALL CELL NETWORKS

Deployment, Management, and Optimization

Holger Claussen
David López-Pérez
Lester Ho
Rouzbeh Razavi
Stepan Kucera

IEEE Press
Series on
Networks and
Services Management

Thomas Plevyak and
Veli Sahin, *Series Editors*

IEEE PRESS

WILEY

Published by John Wiley & Sons, Inc., Hoboken, New Jersey.
Published simultaneously in Canada.

For general information on our other products and services or for technical support, please contact our Customer Care Department within the United States at (800) 762-2974, outside the United States at (317) 572-3993 or fax (317) 572-4002.

Wiley also publishes its books in a variety of electronic formats. Some content that appears in print may not be available in electronic formats. For more information about Wiley products, visit our web site at www.wiley.com.

Library of Congress Cataloging-in-Publication Data is available.

ISBN: 978-1-118-85434-1

Printed in the United States of America.

10 9 8 7 6 5 4 3 2 1

CONTENTS

ABOUT THE AUTHORS

Dr. Holger Claussen is the leader of the Small Cells Research Department of Nokia Bell Labs located in Ireland and the United States. In this role, he and his team are innovating in all areas related to future evolution, deployment, and operation of small cell networks to enable exponential growth in mobile data traffic. His research in this domain has been commercialized in Nokia's (formerly Alcatel-Lucent's) Small Cell product portfolio and continues to have significant impact. He received the 2014 World Technology Award in the individual category Communications Technology for innovative work of the "greatest likely long-term significance". Prior to this, Holger was the head of the Autonomous Networks and Systems Research Department at Bell Labs Ireland, where he directed research in the area of self-managing networks to enable the first large-scale femtocell deployments from 2009 onward. Holger joined Bell Labs in 2004, where he began his research in the areas of network optimization, cellular architectures, and improving energy efficiency of networks. Holger received his Ph.D. degree in signal processing for digital communications from the University of Edinburgh, UK, in 2004. He is the author of more than 100 publications and 120 filed patent families. He is a fellow of the World Technology Network, a senior member of the IEEE, and a member of the IET.

Dr. David López-Pérez is a member of technical staff in the Small Cells Research Department of Nokia Bell Labs in Ireland. Prior to this, David received B.Sc. and M.Sc. degrees in telecommunication from Miguel Hernandez University, Spain, in 2003 and 2006, respectively, and Ph.D. degree in wireless networking from the University of Bedfordshire, UK, in 2011. David was also an RF engineer with Vodafone, Spain, from 2005 to 2006, and a research associate with King's College London, UK, from 2010 to 2011. David has authored the book *Heterogeneous Cellular Networks: Theory, Simulation and*

Deployment (Cambridge University Press, 2012), as well as more than 100 book chapters, journal articles, and conference papers, all in recognized venues. He also holds more than 35 patent applications. David received the Ph.D. Marie-Curie Fellowship in 2007 and the IEEE ComSoc Best Young Professional Industry Award in 2016. He was also a finalist for the Scientist of the Year prize in the Irish Laboratory Awards in 2013 and 2015. He is an editor of *IEEE Transactions on Wireless Communications* since 2016 and he was awarded Exemplary Reviewer of *IEEE Communications Letters* in 2011. He is or has also been a guest editor of a number of journals, for example, *IEEE Journal on Selected Areas in Communications* and *IEEE Communication Magazine*.

Dr. Lester Ho is a distinguished member of technical staff in the Small Cells Research department of Nokia Bell Labs in Ireland. He obtained his B.Eng. degree in electronic engineering in 1999, and his Ph.D. degree on self-organizing wireless networks in 2003, both from the University of London. He joined Bell Labs, which was then part of Lucent Technologies, in 2003, where he performed research in various topics in wireless communications, particularly in small cells, self-organizing networks, and network optimization techniques, many of which can be found in commercial deployments today. He has over 40 patents granted, as well as more than 40 publications in journals, conference papers, and book chapters. He is a senior member of the IEEE, and the recipient of the Eckermann-TJA Prize in 2008, and a finalist for the Scientist of the Year prize in the Irish Laboratory Awards in 2014.

Dr. Rouzbeh Razavi received a master's degree with distinction in information systems and a Ph.D. in computer science, both from the University of Essex, UK. He is currently a faculty member in the Department of Management and Information Systems at the Kent State University, OH, USA. Prior to this, Dr. Razavi was a director in the Group Decision Sciences at the Commonwealth Bank of Australia in New York. Before joining the Commonwealth Bank of Australia, Rouzbeh served as a senior research data scientist at SAP and as a member of technical staff in the Small Cells Research Department of Nokia Bell Labs in Ireland for several years. He has been supervising a number of Ph.D. students and post-doctoral researchers and has published more than 60 technical papers in peer-reviewed journals and conferences. In addition, he has authored five book chapters and filed more than 30 patent applications. Rouzbeh is a senior member of the IEEE and recipient of best paper awards from two international conferences.

 **Dr. Stepan Kucera** is a member of technical staff in the Small Cells Research Department of Nokia Bell Labs in Ireland. In this role, he serves as a principal investigator and technical lead of multiple projects aiming to create novel disruptive technologies for wireless networking with current focus on multi-connectivity and gigabit wireless technologies. His expertise lies mainly in the area of wireless and IP networking technology, including both proprietary solutions and 3GPP/IEEE/IETF standards. Stepan has filed more than 50 patents and published over 30 book chapters, transactions, and conference papers in peer-reviewed IEEE venues. He is also the (co)-recipient of several professional awards. Between 2008 and 2011, he was a research scientist at the New Generation Wireless Communications Research Center in the Keihanna Research Laboratories, NICT, Japan. He received his Ph.D. degree in informatics from the Graduate School of Informatics at Kyoto University, Kyoto, Japan, in 2008. He is a senior member of the IEEE, and actively serves on technical boards of major IEEE journals and conferences.

FOREWORD

Over just the last 3 decades communication networks have evolved from simply delivering voice over copper wires to a set of fixed locations (homes, businesses, phone booths, etc.) to connecting over 3.7 billion people wirelessly to a seemingly infinite amount of information accessible in any location, anywhere and at any time.

Despite this remarkable transformation, today we are on the brink of another networking revolution wherein not only all people, but also all machines, systems, processes and devices will be wirelessly connected, with optimised access to a near-unlimited pool of computing and processing resources. This will allow people and automata to digitise, communicate with, and control much of the physical world, which, in turn, will allow a manifest simplification of many aspects of work or personal life. In essence, this simplification by automation will augment human perception and capabilities, to increase the ability to perform tasks and, in effect, to save time.

In order to accomplish this remarkable human transformation, over the next 10 years the capacity of the enabling networks will have to increase by a factor of 100×. To enable such growth, fundamental changes in the way we design, deploy, and operate future networks are required. Future networks will have to be built from a dense array of small cells to allow the massive re-use of the limited low frequency spectrum (<6GHz frequency) and the widespread use of limited-range high frequency (>6GHz) spectrum. In turn, these small cells will have to be connected via an ultra-fast, and low latency backhaul network to a massively-distributed 'cloud' of servers, that will host the set of critical 'life-enabling' applications, and provide real time analytics, with the requisite privacy and security protections. This book focuses on describing the creation and deployment of these small cells, and the many associated challenges in terms of network design, deployment and optimisation.

The authors of this book have played a pioneering role in the fundamental understanding of small cells and heterogeneous networks, and invented many of the key technologies that underpin small cells. For example, they have pioneered scalable network architectures, and self-configuration and optimisation techniques that have allowed small cells to be deployed with high efficiency and sustainable economics. As a result, for the first time, cellular small cell networks can now be simply deployed without extensive (and expensive) planning, manual configuration or a specialised field force. Their work has paved the way for the commercial small cell deployments, which now number more than 13 million cells, exceeding already the number of conventional macrocells worldwide by more than a factor of 2. Today, it is recognised that small cells are an essential component of future networks, and will form the foundation of all ultra-high capacity (>1Gbps) wireless networks going forward.

This book provides a comprehensive view of the network evolution towards future networks where the majority of capacity will be provided by small cells. It begins by describing the fundamental challenges of enabling a 100× scaling of cellular capacity, and key requirements such as full network automation to enable cost effective deployment and operation. The authors then discuss critical technical elements such as frequency assignment and access methods, coverage and capacity optimisation, interference management, mobility management, energy efficiency and idle modes, backhaul, deployment planning, the management of ultra-dense and heterogeneous networks, and future ultra-high capacity applications. Finally, they provide the detailed model and methodology for simulating and evaluating small cell networks that is used throughout the book.

The innovations described in this book will have a lasting impact on how future wireless networks are architected, deployed and used, not only in the near term (legacy 3G and current 4G networks), but as an essential part of the next generation 5G networks that will be deployed in 2020 and beyond. Researchers, network designers and operators will find this book provides invaluable insights and an in-depth understanding of the foundation of future networks – networks that will form a new digital fabric that will redefine human existence and transform societies and economics. As such, this book will allow the reader to not only understand the past and present, but also the future – and that is undoubtedly time well spent!

DR. MARCUS WELDON
CTO of Nokia and President of Nokia Bell Labs

ACRONYMS

3G	third-generation
3GPP	Third-Generation Partnership Project
4G	fourth-generation
5G	fifth-generation
AAA	authentication, authorisation and accounting
ABS	almost blank subframe
ACK	acknowledgment
ADC	analog-to-digital converter
AGG	aggressor cell
ANR	automatic neighbor relation
AoA	angle of arrival
API	application programming interface
AR	augmented reality
AWGN	additive white Gaussian noise
BER	bit error rate
BLER	block error rate
BS	base station
CA	closed access
CaCo	carrier component
CAG	carrier aggregation
CAPEX	capital expenditure
CCE	control channel element
CCO	coverage and capacity optimization
CDF	cumulative distribution function
CDMA	code division multiple access
CESM	capacity effective SINR mapping
CFI	control format indicator
CGI	cell global identity
CIF	carrier indicator field
CIO	cell individual offset
CIR	channel impulse response
COC	cell outage compensation
COD	cell outage detection

CoMP	coordinated multi-point
CP	cycle prefix
CPICH	common pilot channel
CQI	channel quality indicator
C-RAN	cloud radio access network
CRE	cell range expansion
C-RNTI	cell radio network temporary identifier
CRS	cell-specific reference symbol
CRT	cell re-selection threshold
CSG	closed subscriber group
CSI	channel state information
CSI-RS	channel state information-reference signals
CSO	cell selection offset
CV	cross-validation
DAC	digital-to-analog converter
DC	direct current
DC-HSPA	dual-carrier high-speed packet access
DCI	downlink control information
DFT	discrete Fourier transform
DFTS	discrete Fourier transform spread
DHCP	dynamic host control protocol
DL	downlink
DRS	discovery reference signal
DRX	discontinuous reception
DSL	digital subscriber line
DSLAM	digital subscriber line access multiplexer
DSP	digital signal processor
DT	decision tree
DTX	discontinuous transmission
EA	evolutionary algorithm
EESM	exponential effective SINR mapping
eICIC	enhanced inter-cell interference coordination
ELF	evolutionary learning of fuzzy rules
EMR	electromagnetic radiation
eNodeB	evolved NodeB
EPB	equal path-loss boundary
EPC	evolved packet core
EPDCCH	enhanced physical downlink control channel
EPS	evolved packet system
ESB	equal downlink received signal strength boundary
E-UTRA	evolved UTRA
FARL	fuzzy-assisted reinforcement learning
FCC	Federal Communications Commission

FDD	frequency division duplexing
FDTD	finite-difference time-domain
FFR	fractional frequency reuse
FFT	fast Fourier transform
FPC	fractional power control
FPGA	field-programmable gate array
FTP	file transfer protocol
FTTx	fiber to the x
GA	genetic algorithm
GBR	guaranteed bitrate
GCI	global cell identity
GNSS	global navigation satellite system
GP	genetic programming
GPON	gigabit passive optical network
GPS	global positioning system
GSM	global system for mobile communication
GTP	GPRS tunnel protocol
GTP-U	GPRS tunnel protocol—user plane
HARQ	hybrid automatic repeat request
HCN	heterogeneous cellular network
HetNet	heterogeneous network
HiFi	high-fidelity
HII	high-interference indicator
HO	handover
HOF	handover failure
HSB	hotspot boundary
HSDPA	high-speed downlink packet access
HSPA	high-speed packet access
HVAC	heating, ventilating, and air conditioning
ICIC	inter-cell interference coordination
ICT	information communication technology
IDFT	inverse discrete Fourier transform
IE	information element
IEEE	Institute of Electrical and Electronics Engineers
IFA	Inverted-F-antennas
IFFT	inverse fast Fourier transform
IIR	infinite impulse response
IOI	interference overload indicator
IP	Internet protocol
ISD	inter-site distance
ITU	International Telecommunication Union
JFI	Jain's fairness index
KPI	key performance indicator

KNN	k-nearest neighbors
LDA	linear discriminant analysis
LOS	line-of-sight
LPC	logical PDCCH candidate
LSAS	large-scale antenna system
LTE	long-term evolution
LTE-A	long-term evolution advanced
LUT	look-up table
MAC	medium access control
MBMS	multicast-broadcast multimedia service
MBSFN	multicast-broadcast single-frequency network
MCB	main circuit board
MCS	modulation and coding scheme
MCSR	multi-carrier soft reuse
MDT	minimization of drive tests
MEA	multi-element antenna
MeNodeB	Master eNodeB
MIESM	mutual information effective SINR mapping
MIMO	multiple-input multiple-output
MLB	mobility load balancing
MME	mobility management entity
MNO	mobile network operator
MPC	multi-path components
MR	measurement report
MRC	maximal ratio combining
MRO	mobility robustness optimization
MRT	maximum ratio transmission
MSE	mobility state estimation
MUE	macrocell user equipment
MU-MIMO	multi-user MIMO
MVNO	mobile virtual network operators
NACK	negative acknowledgment
NAS	nonaccess stratum
NB	Naive Bayes
NGMN	next-generation mobile networks
NLOS	non-line-of-sight
NN	nearest neighbor
OAM	operation, administration, and maintenance
ODA	omni-directional antenna
ODU	optical distribution unit
OFDM	orthogonal frequency division multiplexing
OFDMA	orthogonal frequency division multiple access
OFS	orthogonally filled subframe

OLT	optical line termination
OPEX	operational expenditure
OTT	over-the-top
PAPR	peak-to-average power ratio
PBCH	physical broadcast channel
PC	power control
PCB	printed circuit board
PCC	primary carrier component
PCell	primary cell
PCFICH	physical control format indicator channel
PCI	physical layer cell identity
PDCCH	physical downlink control channel
PDCP	packet data convergence protocol
PDF	probability density function
PDSCH	physical downlink shared channel
PF	proportional fair
PhD	doctor of philosophy
PHICH	physical HARQ indicator channel
PHY	physical
PMP	point-to-multi-point
PON	passive optical network
POP	point of presence
PRACH	physical random access channel
PRB	physical resource block
PSD	power spectral density
PSS	primary synchronization channel
PTP	point-to-point
PUCCH	physical uplink control channel
PUE	picocell user equipment
PUSCH	physical uplink shared channel
QAM	quadrature amplitude modulation
QoE	quality of experience
QoS	quality of service
QPSK	quadrature phase-shift keying
RACH	random access channel
RAN	radio access network
RAT	radio access technology
RB	resource block
RE	range expansion
REB	range expansion bias
RF	radio frequency
RFID	radio frequency identification
RLC	radio link control

RLF	radio link failure
RLM	radio link monitoring
RMS	root mean square
RNC	radio network controller
RNP	radio network planning
RNTP	relative narrowband transmit power
RR	round robin
RRC	radio resource control
RRH	remote radio head
RRM	radio resource management
RS	reference signal
RSQ	reference signal quality
RSRP	reference signal received power
RSRQ	reference signal received quality
RSS	reference signal strength
RSSI	received signal strength indicator
RTT	round trip time
SCell	secondary cell
SCFDMA	single-carrier FDMA
SCTP	stream control transmission protocol
SeNodeB	secondary eNodeB
SIB	system information block
SINR	signal-to-interference-plus-noise ratio
SMB	small and medium-sized businesses
SMS	short message service
SNR	signal-to-noise ratio
SON	self-organizing network
SRS	sounding reference signals
SSS	secondary synchronization channel
STA	steepest ascent
TCO	total cost of ownership
TDD	time division duplexing
TDoA	time difference of arrival
TNL	transport network layer
ToA	time of arrival
TPC	transmit power control
TTI	transmission time interval
TTT	time-to-trigger
TU	typical urban
UDP	user datagram protocol
UE	user equipment
UL	uplink
UMTS	universal mobile telecommunication system

VDSL	very-high-bit-rate digital subscriber line
VIC	victim cell
VoIP	voice over IP
VoLTE	voice over LTE
WCDMA	wideband code division multiple access
Wi-Fi	wireless fidelity
WiSE	wireless system engineering
WLAN	wireless local area network

PART I

INTRODUCTION

1

SMALL CELLS—THE FUTURE OF CELLULAR NETWORKS

1.1 INTRODUCTION

"Any sufficiently advanced technology is indistinguishable from magic."
—Arthur C. Clarke, *Profiles of the Future*

One of the most widely used devices that appears "magical" today is the smartphone, which allows the user to connect instantaneously with people anywhere on the planet, can provide professional answers to any question, has access to a map of the entire world, and can guide the user to any desired destination.

However, the "magic" does not occur in the smartphone but in the network, which enables its functionality, provides ultra-broadband wireless access, and processes information to deliver voice and data services, invisible to the user.

Popular user demand has thus fueled a remarkable growth of cellular network infrastructure and mobile devices. In 2014, the number of connected mobile devices for the first time exceeded the number of people on Earth, increasing rapidly from zero to 7.6 billion connected devices and 3.7 billion unique subscribers in only three decades [1, 2]. This has fundamentally transformed the way we communicate and access information.

Small Cell Networks: Deployment, Management, and Optimization, First Edition. Holger Claussen, David López-Pérez, Lester Ho, Rouzbeh Razavi, and Stepan Kucera.
© 2017 by The Institute of Electrical and Electronic Engineers, Inc. Published 2017 by John Wiley & Sons, Inc.

Today, we are on the brink of another significant change. While up to now the network mainly served humans, in the future, this capability will increasingly be used by machines as well. The emergence of machine-originated data traffic not only drives further the demand for network capacity but also imposes additional requirements on network performance, mainly in the area of end-to-end latency, which currently is the limiting factor for many new applications.

Nowadays, most of the data services reside in the Internet, far away from the user where the speed of light becomes one of the main factors limiting latency. To address this problem, processing will have to move closer to the user into a cloud computing infrastructure as part of the network. In addition, adaptive network management and well-designed congestion control can help to control latencies and enable new real-time applications such as augmented reality or efficient machine communication.

With these changes, the future network is evolving to become our main interface with the virtual world, and increasingly also with the physical world, to simplify and automate much of life. This will allow us to effectively "create time" by improving the efficiency in everything we do [3].

Making this vision of the future network a reality will require both:

1. Ultra broadband wireless access, providing orders of magnitude improved performance and quality-of-service control, as well as
2. A flexible and programmable cloud computing infrastructure located close to the edge of the wireless network.

Throughout this book, we argue that small cells are the answer to the technological challenges of creating a wireless access network that connects the mobile devices, machines and objects to a processing cloud engine.

As an introduction to our technological philosophy and the content of the book, the remainder of this chapter first summarizes the industry challenge, followed by an overview of the small cell technology and its history. Then, individual parts and chapters of the book are introduced as well as their relationship to various aspects of deploying and operating small cell networks.

1.2 THE INDUSTRY CHALLENGE

The proliferation of highly capable mobile devices as well as the user expectation to be fully connected and have access to all services anywhere and anytime has resulted in an exponential increase in cellular capacity demand over the past few years. With the addition of wirelessly connected machines, which can send, receive, and process massive amounts of data, this trend will continue and is driving an explosion of cellular capacity demand. The expectations are that machines will significantly outnumber human users in the future.

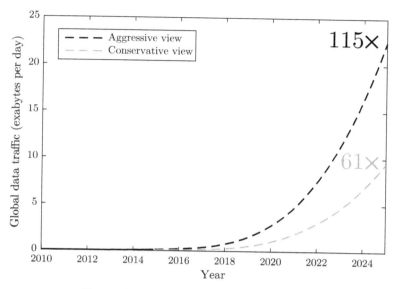

Figure 1.1. Growth in capacity demand [7].

Figure 1.1 shows the predicted increase in data traffic until 2025, taken from an analysis done by Bell Labs Consulting in 2014 based on LTE traffic models and drawing from multiple data sources, including Alcatel-Lucent field data and [1,4–6]. It is shown that the global bearer traffic is expected to grow by a factor between 61× and 115× over the next decade to 22.5 Exabytes per year [7].

Moreover, the control plane demand is predicted to increase proportionally to support an increasing number of short traffic messages generated by machines [7].

However, although the demand for capacity is increasing, users are not willing to pay substantially more for higher data rates, and the average revenues per unique subscriber in recent years have been stagnating [8]. This means we have to provide exponentially more capacity for the same costs as today, adding a significant commercial challenge to the already difficult physical challenge of scaling capacity by orders of magnitude.

A further difficulty is the energy consumption of networks. A report from Ofcom suggests that information communication technology (ICT) accounts for 2% of global CO_2 emission with 0.7% contribution from mobile and fixed communication devices [9]. For example, British Telecom consumes 0.7% of all electricity usage in the United Kingdom [10]. The energy consumption already accounts for 7–15% of operational expenditure (OPEX), reaching up to 50% in developing countries. As a result, we cannot scale capacity using traditional macrocellular network technology, since this would quickly become unsustainable both from a commercial and environmental point of view.

In summary, we can state the industry challenge as *enabling orders of magnitude increase in wireless capacity without increasing costs.*

1.3 ARE SMALL CELLS THE ANSWER?

1.3.1 Dimensions for Capacity Scaling and Historic Capacity Gains

When aiming at orders of magnitude improvements of network capacity, it is important to understand the different dimensions of how capacity can be improved. In a simplified form, the Shannon–Hartley theorem

$$C = B \log_2 \left(1 + \frac{S}{N} \right) \tag{1.1}$$

provides an insight into what are the variables that influence the amount of information (capacity C) one can transmit over a communication channel of a specified bandwidth B with a signal received with power S in the presence of white Gaussian noise with power N [11].

Capacity C can be scaled by increasing the bandwidth B per user, and by increasing the signal-to-noise ratio S/N, or in a multi-user network the signal-to-interference-plus-noise ratio (SINR). Addressing the bandwidth is a more promising approach since this results in a linear scaling compared to the logarithmic scaling when increasing spectral efficiency by improving the SINR. In a network with multiple users, the bandwidth per user can be scaled by either increasing the frequency resources, or by network densification based on the reduction of cell size. This measure results in improved spatial reuse of frequency resources and less sharing of the available bandwidth between users.

An overview of how each of the three degrees of freedom governing wireless channel capacity—densification, bandwidth, and spectral efficiency—contributed to capacity gains between 1950 and 2000 is provided in [12]:

- 2700× from densifying to smaller cells,
- 15× by using more spectrum bandwidth (3 GHz vs. 150 MHz),
- 10× by improving spectral efficiency (coding, medium access control (MAC) and modulation methods).

From this, it is clear that the majority of the gain was achieved by increasing the spatial frequency reuse though densifying the network to small cells. This leaves us with the question: How much further can we increase the spatial reuse by reducing the cell size?

1.3.2 Densification

In a multi-user network, users in the coverage of a cell share the available bandwidth. Thus, reducing the cell size and deploying more cells also reduces the number of users per cell, and in turn increases the bandwidth available to each user. Through this approach, the bandwidth per user can be increased until each cell serves only a single user. When densifying further, only the SINR is improved by reducing the distance between the base station and user. As a result, the capacity scales linearly until the single user limit is reached by increasing the bandwidth per user, after which the scaling becomes logarithmic (see Eq. 1.1). This is illustrated in Fig. 1.2 showing that with increasing cell densities the capacity initially increases very quickly through cell splitting gains, but then slows down when the gains are mainly achieved via improving SINR through proximity gains. Note that if base stations (BSs) are not deployed following the user distribution, a large number of cells are required to achieve the one user limit.

A second aspect of densification is that the required transmit power reduces to an extent where its contribution to the total energy consumption becomes insignificant, and the processing power becomes the dominant factor. In addition, with smaller cell sizes the required number of cells increases such that many of them do not serve active users for most of the time. However, they still consume power and transmit unnecessary pilot signals which cause interference. This can be addressed by introducing

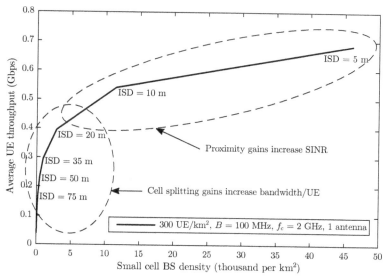

Figure 1.2. Capacity scaling with densification for different inter-site distances (ISDs).

idle modes, where cells are only woken up when actively serving users. With efficiently controlled idle modes, the SINR and network energy consumption improves significantly.

The main challenge of network densification is the issue of increasing costs for equipment, deployment and operational expenses. In this light, it is important to note that the costs of a small cell BS only accounts for approximately 20% of the total deployment costs associated with outdoor small cells in 2015. The majority of the costs are site leasing (26%), backhaul (26%), planning (12%), and installation (8%) [13], which can be addressed by changing the deployment model from operator deployment to a "drop and forget" user deployment, and reusing of existing power and backhaul infrastructure. The end user simply connects the base station to power and backhaul, which triggers a fully automatic configuration, and a continuous self-optimization process during operation. This user deployment model is feasible for both the residential and enterprise market. In addition, it becomes increasingly important to deploy the cells where the users are, since small cells cannot compensate for misplacement as well as larger cells. However, accurate user demand distributions are hard to derive today because the accuracy of conventional localization techniques in cellular networks such as triangulation is poor and the use of more accurate techniques such as the global positioning system (GPS) is not available indoors, where 80% [14] of the traffic demand is located.

In summary, densification continues to have a high potential to increase capacity until reaching one user per cell. In order to maintain high performance and energy efficiency, idle modes that switch off cells when they are not serving users are necessary. Transitioning to a "drop and forget" deployment by the user has a high potential to reduce the deployment and operational costs.

1.3.3 Bandwidth

Another dimension we can use to increase capacity is increasing the bandwidth, where we can achieve linear scaling (see Shannon–Hartley theorem, Eq. 1.1).

However, there are several challenges with this approach as well. First, the available bandwidth at lower frequencies is limited which implies that increasing bandwidth often requires an increase in the carrier frequency band as well, and higher frequencies lead to increased path loss (even line-of-sight requirements when moving to mm-wavelengths). Second, the required transmit power increases significantly when increasing bandwidth due to the higher path loss at higher frequency bands and the fact that more carriers need to be allocated. This is illustrated in Fig. 1.3, which shows that with increasing bandwidth, the transmit power also increases quickly, making higher bandwidths only usable for small cells.

Although increasing spectrum availability can provide high capacity gains, bandwidth is already used up at lower carrier frequencies. More bandwidth is available at higher carrier frequencies, but is mainly applicable to smaller cells due to the increasing transmit power requirements.

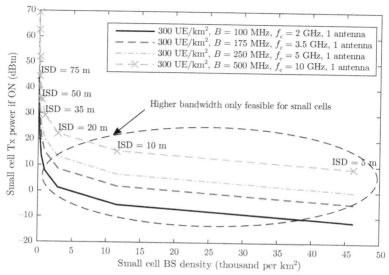

Figure 1.3. Transmit power for increasing bandwidth, small cell densities, and inter-site distances (ISDs).

1.3.4 Spectral Efficiency

The third dimension for increasing capacity is increasing the spectral efficiency, for example, by signal processing through error correction coding, increasing the SINR using interference mitigation, or with multiple antennas.

The progress in signal processing has already led to near-saturation of gains in this dimension. Current coding schemes already operate close to the Shannon–Hartley capacity limit, and further signal processing gains require significant overhead (e.g., Network multiple-input multiple-output (MIMO)). Multiple antennas can be used to increase SINR through beamforming or for spatial multiplexing, but the low number of antennas at the user equipment (UE) at cellular bands and issues with channel state information acquisition limit the gain for traditional MIMO systems.

Figure 1.4 shows the achievable gains exploiting densification, high bandwidth and different numbers of antennas used for beamforming. It is shown that increasing the number of antennas results in gains, but they are lower compared to densification and increasing bandwidth. This results from the fact that improving the SINR only leads to logarithmic gains (see Shannon–Hartley theorem, Eq. 1.1). Consequently, the gains in this dimension are more limited compared to densification and increasing bandwidth. A further challenge is that the gains from improving SINR are not fully complementary with densification. For example in an ultra-dense small cell network, with on average less than 1 user per cell, the SINR is already high due to low

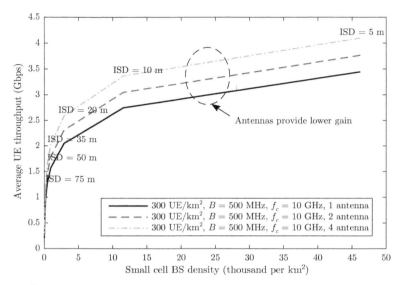

Figure 1.4. Increasing spectral efficiency through beamforming with multiple antennas for different small cell densities, and inter-site distances (ISDs).

interference resulting from empty neighboring cells. Further increasing the SINR leads to diminishing gains due to the logarithmic scaling.

As a concluding remark, techniques improving spectral efficiency should be exploited but any improvement in SINR leads to only logarithmic gains. Thus, the gains in this dimension are more limited compared to densification and increasing bandwidth.

1.3.5 The Answer

Figure 1.5 illustrates the three dimensions for scaling cellular capacity, summarizes their characteristics, and outlines the expected future gains in cellular bands. More spatial reuse by densifying the network with smaller cells continues to provide the highest potential for capacity gains with an expected increase of 18×. Increasing bandwidth can lead to 5× improvements, and improving spectral efficiency with multiple antennas can increase capacity by 2× at cell edges. The challenge of capacity scaling and all assumptions for these results are discussed in more detail in Chapter 2. By exploiting all dimensions, cellular capacity can be increased on the order of 180×. Beyond that we need to explore higher frequencies such as the unlicensed bands and mm-wave. In addition, a high increase in user density due to machine traffic will allow further gains through densification.

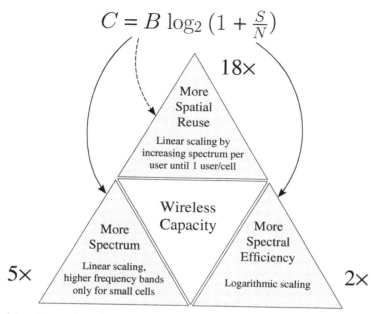

Figure 1.5. Dimensions for scaling wireless capacity and achievable future gains at cellular bands.

The discussion above outlines the physical feasibility of achieving orders of magnitude capacity gains. However, to make this commercially viable, it is necessary to be able to implement and deploy this future network at a cost point not much different from where we are today. This requires a significant increase in automation of the network, including all aspects from planning, deployment, configuration, and optimization.

In summary, orders of magnitude increases in wireless network capacity are possible. Small cells are a necessary topology evolution and will be the main solution to providing the future capacity growth until reaching one user per cell. The other dimensions of frequency and spectral efficiency should also be exploited. To make this future network commercially viable, significant changes are required in the way we deploy and operate the network. Many of these topics are covered in this book.

1.4 A BRIEF HISTORY OF SMALL CELLS

The idea of small cells has been around for over three decades [15]. Initially, "small cells" was the term used to describe the cell size in a metropolitan area, where a macrocell (on the order of kilometers in diameter) would be split into a larger number of smaller cells with reduced transmit power, known today as metropolitan macrocells or microcells. These cells have a radius of a few hundreds of meters.

In the 1990's picocells appeared [16] with a cell size between a few tens of meters and around a hundred meters. These "traditional" small cells are used for capacity and coverage infill, that is, where macrocell penetration was insufficient to give a good connection or where the macrocell was at its capacity limit. Moreover, these types of small cell BSs are essentially a smaller version of the macrocell BS. They had to be planned, managed, and interfaced with the network the same way as macrocell BSs. This last point alludes to why small cells (other than metropolitan microcells) have not gained in popularity for some time. Essentially, the costs associated with deploying and running a large number of small cells outweighed the advantage that this kind of cellular topology provided.

New thinking on the deployment and configuration of cellular systems began to address the operational and cost aspects of small cell deployment [17, 18], which enabled cost-effective deployment of even smaller cells. Early simulation results for femto- and small cells were presented by the authors [19–21], which were extended to self-optimization strategies, multiple antennas, energy efficiency, and offloading benefits shortly afterward [22–29]. On the academic side, early work included new mathematical models and analysis, specifically looking at the uplink interference problem in code division multiple access (CDMA)-based networks with closed access [30, 31].

Femtocells [22, 32, 33] emerged as the first step toward a heterogeneous network deployment model. Femtocells are low-cost cellular BSs with advanced auto-configuration and self-optimization capability, which allows them to be deployed by the user in the home in a plug-and-play manner. They use a broadband Internet connection as backhaul and connect to the cellular network via gateways, which allows better scaling to millions of BSs.

Following early research, the number of publications including femtocell or femtocells in the topic registered in the IEEE database [34] have increased from 3 in 2007 to 11 (2008), 52 (2009), 117 (2010), and a total of 1088 publications registered by the end of 2015.

In addition, the European Union has started funding research on femtocells, for example the ICT-4-248523 BeFEMTO project, which focuses on the analysis and development of long-term evolution (LTE)-compliant femtocell technologies [35].

In 2007, several industry players advocating small cell technology formed the Femto Forum, rebranded as the Small Cell Forum in 2012, to create a venue for promotion, standardization, and regulation.

The first commercial deployments of residential femtocells started in 2008 when Sprint launched a nationwide service in the US, followed in 2009 by Vodafone in Europe, and Softbank in Japan. Since then, small cell technology has proliferated quickly. The number of deployed 3G small cells for the first time exceeded those of macrocells in 2011 [36], and in 2015, over 77 operators used small cells worldwide [37]. The business impact of small cells with 4G LTE-Advanced is now considered by far the most important, but multi-mode cells with additional 3G as well as wireless fidelity (Wi-Fi) capability are becoming increasingly available [38].

The scope of femtocells was then extended from residential deployments to public indoor spaces, which generate most of the cellular traffic. By 2015, 71 carriers operated in-building small cells, known as pico- or microcells, in enterprise or public buildings [37]. Their design is tailored to reduce planning and deployment costs, decrease the need for large customer support teams, and eliminate the need for massive reprovisioning. The generally good availability of Internet protocol (IP) backhaul such as Ethernet in indoor spaces is an important deployment advantage. The overall system configuration as well as the overlap with outdoor macro- and metrocells must be monitored and well managed. To this end, in-building small cell deployments are equipped with quality of service (QoS) and per-call analytics, as well as self-organizing network (SON) features.

The expansion to the outdoor space is more difficult due to barriers such as backhaul availability and cost, site rental, network provisioning and management as well as monetization challenges. Nevertheless, even in this area, small cells proved to be a viable approach in the form of metrocells—smaller and more flexible versions of macrocells with whom they share many hardware and software features, most notably the support of a high number of simultaneous active users and the SON capability for self-configuration and self-optimization of inter-cell interference mitigation, neighbor relation management and handover parameter configuration.

By 2014, AT&T has deployed small cells in 30 US states, mainly to fill coverage gaps in areas with complex landscape or poor signal propagation [39]. Outdoor capacity-oriented trials using cloud radio access network (C-RAN) technology were announced in 2015 [40]. In Europe, the trend is similar with Vodafone rolling out metrocells for years and small cells being part of Orange's strategy, although the overall small cell penetration is still moderate as the available spectrum has not yet been fully utilized [41]. As of 2015, 23 operators have deployed metrocells in outdoor urban zones and large venues such as stadiums and shopping malls, while over 32 operators have used them in rural or remote scenarios [37].

By February 2016, 13.3 million small cells were shipped, and the small cell market size increased to more than $1 billion annually [42]. However, the growth of the small cell technology evolution is nowhere near saturation given the ongoing capacity crunch. It is predicted that small cell densification will continue at a fast pace for both residential and enterprise networks, and the market revenue will reach $6.7 billion by 2020 [42].

In addition, small cells are expected to play a big role in 5G, especially in terms of low latency and high data rates due to the close proximity to the user. Small cells also drive the discussion around long-term evolution (LTE) Unlicensed, a technology that may coexist with Wi-Fi in unlicensed bands and could substantially increase the LTE capacity.

1.5 SMALL CELL CHALLENGES AND OUTLINE OF THE BOOK

1.5.1 Part I: Introduction

Following the introduction and motivation provided in this chapter, the remainder of Part I focuses on the fundamental challenges of enabling 100× scaling of cellular capacity, and automating cellular networks, the key to cost-effective deployment and operation.

Chapter 2 extends the high-level discussion on scaling capacity, and provides an in-depth analysis on how to make it physically possible to scale the cellular capacity by orders of magnitude. The three dimensions of densification, bandwidth, and spectral efficiency are discussed in detail. The one user per cell concept is introduced as the sweet spot for densification. The importance of idle modes, both for reducing interference and improving the energy efficiency in small cell networks is discussed. The impact of increased bandwidth on the power consumption is evaluated and improving spectral efficiency with beamforming techniques is explored. In addition, the impact on scheduling and the resulting network energy efficiency is evaluated. Finally, the technology mix that can enable an average capacity of 1 Gbps per UE is derived.

In addition to making it physically possible to scale the network capacity, making such networks commercially viable is the second critical challenge. To achieve this, significant changes are required in the way networks are planned, deployed, configured, and optimized. Chapter 3 provides an overview of what parts of the network can be automated today. The main SON use cases introduced by Third-Generation Partnership Project (3GPP) and next-generation mobile networks (NGMN) are discussed. Then, key requirements for SON are described and different mechanisms for implementing intelligence across the network are explained. Finally, two examples are given as case studies for SON functionality. First, an antenna tilt angle optimization in a heterogeneous network with macrocells and small cells based on fuzzy-reinforcement learning is presented. Second, the online evolution of femtocell coverage algorithms using genetic programming shows how networks can become truly intelligent in the future and can even rewrite their own optimization code while operating.

The main optimization challenges in small cell networks are addressed in the following parts in more detail.

1.5.2 Part II: Coverage and Capacity Optimization

One of the initial challenges when deploying small cell networks is selecting the carrier frequency and access method, and optimizing the coverage of the cells.

One of the first choices an operator needs to make when deploying small cells is on the frequency allocation and the user access model. In Chapter 4, the different options for frequency allocation: separate, shared, and partially shared frequency resources are presented and their trade-offs and applicability to private and public access models discussed. Then, some of these concepts are illustrated using two examples. First, the feasibility of co-channel deployment of residential femtocells is demonstrated showing that co-channel femtocells with public access can achieve high performance while only causing negligible performance degradation for the macro-cellular network. As second example, the optimization of a partially shared spectrum configuration using multi-carrier soft reuse (MCSR) for outdoor small cells is investigated. It is shown that correct optimization of macrocell power in the reduced power band is critically important to maximize offloading to small cells and overall network performance.

Once the initial parameters such as frequency and access model are configured, coverage optimization is a critical step that has significant impact on the overall network performance. Chapter 5 focuses on coverage and capacity optimization for both residential and enterprise small cells. For residential femtocells, the main objective is to minimize leakage of the coverage into busy public spaces, which can cause a significant amount of handover traffic and signaling for passing users, while maximizing indoor coverage of the home. Coverage optimization algorithms are presented that can reduce the number of mobility events by 90% for a single antenna configuration and by a further 31% for a multiple antenna configuration. For enterprise femtocells, the objective is different. Small cells need to jointly work together to provide full coverage and balance load, with the minimum necessary power to minimize interference. A distributed algorithm is presented that achieves this and improves capacity by 18% compared to a static configuration. An alternative approach to automatically evolving optimization algorithms for joint coverage optimization is already covered in Chapter 3.

Chapter 6 focuses on coverage optimization for outdoor small cells. Here, the problem is different again, since outdoor small cells are typically deployed to cover a hotspot area. When small cells reuse the macrocell frequencies, their coverage is often limited by their lower transmit power. This power imbalance can lead to insufficient traffic offloading particularly when small cell BSs are deployed close to a macrocell BS and also to high interference in the uplink. To address this problem, range expansion using a bias value is discussed and optimization strategies to calculate the bias value for the expanded coverage are presented. A second problem with outdoor small cells is that their placement is often sub-optimal due to deployment limitations and availability of backhaul and power. This problem is addressed using a switched multi-element antenna system, which allows better adaptation of the coverage to the user locations. The proposed solution can achieve an average SINR improvement of 5.1 dB, translating in a peak throughput improvement of 72% in the investigated scenario.

1.5.3 Part III: Interference Management

To achieve high spectral efficiency, small cells need to reuse the frequency resources of the macrocellular network in a shared or partially shared configuration, as discussed in Chapter 4. However, this can create high interference between macro-cells and small cells resulting in degraded network performance at the cell boundaries. In this part, different interference management strategies for both downlink and uplink are presented and their trade-offs discussed.

Chapter 7 focuses on inter-cell interference management by frequency-domain multiplexing. Here, potentially interfering transmissions can be transmitted in different frequency bands using carrier aggregation. Different configuration options and aspects such as signaling, cell activation, power control, and mobility management are discussed. The main benefit of this approach is backward compatibility and its relative simplicity. This comes at the cost of limited flexibility and overhead for separate reference, synchronization and control signals on each component carrier.

In Chapter 8, time-domain interference coordination is discussed. In particular, the technique of almost blank subframes (ABSs) is examined, which allows an interfering cell to mute some of its non-critical transmissions in selected subframes to reduce interference in these resources to other cells. The benefit over carrier aggregation-based approaches discussed in Chapter 7 is the finer granularity of the blanking, which results in improved performance. Methods for optimizing ABS pattern, range expansion biases (REBs) and transmit power are presented, which achieve improvements of the median capacity of almost 60% in a simulated scenario with 132 macrocells and 72 metrocells.

In Chapter 9, a novel sector offset configuration for the macrocell tier is proposed, which can achieve interference coordination without reducing the frequency reuse factor. As a result, the approach can significantly enhance both macrocell and small cell UE throughput, and also improve mobility performance. For a single carrier LTE heterogeneous network (HetNet), cell-edge UE throughput has been shown to improve by 50%, while average and cell-edge macrocell UE throughput have been improved by 17% and 32%, respectively. The handover failure (HOF) rate decreased by 46% when adopting offset sectorization. This comes at the cost of an increased number of antennas at the macrocell site to achieve the sector offset configuration. The concept is fully compatible with current mobile devices and cellular standards, and is in general easy to implement.

Frequency dependent interference coordination, as described in Chapter 7, can be used to improve the quality of LTE data channels. However, the LTE standard does not allow this flexibility in the control channels. As a result, the performance of LTE control signaling under inter-cell interference can become the coexistence bottleneck in co-channel HetNets that limits the effective small cell coverage range and user capacity. In Chapter 10, the concept of orthogonally filled subframes (OFSs) is proposed where control channel interference is minimized by optimizing the pseudo-random allocation of control channel resources. It is shown that this method can either be used to double the small cell radius, enabling a significantly improved offload of the macrocell traffic by the small cell, or triple the control channel capacity—a feature demanded by the emerging high-load voice over LTE (VoLTE) applications.

While previous chapters have mainly focused on interference management in the downlink, Chapter 11 focuses in more detail on uplink-oriented optimization. In particular, conditions under which secondary base stations can share the same communication channel with the primary base stations are examined, assuming constraints such as common interference-limited channels and predefined quality of service such as a minimum SINR for each active transmission. Optimization methods for both CDMA and LTE networks are presented with up to 3× gains in mean uplink user data rates.

1.5.4 Part IV: Mobility Management and Energy Efficiency

Mobility management and energy efficiency are important challenges that highly impact the quality of experience for the user and the operating expenses for the operator. These problems are related since idle modes are the key to energy efficiency in small cell networks, and wake up procedures need to be able to activate idle cells in time that a handover can take place.

Due to the power imbalance and high number of small cells in HetNets, mobility issues are amplified, and if not addressed, they can result in a significantly increased number of dropped calls. In Chapter 12, the handover process in LTE is reviewed as well as the different mobility challenges in HetNets. Special attention is given to its main four stages: measurements and processing, triggering, preparation, and execution, together with important concepts such as radio link failure (RLF), HOF, and ping-pongs. Based on the significant impact of UE velocity on the handover (HO) performance in HetNets, a UE velocity-dependent HO scheme is presented. The need for a better mobility state estimation and new network architectures that consider small cells and the mobility management issue as a key design driver is also discussed.

Chapter 13 presents a framework that is used to quantify the energy saving gains of deploying small cells using a parametric model. It is shown that a power reduction gain of 46× can be achieved when compared to the baseline macrocell-only scenario, with traffic demand assumptions for 2016. With increasing traffic demand in the future, this gain will increase further significantly. Obtaining those gains in practice requires efficient idle mode control and a low idle power consumption. Two idle mode control methods are presented. A simple distributed approach is based on sniffing uplink transmissions to the macrocell and waking the cell up when the measured power exceeds an auto-configured threshold. This simple method can achieve the majority of the efficiency gains, but wakes up more cells than necessary in multitier HetNets. This is addressed by a centralized approach based on RF-fingerprinting which can overcome this problem.

1.5.5 Part V: Small Cell Deployment

The correct placement of small cells is important to maximize the gains of small cell deployments. This part focuses on providing backhaul and on placement optimization to maximize the return of investment.

One significant challenge for the deployment of small cells is the provisioning of backhaul, in particular for outdoor deployments where this represents a high cost factor. In Chapter 14, different backhaul options such as point-to-point and point-to-multipoint wireless backhaul, and wired backhaul options are presented and their capabilities and costs discussed.

Chapter 15 is focused on the placement optimization of small cells. The correct placement is important to maximize the gains of small cell deployments. Placing small cells in the wrong location can not only result in low gains in capacity but, in the worst case, can even cause a degradation due to increased interference. There are many factors which contribute to a good deployment location, but the main ones to consider are interference, offload, backhaul, and costs. In order to maximize the benefits of small cell deployments, they should ideally be placed in locations where the received signal strength from the macrocell underlay is low in order to minimize co-channel interference and maximize the small cell coverage area, while minimizing the costs for site and backhaul. A solution to achieve this is presented including the collection and processing of input data, together with different algorithms for solving the multi-objective placement optimization problem.

1.5.6 Part VI: Future Trends and Applications

This part provides a view into the future and discusses how a transition to ultra-dense networks could be realized, and what new applications could be enabled by the finer spatial information that small cells provide.

For scaling of capacity, densification is one of the most promising approaches as discussed in Chapters 1 and 2. When densifying to ultra-dense networks, one challenge is the transition to line-of-sight path loss resulting in increased inter-cell interference between small cells. Chapter 16 addresses this challenge by employing an attocell deployed at the ceiling with a downward facing patch antenna. It is shown that this significantly reduces inter-cell interference and increases capacity by 4× compared to a traditional small cell with an omni-directional antenna. In addition, aspects such as mobility in ultra-dense networks, and the applicability to higher frequency bands are discussed.

Owing to their special characteristics, such as their short coverage range, small cells can be used as the technology enabler for a number of other applications. Chapter 17 presents different application areas that can be addressed by small cells that can potentially reverse the trend of declining revenues per user for operators. Localized services can provide information and enhanced experience in areas such as museums and for public transport. Proximity detection can provide location-based notifications, for example, when children leave or enter their home. Small cells enable accurate indoor localization where GPS is not available, which can enable a variety of enterprise applications, navigation, tracking, interaction detection, and support E911 emergency call localization. Small cells could also connect devices with applications in home automation, light control, remote control, smart energy, remote care, security, and safety. A further application is enabling selective localized network access control, allowing to restrict the mobile service for specific user groups in hospitals, libraries, theatres, and cinemas without intrusive jamming that causes interference in neighboring areas.

1.5.7 Appendix A: Simulating HetNets

Finally, Appendix A explains how HetNets can be simulated, and describes how the results in this book were derived. Modeling of macrocell and small cell layouts are described, followed by modeling of antennas, path loss, fading, and the calculation of the received signal strengths and SINRs. Scheduling algorithms and the mapping SINR to throughput is also discussed. Then, user traffic and mobility models are presented. This provides the reader with a guide of how to analyze HetNets using computer simulations.

REFERENCES

1. GSMA Intelligence. (2016). Global data. [Online]. Available: https://gsmaintelligence.com/

2. United States Census Bureau. (2016). U.S. and World Population Clock. [Online]. Available: http://www.census.gov/popclock/

3. M. K. Weldon, Ed., *The Future X Network: A Bell Labs Perspective*. CRC Press, 2015.

4. Beecham, "Internet of things: worldwide M2M market forecast," Tech. Rep., 2008.

5. Machina, "M2M communications in automotive 2010–2020," "M2M communications in healthcare 2010–2020," "M2M communications in consumer electronics 2010–2020," "M2M communications in utilities 2010–2020," "M2M global forecast and analysis 2010–2020," Tech. Rep., 2008.

6. Pyramid Research, "Mobile data total region for NA, EMEA, CALA and APAC," Tech. Rep., 2011.

7. H. Viswanathan and T. Sizer II, "The future of wireless access," in *The Future X Network: A Bell Labs Perspective*, M. K. Weldon, Ed. CRC Press, 2015, Ch. 6, pp. 197–224.

8. GSMA Intelligence. (2014). "Evaluating consumer spending: the need for a revised ARPU metric. How multi-SIM offers and bundles dilute the relevance of ARPU" [Online]. Available: https://gsmaintelligence.com/research/2014/10/evaluating-consumer-spending-the-need-for-a-revised-arpu-metric/448/

9. "Understanding the environmental impact of communication systems," Ofcom, White Paper, Apr. 2009.

10. "The line goes green," Delloite, White Paper, May 2010.

11. C. E. Shannon, "Communication in the presence of noise," *Proc. Inst. Radio Eng.*, vol. 37, no. 1, pp. 10–21, Jan. 1949.

12. W. Webb, Ed., *Wireless Communications: The Future*. John Wiley & Sons, 2007.

13. "Game changing economics for small cell deployment," Amdocs, White Paper, Oct. 2014.

14. "Indoor deployment strategies," Nokia, White Paper, Jun. 2014.

15. A. C. Stocker, "Enhanced intercell interference coordination challenges in heterogeneous networks," *IEEE Trans. Veh. Technol.*, vol. 33, no. 4, pp. 269–275, 1984.

16. R. Iyer, J. Parker, and P. Sood, "Intelligent networking for digital cellular systems and the wireless world," *Proc. IEEE Global Telecommun. Conf. (GLOBECOM)*, vol. 1, pp. 475–479, 1990.

17. L. T. W. Ho, "Self-organising algorithms for fourth generation wireless networks and its analysis using complexity metrics," Ph.D. thesis, Queen Mary College, University of London, Jun. 2003.

18. H. Claussen, L. T. W. Ho, H. R. Karimi, F. J. Mullany, and L. G. Samuel, "I, base station: Cognisant robots and future wireless access networks," in *Proc. 3rd IEEE Cons. Commun. Netw. Conf. (CCNC)*, Las Vegas, USA, Jan. 2006, pp. 595–599.

19. H. Claussen, L. T. W. Ho, and L. G. Samuel, "Financial analysis of a pico-cellular home network deployment," in *Proc. IEEE Int. Conf. Commun. (ICC)*, Glasgow, UK, Jun. 2007, pp. 5604–5609.

20. H. Claussen, "Performance of macro- and co-channel femtocells in a hierarchical cell structure," in *Proc. 18th IEEE Int. Symp. Pers., Indoor Mobile Radio Commun. (PIMRC)*, Athens, Greece, Sep. 2007.

21. L. T. W. Ho and H. Claussen, "Effects of user-deployed, co-channel femtocells on the call drop probability in a residential scenario," in *Proc. 18th IEEE Int. Symp. Pers., Indoor Mobile Radio Commun. (PIMRC)*, Athens, Greece, Sep. 2007.

22. H. Claussen, L. T. W. Ho, and L. G. Samuel, "An overview of the femtocell concept," *Bell Labs Tech. J.*, vol. 15, no. 3, pp. 137–147, Dec. 2008.

23. H. Claussen, "Co-channel operation of macro- and femtocells in a hierarchical cell structure," *Int. J. Wireless Inf. Netw.*, vol. 15, no. 3, pp. 137–147, Dec. 2008.

24. H. Claussen, L. T. W. Ho, and L. G. Samuel, "Self-optimization of coverage for femtocell deployments," in *Proc. Wireless Telecommun. Symp. (WTS)*, Los Angeles, USA, Apr. 2008, pp. 278–285.

25. H. Claussen and F. Pivit, "Femtocell coverage optimization using switched multi-element antennas," in *Proc. IEEE Int. Conf. Commun. (ICC)*, Dresden, Germany, Jun. 2009.

26. H. Claussen, L. T. W. Ho, and F. Pivit, "Self-optimization of femtocell coverage to minimize the increase in core network mobility signalling," *Bell Labs Tech. J.*, vol. 14, no. 2, pp. 155–184, Aug. 2009.

27. H. Claussen, L. T. W. Ho, and F. Pivit, "Effects of joint macrocell and residential picocell deployment on the network energy efficiency," in *Proc. 19th IEEE Int. Symp. Pers., Indoor Mobile Radio Commun. (PIMRC)*, Cannes, France, Sep. 2008.

28. H. Claussen, L. T. W. Ho, and F. Pivit, "Leveraging advances in mobile broadband technology to improve environmental sustainability," *Telecommun. J. Australia*, vol. 59, no. 1, pp. 4.1–4.18, Feb. 2009.

29. H. Claussen and D. Calin, "Macrocell offloading benefits in joint macro- and femtocell deployments," in *Proc. 20th IEEE Int. Symp. Pers., Indoor Mobile Radio Commun. (PIMRC)*, Tokyo, Japan, Sep. 2009, pp. 350–354.

30. V. Chandrasekhar and J. Andrews, "Uplink capacity and interference avoidance for two-tier cellular networks," in *Proc. Global Telecommun. Conf. (GLOBECOM)*, Nov. 2007, pp. 3322–3326.

31. V. Chandrasekhar and J. Andrews, "Uplink capacity and interference avoidance for two-tier femtocell networks," *IEEE Trans. Wireless Commun.*, vol. 8, no. 7, pp. 3498–3509, Jul. 2009.

32. V. Chandrasekhar, J. Andrews, and A. Gatherer, "Femtocell networks: a survey," *IEEE Commun. Mag.*, vol. 46, no. 9, pp. 59–67, Sep. 2008.

33. J. G. Andrews, H. Claussen, M. Dohler, S. Ragan, and M. C. Reed, "Femtocells: Past, present, and future," *IEEE J. Sel. Areas Commun. (JSAC), Spec. Issue: Femtocell Netw.*, vol. 30, no. 3, pp. 497–508, Apr. 2012.

34. IEEE. (2016). IEEE Xplore Digital Library. [Online]. Available: http://ieeexplore.ieee.org/

35. BeFEMTO. (2016). Broadband Evolved Femto Networks. [Online]. Available: http://www.ict-befemto.eu/

36. Small Cell Forum Press Release. (2011, Jun.). "3G femtocells now outnumber conventional 3G basestations globally" [Online]. Available: http://www.smallcellforum.org/press-releases/3g-femtocells-now-outnumber-conventional-3g-basestations-globally/

37. Small Cell Forum Press Release. (2015, Jun.). "Market status statistics June 2015 – Mobile Experts" [Online]. Available: http://scf.io/en/documents/050_-_Market_status_report_June_2015_-_Mobile_Experts.php

38. S. Reedy. (2013, Sep.). "Multimode small cells get stalled in labs" [Online]. Available: http://www.lightreading.com/mobile/small-cells/multimode-small-cells-get-stalled-in-labs/d/d-id/703334

39. S. Reedy. (2014, Dec.). "LTE small cells set to be big in 2015" [Online]. Available: http://www.lightreading.com/mobile/small-cells/lte-small-cells-set-to-be-big-in-2015/d/d-id/712704

40. M. Degrasse. (2015, Jul.). "Super Bowl, usage growth prompt Verizon rollout of small cells in San Francisco; carrier uses C-RAN architecture" [Online]. Available: http://www.rcrwireless.com/20150729/network-infrastructure/verizon-explains-small-cell-rollout-tag4

41. M. Donegan. (2014, Jun.). "Lessons from your friendly neighborhood small cells" [Online]. Available: http://www.lightreading.com/mobile/small-cells/lessons-from-your-friendly-neighborhood-small-cells/d/d-id/709456

42. Small Cell Forum Press Release. (2016, Feb.). "Market status statistics February 2016 – Mobile Experts" [Online]. Available: http://scf.io/en/documents/050_-_Market_status_report_Feb_2016_-_Mobile_Experts.php

2

100× CAPACITY SCALING OF CELLULAR NETWORKS

2.1 INTRODUCTION

Even though there is uncertainty on how future networks will look like, it is expected that they follow the same trends as previous communication systems and technology breakthroughs, and provide increasingly more capacity as time goes by. Voice services [1] were the killer applications at the beginning of this century demanding tens of kbps per user equipment (UE), while high-quality video streaming [2] is the most popular one today requiring tens of Mbps per UE [3]. Future services such as augmented reality, 3D visualization, and online gaming may use multiple displays requiring hundreds of Mbps each, resulting in a total sum of up to 1 Gbps per UE, and who knows what else tomorrow will bring?

In view of the significant future traffic demands, the mobile industry has set its targets high, and has decided to improve the capacity of today's networks by a factor of 100× or more over the next 10 years (see Chapter 1).

In order to achieve this goal, vendors and operators are currently looking at using every tool they have at hand, where the existing tools can be classified within the following three paradigms, as illustrated in Fig. 1.5:

Small Cell Networks: Deployment, Management, and Optimization, First Edition. Holger Claussen,
David López-Pérez, Lester Ho, Rouzbeh Razavi, and Stepan Kucera.
© 2017 by The Institute of Electrical and Electronic Engineers, Inc. Published 2017 by John Wiley & Sons, Inc.

- Enhance spatial reuse through network densification, that is, heterogeneous networks (HetNets) and small cells [4–7].
- Use of larger bandwidths, exploiting higher carrier frequencies, both in licensed and unlicensed spectrum [8–10].
- Enhance spectral efficiency through multi-antenna transmissions [11], cooperative communications [12], dynamic time division duplexing (TDD) techniques [13–15], etc.

These techniques provide vendors and operators with choice and flexibility to improve network capacity, but open new questions on how operators should divide their investments to meet capacity demands. Operators differ in financial means, customer market segmentation, existing network assets and technical expertise, and may therefore require different solutions. Network densification is challenging in terms of deployment planning, backhaul provisioning, and mobility management, while the usage of higher carrier frequencies is characterized by larger path losses typically implying more expensive equipment. Technologies offering spectral efficiency enhancement depend on a tight synchronization as well as relatively complex signal processing capabilities and may be compromized due to inaccuracies in channel state information (CSI). Furthermore, the three paradigms above have their own fundamental limitations and thus cannot be infinitely exploited, and capital expenditure (CAPEX) and operational expenditure (OPEX) are also major concerns. Finding the right portfolio of tools to meet the key requirements is vital for both vendors and operators.

In this context, this chapter aims at bringing further common understanding and analysis of the potential gains and limitations of network densification, in combination with the use of higher frequency bands and spectral efficiency enhancement techniques. The impact of idle mode capabilities at the small cells (see Chapter 13) as well as UE density and distribution are also studied since they are foreseen to have a large impact. This chapter sheds new light on the Pareto set of network configurations in terms of small cell density, frequency band of operation, and number of antennas per small cell base station (BS) that can meet future traffic demands, achieving an average throughput exceeding 1 Gbps per UE [16].

The rest of the chapter is organized as follows: In Section 2.2, the system model used in this chapter to analyze ultra-dense small cell deployments is presented. In Sections 2.3, 2.4 and 2.5 the three paradigms to enhance network performance are discussed, that is, network densification together with the usage of higher carrier frequencies and multi-antenna transmissions, respectively. Then, in Section 2.6, the impact of network densification on the performance of small cell BSs is investigated, followed by the impact on the energy efficiency in Section 2.7. In Section 2.8, the main differences between regular HetNets and ultra-dense HetNets are highlighted, and finally conclusions are drawn in Section 2.9.

2.2 SYSTEM MODEL

To achieve a good performance in a co-channel deployment of small cells with the macrocell tier, a joint optimization of range expansion bias (REB), almost blank subframe (ABS) patterns, power reduction factors in reduced power subframes, scheduling thresholds, and frequent inter-BS coordination may be required, as discussed further in Chapters 4 to 10. The complexity of these optimization procedures is aggravated with the number of cells, and thus we anticipate that co-channel deployment of small cells and macrocells is not suitable for ultra-dense small cell deployments, where network planning should be completely avoided. This suggests that co-channel deployments of small cells with the macrocell tier should be aimed at hotspot locations with a reasonable low number of small cells (e.g., less than 10 per macrocell sector), while ultra-dense deployments should use dedicated carriers to circumvent mobility and interference problems, and thus simplify the planning stage.

In this chapter, the focus is on non-co-channel mid-frequency small cell deployments in combination with a macrocell tier for area coverage. This cell type has the potential to significantly enhance network performance through high network densification and the usage of relatively high frequency bands, while avoiding coordination issues with the macrocell tier designed to support fast moving UEs. The relatively small size of antennas at mid-frequency bands around 10 GHz is also an appealing feature in order to exploit multi-antenna transmission techniques.

In the following, the system model used to analyze ultra-dense small cell networks is introduced. A wrap-around 500×500 m scenario is used, and small cells are placed outdoors in a uniform hexagonal grid with different inter-site distances (ISDs) of 200, 150, 100, 75, 50, 35, 20, 10, or 5 m, which result in 29, 52, 116, 206, 462, 943, 2887, 11548, or 46189 small cell BSs per square km deployed in the scenario, respectively.

Two scenarios for UE distribution are considered, each with three different UE densities of 600, 300, or 100 active UEs per square km:

- Uniform: UEs are uniformly distributed within the scenario.
- Non-uniform: Half of the UEs are uniformly distributed within the scenario, while the other half are uniformly distributed within circular hotspots of 40 m radius with 20 UEs each. Hotspots are uniformly distributed, and the minimum distance between two hotspot centers is 40 m.

Note that 300 active UEs per square km is the density usually considered in dense urban scenarios, such as Manhattan [17].

In terms of frequency bands, four carrier frequencies are considered, with 2.0, 3.5, 5.0, and 10 GHz, where the available bandwidth is 5% of the carrier frequency, with 100, 175, 250, and 500 MHz, respectively.

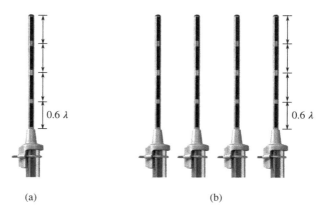

(a)   (b)

Figure 2.1. Horizontal dipole array.

In terms of antenna implementation and operation, each small cell BS has 1, 2, or 4 antennas, where each antenna consists of a vertical dipole array as presented in more detail in Section A.3.2.2 and shown in Fig. 2.1a. These antennas are deployed in a horizontal array, shown in Fig. 2.1b. The UE has only 1 antenna. The standardized long-term evolution (LTE) code book beamforming is adopted, targeted in this case at maximizing the received signal strength of the intended UE (quantized maximum ratio transmission (MRT) beamforming with no inter-BS coordination required). It is important to note that the transmit power of the small cell BS is equally divided among antennas, and that beamforming is only applied to the data channels and not to the control channels, which define the small cell coverage and UE cell association.

The transmit power of each active BS is configured such that it provides a signal-to-noise ratio (SNR) of 9, 12, or 15 dB at the targeted coverage range, which is $\sqrt{3}/2$ of the ISD [18].

UEs are not deployed within a 0.5 m range of any BS, and all UEs are served by the BS from which they can receive the strongest received pilot signal strength, provided that the pilot signal-to-interference-plus-noise ratio (SINR) is larger than −6.5 dB.

A BS with no associated active UE is switched off. This idle mode capability is presented in Section 13.3 [19], and adopted hereafter.

With regard to scheduling, we envision that because of the use of mid- to high-frequency bands and due to the importance of the line-of-sight (LOS) component in small cells, the time spread of the channel impulse response (CIR) will be very small, typically on the order of several μs, and hence the impact of multi-path fading will become less significant in this type of deployment. Therefore, multi-path fading is not considered in our analysis, and a round robin scheduler designed for simple implementation is adopted. This argument is supported by the analysis in Section 13.3, where the impact of network densification in the performance of the small cell BS scheduler is investigated, while considering multi-path fading [20].

For details on path loss, antenna gain, shadow fading, SINR computation, and capacity mapping, please refer to Appendix A. Simulation details are also summarized in [16].

For the sake of clarity, it is also important to mention that in the legend of the figures of this chapter, i indicates the ISD of the scenario in meter, d indicates the density of UEs per square km, ud indicates the UE distribution ($ud = 0$ uniform; $ud = 1$ non-uniform), s indicates whether the idle mode capability of the small cell BS is deactivated or activated ($s = 0$ deactivated; $s = 1$ activated), sm indicates the index of the idle mode used in the energy efficiency analysis, f indicates the carrier frequency in GHz, a indicates the number of antennas, and t indicates the SNR target of the small cell BS at its cell edge, which is located at $\sqrt{3}/2$ of the ISD.

2.3 NETWORK DENSIFICATION

Network densification has the potential to significantly increase the capacity of the network with the number of deployed cells through spatial spectrum reuse, and is considered to be the key enabler to provide most of the capacity gains in future networks.

In order to better understand the implications of network densification on network capacity, let us define the network capacity C[bps] based on the framework developed by Claude Shannon [21] as

$$C = \sum_{m}^{M} \sum_{u}^{U_m} B_{m,u} \log_2(1 + \gamma_{m,u}), \tag{2.1}$$

where $\{1, \ldots, m, \ldots, M\}$ is the set of BSs deployed in the network, $\{1, \ldots, u, \ldots, U_m\}$ is the set of UEs connected to BS m, B[Hz] is the total available bandwidth, and $B_{m,u}$[Hz] and $\gamma_{m,u}$ are the bandwidth granted to and the SINR experienced by UE u when connected to BS m. This model assumes Gaussian interference.

At the network level, network densification increases the number of geographically separated BSs M that can simultaneously reuse the available bandwidth B, thus linearly improving spatial reuse and increasing network capacity with M.

At the cell level, a consequence of network densification is cell size reduction, which directly translates into a lower number of UEs U_m connected per BS m and thus a larger bandwidth $B_{m,u}$ available per UE. In this way, the network capacity increases linearly with the number of offloaded UEs.

Moreover, at the cell level too, the average distance between a UE and its serving BS reduces, while the distance to its interfering BSs does not necessarily reduce at the same pace assuming idle mode capabilities. This leads to an increased UE signal quality $\gamma_{m,u}$, and thus the network capacity logarithmically increases with the $\gamma_{m,u}$.

As can be derived from the above discussion, network densification increases M and in turn improves both $B_{m,u}$ and $\gamma_{m,u}$, resulting in an increase of the network capacity—see (Section 2.1). However, the impact of UE density and distribution should not be forgotten. If network densification is taken to an extreme, and the number of deployed BSs is larger than the number of existing active UEs, this ultra-dense small cell deployment may reach a fundamental limit in which the number of average active UEs per cell is equal or lower than one, that is, $U_m \leq 1$. At this point, the bandwidth $B_{m,u}$ available per UE cannot be further increased through cell splitting, and thus network densification can only enhance network capacity at a lower pace in a logarithmic manner by bringing the network closer to the UE and improving the UE signal quality $\gamma_{m,u}$, which may not be as cost-effective. As a result, *one UE per cell* may be the operational sweet spot from a densification view point.

An important note is that even if the supply of UEs is infinite, and thus the spatial reuse gain increases linearly with the small cell density, it is still premature to claim that network capacity will increase linearly with the BS density. In some recent studies [22,23], the authors introduced a sophisticated path-loss model into the stochastic geometry analysis incorporating both LOS and non-line-of-sight (NLOS) transmissions to study their performance impact in dense small cell networks. The performance impact of LOS and NLOS transmissions in small cell networks in terms of network capacity was shown to be significant both quantitatively and qualitatively, compared with previous works. In particular, when the density of small cells is larger than a threshold, $\gamma_{m,u}$ will *decrease* as small cells become denser due to NLOS to LOS transition of the interfering signals, which in turn makes the network capacity suffer from a slow growth or can even result in a capacity *decrease*, as shown in [22,23].

In the following, we discuss the *one UE per cell* concept.

2.3.1 Idle Mode Capability and the One UE per Cell Concept

One important advantage of having a surplus of cells in the network is that a large number of them could be switched off if there is no active UE within their coverage area, which reduces interference to neighboring UEs as well as energy consumption.

Provided that a surplus of cells exists and as a result of an optimal idle mode capability [19], the network would have the key ability of adapting the distribution of active BSs to the distribution of active UEs, and thus the number of active cells, transmit power of the network, and interference conditions would strongly depend on the UE density and distribution.

According to the system model, Fig. 2.2 shows the average number of active BSs in the network per square km and the average number of active UEs per active BS, with the aforementioned idle mode capability (a BS with no associated UE is switched off). Different UE densities and distributions are considered. From Fig. 2.2a, the following conclusions can be drawn. The lower the UE density, the lower the average

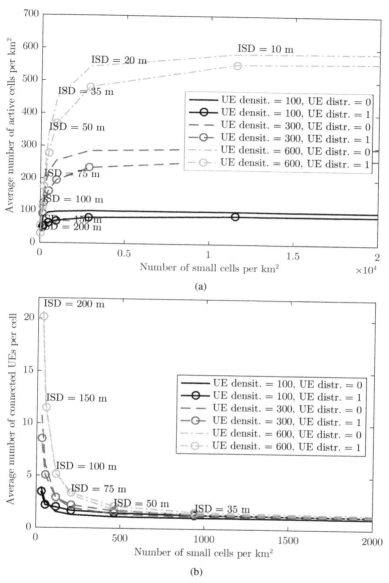

Figure 2.2. (a) Average number of active BSs in the network per square km and (b) average number of active UEs per active BS. The UE densities are 600, 300, and 100 active UEs per km². The UE distributions are uniform and non-uniform within the scenario. The rest of the parameters are $s = 1$, $f = 2$ GHz, $a = 1$ and $t = 12$ dB.

number of active BSs in the scenario and the lower the average number of active UEs per active BS. With regard to UE distribution, a uniform distribution requires more active BSs than a non-uniform distribution to provide full coverage to a given UEs density. This is because UEs are more widely spread in the former. In contrast, due to the more active BSs, the uniform distribution results in a lower number of active UEs per active BS for a given UEs density, which tends to provide a better UE performance at the expense of an increased number of deployed active cells and thus cost. This indicates the importance of understanding the UE density and distribution in specific scenarios to realize efficient deployments.

An important result that can be extracted from Fig. 2.2b is that an ISDs of 35 m can already achieve an average of 1.1 active UEs per active BS or smaller, approaching the fundamental limit of spatial reuse. Reaching such limit has implications for the network and UE performance, as it has been qualitatively explained before, and as it will be quantitatively shown in Section 2.4. Moreover, it also has implications for the cost-effectiveness of the deployment, since densifying further than the 1 UE per cell sweet spot requires an exponential increase in investment to achieve a diminishing logarithmic capacity gain through signal quality enhancement. In other words, when the network is dense enough, a large number of cells have to be added to the network to enhance the UE throughput in a noticeable manner, and this may not be desired since the operator may have to pay exponentially more money to carry on with the deployment and to achieve further gains from densification.

2.3.2 Transmit Power and UE SINR Distribution

Combining ultra-dense small cell deployments together with an efficient idle mode capability has the potential to significantly reduce the transmit power of the network. This is because active cells transmit to UEs with a lower power due to their reduced cell size and empty cells with no active UE can be put into idle mode until a UE becomes active [19].

Moreover, by turning off empty cells, the interference suffered by UEs from always on channels, for example, synchronization, reference, and broadcast channels, can also be removed, neutralizing some neighboring cells and thus improving UE SINR distributions.

Working in this direction, LTE Release 12 networks have defined periodic discovery reference signals (DRSs) to facilitate UEs the discovery of small cells that are turned off [24]. DRSs are transmitted sparsely in the time domain and they consist of multiple types of reference signals (RSs), based on which UEs are able to perform synchronization, detect cell identity, and acquire coarse CSI, etc. Due to the low periodicity of DRSs, the impact of DRSs on UE SINR distribution is marginal, and thus it can be ignored.

2.3.2.1 Transmit Power In terms of transmit power, Fig. 2.3a shows how the transmit power per active BS significantly reduces with the small cell BS density

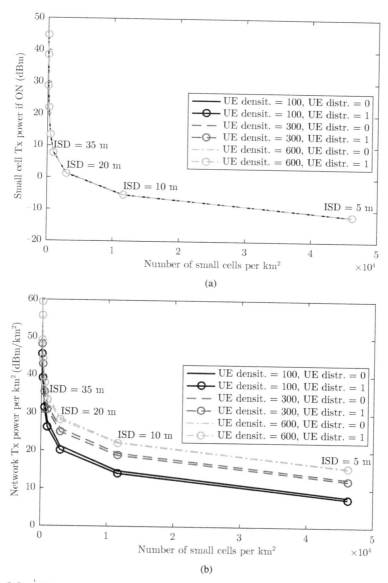

Figure 2.3. (a) Transmit power per active BS and (b) transmit power of the network per km². The UE densities are 600, 300, and 100 active UEs per km². The UE distributions are uniform and non-uniform within the scenario. The rest of the parameters are $s = 1$, $f = 2$ GHz, $a = 1$, and $t = 12$ dB. It is important to note that in Fig. 2.3a there are six overlapping curves. This is because the power used by the cell if it is activated does not depend on the UE density.

in the studied scenario. In this case, the transmit power of each active BS is configured such that it provides a SNR of 12 dB at the targeted coverage range, which is $\sqrt{3}/2$ of the ISD. Note that here the required transmit powers are significantly lower compared to co-channel deployments where the small cells should exceed the received macrocell power in the intended coverage area. In addition, when the idle mode capability is considered, Fig. 2.3b shows how the overall transmit power used by the network also significantly reduces with the small cell BS density in the studied scenario. This is because the reduction of transmit power per cell outweighs the increased number of active cells, and this fact holds true for both the uniform and non-uniform UE distributions, with network transmit power reductions of up to 43 dB. Note that this discussion only considers transmit powers, and that the overall power consumption of the network when considering the power consumed by the BSs in idle mode will be analyzed in Section 2.7.

2.3.2.2 UE SINR Distribution   Traditional understanding has led to the conclusion that the UE SINR distribution and thus outage probability is independent of BS density. The intuition behind this phenomenon is that the increase in signal power is exactly counter-balanced by the increase in interference power, and thus increasing the number of BSs does not affect the coverage probability. This is a major result in the literature [25, 26], which only holds for sparse networks and/or under the assumption of a single-slope path-loss model, more suited for rural areas. However, conclusions may be different for ultra-dense and urban scenarios where NLOS to LOS transitions may occur [22, 23].

In order to show this, Figs. 2.4 and 2.5 respectively show the SINR spatial distribution and UE SINR cumulative distribution functions (CDFs) for several ultra-dense small cell deployments with different small cell densities with and without idle mode capability. Recall that in our path-loss models, the NLOS to LOS transition is modeled with the probabilistic function defined for urban microcell environments in [27]. Let us now analyze the results.

When the idle mode capability is deactivated, our results show that the UE SINR CDF degrades with the small cell density, contradicting the results in [25, 26] where it is independent of BS density. While in traditional networks with large ISDs the carrier signal may be subject to LOS depending on the distance between the UE and its serving BS, the interfering signal is not usually subject to LOS due to the large distance between the serving BS and its interfering BSs. However, with the smaller ISDs, LOS starts dominating the interfering signal too and this brings down the UE SINR, thus lowering the UE and cell throughputs. In other words, the interference power increases faster than the signal power with densification due to the transition of the former from NLOS to LOS, and thus the BS density matters [22, 23]. This new conclusion should significantly impact network deployment strategies, since the network capacity no longer grows linearly with the number of cells as indicated in previous works. Instead, it increases at a lower pace as the BS density increases, and

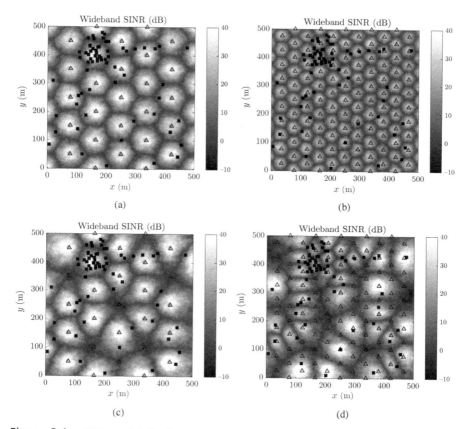

Figure 2.4. SINR spatial distributions of several ultra-dense small cell deployments with ISD among small cell BSs of 100 and 50 m with or without idle mode capabilities. The rest of the parameters are $d = 300$ UE/km^2, $ud = 1$, $s = 1$, $f = 2$ GHz, $a = 1$ and $t = 12$ dB. The triangles represent BSs and the squares represent active UEs. (a) 100 ISD and idle mode capability deactivated, (b) 50 ISD and idle mode capability deactivated, (c) 100 ISD and idle mode capability activated and (d) 50 ISD and idle mode capability activated.

thus operators may find themselves paying exponentially more money for diminishing gains when heavily densifying their networks.

However, the good news is that when the idle mode capability is activated, the trend is reversed, and the UE SINR CDF is significantly boosted with the cell density as a result of interference mitigation. The denser the BS deployments, the larger the capacity increase, since more BS can be turned off, which reduces interference. When the ISD among BSs is 35 m, the median SINR improvement compared to the case when the idle mode capability is deactivated is around 8.76 dB, while for an ISD of 10 m, the median SINR improvement is around 20.62 dB. This is a significant improvement.

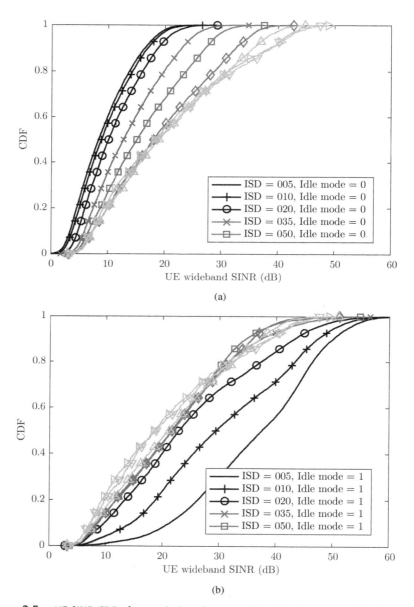

Figure 2.5. UE SINR CDF of several ultra-dense small cell deployments with ISD among small cell BSs of 200, 75, 50, 20, 10, and 5 m with or without idle mode capabilities. The rest of the parameters are $d = 300$ UE/km^2, $ud = 1$, $s = 1$, $f = 2$ GHz, $a = 1$ and $t = 12$ dB. (a) CDF of SINR for different ISDs with idle mode capability deactivated and (b) CDF of SINR for different ISDs with idle mode capability activated.

As conclusion, it is important to note that an optimum idle mode capability not only plays a significant role in transmit power savings, but also acts as an interference mitigation technique.

2.3.3 Transition from the Interference-Limited Regime to the Noise-Limited Regime

As shown in Fig. 2.2, in ultra-dense small cell deployments, a large number of cells can be switched off when the BS density is large and the UE density is low; this combination leads to the most effective interference mitigation. However, is this interference mitigation through small cell deactivation large enough to transition from an interference-limited scenario to a noise-limited scenario? In an interference-limited scenario, the signal quality of a UE is independent of the transmit power of the serving and the interfering BSs, provided that they all use the same transmit power. However, this is not the case in a noise-limited scenario where the signal quality of a UE improves with the transmit power of the serving BS, as the noise remains constant. Therefore, if such transition takes place, the tuning of the small cell BS transmit power becomes even more important, since the transmit power will not only determine the coverage radius of the cell but will also affect the capacity of the network.

In order to answer this question, Fig. 2.6 shows the UE SINR CDF in different ultra-dense small cell deployments, while considering different transmit power for the small cell BSs. In more detail, the targeted SNRs at $\sqrt{3}/2$ of the ISD are set to 9, 12, or 15 dB. Different UE densities are also considered, 100 and 600 UEs per square km. The results show that the change in transmit power only has an impact on the SINR distribution of the scenario with the largest BS densities, ISD = 5 m and ISD = 10 m, and the lowest UE densities, 100 active UEs per km^2. Otherwise, the SINR distribution is independent of the transmit power, indicating that this transition does not occur in representative scenarios. It is important to note that even in the indicated extreme cases, the decoupling of the UE SINR CDF only happens at the high SINR regime, whose SINRs belong to non-cluster UEs suffering from low interference. As a result, since the decoupling only happens for a very extreme BS density, it can be concluded that such transition from interference-limited to noise-limited does not occur in realistic deployments, and that the small cell BS transmit power should be configured to guarantee a targeted range.

2.4 HIGHER FREQUENCY BANDS

As shown by the Shannon–Hartley theorem presented in (2.1), the network capacity linearly increases with the available bandwidth. Therefore, increasing the available bandwidth is an appealing proposition to enhance network capacity. However, spectrum is a scare resource, especially at the lower frequency bands, 500–2600 MHz,

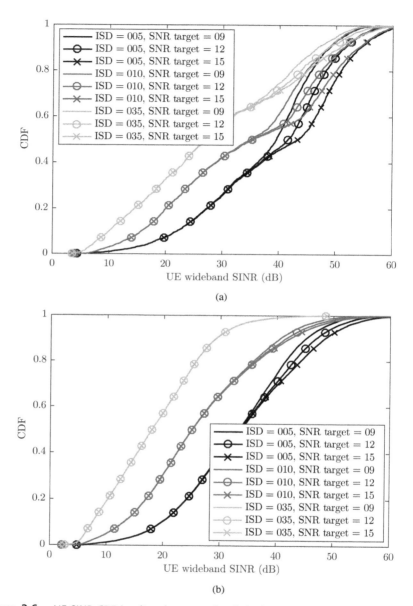

(a)

(b)

Figure 2.6. UE SINR CDF in ultra-dense small cell deployments with large BS densities, 35, 10, and 5 m, low UE densities, 100 and 600 active UE per square km and SNR targets, 15, 12, and 9. The rest of the parameters are $ud = 1$, $s = 1$, $f = 2$ GHz, and $a = 1$. (a) 100 active UEs per km^2 and (b) 600 active UEs per km^2.

which are in use today by radio and TV stations as well as the first wireless communication systems due to their good propagation properties. These frequency bands are thus heavily regulated, and it is unlikely that large bandwidths become available from them in the near future.

Due to the higher path losses at higher frequency bands, these bands were never appealing for large macrocells but now suit well the operation of small cells, targeted at short ranges. These higher path losses should not only be considered as a disadvantage but also as an opportunity, since they effectively mitigate interference from neighboring cells, and thus allow a better spatial reuse. Moreover, it allows using smaller antennas and packing more of them per unit of area, which benefits multi-antenna techniques. In the following, the capacity gains provided by and the challenges faced when using higher frequency bands are discussed.

In order to assess the capacity gains provided by ultra-small cell deployments and the use of higher frequency bands, Fig. 2.7 shows the average and 5th percentile UE throughput for different densification levels and four different network configurations where the carrier frequencies are 2.0, 3.5, 5.0, and 10 GHz and the available bandwidth is 5% of the carrier frequency, that is, 100, 175, 250, and 500 MHz, respectively. In this case, the targeted SNRs by the small cell BSs at $\sqrt{3}/2$ of the ISD is 12 dB, thus assuring a constant coverage area regardless of the frequency band. From the results, different observations can be made. In the following and based on Fig. 2.7, we first analyze the impact of densification on the network capacity and then improvements brought by the use of higher frequency bands.

In terms of densification, the average and 5th percentile UE throughput do not increase linearly with the number of deployed BSs, but with diminishing gains. This is due to the finite nature of UE density and the characteristic of its distribution. In a first phase, with a low BS density and up to an ISD of 35 m, the UE throughput grows rapidly, almost linearly with the number of cells, due to cell splitting gains and spatial reuse. Subsequently, in a second phase, once the fundamental limit of spatial reuse—*one UE per cell*—is reached, the UE throughput continues growing with the network densification but at a lower pace. This is due to the combined effect of both, bringing the UE closer to the serving cell BS through densification and further from the interfering BSs through idle modes. These two effects together result in a SINR enhancement, which improves network performance in a logarithmic manner (see Eq. 2.1). This transition and the two regimes are more obvious for the average than for the 5th percentile UE throughput, since cell-edge UEs benefit more from proximity gains and interference mitigation.

Looking at the average and 5th percentile UE throughput using the 200 m ISD case with a 100 MHz bandwidth as a baseline, the 35 m ISD case can provide an average and cell-edge gain of 7.56× and 5.80×, respectively, while the gains provided by the 5 m ISD case are 17.56× and 48.00×, respectively. This shows that a significant increase in network performance can be achieved through network densification.

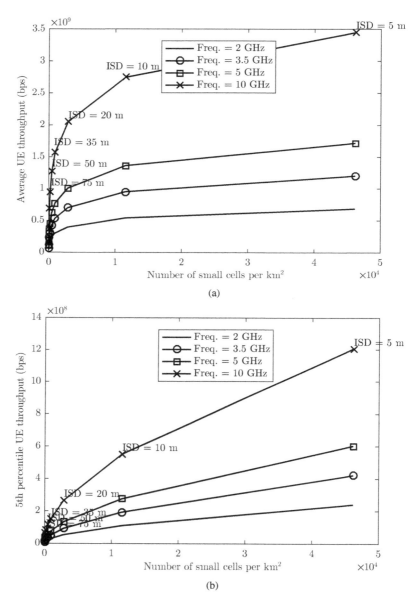

Figure 2.7. Average (a) and 5th percentile UE (b) throughput for four different network configurations where the carrier frequencies are 2.0, 3.5, 5.0, or 10 GHz. The rest of the parameters are d = 300 UE/km², ud = 1, s = 1, a = 1 and t = 12 dB.

In terms of frequency bands, the average and 5th percentile UE throughput increase linearly with the carrier frequency due to the larger available bandwidth, showing the use of larger bandwidths as a key to achieve very high UE throughputs.

Looking at the average and 5th percentile UE throughput with the 35 m ISD case with a 100 MHz bandwidth as a baseline, a bandwidth of 250 MHz can provide an average and cell-edge gain of 2.59× and 2.58×, respectively, while the gains provided by a bandwidth of 500 MHz are 5.31× and 5.17×, respectively. Although not as large as the gains provided by network densification, this linear gains also represent a significant increase in network performance.

It is important to highlight that the capacity gains seen through using higher frequency bands do not come for free. The cost of the UE and BS equipment increases with the carrier frequency, as more sophisticated analogue circuit components are needed. Moreover, a larger transmit power is required for both activating the more subcarriers existing in a wider bandwidth and compensating for the higher path losses at higher frequency bands [28]. Figure 2.8 shows how both the transmit power per active BS and the overall transmit power used by the network increase with the wider bandwidth and the larger path losses, where this increase is not negligible and up to 24 dB. This transmit power increase is prohibitive in macrocell BSs where the required power would be up to 70 dBm, but it is still well-suited to ultra-dense small cell deployments where BSs could operate this large bandwidth with less than 20 dBm of transmit power. In order to reduce transmit power, the identification and characterization of hotspots becomes critical [29,30]. Deploying BSs where they are most needed, where the UEs are, for example, in the middle of a hotspot, will decrease the average UE path loss to the serving BS, and reduce this transmit power. Activating only those subcarriers with good channel quality that are necessary to achieve the required UEs throughput targets will also help to reduce transmit power.

From these results, one can see that, when combining network densification with increased bandwidth, the targeted average 1 Gbps per UE is reachable with an ISD of 50 m and 500 MHz bandwidth, or 35 m ISD and 250 MHz bandwidth. The first and the second combination resulting in an average of 1.27 Gbps and 1.01 Gbps per UE, respectively. Thus, it can be concluded that the usage of larger cellular bandwidth than today's 100 MHz is required to meet the targeted high data rates of around 1 Gbps or larger.

Taking the usage of higher frequency bands to an extreme, vendors and operators have also started to look at exploiting mm-wave with carrier frequencies on 22, 60, and 77 GHz, where the available bandwidth is enormous, ≥1 GHz [31]. However, diffraction and penetration through obstacles are hardly possible at these high frequency bands, and thus only LOS or near-LOS links seem feasible. In addition, the range of the cell may be confined by high atmospheric phenomena. Water and oxygen absorption significantly increase path losses, especially at 22 GHz and 60 GHz, respectively. As a result, providing the required coverage range through larger transmit powers is not feasible anymore, as it is in the sub-10 GHz bands.

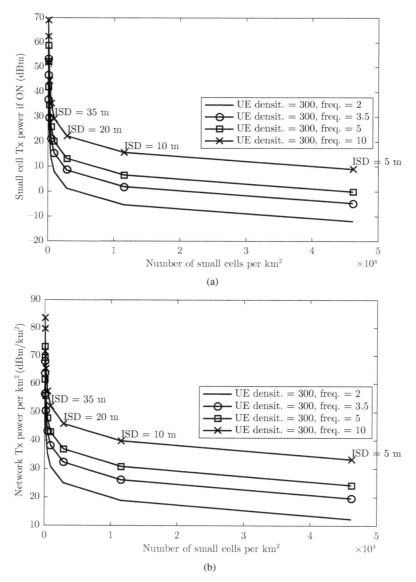

Figure 2.8. Transmit power per active BS (a) and transmit power of the network per km² (b) for different network configurations where the carrier frequencies are 2.0, 3.5, 5.0, or 10 GHz. The rest of the parameters are d = 300 UE/km², ud = 1, s = 1, f = 2 GHz, a = 1 and t = 12 dB.

Thus, it is expected that active antenna arrays and beamforming techniques become essential to overcome the increased path losses at these high frequency bands [32].

2.5 MULTI-ANTENNA TECHNIQUES AND BEAMFORMING

Previous results showed that for a single-antenna small cell BS, the targeted average 1 Gbps per UE is only reachable with an ISD of at least 50 m and a bandwidth of 500 MHz (or 35 m and a bandwidth of 250 MHz). In order to enhance UE performance and bring down this still large BS density, the usage of multiple antennas at the small cell BS is explored in the following.

Multiple antenna systems provide a number of degrees of freedom for transmitting information, which may vary from one to a number upper bounded by the number of antennas. The higher the number of degrees of freedom available, the better spectral efficiency can be expected. However, how many effective degrees of freedom are available is mostly related to the spatial correlation of the channels [33]. Given a number of degrees of freedom available, two multi-antenna techniques stand out. Beamforming makes use of only one of the degrees of freedom available, while spatial multiplexing may use all of them. Beamforming benefits from a low-complexity implementation, together with its ability to extend the cell range by focusing the transmit power in a certain direction. However, it may be suboptimal in terms of spectral efficiency. In contrast, spatial multiplexing has the potential to approach the maximum channel capacity, linearly increasing the capacity of the channel with the minimum of number of transmit and receive antennas. However, it is significantly more complex to implement [33].

Partly due to its simplicity and partly due to the fact that the small cell sizes in an ultra-dense deployment may suffer from a large spatial correlation of the channels, which limits the degrees of freedom available and renders spatial multiplexing unsuitable, this section focuses on beamforming. In more detail, it focuses on quantized maximal ratio transmission (MRT) beamforming, using the standardized LTE code book beamforming approach [34]. The effectiveness of spatial multiplexing in ultra-dense small cell deployments is left for future analysis.

In the considered model, it is assumed that each BS is equipped with a horizontal antenna array comprised of 1, 2, or 4 vertical dipole arrays, and each UE has a single antenna. The existing transmit power is equally distributed among antennas. The characteristics of the vertical dipole array that serves as antenna unit in this analysis are described in Appendix Section A.3.2.2. Based on measurements over BS pilots signals, the UE suggests to the BS the precoding weights specified in the LTE code book [35] that maximize its received signal strength, and the BS follows such suggestion. This horizontal beamforming helps to shape the horizontal antenna pattern at the BS and thus focus the energy toward the UE in the horizontal plane. Interference mitigation is also achieved in an opportunistic manner [36]. Real-time inter-BS coordination required for cooperation is not supported.

Figure 2.9 shows the average and 5th percentile UE throughput for different network configurations where the carrier frequencies are 2.0 and 10 GHz and the number of antennas per BS are 1, 2, or 4. Control channels are not beamformed. The targeted SNRs by the small cell BS at $\sqrt{3}/2$ of the ISD is 12 dB, thus assuring a constant coverage area regardless of the frequency band. The constant coverage is maintained at the expense of larger transmit power, as explained earlier. Different remarks can be made based on the results.

Beamforming gains increase in a diminishing manner with the number of antennas. This is in line with the linear antenna array theory, which indicates that beamforming antenna gains increase logarithmically with the number of antennas [37]. Looking at the average UE throughput for the 35 m ISD case with a 500 MHz bandwidth as a base line, the average gain of using 2 antennas over 1 antenna is 13.77% (1 more antenna is needed), while the gain of using 4 antennas over 2 antenna is 12.61% (2 more antennas are needed). The overall average gain from 1 to 4 antennas is 28.11%.

Beamforming gains are larger at the cell edge. This is because of the interference mitigation provided by the beamforming, which is more noticeable at the cell edge. Looking at the cell-edge UE throughput for the 35 m ISD case with a 500 MHz bandwidth as baseline, the cell-edge gain of using 2 antennas over 1 antenna is 46.67% (1 antenna increase), while the gain of using 4 antennas over 2 antennas is 38.64% (2 antennas increase). The overall cell-edge gain from 1 to 4 antennas is 103%, that is, 2×.

Beamforming gains are larger for larger cell sizes. This is because the beamforming helps improving the received signal strength of UEs, which may receive only a poor received signal strength if beamforming is not in place. For the 200 m ISD and the 5 m ISD cases with a 500 MHz bandwidth, the average gains of using 2 antennas over 1 antenna are 19.05% and 9.23%, respectively (1 antenna increase), while the gains of using 4 antennas over 2 antenna are 17.37% and 8.87%, respectively (2 antennas increase).

From the results, one can see that the targeted average 1 Gbps per UE is now reachable with an ISD of 75 m, 500 MHz bandwidth and 4 antennas (or 35 m, 250 MHz bandwidth and 4 antennas), showing that multi-antenna techniques enhance UE performance and reduce the required BS density to meet the targeted UE demands. However, overall, beamforming gains are estimated to be on the order of up to 2×, which are low compared to the larger gains provided by network densification and use of higher frequency bands. Therefore, since the larger path loss at the frequency bands between 2 and 10 GHz could be compensated via larger transmit powers, the existing antennas at the small cell may be better exploited by using spatial multiplexing, which has the potential to linearly increase performance with the number of antennas provided that the required degrees of freedom exist in the channel. Spatial multiplexing in ultra-dense small cell networks is one of our future lines of research.

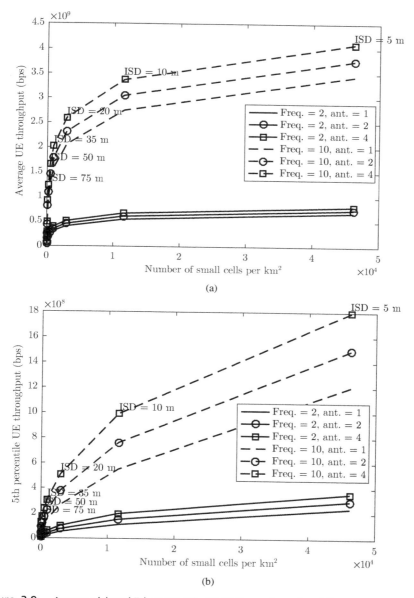

Figure 2.9. Average (a) and 5th percentile UE (b) throughput for different network configurations where the carrier frequencies are 2.0 or 10 GHz and the number of antennas per BS are 1, 2 or 4. The rest of the parameters are $d = 300$ UE/km^2, $ud = 1$, $s = 1$, $f = 2$ GHz and $t = 12$ dB.

Finally, it is important to note that although the beamforming gains are the lowest in absolute values, it may be the most cost-effective solution since implementing more antennas at the small cell BS is cheaper than deploying more BSs or purchasing more spectrum.

2.6 SCHEDULING

In LTE, a resource block (RB) pair, here also called RB for simplicity, refers to the basic time/frequency scheduling resource unit to which a UE can be allocated. Each RB expands 180 kHz in the frequency domain and its 1 ms transmission time interval (TTI) is referred to as a subframe. Depending on the bandwidth of the system, a different number of RBs are available for allocation. For example, a network with 10 MHz has 50 RBs and one with 20 MHz has 100 RBs.

One of the principal enhancements of LTE over the universal mobile telecommunication system (UMTS) is its channel-dependent scheduling, which takes advantage of multi-path fading by allocating to each RB the UE having the best conditions according to a given metric [38], for example, *proportional fair (PF)*. Such type of scheduling leads to multi-user diversity gains, which have been shown to roughly follow a double logarithmic scaling law in terms of capacity with regard to the number of UEs per BS in macrocell scenarios [39]. However, it is important to note that these gains come at the cost of complexity and overhead compared to schedulers that do not consider channel conditions, for example, *round robin (RR)*. In order to aid the channel-sensitive scheduling and exploit multi-user diversity gains, UEs need to report downlink (DL) channel quality indicator (CQI) back to their serving BSs, which allows the scheduler to assess the UE channel quality and perform the scheduling according to the specified metric.

In the following, the performance of RR and PF schedulers is analyzed under different densification levels. For a more detailed description of PF and RR schedulers, please refer to Appendix Section A.9.1. Moreover, it is important to note that in order to consider the multi-path fast-fading effect in this section, the BS to UE channel gain $G_{m,u}$[dB] also comprises the multi-path fast fading gain, which is modeled using a distant dependent Rician channel model. The lower the UE to serving BS distance, the lower the channel fluctuations due to multi-path fading, as the probability of LOS is higher [20].

Figure 2.10 shows the performances of RR and PF schedulers in terms of cell throughput with respect to the number of served UEs per BS for different ISDs. Let us look first at the trends. When using RR, the number of served UEs per BS does not impact the cell throughput, since RR does not take into account the UE channel quality, and therefore does not take advantage of multi-user diversity. In contrast, PF is able to benefit from multi-user diversity, and the cell throughput increases with the number of served UEs per BS up to a certain extent. There is a point in which having more UE per BS does not bring more cell throughput gains, and this point is lower

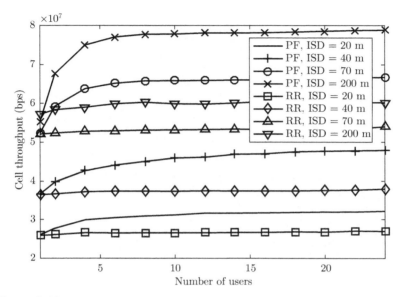

Figure 2.10. Average cell throughput for various ISDs with RR and PF schedulers.

with the network density. For example, for an ISD of 200 m, having more than 8 UEs per cell does not noticeably increase cell throughout, while for an ISD of 20 m, this number is reduced to 6 UEs. But, why is this point lower with the network density? This is because the more cells, the stronger LOS propagation and less fluctuating channel conditions (diversity decreases), which make multi-user diversity gains less noticeable. It order to corroborate this, note that although PF always outperforms RR in terms of cell throughput (see Fig. 2.10), it starts losing its advantage with respect to RR in terms of cell throughput with the reduced cell size. For a given number of UE per BS, for example, 4, the PF gain over RR reduces from 21.2% to 12.4% and 10.5% for ISDs of 150, 40 and 20 m, respectively.

Shrinking the cell size not only reduces cell throughput but also reduces UE throughput. Using a PF scheduler, Fig. 2.11 shows that as a result of both interference enhancement due to NLoS to LoS transition as well as multi-user diversity loss due to low channel fluctuations, the UE throughput for a given number of served UEs per BS drops down with network densification. For instance, reducing the ISD from 150 m to 40 m and 20 m, the UE 5th percentile throughput drops by ∼40.8% and ∼36.7%, respectively. Comparing the UE 5th percentile throughput of PF and RR for an ISD of 20m, the gain of the former is almost negligible, around ∼9%.

The minor gains of PF scheduler over the RR one at low ISDs in terms of cell and UE throughput suggests that RR scheduler may be a good alternative choice in dense small cell deployments considering its lower complexity. The complexity of PF

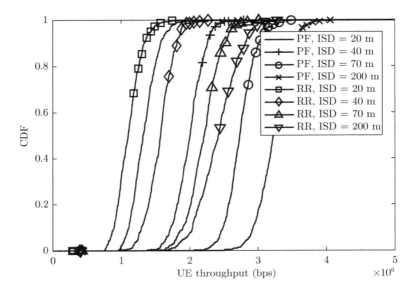

Figure 2.11. CDF of UE throughput for different ISDs.

lies in the evaluation of each UE on each RB considering a greedy PF that operates on a per RB. The PF complexity with exhaustive search is considerably higher. This conclusion may have a significant impact in the manufacturing of small cell BSs where the digital signal processor (DSP) cycles saved due to the adoption of RR scheduling can be used to enhance the performance of other embedded technologies.

2.7 ENERGY EFFICIENCY

Deploying a large number of small cells—millions, or tens of millions—poses some concerns in terms of energy consumption. For example, the deployment of 50 million femtocells consuming 12 Watts each will lead to an energy consumption of 5.2 TWh/a, which is equivalent to half the power produced by a nuclear plant [40]. This approach does not scale, and thus the energy efficiency of ultra-dense small cell deployments should be carefully considered to allow the deployment of sustainable networks.

In Section 2.3, Fig. 2.3b showed how the overall transmit power used by the network significantly reduces with the small cell BS density when an efficient idle mode capability is used. This is because the reduction of transmit power per cell outweighs the increased number of active cells. However, this observation may not hold when considering the total power consumption of each small cell BS, since a BS in idle mode may still consume a non-negligible amount of energy, thus impacting the energy efficiency of the network. In order to better understand the impact of network densification on the power consumption, the energy efficiency of ultra-dense small cell networks in terms of throughput per Watt (bps/W) is analyzed in the following. This study uses the power model developed in the GreenTouch project [41]. This power model estimates the power consumption of a cellular BS, and is based on tailored modeling principles and scaling rules for each BS component, that is, power amplifier, analogue front-end, digital base band, digital control and backhaul interface and power supply. Moreover, it provides a large flexibility, that is, multiple BS types are available, which can be configured with multiple parameters (bandwidth, transmit power, number of antenna chains, system load, etc.) and includes different optimized idle modes. The idle mode profile is indicated by sm. Among the provided idle modes, we considered the GreenTouch slow idle mode ($sm = 1$) and the GreenTouch shut-down state ($sm = 2$), where most or all components of the BS are deactivated, respectively. These two modes are the most efficient ones provided by the model, that is, result in the lowest power consumption while in idle mode.

Tables 2.1, 2.2, and 2.3 show the estimated power consumption for different small cell BS types with different transmit power, where such power consumption is given for the active mode as well as the slow idle mode ($sm = 1$) and the shut-down state ($sm = 2$), and for BSs with 1, 2, or 4 antennas. Note that we used the 2020 small cell BS type in the model and that the presented analysis considers a 20 MHz bandwidth. In contrast to macrocell BSs, it is important to note that the power consumption of a small cell BS linearly scales with the number of antenna chains, since it is the most contributing component to power consumption in a small cell BS.

Based on these values and the throughput analysis in previous sections, Fig. 2.12 shows the energy efficiency in bps/W for different network configurations with 1, 2, or 4 antennas per BS. In addition to the slow idle mode ($sm = 1$) and the shut-down mode ($sm = 2$) models provided by the GreenTouch project, two futuristic idle modes

TABLE 2.1 Power Consumption with All BS in Active Mode

Small Cell ISD (m)	Tx Power (dBm/mW)	Consumed Power (W)		
		1 Antenna	2 Antenna	4 Antenna
200	23.27/212.32	1.8923	2.5848	4.4560
150	20.52/112.72	1.3405	2.0316	3.9015
100	16.64/46.13	0.9793	1.6696	3.5386
75	13.90/24.55	0.8643	1.5544	3.4231
50	10.02/10.05	0.7853	1.4752	3.3437
35	6.61/4.58	0.7558	1.4456	3.3141
20	1.27/1.34	0.7383	1.4281	3.2965
10	−5.20/0.30	0.7326	1.4224	3.2908
5	−11.89/0.06	0.7314	1.4211	3.2895

TABLE 2.2   Power Consumption with Idle Mode 1 ($sm = 1$)

Small Cell ISD (m)	Tx Power (dBm/mW)	Consumed Power (W)		
		1 Antenna	2 Antenna	4 Antenna
200	23.27/212.32	0.2324	0.3105	0.4959
150	20.52/112.72	0.2191	0.2971	0.4825
100	16.64/46.13	0.2104	0.2884	0.4738
75	13.90/24.55	0.2076	0.2856	0.4710
50	10.02/10.05	0.2057	0.2837	0.4691
35	6.61/4.58	0.2050	0.2830	0.4683
20	1.27/1.34	0.2046	0.2826	0.4679
10	−5.20/0.30	0.2044	0.2824	0.4678
5	−11.89/0.06	0.2044	0.2824	0.4678

are considered, where their energy consumption is 15% ($sm = 3$) or 0% ($sm = 4$) of the GreenTouch slow idle mode power consumption model ($sm = 1$). The former model accounts less energy consumption than the shut-down mode ($sm = 2$), and the latter model assumes that the BS consumes nothing in idle mode.

For any given idle mode, results show that increasing the number of antennas at the small cell BS always decreases the energy efficiency. This is because the performance gain provided by adding a new antenna through beamforming is not large enough to cope with the increase in power consumption resulting from adding a new antenna chain at the small cell BS. A different conclusion may be obtained for the macrocell case, where adding a new antenna chain does not lead to a large increase in the total power consumption of the BS, since other components consume much more energy. Conclusions for the small cell BS case may also be different when considering spatial multiplexing instead of beamforming. Indeed, provided that the required

TABLE 2.3   Power Consumption with Idle Mode 2 ($sm = 2$)

Small Cell ISD (m)	Tx Power (dBm/mW)	Consumed Power (W)		
		1 Antenna	2 Antenna	4 Antenna
200	23.27/212.32	0.1881	0.2478	0.4073
150	20.52/112.72	0.1748	0.2345	0.3939
100	16.64/46.13	0.1661	0.2257	0.3852
75	13.90/24.55	0.1633	0.2230	0.3824
50	10.02/10.05	0.1614	0.2210	0.3804
35	6.61/4.58	0.1607	0.2203	0.3797
20	1.27/1.34	0.1603	0.2199	0.3793
10	−5.20/0.30	0.1601	0.2198	0.3792
5	−11.89/0.06	0.1601	0.2197	0.3791

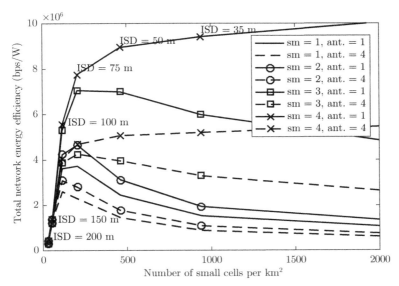

<u>Figure 2.12.</u> Energy efficiency in bps/W for different network configurations where the carrier frequency is 2.0 GHz, the bandwidth is 20 MHz, and the number of antennas per BS are 1, 2, or 4. The rest of the parameters are d = 300 UE/km², ud = 1, s = 1, f = 2 GHz and t = 12 dB.

degrees of freedom exist in the channel, the performance of spatial multiplexing in terms of capacity can follow a linear scaling law with the minimum number of transmit and receive antennas, and this capacity boost will enhance energy efficiency. However, it is important to note that in ultra-dense deployments, the strong LOS limits the multiplexing gain too.

When comparing the performance of the different idle modes in terms of energy efficiency, it can be seen that the lower the power consumption in the idle mode, the larger the energy efficiency of the network. This is because less energy is required to transmit the same amount of bits at the network level. When using the idle mode models provided by the GreenTouch project, idle modes 1 and 2, energy efficiency decreases with densification. This is because the increase in throughput provided by adding more cells is not large enough compared to the increase in their power consumption, mostly because idle cells following the GreenTouch project model still consume a non-negligible amount of energy. When considering the energy consumption of the futuristic idle mode 3, this trend starts changing. First, the energy efficiency increases with densification, and then starts decreasing when the number of deployed cells becomes large and many cells are empty. When considering the energy consumption of the new idle mode 4, idle cells do not consume anything, and the trend significantly changes again. Energy efficiency always increases with densification, and it does it in a significant manner. This is because the increase in throughput provided by adding more cells comes now at no or very low energy cost since idle cells do not consume power from the energy grid.

This last observation shows the need for the development of more advanced idle mode capabilities, where the power consumption of a small cell BS from the energy grid in idle mode is zero. This can be optimally realized by using energy harvesting approaches that are able to supply sufficient power to keep the small cell BS alive when it is in idle mode. For example, assuming the use of idle mode 2 and 4 antennas at the small cell BS, the energy harvesting mechanism would need to provide the affordable amount of 0.4073 W per cell to make ultra-dense small cell deployments highly energy efficient. In this area, researchers are looking at both thermoelectric as well as mechanical energy harvesting techniques. An example of the latter is the research in [42] and [43], based on vibration energy harvesting techniques for powering wireless sensors and small cell infrastructure. Moreover, an issue with all ambient environment energy sources, to varying degrees, is their intermittent nature. For example, harvesting useable levels of solar and wind power is dependent on the time of the day, the season, and the local weather conditions. For this reason, an efficient and cost-effective energy storage system is also key [44]. These are important areas of research, whose results are vital to deploy energy efficient and green telecommunications networks that have a minimal impact on the ecosystem.

2.8 WHAT IS DIFFERENT IN ULTRA-DENSE SMALL CELL DEPLOYMENTS

In this section, the main differences between regular HetNets and ultra-dense HetNets are highlighted, with their implications in network modeling.

Difference 1: BS to UE density. In regular HetNets, the UE density is larger than the BS density, while in ultra-dense HetNets, the UE density is smaller than the BS density. As a result, and in contrast to regular HetNets where the BS are powered on most of the time, in ultra-dense HetNets, BSs with no active UEs should be powered off to reduce unnecessary interference and save power.

Difference 2: Propagation conditions. In regular HetNets, NLoS interferers count for most cases, while in ultra-dense HetNets, LoS interferers count for most cases. As a result, and in contrast to regular HetNets where simple single-slope path-loss model is usually assumed to obtain the numerical and analytic results, in ultra-dense HetNets, more sophisticated and practical path-loss models should be considered.

Difference 3: Diversity loss. In regular HetNets, there is a rich UE diversity, while in ultra-dense HetNets, there is a limited UE diversity. As a result, and in contrast to regular HetNets, where independent shadowing and multi-path fading among UEs in one cell is usually assumed to obtain the numerical and analytic results, in the ultra-dense HetNets, such assumption is not true. Thus, new results for ultra-dense HetNets should be derived.

Because of these differences, the numerical and analytic results for regular HetNets cannot be directly applied to ultra-dense HetNets due to different assumptions originated from the fundamental characteristics of the networks.

2.9 SUMMARY AND CONCLUSIONS

In this chapter, the gains and limitations of network densification, use of higher frequency bands and multi-antenna techniques have been analyzed. Network densification has been shown to provide the highest UE throughput gains of up to 18× in average and 48× at the cell edge. The usage of higher frequency bands up to 10 GHz has been shown to provide an appealing linear scaling of the network capacity with the available bandwidth of up to 5×. However, this linear scaling comes at the expense of more transit power in order to light up the higher number of subcarriers and to overcome the larger path losses. Beamforming gains are on the order of 2× at the cell edge, and are low compared to the gains achievable by network densification and the use of higher frequency bands. Therefore, using the antennas for spatial multiplexing seems more appealing compared to beamforming, provided that the channels between the antennas are not too correlated. It is important to note that although the beamforming gains are the lowest in absolute values, it may be the most cost-effective solution since implementing more antennas at the small cell BS is cheaper than deploying more BSs or purchasing more spectrum.

Efficient idle mode capabilities at the small cells have been shown to be key to mitigate interference and save energy. Moreover, one UE per cell has been shown to be the fundamental limit of spatial reuse and may drive the cost-effectiveness of ultra-dense small cell deployments. As a result, UE density and distribution have to be well understood before any rollout.

Network densification has also been shown to reduce multi-user diversity, and thus PF alike schedulers lose their advantages with respect to RR ones. For an ISD of 20 m, the difference in average UE throughput of schedulers over RR is only of around 5%, indicating that RR may be a good alternative for ultra-dense small cell deployments due to their reduced complexity.

As a bottom line, simulation results have shown that for a realistic non-uniformly distributed UE density of 300 active UE per square km, it is possible to achieve an average 1 Gbps per UE with roughly an ISD of 35 m, 250 MHz bandwidth and 4 antennas per small cell BS. Today, these deployments are neither cost-effective nor energy efficient. However, this may change in the future with new deployment models, lower cost and more efficient equipment.

REFERENCES

1. T. Halonen, J. Romero, and J. Melero, *GSM, GPRS and EDGE Performance: Evolution Towards 3G/UMTS*. John Wiley & Sons, 2004.

2. D. Wu, Y. T. Hou, W. Zhu, Y. Zhang, and J. M. Peha, "Streaming video over the Internet: approaches and directions," *IEEE Trans. Circuits Systems Video Technol.*, vol. 11, pp. 282–300, 2001.

3. E. Setton, T. Yoo, X. Zhu, A. Goldsmith, and B. Girod, "Cross-layer design of ad-hoc networks for real-time video streaming," *IEEE Wireless Commun.*, vol. 12, no. 4, pp. 59–65, Aug. 2005.

4. X. Chu, D. López-Pérez, Y. Yang, and F. Gunnarsson, *Heterogeneous Cellular Networks: Theory, Simulation and Deployment*. University Cambridge Press, 2013.

5. H. Claussen, L. T. W. Ho, and L. G. Samuel, "An overview of the femtocell concept," *Bell Labs Tech. J.*, vol. 13, no. 1, pp. 221–245, May 2008.

6. V. Chandrasekhar, J. G. Andrews, and A. Gatherer, "Femtocell networks: A survey," *IEEE Commun. Mag.*, vol. 46, no. 9, pp. 59–67, Sep. 2008.

7. J. G. Andrews, H. Claussen, M. Dohler, S. Ragan, and M. C. Reed, "Femtocells: Past, present, and future," *IEEE J. Sel. Areas Commun. (JSAC), Spec. Issue: Femtocell Netw.*, vol. 30, no. 3, pp. 497–508, Apr. 2012.

8. 3GPP TR 36.808, "Evolved universal terrestrial radio access (E-UTRA); carrier aggregation; base station (BS) radio transmission and reception," 3GPP, Tech. Rep., 2013.

9. 3GPP TR 36.889, "Feasibility study on licensed-assisted access to unlicensed spectrum," 3GPP, Tech. Rep., 2014.

10. T. S. Rappaport, R. W. Heath Jr., R. C. Daniels, and J. N. Murdock, *Millimeter Wave Wireless Communications*. Prentice Hall, 2014.

11. L. Hanzo, Y. Akhtman, L. Wang, and M. Jiang, *MIMO-OFDM for LTE, WiFi and WiMAX: Coherent versus Non-coherent and Cooperative Turbo Transceivers*. John Wiley & Sons, 2011.

12. M. Ding and L. Hanwen, *Multi-point Cooperative Communication Systems: Theory and Applications*. Springer, 2013.

13. Z. Shen, A. Khoryaev, E. Eriksson, and X. Pan, "Dynamic uplink-downlink configuration and interference management in TD-LTE," *IEEE Commun. Mag.*, vol. 50, no. 11, pp. 51–59, Nov. 2012.

14. M. Ding, D. López-Pérez, W. Chen, and A. Vasilakos, "Dynamic TDD transmissions in homogeneous small cell networks," in *IEEE Int. Conf. Commun. (ICC)*, Sydney, Australia, Jun. 2014.

15. M. Ding, D. López-Pérez, R. Xue, W. Chen, and A. Vasilakos, "Small cell dynamic TDD transmissions in heterogeneous networks," in *IEEE Int. Conf. Commun. (ICC)*, Sydney, Australia, Jun. 2014.

16. D. López-Pérez, M. Ding, H. Claussen, and A. H. Jafari, "Towards 1 Gbps/UE in cellular systems: Understanding ultra-dense small cell deployments," *IEEE Commun. Surveys Tuts.*, vol. 18, no. 6, pp. 983–986, 2015.

17. J. Deissner and G. Fettweis, "A study on hierarchical cellular structures with inter-layer reuse in an enhanced GSM radio network," in *IEEE Int. Workshop Mobile Multimedia Commun. (MoMuC '99)*, 1999, pp. 243–251.

18. H. Claussen, L. T. W. Ho, and L. G. Samuel, "An overview of the femtocell concept," *Bell Labs Tech. J.*, vol. 13, no. 1, pp. 221–245, May 2008.

19. I. Ashraf, L. T. W. Ho, and H. Claussen, "Improving energy efficiency of femtocell base stations via user activity detection," in *Proc. IEEE Wireless Commun. Netw. Conf. (WCNC)*, Sydney, Australia, Apr. 2010.

20. A. Jafari, D. López-Pérez, M. Ding, and J. Zhang, "Study on scheduling techniques for ultra dense small cell networks," in *IEEE Veh. Technol. Conf. (VTC-Fall)*, Boston, USA, Sep. 2015, pp. 1–6.

21. C. E. Shannon, "A mathematical theory of communication," *Bell System Tech. J.*, vol. 27, pp. 379–423, Jul. 1948.

22. M. Ding, P. Wang, D. López-Pérez, G. Mao, and Z. Lin, "Performance Impact of LoS and NLoS Transmissions in Dense Cellular Networks," *IEEE Trans. Wireless Commun.*, vol. 15, no 3, Page(s):1536–1276, 2015.

23. M. Ding, D. López-Pérez, G. Mao, P. Wang, and Z. Lin, "Will the area spectral efficiency monotonically grow as small cells go dense?" in *IEEE Global Telecommun. Conf. (GLOBECOM)*, San Diego, USA, Dec. 2015, pp. 1–7.

24. ETSI MCC, "Draft Report of 3GPP TSG RAN WG1 #77," 3GPP TSG RAN WG1 Meeting #77, Seoul, Korea, Tech. Rep., May 2014.

25. J. G. Andrews, F. Baccelli, and R. K. Ganti, "A tractable approach to coverage and rate in cellular networks," *IEEE Trans. Commun.*, vol. 59, no. 11, pp. 3122–3134, Nov. 2011.

26. S. Mukherjee, "Distribution of downlink SINR in heterogeneous cellular networks," *IEEE J. Sel. Areas Commun. (JSAC)*, vol. 30, no. 3, pp. 575–585, Apr. 2012.

27. 3GPP TR 36.814, "Evolved universal terrestrial radio access (E-UTRA); further advancements for E-UTRA physical layer aspects," 3GPP-TSG R1, Tech. Rep. v 1.0.0.

28. Y. Kishiyama, A. Benjebbour, T. Nakamura, and H. Ishii, "Future steps of LTE-A: Evolution toward integration of local area and wide area systems," *IEEE Wireless Commun.*, vol. 20, no. 1, pp. 12–18, Feb. 2013.

29. F. Gustafsson and F. Gunnarsson, "Mobile positioning using wireless networks: possibilities and fundamental limitations based on available wireless network measurements," *IEEE Signal Process. Mag.*, vol. 22, no. 4, pp. 41–53, Jul. 2005.

30. K. Klessig and V. Suryaprakash, "Spatial traffic modelling for architecture evolution," Green Touch, Tech. Rep., 2014.

31. Z. Pi and F. Khan, "An introduction to millimeter-wave mobile broadband systems," *IEEE Commun. Mag.*, vol. 49, no. 6, pp. 101–107, Jun. 2011.

32. E. Larsson, O. Edfors, F. Tufvesson, and T. Marzetta, "Massive MIMO for next generation wireless systems," *IEEE Commun. Mag.*, vol. 52, no. 2, pp. 186–195, Feb. 2014.

33. M. Sanchez-Fernandez, S. Zazo, and R. Valenzuela, "Performance comparison between beamforming and spatial multiplexing for the downlink in wireless cellular systems," *IEEE Trans. Wireless Commun.*, vol. 6, no. 7, pp. 2427–2431, Jul. 2007.

34. F. Khan, *LTE for 4G Mobile Broadband: Air Interface Technologies and Performance.* University Cambridge Press, 2009.

35. 3GPP TS 36.211, "Evolved universal terrestrial radio access (E-UTRA); physical channels and modulation," v.13.5.0, Mar. 2017. [Online]. Available: https://portal.3gpp.org/desktopmodules/Specifications/SpecificationDetails.aspx?specificationId=2425

36. P. Viswanath, D. N. C. Tse, and R. Laroia, "Opportunistic beamforming using dumb antennas," *IEEE Trans. Inf. Theory*, vol. 48, no. 5, pp. 1277–1294, Jun. 2002.

37. C. A. Balanis, *Antenna Theory: Analysis and Design*. John Wiley & Sons Ltd, 2005.

38. S. Sesia, I. Toufik, and M. Baker, *LTE: The UMTS Long Term Evolution, from Theory to Practice*. John Wiley & Sons Ltd, Feb. 2009.

39. M. Sharif and B. Hassibi, "A comparison of time-sharing, DPC, and beamforming for MIMO broadcast channels with many users," *IEEE Trans. Commun.*, vol. 55, no. 1, pp. 11–15, Jan. 2007.

40. H. Claussen, "Future Cellular Networks," *in IEEE Wireless Communications and Networking Conference (WCNC)*, Paris, France, Apr. 2012.

41. B. D. Claude Desset and F. Louagie, "A flexible and future-proof power model for cellular base stations," in *IEEE Veh. Technol. Conf. (VTC-Spring)*, Nanjing, China, May 2015, pp. 1–7.

42. D. O'Donoghue, V. Nico, R. Frizzell, G. Kelly, and J. Punch, "A multiple-degree-of-freedom velocity-amplified vibrational energy harvester: Part A – experimental analysis," *Proc. Amer. Soc. Mech. Eng.*, vol. 2, Sep. 2014.

43. V. Nico, D. O'Donoghue, R. Frizzell, G. Kelly, and J. Punch, "A multiple degree-of-freedom velocity-amplified vibrational energy harvester: Part B – modelling," *Proc. Amer. Soc. Mech. Eng.*, vol. 8, Sep. 2014.

44. K. Divya and J. Oestergaard, "Battery energy storage technology for power systems – an overview," *Elect. Power Syst. Res. J.*, vol. 79, no. 4, pp. 511–520, 2009.

3

AUTOMATION OF CELLULAR NETWORKS

3.1 INTRODUCTION

The increasingly rapid evolution of the telecommunication industry and complex technology migration implies that the traditional revenue avenues of mobile network operators (MNOs) are increasingly challenging. Conventionally, voice services and, more recently, provisioning of data are considered as key revenue streams for most operators. The advances in consumer electronics and smart devices have resulted in the fast development of over-the-top (OTT) applications too, which have shaped a new landscape posing a significant challenge for the operators. This, together with growth of mobile virtual network operators (MVNOs), has intensified the competition and has resulted in reduced margins for the operators. In addition to this, consumers have become much more demanding in terms of the required quality of experience (QoE), and the required network capacity to fulfill this will need to increase by a factor of 100× or more over the next 10 years (see Chapter 1). Finally, regulatory issues have obliged operators to provide coverage in areas which might not be necessarily appealing from the business perspective. With all these pressures and limited opportunities for new revenue streams, reducing capital expenditure (CAPEX) and operational expenditure (OPEX) can be envisioned as a key way for operators to improve their cost position.

Small Cell Networks: Deployment, Management, and Optimization, First Edition. Holger Claussen, David López-Pérez, Lester Ho, Rouzbeh Razavi, and Stepan Kucera.
© 2017 by The Institute of Electrical and Electronic Engineers, Inc. Published 2017 by John Wiley & Sons, Inc.

Moving toward automation and self-managed networks is a big step toward reducing OPEX. This is especially true considering the migration of the network deployments toward a more distributed architecture, in which small cells are expected to play a critical role to improve network capacity and coverage. In addition, the increased competition and consumers demands, together with the rapid evolution of the technology and introduction of new services as a result, require operators to maximize the pace of their network rollout, while minimizing service disruptions and poor user equipment (UE)s' QoE during the deployment phase. With this objective in mind, network automation can significantly facilitate and enhance the network rollout process.

The introduction of self-organizing networks (SONs) by the next-generation mobile networks (NGMN) alliance has made the realization of network automation to become more closer to the reality. The main objective of such algorithms is to simultaneously eliminate manual configuration and to improve performance, hence reducing costs while increasing revenue opportunities by improving UEs' QoE. In addition to the cost and market considerations, the ever-increasing trend of mobile data usage has resulted in the evolution of the mobile network architecture toward more complex multi-tier, multi-radio access technology (RAT) and multi-frequency deployments. Subsequently, network automation and SON functionality are critically important for effective management of such complex future mobile networks.

This chapter provides an overview of the areas where SON functions can be applied. The main use cases introduced by the Third-Generation Partnership Project (3GPP) are briefly described and discussed. In addition, key requirements for SON mechanisms and various methods to implement intelligence across the network are explained. Finally, two SON case studies are presented and studied.

3.2 SELF-ORGANIZING NETWORK USE CASES AND STANDARDIZATION

While some of the SON functionality can be implemented in a fully vendor-specific fashion, in many scenarios, network automation requires a number of network elements to communicate and exchange information. This implies that there is a need for active standardization of such processes. As a result, shortly after the introduction of SON use cases by NGMN in 2007, the 3GPP standardization activities started in order to assure unified ways of implementing interfaces, signaling and measurements amongst different vendors, while the core logic behind such SON functionality still remains vendor-specific. Later on, the NGMN released operation efficiency recommendations in 2010. Generally, the 3GPP SON use cases are closely related to NGMN recommendations. This section briefly introduces these use cases and describes why they are important for network operators.

3.2.1 Base Station Self-Configuration

When a new base station (BS) is installed, many different aspects of its configuration must be provided. This includes establishing and maintaining the transport links and connection to the core network, antenna and power level configurations, as well as neighbor relations. Performing these tasks manually is usually very time-consuming, and there is a possibility of human errors [1].

Self-configuration of BSs reduces the effort required for network deployment and significantly improves the rollout time. Self-configuration is rather a broad concept, which involves various functions including automatic software management, self-testing and automatic configuration of neighbor relations. The 3GPP has considered the latter as a separate use case. In order to have self-configuration functionality, the BS must have the following elements [2]:

- A process control mechanism to assure commissioning sequence of steps
- A dynamic host control protocol (DHCP) client to obtain the initial IP address configuration
- A certificate client to obtain and store relevant certificates
- An auto-connection client to establish the connection to the auto-connection server

3.2.2 Automatic Neighbor Relation

One of the most labor-intense tasks of operators is to handle the neighbor relations of cells to assure successful handovers. Migrating to a heterogeneous network (Het-Net) architecture consisting of small cells, which are most of the time deployed in an *ad hoc* manner, has made this problem significantly more complex [3]. Consequently, automatic neighbor relation (ANR) is considered an important SON feature, and one of the first functions considered for 3GPP standardization.

In summary, ANR enables the automatic discovery and setup of neighbor relations using the help of UE measurements. More specifically, once a radio resource control (RRC) connection between a UE and the serving cell is established, the UE reports all detected physical layer cell identities (PCIs) of neighboring cells that satisfy the measurement criteria back to the serving cell.

If a reported PCI does not correspond to a known neighbor of the serving cell, it requests the UE to report the global cell identity (GCI) of the cell with the unknown PCI. GCI information is broadcasted by all cells, hence the UE can retrieve and report this information to the serving cell. Thereafter, the serving cell contacts the mobility management entity (MME) to retrieve the Internet protocol (IP) address of the unknown cell. Once the IP address of the unknown cell is known, the serving cell sends a request to establish a connection via the X2 interface to the unknown cell. In addition, the UE forwards all the necessary information, including the PCI and the frequency information, to create a neighbor relation to the unknown cell. Upon reception of this request, the unknown cell creates the neighbor relation for the serving cell and forwards its own related information to the serving cell such that a corresponding neighbor relation record can be created at the serving cell too.

3.2.3 Physical Cell Identity Planning

PCI can be envisioned as the unique signature of a cell, which can be used by UEs for various operation tasks including signal measurements for handover purposes. According to the 3GPP [4], a total of 504 unique PCI codes are available for a cell to select from.

These identity codes are grouped into 168 code groups each consisting of three unique codes. The PCI is allocated to the cell when it is installed. However, considering dense cell deployments, it is possible for two neighboring cells to use the same PCI [5]. This is referred to as physical layer cell identity (PCI) collision. In addition, there is a possibility of PCI confusion, which refers to the scenario when a cell has two neighbors with identical PCIs. In the latter case, the serving cell may not be able to distinguish between the two neighboring cells, hence handover procedures may not be completed successfully. Furthermore PCIs of neighboring cells can be transmitted in the same resource blocks, which causes interference and can prevent UEs from detecting neighboring cells. The problem of identity code allocation also exists in third-generation (3G) wideband code division multiple access (WCDMA) networks in the form of scrambling code allocation [6].

PCI assignment is traditionally performed using offline planning tools. However, apart from the tedious tasks involved in using such tools and potential human errors involved, a common feature of such tools is that the neighbor relations are derived based on estimated coverage of cells which may not be accurate.

3.2.4 Mobility Robustness Optimization

This use case was introduced in 3GPP TR36.902 [7] and considers self-optimization of handover parameters. With small cells being deployed as an integrated part of the operator's network, seamless handovers between small cells as well as handovers from and to the underlying macrocell network is a key advantage compared to alternative offloading solutions, which are not integrated into the operator's network, for example, most offload methods based on wireless fidelity (Wi-Fi). However, this simultaneously implies that the handover process should be adapted to account for new constraints imposed as a result of migrating to such a heterogeneous architecture. More specifically, with dense small cell deployments, where each small cell is serving only a limited area, not only is the total number of handovers expected to increase, but such handovers have to be completed in a timely manner, or otherwise there will be a risk of dropped calls as a result of radio link failures (RLFs). This is because the fluctuations of the signal strength with respect to distance are much more significant at the edge of a small cell compared to that at the edge of a macrocell [3]. For example, if an inbound handover to a co-channel deployed small cell is not completed on time and the UE moves toward the center of the small cell, it will be very likely that the call is dropped as a result of the significant increase of interference from the small cell. In addition, since small cells are principally deployed to offload static UEs, it would be ideal not to trigger a handover for a passing-by UE if possible. This would decrease potential call drops and unnecessary handover signaling overheads. In addition, this can increase the performance robustness by avoiding ping-pong effects between cells.

Generally, depending on the connection mode of the UE and carrier frequency of the source and the target cells, four possible categories of mobility events can be identified:

- Intra-carrier handover in connected mode
- Inter-carrier handover in connected mode
- Intra-carrier cell reselection in idle mode
- Inter-carrier cell reselection in idle mode

From an operator's perspective, the connected mode performance is the most important. Additionally, intra-carrier handovers are more vital since there is a higher risk of dropped calls as a consequence of RLFs due to interference. With this in mind, from the interference perspective, inter-RAT handovers can be considered similar to inter-carrier handovers. In fact, the primary objective of mobility robustness optimization (MRO) is to minimize the number of RLFs that are related to handovers. As an example, the target drop call rate for the voice over LTE (VoLTE) is 10^{-4}/s, equivalent to that of circuit switch (CS) services in 3G networks. As a secondary objective, MRO seeks to additionally minimize the inefficient use of network resources and unnecessary handover overheads.

Following the 3GPP notation, RLFs can occur due to late handovers (the most common cause), early handovers and handover to an inappropriate target cell. The parameters considered to be optimized in order to avoid such failures include the time-to-trigger (TTT), the Layer 3 filtering coefficient (k), the cell individual offset (CIO), and the hysteresis. In the case of inter-carrier handovers, instead of the CIO, two threshold quantities (referred to as threshold 1 and threshold 2) are used to trigger an $A5$ event. That is when the signal quality of the current serving cell falls below threshold 1 and the signal quality of the target cell exceeds threshold 2. In addition to the connected mode optimization, appropriate cell reselection mechanisms are essential to ensure that a UE is always camped to the right cell, hence a connection can be set up immediately at any time.

Generally, the MRO addresses two sub-tasks, namely, identification of the handover problems and correction of the relevant mobility parameters [2]. The 3GPP has been mainly focused on the standardization of processes required to detect and report the causes of handover-related RLFs whereas mechanisms to improve the performance through adjustment of related parameters are mainly left to vendor-specific solutions.

As for the legacy macrocell networks, MRO can be realized in either centralized or distributed fashion. In the centralized scheme, the adjustment of parameters occurs at the Domain Manger (DM) level. However, considering HetNets consisting of small cells, distributed algorithms are more appropriate since different DMs may be responsible for a set of neighboring cells [2].

3.2.5 Mobility Load Balancing

As the name implies, the key objective of the mobility load balancing (MLB) use case is to provide intelligent mechanisms whereby highly loaded cells can offload some of their UEs to their neighboring cells, while minimizing the number of handovers as a result. This will result in a much better way of utilizing network-wide radio resources and improves the UEs' QoE. One way to re-distribute the load amongst the neighboring cells is to change the coverage area of the cells adaptively. However, this may result in the creation of coverage holes and should be preferably avoided if such changes are temporary. The main idea behind MLB is to adjust the handover region by modifying related parameters. This can be achieved either in a centralized or a distributed manner.

In case of a decentralized scheme, the algorithms are run at the BS and the load information, which is the key input parameter, is being exchanged amongst the neighboring BSs. More specifically, the reported information elements (IEs) consist of guaranteed bitrate (GBR) and non-GBR physical resource block (PRB) usage, the total available PRBs, hardware load and optionally transport network layer (TNL) load. In the centralized MLB schemes, however, input load information from all BSs is communicated to a central entity and appropriate parameter values are calculated and communicated back to the BSs. Regardless of how the functionality is realized, the 3GPP specifies the following four steps as part of the MLB procedure:

- Active monitoring of cell load
- Detection of overloaded cells
- Selecting a subset of UEs to be handed over to neighbor cells and identification of appropriate target cell(s)
- Configuration of handover parameters

In addition to the active UEs, the MLB framework is additionally applicable to idle UEs. This can be realized by modifying the cell reselection parameters. However, this has less critical performance impact since idle UEs have no direct effect on the air interface system capacity, but may affect the signaling overhead of the core network.

3.2.6 Inter-cell Interference Coordination

The primary objective of inter-cell interference coordination (ICIC) is to shape the inter-cell interference by scheduling UEs over specific portions of the carrier frequency to avoid interference to the neighboring cell UEs. In other words, each BS gives up a certain portion of the bandwidth in a coordinated fashion to improve the performance of cell-edge UEs that are most vulnerable to the inter-cell interference. To achieve this, the following 3GPP-defined metrics are exchanged between BSs:

- High-interference indicator
- Overload indicator
- Relative narrowband transmit power

In addition, Release 10 of the long-term evolution (LTE) standard has introduced enhanced ICIC (enhanced inter-cell interference coordination (eICIC)) to address issues related to HetNets consisting of macrocells and small cells. An effective technique to maximize the offload gain of small cells is cell range expansion (CRE), where the expanded region refers to the area where a UE is being served by the small cell whilst the small cell may not have the strongest signal. The issue of increased interference in the expanded region can be addressed by introducing the concept of almost blank subframe (ABS) where macrocell transmissions in part of the spectrum are muted.

3.2.7 Coverage and Capacity Optimization

Coverage and capacity are considered the two key metrics for any network operator, which directly impact the UEs' QoE. Optimizing network parameters such as BS antenna configurations and power to improve the network coverage and capacity has always been a priority task for operators. While this task has been to some extent facilitated by the introduction of software tools for cell planning and optimization, the need for extensive measurement data from the network is still there. In addition, the dynamic and rapidly changing configuration of the network topology and load as a result of introducing small cells that are deployed in a rather unplanned fashion, strictly requires coverage and capacity optimization (CCO) tasks to be automated.

In 3GPP TS 32.521 [8], the requirements for CCO SON mechanisms are specified. According to this, CCO should be implemented with minimal human intervention, and operators should be able to define different objectives and targets for different parts of the network. Finally, the data collection process used for CCO should be automated to the maximum extent. Generally, it should be noted that due to transmit power limitations, there usually exists a trade-off between the cell coverage and capacity where larger coverage implies lower spectral efficiency hence lower effective capacity. Subsequently, maximizing both quantities simultaneously is not feasible and the trade-offs should be taken into account by operators when implementing

CCO mechanisms. Moreover, different metrics (e.g., mean/median cell throughput or *x*-percentile of the UE throughput) can be used to describe the network capacity. A common approach is to define an objective function, which consists of different metrics that can be customized according to the MNO's policies and priorities [9, 10].

3.2.8 Random Access Channel Optimization

In LTE, the random access channel (RACH) provides a means for a UE to establish uplink time synchronization. The procedure is performed in the following scenarios [11]:

- Connecting an idle UE to the BS
- Re-establishing a connection after RLF
- Handover to a neighboring BS
- Downlink data transmission to an out of synchronization UE
- Uplink data transmission from an out of synchronization UE

Considering the first scenario, a UE is unknown to the BS when it is in idle mode. In order to establish a connection, the UE searches for a suitable serving BS and retrieves the cell-specific random access format and procedures from the broadcast system information. This provides the preamble format and the initial power configuration for the UE. Subsequently, the UE selects one of the preambles at random in order to establish a relation with the BS. A preamble collision refers to the scenario when a selected preamble is currently being used by another UE. In cases of handover and downlink or uplink data transmission to or from a UE that is out of synchronization, the BS can control the random access of the UE to avoid such collisions. This is usually referred to as the contention-free random access procedure. In case of a preamble collision in the traditional case, that is, contention-based random access procedure, the UE is required to select another preamble until a free one is found. Such preambles have a high auto-correlation property such that the arrival time can be estimated accurately while the correlation with other preambles are low such that different UEs can be easily distinguished. In addition, the preamble should be transmitted with high enough power such that it can be detected by the serving BS, while keeping the uplink interference to the neighboring cells at a minimum level. One way to improve the detectability of UEs is to increase the length of the random access transmission. To enable this, the 3GPP has defined different random access formats corresponding to different length of random access transmission as well as cyclic prefixes to account for transmission delay spread [4].

3.2.9 Energy Savings

Considering the rapid network growth to accommodate the increased traffic demands, operators are looking for opportunities to enhance the energy efficiency of their network to reduce their OPEX. The introduction of small cells has resulted in significant energy saving gains since such cells provide high capacity at a low transmit power and normally serve few UEs due to their limited coverage range. Therefore, they can be brought to sleep mode more often [12]. In addition, with more legislation being introduced to protect the environment, improving green credentials is now becoming an important aspect for operators.

Introducing effective idle modes, as well as switching BSs off and on, as indicated in Chapter 2, is considered as practical approaches to achieve capacity and energy savings. These functions can be realized in different ways; for example, by the operator and through the operation, administration, and maintenance (OAM) manager or by defining criteria where cells can decide to turn off and on upon the fulfilment of such criteria. However, such energy-saving operations should not lead to compromising other performance metrics. In this line, the 3GPP has indicated that the following criteria should be met when considering energy-saving mechanisms [1]:

- Accessibility of UEs to the network should not be impacted as a result of transferring cells into the energy saving mode.
- Backward compatibility and ability to provide energy savings for Release 10 LTE network deployment should be met.
- The energy saving solution should not impact the physical layer.
- The energy saving solution should not imply higher power consumption for the UE.

3.2.10 Cell Outage Detection and Compensation

Self-healing is another important aspect of SON functions. This refers to the ability of the system to autonomously detect abnormality and failures in the system, and take appropriate actions to mitigate such effects. A BS may fail without the OAM being aware of such failure for a relatively long time. This can have significant impact on the operator's revenue as a result of service unavailability in the affected areas as well as the impact on UEs' QoE and an increased churn rate.

Proposed by the 3GPP [7], cell outage detection (COD) refers to a mechanism whereby performance statistics and evidence information are collected to determine whether or not a cell is operating properly. In addition, COD mechanisms are normally capable of detecting the type of the failure. This can be achieved by collecting comprehensive performance statistics over long enough time intervals. A significant deviation from values seen when a cell is operating normally can be an indication of a fault. This implies that failure detection is far more challenging when considering small cells with sparse UE statistics [13].

The cell outage compensation (COC) framework refers to the actions taken after the COD stage in order to compensate the effect of the outage to the maximum possible extent. Both network coverage and capacity metrics are impacted as a result of a cell outage. Amongst all possible parameters, adjusting the tilt angle of the neighboring cells and power control parameters are shown to have a key impact on mitigating outage the effects [10, 14].

3.2.11 Minimization of Drive Tests

In order to assure acceptable QoE for UEs, network operators have to conduct drive tests to detect areas with coverage, capacity, or handover issues. Evidently, this is considered as a very labor-intensive task, which needs to be performed regularly considering the fast growth of the operator's networks and the high level of network dynamics as a result of deploying small cells. Moreover, such drive tests are far more complicated for indoor areas where there are also higher risks of coverage holes and poor services. As a result, the 3GPP [15] has proposed to collect UE measurements to enable more efficient optimization of the network. The proposal aims at both LTE and high-speed packet access (HSPA) access technologies. Generally, the minimization of drive tests (MDT) should be envisioned as a supporting tool for other SON functions rather than a separate SON function on its own. The MDT information can be collected in two different ways:

- Immediate MDT where a UE in a connected state collects measurement information and reports them to the BS in an online fashion.
- Logged MDT where a UE in an idle state collects measurement information when configured criteria are satisfied and store them locally. This information will be communicated to the BS later on.

Additionally, it should be possible for an operator to configure the geographical area where the defined set of MDT measurements should be collected. Moreover, the MDT measurements should contain time, location, and information when possible. Since different SON functions may use MDT information, MDT should be performed independently of particular SON functions.

3.3 INTELLIGENT TECHNIQUES FOR SELF-ORGANIZING NETWORKS

While the aim of the 3GPP is to provide a unified framework for implementing SON functions to assure interoperability amongst different vendors by defining common interfaces and signaling procedures between the network elements, the core logic behind such SON functions is left open for vendor specific solutions. Many of the SON functions are based on intelligent algorithms. In this section, the key characteristic requirements for SON algorithms are described, and then three categories of intelligent techniques applicable to SON functions, namely rule-based expert systems, machine learning, and evolutionary computing are discussed.

3.3.1 Key Requirements for Self-Organizing Mechanisms

Irrespective of the specific operation or configuration aspect to which a SON function is applied, the successful and practical deployment of SON mechanisms would not be possible unless a number of generic requirements are addressed. These requirements are in fact necessary to address the concerns of network operators when it comes to SON deployments.

3.3.1.1 Scalability Scalability implies that the complexity of a SON mechanism must not increase unboundedly as the size and scale of the system starts to grow [16]. This is especially important when considering the evolution of the network toward a heterogeneous architecture where there are a large number of small cells. To assure such scalability, the complexity of the SON mechanism should be kept at the minimum possible level. Distributed algorithms which often require local interactions, signaling and coordination between entities usually scale better compared to those which need global state information. However, with more modern and computationally powerful processing capability being available, scalability has become a relative notation and many of the centralized SON mechanisms are now feasible. Clustering is a simple way to avoid scalability problems related to centralized schemes whereby a subset of relevant network elements (e.g., BSs) are selected and jointly controlled. There are currently numerous debates in the industry over the choice of decentralized and centralized architectures for SON mechanisms. Each architecture has its own advantages and drawbacks. In summary, while distributed algorithms may scale better than centralized ones, they may result in sub-optimal configurations as a consequence of lacking the global state information. A hybrid solution refers to a case where part of the SON solution is executed in a distributed fashion and the other part centrally.

3.3.1.2 Stability and Robustness Stability and robustness is a very important aspect of a SON mechanism. A major concern of network operators is to give control of their network to SON functions, which may fail unexpectedly or force the system into undesirable states. Considering the co-existence of different SON mechanisms in the network, it is not only sufficient to test and evaluate the response of a given SON mechanism individually under different conditions, but it is also essential to take the interactions of different SON functions into account. For example, there might be a conflict of output parameters between different SON functions. If not properly addressed, such unresolved conflicts may severely impact the network performance and stability. It is important to note that not only concurrently active mechanisms can be in conflict, but additionally, dependency chains that originate in two different functions can cause conflicting behaviors [2]. Therefore, SON coordination and co-design is of particular importance to assure network stability.

3.3.1.3 Convergence A SON mechanism should be able to shift the system into the desired state within an acceptable time interval. Evidently, the definition of the acceptable time interval very much depends on the underlying task. While some of the parameters need to be adjusted in a microsecond time scale (e.g., scheduling decisions), there are others which are normally adjusted on a weekly or daily basis (e.g., load balancing by modifying coverage). If a SON mechanism takes too much time to converge, it is not useful for the operator. In addition, there are scenarios where the system is forced to oscillate between different states indefinitely (ping-pong effect). Again, SON coordination is essential to assure such oscillations are avoided as a result of conflict between different SON functions.

3.3.2 Rule-Based Expert Systems

Expert systems are currently envisioned as a popular approach of representing large bodies of knowledge for specific fields of expertise and for using this knowledge in order to solve problems. An expert system normally consists of two main components namely a knowledge base and an inference engine. The knowledge base represents facts about the field. While in the early expert systems (e.g., [17]), these facts were represented mainly as assertions about variables, modern expert systems deploy object oriented structures to represent such facts in a more complex and accurate arrangement by defining classes, sub-classes and instances. The inference engine, on the other hand, is an automated reasoning system, which evaluates the current state of the knowledge base and applies relevant rules. The inference engine may also have the capability of explanation. In other words, it may show the chain of reasoning used to arrive at a particular conclusion by tracing back over the rules which have been triggered and resulted in the assertion [18].

Rule-based expert systems are one of the most practical and popular forms of expert systems where the expert knowledge is encoded in a rule base. A rule base refers to the set of rules that occur in sequences and often are expressed in the form of "if...then" statements. When these rules are examined by the inference engine, the satisfaction of the rule antecedent leads to the execution of the consequent and the corresponding action is executed. Considering SON mechanisms, rule-based expert systems allow the existing knowledge of the field experts to be integrated into SON functions, while still benefiting from the automation, which is inherently provided by the SON function. In addition, this provides the operator the flexibility to define the SON mechanisms according to their needs.

Fuzzy rule-based expert systems are considered as a major extension of classical rule-based expert system by considering fuzzy sets instead of classical sets. They are considered as the most important application of the fuzzy set theory by Zadeh in 1965 [19]. According to Zadeh's definition, a fuzzy set is a class of objects with continuous grades of membership. In this case, a set is characterized by a membership function, which assigns to each object a grade (or degree) of membership valued in the real unit interval [0, 1]. In contrast to the classical set theory where the membership of an object in a set is defined in binary terms (i.e., an object is either a member of the set or not), fuzzy membership functions allow partial membership in a set.

A rule-based fuzzy inference system commonly consists of three steps, namely: fuzzification, rule-based inference, and finally defuzzification. The process of mapping the crisp input values into membership functions is referred to as fuzzification. Moreover, membership functions may be combined in fuzzy "if ... then" rules to make inferences, such as "if x is high and y is low, then z is normal," in which "high," "low," and "normal" are membership functions of the matching fuzzy subsets. In fact, a fuzzy inference system is used to map from fuzzy inputs to a fuzzy output. Finally, defuzzification refers to computing a crisp value from fuzzy values. There are several methods for defuzzification. One of the most popular methods is the centroid method, in which the "center of mass" of the result provides a crisp (defuzzified) output value. In this method, the crisp output value is calculated such that half of the area under the resulting output fuzzy set is on each side of the output value. Figure 3.1 illustrates different steps and components involved in a typical fuzzy rule-based system.

With respect to SON functions, fuzzy rule-based expert systems have been applied to a number of scenarios. In [20], two neural encoded fuzzy models for load balancing in LTE networks were proposed. Both models use the same load indicators but utilize different key performance indicators (KPIs) for load balancing. The load indicators used in this study were the virtual load of BSs and their overall load state. Moreover, in [21], a rule-based fuzzy logic controller is used to adjust the soft handover parameters in a HSPA network.

To summarize, fuzzy rule-based expert systems are well capable of coping with uncertainty and dealing with inaccurate data. In addition, they can be used to reconcile conflicting objectives and to simplify knowledge representation by means of membership functions and the rule base. Furthermore, the use of linguistic terms for input and output variables makes their understanding easier, as they try to mimic the human reasoning process. Despite of all these advantages, a main drawback related to such systems is that they heavily rely on the field expert knowledge, which may not be always available or may be sub-optimal in some cases.

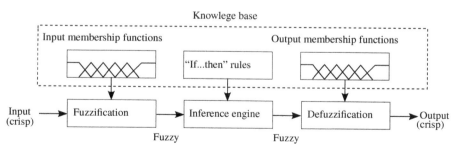

Figure 3.1. Block diagram of a typical fuzzy rule-based system.

3.3.3 Machine Learning

The main goal of machine learning is to build intelligent systems capable of adapting and learning from their past experience. Subsequently, such systems can act without being explicitly programmed. Machine learning has found several applications in different fields over the past two decades. As a matter of fact, most of the real problems are extremely complex, ill-defined, and are hard to model. Machine learning techniques seek to deploy statistical reasoning to find approximate solutions to address these difficulties. Such techniques can be classified into three categories, namely supervised learning, unsupervised learning, and reinforcement learning methods. The following subsections provide a brief overview of these categories.

3.3.3.1 Supervised Learning   In supervised learning, the system is provided with a set of labeled training data and the main task is to analyze this training data in a way that an inferred function can be produced in order to label new unseen examples most accurately. In other words, a supervised learning algorithm tries to generalize from the training data to unseen situations. Supervised learning techniques can be broadly divided into two categories of generative and discriminative methods.

Generative methods try to learn the joint probability of input and class labels (output). They will then rely on the Bayes rule [22] to compute the conditional probability of input and class labels. The class label with highest conditional probability will then be selected. The term generative refers to the fact that they capture the underlying generation process of a data population of interest [23]. The naive Bayes (NB) classifier is a good example of generative methods. This is a simple classifier, which assumes that the presence or absence of a particular feature of a class is completely unrelated to the presence or absence of any other feature of the class. Despite this strong and rather naive assumption, NB classifiers have shown to perform well in many practical applications, especially when the dimensionality of the input is high. The work in [24] is an example of application of NB for SON functions where an automated diagnosis model for universal mobile telecommunication system (UMTS) networks is introduced using a NB classifier.

In contrary to the generative schemes, discriminative methods try to learn directly the conditional probability of input and class labels. Discriminative classifiers are generally simpler, and it has been shown that they result in more accurate predictions than generative ones when the training set is limited [25]. Amongst many other techniques, k-nearest neighbors (k-NN) algorithms are examples of key discriminative classification methods. As the name implies, in the k-NN algorithms, an input is classified by a majority vote of its neighbors. In other words, the input is assigned to the class that is most common among its k-nearest neighbors. There are different ways to quantify the distance of an input instance to the neighbors. k-NN algorithms are very simple and perform relatively well on most basic classification problems. However, instead of learning from the training data, they directly use the training set when making classification decisions (i.e., memorizing instead of learning). This implies that they might not scale well and may perform poorly when the training data are noisy. k-NN classification has been applied in [26] for dynamically adapting the modulation and coding scheme in a multiple-input multiple-output (MIMO) orthogonal frequency division multiplexing (OFDM) system. Moreover, in [27] a k-NN classifier was deployed intrusion detection in IP multimedia subsystems.

Despite their wide application, a common problem in all supervised learning techniques is their dependency on the availability of a training set. While such training data might be available for some SON functions, there are numerous cases where such data are not available or is inaccurate.

3.3.3.2 Unsupervised Learning

Unsupervised learning is a general term, which covers several types of learning scenarios, where the input data set contains the data points themselves without any additional information (e.g., corresponding classification labels). Unsupervised learning methods can be broadly divided into two categories namely data clustering algorithms and feature extraction methods.

The main objective of clustering is to unravel the structure of the provided data set by partitioning the input data instances into groups such that the instances within each group are similar to each other to the maximum possible extent while are different to members of other groups. Such input instances are usually in a high-dimensional space, hence similarity is defined using a measure of distance between instances. A typical pattern clustering process involves the following stages [28]:

- Pattern representation
- Definition of a suitable pattern proximity measure
- Clustering or grouping
- Data abstraction (if needed)
- Assessment of output (if needed)

Pattern representation refers to the number of classes, the number of patterns available, and the number, type, and scale of the features that are considered by the clustering algorithm. Pattern proximity is usually measured by a distance function defined on pairs of patterns. There are a variety of approaches available for implementing a clustering task. Examples of such techniques are probabilistic clustering methods [29] and graph-theoretic approaches [30]. Data abstraction refers to the process of extracting a compact representation of a data set. While this is not a core part of the clustering process itself, it can be very beneficial in many applications. Finally, the result of the clustering process can be assessed and validated. The work in [31] is an example of application of clustering algorithms for SON functions, where a method is proposed to cluster small cells with similar soft frequency reuse patterns. As a result of this, the cross-tier interference in the downlink is mitigated and throughput of UEs is improved.

Feature extraction involves reducing the amount of information required to describe a large set of data accurately. A major problem that commonly occurs when dealing with complex data is related to the high number of variables. Analysis of such data normally requires a large amount of memory and computing power. Feature extraction is used as a general term that refers to techniques which can construct combinations of the variables to address these issues without compromising accuracy significantly.

3.3.3.3 Reinforcement Learning   Reinforcement learning is a sub-area of machine learning that is concerned with how a learning agent should take actions in an environment to maximize its long-term reward [32]. In this scope, the learner is not explicitly commanded (which actions to take) as in supervised learning, but instead must discover which actions yield the most reward by trying them. Reinforcement learning algorithms attempt to find a policy that maps states of the world to the actions the agent have to take in those states. Therefore, it is different from supervised learning, in which knowledge is acquired via examples provided by an external supervisor. For interactive problems, it is often impractical to obtain examples of desired behavior that are both correct and representative of all the situations in which the agent has to act. Reinforcement learning consists of the following components [33]:

- A discrete set of environment states, S
- A discrete set of agent actions, A
- A set of scalar reinforcement signals, R

The learning occurs iteratively where in each step of the interaction the agent senses the current state, $s \in S$, of the environment and then executes an action, $a \in A$. The action changes the state of the environment and the value of this state transition is communicated to the agent through a scalar reinforcement signal, $r \in R$. The agent should ideally execute actions that tend to maximize the long-run sum of values of the reinforcement signal. This can be achieved through a systematic trial and error scheme, which is the underlying principle of reinforcement learning.

There are three fundamental methods for solving reinforcement learning problems namely: dynamic programming, Monte Carlo methods, and Temporal Difference (TD) learning. Dynamic programming approaches are very well developed mathematically but require a complete and accurate model of the environment, which unfortunately does not exist in many application scenarios. Monte Carlo methods do not require a model and are very simple conceptually, but are not suited for step-by-step incremental computation. Finally, TD methods require no model of the environment and are naturally implemented in an online, fully incremental fashion. TD methods learn their estimates in part on the basis of other estimates.

Q-learning is a TD technique that works by learning an action-value function. In the Q-learning context, one can identify three main sub-elements of the system, beyond the learning agent and the environment: a policy, a reward function, and a value function. A policy, π, is a mapping from states, s, and actions, a, to the probability of taking action a when in state s. The value of state s under a policy, denoted by $V^{\pi}(s)$, is the expected return when starting in state s, and thereafter following the policy π. Additionally, the value of taking action a in state s under the policy π, denoted $Q^{\pi}(a, s)$, is the expected return reward when starting in state s, taking the action a, and thereafter following the policy π. The expected return reward in its simplest form is just the sum of the rewards. This approach makes sense in applications in which there is a natural notion of finite time steps, that is, when the agent-environment interaction breaks naturally into sub-sequences. But this representation becomes problematic in case of continuous tasks where the agent-environment interaction does not break into identifiable episodes. For such a case, an additional discount rate can be applied. Under this concept, the overall return can be defined as

$$Re_t = r_{t+1} + \gamma \, r_{t+2} + \gamma^2 \, r_{t+3} + \cdots = \sum_{k=0}^{\infty} \gamma^k \, r_{t+k+1}, \tag{3.1}$$

where Re_t is the return value after time step t, r_{t+n} is the immediate reward after time step t, and γ is a discounting factor ($\gamma < 1$). The discount rate is a relative value of delayed versus immediate rewards and determines the present value of future rewards (i.e., a reward received k time steps in the future is worth only γ^k times what it would be worth if it were received immediately). Then, the Q-values can be iteratively updated as

$$Q(s_t, a_t) \leftarrow Q(s_t, a_t) + \alpha(s_t, a_t) \, [Re_t - Q(s_t, a_t)], \tag{3.2}$$

where $\alpha(s_t, a_t)$ is the learning rate associated to state s and action a in time step t, which can be also set to a fixed value. One issue is that future rewards may not be known in advance. For this reason, the expected return can be envisioned as the sum of the immediate reward and the maximum of the possible future reward. The latter can be estimated through the Q-value of the next state as

$$Re_t = r_{t+1} + \gamma \, \max_a Q(s_{t+1}, a). \tag{3.3}$$

Combining 3.2 and 3.2 yields

$$Q(s_t, a_t) \leftarrow Q(s_t, a_t) + \alpha(s_t, a_t) \left[r_{t+1} + \gamma \, \max_a Q(s_{t+1}, a) - Q(s_t, a_t) \right]. \tag{3.4}$$

In this case, the learned action-value function, $Q(s_t, a_t)$, directly approximates the optimal action-value function, independently of the policy followed. This dramatically simplifies algorithm analysis and enables early convergence proofs. If $\gamma = 0$, the agent is said to be myopic in that it is only concerned with maximizing immediate rewards.

Action selection is an important aspect of reinforcement learning. At each state, the agent is expected to select an action. Evidently, this can be done in many different ways. In its simplest form, an agent can always select the action with the highest state-action value. This is referred to as exploitation. In addition, in order to enhance the learning process and to construct a better estimate of the optimal Q-function, the agent can learn by trying new actions that are expected to be sub-optimal by the agent. This process is called exploration. Balancing between the exploitation and exploration is an active field of research on its own [32].

Q-learning is generally considered in the case that states and actions are both discrete. Unfortunately, in most real-word scenarios, states and actions are continuous. Dividing a continuous space into a set of non-overlapping subsets and considering each subset as a discrete quantity is an option. However, large quantization errors as the result of this process may lead to inefficient learning. On the other hand, generating too many states where the quantization error is small will result in the curse of dimensionality problems where the agent requires a significant number of learning iterations before it converges. In addition to how such states and actions are defined, for most real-world problems, it is expected that actions vary smoothly in response to smooth changes in state. The very discrete nature of states and actions definition makes this a challenging task [34]. Another issue with the classic reinforcement learning relates to the fact that integration of existing knowledge into the system is not always a very straightforward task. While for most applications, domain knowledge might be available, it is not easy to benefit from that in the context of the reinforcement learning, as the learning agent seeks to learn the value of each action and state pair by trying them in a systematic trial-and-error fashion.

3.3.4 Evolutionary Computing

Evolutionary computing (EC) is an umbrella term for a wide range of problem-solving techniques based on principles of biological evolution, such as natural selection and genetic inheritance [35]. Such methods have recently found a large number of applications in the industry. An evolutionary algorithm (EA) is in fact a population-based optimization algorithm that uses genetic operations such as mutation and recombination on candidate solutions over multiple generations in order to produce a solution with a better performance. The main steps involved in setting up a problem for EA are:

1. The definition of a fitness (or objective) function that is used when assessing the optimality and performance of an individual solution. This allows considering multi-objective optimization, where the fitness can be constructed by aggregating different metrics. Moreover, appropriate weights can be applied to establish the priority of some metrics over others.

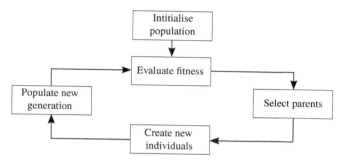

Figure 3.2. Steps of an EA scheme.

2. Creating a representation of solutions in a form that can be manipulated by the genetic operators. These are called chromosomes, and depending on the problem being addressed, they can appear in many different forms. An example is a binary string chromosome, representing a particular numerical value.

3. The specification of the EA parameters to be used. This includes the parent selection procedure, probabilities of mutation and recombination, population size, and the termination condition. The termination condition can be defined as when a certain level of fitness is achieved or after a predefined number of generations.

The process of evolving an individual solution includes finding the right combination or sequence of chromosomes, driven by the fitness function. Figure 3.2 illustrates this process where the steps involved are as follows:

1. Initialize the population with random solutions.

2. Evaluate the performance of individual solutions in the population according to the fitness function.

3. Populate the next generation with offspring by applying genetic operators such as mutation and recombination on individual solutions with high fitness.

4. Repeat steps 2 and 3 until stopping criteria are satisfied.

There exist several different forms of EAs, which differ in terms of the specifics of implementation and their applications. Genetic algorithms (GAs) are arguably the best-known form of EA and are considered as an intelligent exploitation of a random search within a defined search space in order to solve a specific problem. In classical genetic algorithm (GA), the standard representation of each candidate solution is as a stream of bits. Genetic programming (GP) is a variant of GA, in which the hypothesis being manipulated are computer programs rather than bit strings [36].

Generally, EAs are considered as effective tools for problems that cannot be solved using traditional optimization techniques. In addition, as the search process starts over multiple locations in the fitness landscape, the algorithm is able to locate several different optima and is thus less likely to be trapped in a sub-optimal point. However, the application of such algorithms might be a challenging task in some scenarios since evaluation of poor individual solutions might be costly or infeasible.

3.4 CASE STUDIES

In this section, two SON-related case studies are described. In the first case, the problem of antenna downtilt angle optimization for improving coverage and capacity is addressed. A combination of a fuzzy logic controller and a reinforcement learning agent is used to adjust the antenna angle of BSs in a distributed fashion [9, 10]. In the second case, the issue of coverage optimization for cluster of small cells is considered [37]. This has been achieved by optimizing the pilot power of small cells and by an automatically evolved algorithm using genetic programming (GP) as the optimization technique.

3.4.1 Tilt Angle Optimization Using Fuzzy-Reinforcement Learning

Individually adjusting the vertical antenna downtilt angle of every sector allows optimizing the coverage and the capacity of a cellular system. This is especially important in HetNet scenarios where the traffic intensity of the macrocell tier is dynamically changing as a result of deploying small cells for traffic offloading. Such small cells are being deployed at a fast pace and in a rather unplanned fashion. Moreover, in order to achieve higher energy efficiency, small cells can be turned off at certain periods. In addition, the distribution of UEs and their traffic profile may change over the time. As a result, optimizing tilt angle of macrocell BSs is a complex task that can provide consistent performance in terms of coverage and capacity. Techniques such as the continuously adjustable electrical downtilt [38] enable intelligent algorithms to be employed for autonomous configuration and optimization of the antenna downtilt angles.

In terms of the optimization techniques, reinforcement learning is a suitable candidate for the tilt angle configuration issue. In this context, the BS is considered as a learning agent, which takes actions by increasing or decreasing the antenna tilt angle. As result of such actions, the overall network's performance is changed. Such changes can be formulated into a reinforcement signal and can be observed by (or communicated to) the BS. After a number of iterations, the BS is expected to learn the optimal policy. However, as mentioned earlier, major issues arise with reinforcement learning when the input space is continuous. In the reinforcement learning paradigm, this problem is referred to as the curse of dimensionality. Functional approximation and generalization methods seem to be a more feasible solution, but unfortunately, the optimal convergence of functional approximation has not yet been proven. Moreover, using fixed thresholds for partitioning the input space into discrete states leads to sudden state transitions, which may lead to abrupt actions that are naturally undesirable in most learning systems. This is especially true in the case of multi-agent learning systems, in which a number of learning agents interact simultaneously with the environment. In addition, embedding a priori knowledge in reinforcement learning systems is not a straightforward task, if it is possible at all. However, some general knowledge about characteristics of the environment is often available and can be used as a starting point to accelerate the learning process.

By employing fuzzy sets that allow for graded membership of variables in sets, not only can a very good level of input abstraction be achieved, but also smooth transitions from one state to another are feasible. The latter is due to the fact that when the value of a fuzzy variable changes, the degree of membership to a particular set can be gradually decreased, while its membership to another set can be gradually increased. Moreover, as a sub-class of expert systems, a priori knowledge can easily be incorporated into fuzzy systems. In addition, fuzzy systems can cope well with uncertainty and imprecise data, and unlike many other models, fuzzy systems are easily interpretable and their functionality is traceable. However, one of the main drawbacks of classic fuzzy systems is that they significantly rely on the existence of domain expert knowledge.

The above description implies a successful combination of reinforcement learning techniques and fuzzy systems. While the former is necessary for training and tuning the controller when no direct supervisor is available, the latter provides a flexible framework for many control and optimization problems.

One way to implement a fuzzy-reinforcement learning system is to construct the sate variable by combining all input variables in an intersectional fashion using the fuzzy AND operator when there is more than a single input variable. For example, assuming a simple system with two input parameters (e.g., x_1 and x_2), each using three membership functions (e.g., "low," "medium," and "high"), the input space could be classified into nine states where, for example, the case in which x_1 is low AND x_2 is high is considered as one state.

With a single output (e.g., y), again with three membership functions (e.g., "decrease" "no change," and "increase"), the main task is to find the appropriate consequence for each given "if ... then" rule, which corresponds to an input state (e.g., if x_1 is "low" AND x_2 is "high" then y is "increase"). It is important to note that due to the fuzzy nature of the inputs and the overlapping membership functions, more than one "if ... then" rule might be fired at each time. In other words, each input can belong to more than one membership function with different degrees of membership. In the above example, input x_1 could belong to the "low" membership function with a degree of 0.2 and to the "normal" membership function with degree of 0.8, resulting in the triggering of more than one rule. Recall that each rule is defined through an identical combination of input variables. This is not the case in conventional

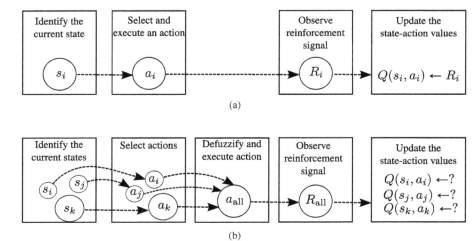

(a)

(b)

Figure 3.3. Block diagram of (a) a classical reinforcement learning system and (b) a fuzzy logic assisted reinforcement learning scheme.

reinforcement learning scenarios, where states are non-overlapping. The implication of this is the fact that once the reinforcement signal is observed or communicated back to the learning agent, the learning agent should deploy a suitable credit assignment technique to relate the reinforcement signal to the appropriate states and actions.

Figure 3.3 compares the learning procedures for conventional and fuzzy-reinforcement learning systems. Here, s_i is the identified state and a_i refers to the action which is selected by the learning agent in state s_i. Note that identical actions can be selected in different states (i.e., $a_i = a_j, s_i \neq s_j$). As illustrated, unlike for classical reinforcement learning schemes, credit assignment is not very straightforward when considering a combination of fuzzy logic and reinforcement learning.

The fuzzy-assisted reinforcement learning (FARL) is a framework that aims at addressing the issue of credit assignment problem. The main concept behind FARL is described in [10]. Ideally, the credit assignment task should consider the firing strength of both states and actions. The firing strength of each rule is the degree to which the antecedent part of a fuzzy rule is satisfied and can be interpreted as a degree of truth of its associated state. Consequently, after executing an action and observing the corresponding reinforcement signal, for each state $s \in S$, the state strength, S_s, can be defined as the normalized firing strength of that state as,

$$\forall s \in S, \quad S_s(s) = \frac{f_s}{\sum_{\forall s' \in S} f_{s'}}, \tag{3.5}$$

where f_s is the firing strength of the state s. Additionally, one can define a similar measure for the action space to determine the contribution of each action to the final collective action after the defuzzification stage. Generally speaking, since defuzzification is not a linear procedure, it is not an easy task to isolate and evaluate the impact of each action on the overall collective result. However, the degree of truth of the state associated to each action can be considered as a fairly relevant measure. Additionally, as explained, it is possible that identical actions are selected by a number of different fired states. The involvement of each individual action in deciding the final collective output would be in proportion to the sum of the firing strength of all states selecting that particular action. Therefore, the action strength can be defined as

$$\forall a \in A \ A_s(a) = \frac{\sum_{\forall s \in S} f_s \, \psi(a, s)}{\sum_{\forall s' \in S} f_{s'}}, \tag{3.6}$$

where $\psi(s, a)$ is a binary parameter indicating whether or not action a is the selected action in state s. That is:

$$\psi(s, a) = \begin{cases} 1 & \text{if action } a \text{ is selected for state } s \\ 0 & \text{otherwise} \end{cases}. \tag{3.7}$$

For every state, s, and action, a, a joint state-action strength, $J(s, a)$, can be defined as the product of the state and the action strength:

$$\forall a \in A, \ \forall s \in S, \ J(s, a) = S_s(s) \, A_s(a). \tag{3.8}$$

With this definition, FARL uses $J(s, a)$ as a measure by which the final reinforcement signal can be related to the individual state and action pairs. Higher values of $J(s, a)$ imply that the final reinforcement is closely related to s and a. Evidently, if state s is not triggered or action a is not selected, $J(s, a)$ will be zero. On the other hand, the learning rate defines how Q-values are updated as a result of a recent reinforcement observation. Subsequently, $J(s, a)$ can be used to modulate the effective learning rate for each state-action pair:

$$\forall a \in A, \ \forall s \in S \ \alpha(s, a) = \alpha_{\max} J(s, a), \tag{3.9}$$

where $\alpha_{\max}$ is the maximum learning rate. A higher learning rate implies faster adaptation of the Q-value, which will be applied when the learning agent believes that the current reinforcement observation is closely related to the given state and action pairs. Figure 3.4 schematically illustrates the work flow of FARL including the credit assignment task.

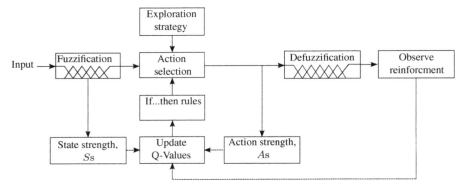

Figure 3.4. Block diagram of FARL scheme.

Evolutionary learning of fuzzy rules (ELF) [39] is a similar framework to FARL, which aims at facilitating the combination of reinforcement learning and rule-based fuzzy logic systems. Despite of their similarities, the credit distribution mechanism is different in FARL and evolutionary learning of fuzzy rules (ELF) schemes. More specifically, ELF considers only state strength when distributing the reinforcement signal between contributing state-action pairs. Moreover, the ELF update procedure is based on the weighted averaging.

There are different ways where a learning agent can select actions. This case study considers an ϵ-greedy algorithm for action selection where the learning agent exploits by choosing the action with the highest associated Q-value with probability of $1 - \epsilon$, and explores by choosing another random action with the probability of ϵ.

In the antenna tilt angle optimization problem of this case study, the state space is defined as the current antenna tilt angle, and actions are the changes that are made to the antenna tilt angle. In fuzzy terminology, the state space corresponds to the input, and the action space corresponds to the output of the fuzzy system. Figure 3.5 illustrates the input and output membership functions used in this case study.

A general fitness function needs to be used to quantify the performance of the system for the experiment. Here, a linear combination of two performance metrics, namely, the mean and the lowest 5th percentile of the UE's spectral efficiency, is used as a performance indicator. The latter is especially important as the UEs on the cell boundary experience significantly poorer performance than the UEs in the cell center. In acknowledgement of this reality, the specification for spectrum efficiency only requires that at least 95 percent of UEs should be served at better than approximately 0.1 bit/s/Hz [40]. The system average, on the other hand, is expected to be 2 bit/s/Hz to 3 bit/s/Hz per UE. The fitness function, F, can be formulated as

$$F = S_{\text{Av}} + cS_{\text{edge}}, \tag{3.10}$$

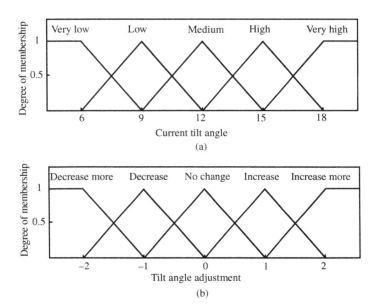

Figure 3.5. (a) Input fuzzy membership functions and (b) output fuzzy membership functions.

where S_{Av} and S_{edge} are the mean and 5th percentile of UE's spectral efficiency in bit/s/Hz respectively. The unit-less coefficient $c(> 1)$ serves to magnify the contribution of the edge UE's spectral efficiency, and was set to two in this case study. Moreover, for each learning agent the immediate reward is defined as the changes of the fitness function as a result of executing the selected action(s) in the given state(s). In other words, in a given state, positive rewards are associated only with those actions which can improve the fitness function.

The simulation scenario consists of 21 cells (seven sites, each with three sectors), in a hexagonal layout and with the inter-site distance of 500 m. Each cell is adjusting its antenna tilt angle independently in a distributed fashion. Mobility of UEs was modeled as random walk with the speed of 3 km/h for pedestrian UEs and 30 km/h for fast-moving UEs. Simulations were performed for 1000 independent runs and the results were averaged.

Starting from randomly configured antenna downtilt angles in the range of 6–10 degree, Fig. 3.6 shows how the average fitness is improved over iterations. It should be noted that the number of iterations on the x-axis refers to the total iterations of all 21 cells, which occur in parallel and without any coordination. In order to provide a comparison benchmark, the global optimum fitness was additionally calculated through an exhaustive search procedure. According to the results, as the learning continues, both FARL and ELF reach close to the optimal performance, while FARL is superior in terms of the convergence speed.

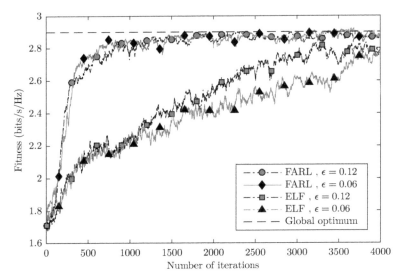

Figure 3.6. Average fitness for FARL and ELF schemes.

The experiments were additionally performed for different values of ϵ. As the results show, variations of the exploration rate do not have a noticeable impact on the convergence time of the FARL scheme, but they do on the convergence time of the ELF scheme. The fact that ELF uses a simple averaging operation (as opposed to exponentially moving average used in FARL) over a number of observations makes it slow to learn; therefore, a further reduction in the exploration rate can intensify this effect and impact the convergence speed.

It should also be noted that the fitness of the FARL scheme has slightly exceeded the average optimal value at some iterations. This is because of the averaging and due to the changing distribution UEs at those particular iterations. The fitness function used here is related to the location of UEs, and despite the fact that the results are averaged over 1000 independent simulation runs, the fitness has slightly exceeded the calculated average optimal value in a few iterations due to the randomness in UEs' distribution and their mobility pattern.

Since in this case study the placement of sites follows a strict hexagonal arrangement and the UE distribution is uniform, the optimal tilt angle of all sectors should converge to a single value, implying that the variance of the tilt angles of different sectors at a given iteration should decrease as the optimization proceeds. This measure can indirectly serve as an indication of the convergence of the algorithm as well. Figure 3.7 shows the variance of tilt angles of different cells over a number of iterations. While FARL converges faster than ELF, ultimately in both schemes, the variance starts to decrease as the learning continues.

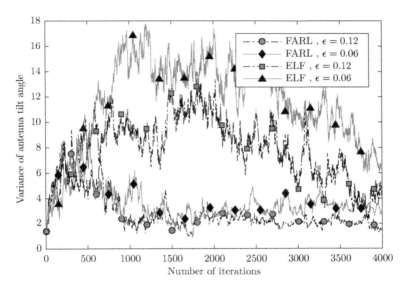

<u>Figure 3.7.</u>   Variance of antenna tilt angles of different BSs over iterations.

Moreover, it is important to take necessary actions when a part of the network fails. To illustrate the self-healing properties, all three sectors of the central site were disabled during the operation of the learning algorithms. A simple way to compensate a site failure is for the neighboring cells to extend their coverage by reducing their antenna downtilt angle. Figure 3.8 shows the response of the FARL and the ELF algorithm when the central site is disabled at the 20th iteration after the network was operating close to its optimal region. Both algorithms are shown to be capable of reacting to the failure, but FARL is more responsive and converges faster than ELF.

In summary, the case study illustrates how the combination of the fuzzy logic and reinforcement learning can be used as a framework for enabling distributed learning in cellular networks where there is a high level of dynamicity. Evidently, tilt angle optimization is only one example application and this learning framework can be applied to provide other SON functions in the network.

3.4.2 Online Evolution of Femtocell Coverage Algorithms Using Genetic Programming

One of the aspects that can benefit from SON functions is the optimization of the femtocell coverage, where the femtocell pilot channel transmit power is adjusted to enhance coverage metrics. This is especially important for enterprise femtocell networks that are deployed to provide enhanced coverage in indoor environments such

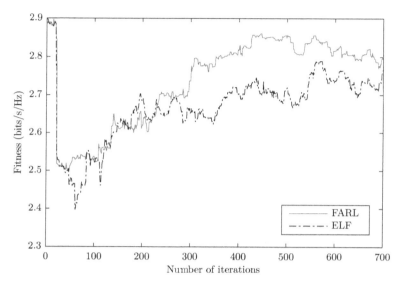

Figure 3.8.   Example of self-healing function in FARL and ELF.

as office and retail buildings. The following three objectives are considered as a part of the optimization process:

- Reduce the amount of coverage gaps
- Perform load balancing
- Reduce the amount of leakage of coverage outside the building of interest

In contrast to some of the pre-defined coverage algorithms that are designed for general cases and often result in sub-optimal performance due to the high variability of the scenarios, this case study uses GP to automatically create distributed enterprise femtocell coverage optimization algorithms for the specific scenario where such cells are deployed in.

Traditionally, the algorithm evolution in GP is performed in an offline manner where a pre-defined model of the network is used, since performing such evolution online and testing evolved algorithm candidates on the live network would result in an unacceptable amount of disruption to the network. For example, testing a highly sub-optimal candidate during the evolution that causes large amounts of coverage gaps in the building would be unacceptable, even for a short period of time. Therefore, creating the network model for testing algorithm candidates is necessary, and requires detailed knowledge of the location of the femtocell BSs, the UE traffic behavior and the radio propagation environment.

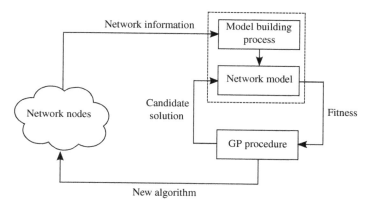

Figure 3.9. High-level overview of centralized online evolution utilizing generalized network model.

To fully automate the algorithm evolution process and run GP in an online manner, the network model for testing evolved algorithm candidates needs to be created in an automated way as well. This negates the need of any manual intervention and introduces the ability for the GP process to dynamically adjust to any changes in the behavior of UEs or the environment.

Automatic network modeling can be achieved by generating hierarchical Markov models that produce statistical models of the network behavior using real-time measurement traces of the femtocells load, dropped calls and mobility of UEs [37]. This model is then used to predict the behavior of the network when different coverage algorithms are used to obtain the performance of potential algorithms generated using the GP procedure. Figure 3.9 illustrates the integration of the model-building process with the GP algorithm.

The followings steps need to be performed prior to the evolution process:

1. Specification of the functions and terminals used to construct the algorithm
2. Formulation of an appropriate fitness function based on the objective
3. Specification and evaluation scenarios that would be used to test the generated algorithms

Table 3.1 describes the functions used in the GP setup, while Table 3.2 illustrates the corresponding terminal names. The parse tree that is constructed is used as an algorithm, which is run locally by each femtocell. The femtocell collects statistics of its load (set as the number of active UEs here), rejected calls due to congestion, and handovers over an update period. After the update period, the femtocell inputs the statistics collected over the update period to the algorithm, which updates its pilot power. The femtocell then starts a new update period and the process is repeated. Here, the update period is set to 20 seconds. Because the update period

TABLE 3.1   List of Functions Used in the GP Procedure

Function Name	Description
if_load_higher	If the maximum load experienced during the update period is higher than a threshold, execute branch 1; else, execute branch 2.
if_call_rejected	If the total number of calls rejected due to overloaded conditions exceeds a threshold, execute branch 1; else, execute branch 2.
if_unreg_ho_higher	If the total number of handover request from unregistered UEs during the update period exceeds a threshold, execute branch 1; else, execute branch 2.
if_ho_higher	If the total number of handovers from a registered UE occurring between a macrocell and a femtocell during the update period exceeds a threshold, execute branch 1; else, execute branch 2.
combine2	Execute branches 1 and 2 consecutively.
if_unreg_ho_higher	Execute each branches 1, 2, and 3 consecutively.

is short, the thresholds used in the if_call_rejected, if_call_rejected, if_unreg_ho_higher, and if_ho_higher functions are set to zero. The if_load_higher threshold is set to seven UEs, which is one UE below the maximum assumed femtocell load of eight UEs.

The fitness function used to evaluate the algorithms during the evolution is calculated using the measurement traces collected whilst the algorithm is running in the network, and consists of fitness components relating to the coverage gaps, coverage leakage, and load:

$$F = P \left[\frac{W_H F_H + W_U F_U + W_L F_L}{W_H + W_U + W_L} \right] . \tag{3.11}$$

F_H is the fitness associated with coverage gaps and is calculated using the total number of mobility events that the femtocells experience. The handovers are

TABLE 3.2   List of Terminals Used in the GP Procedure

Terminal Name	Description
increasepow	Increase the pilot channel power by 1 dB.
increasepow2	Increase the pilot channel power by 2 dB.
decreasepow	Decrease the pilot channel power by 1 dB.
decreasepow2	Decrease the pilot channel power by 2 dB.
donothing	Keep the pilot channel power unchanged.

classified as either handovers between femtocells (H^{sc}), or handovers to or from another macrocell (H^m). Then, F_H is calculated as the proportion of total handovers that involves the macrocell. That is to say:

$$F_H = \frac{H^m}{H^m + H^{sc}}. \tag{3.12}$$

F_L is the fitness associated with load. There are two metrics that are contributing in construction of load related fitness: the amount of load that the femtocell network serves and the total number of calls that were blocked or dropped due to the femtocells being congested. The first part is formulated as the portion of the total load that is supported by the femtocells, L^{sc} over the total load supported by both the macrocells, L^m, and femtocells, L^{sc}. The second part is the portion of calls that were blocked due to a call being initiated or handed over into overloaded femtocell, $C^{sc}_{blocked}$ given the total number of calls or handover attempts, C^{sc}_{total}. F_L is calculated as the product of the two metrics:

$$F_L = \left(\frac{L^{sc}}{L^{sc} + L^m} \right) \left(\frac{C^{sc}_{blocked}}{C^{sc}_{total}} \right). \tag{3.13}$$

F_U is the fitness associated with leakage of coverage, and is calculated as the portion of number of unregistered UEs initiating a mobility request to the femtocells inside the building, N_m, given the total number of unregistered UEs detected, N_{total}:

$$F_U = \frac{N_m}{N_{total}}. \tag{3.14}$$

D denotes a distance from the optimal penalty that is calculated as the Euclidean distance of the solution from the origin point, normalized to 1:

$$D = \frac{\sqrt{F_H^2 + F_L^2 + F_U^2}}{\sqrt{3}}. \tag{3.15}$$

D is used here to avoid solutions that optimize one objective at the detriment of the others, and introduce some bias toward solutions that are more Pareto efficient. W_H, W_U, and W_L are the weights used to modify the impact of each fitness component. Their values can be adjusted to place more importance on a fitness component relative to the others. Here, W_L is set to 2, while W_U and W_H are set to 1 to give higher importance to preventing overload conditions.

The simulation environment is the enterprise building shown in Appendix A. A total of 140 UEs are modeled, and each UE has a voice traffic model such that each UE produces 0.1 Erlangs of traffic. The movement of UEs in the building are modeled using the waypoint-based mobility model introduced in Appendix A. In this case study, the femtocell simulation scenario is used in place of a real network, and an initialization stage is executed to obtain an initial model of the network. After this, the online GP evolution is started where the GP process uses only the generated model to evolve new coverage algorithms. The population size used in the GP process is 30, and a mutation rate of 0.4 and a crossover rate of 0.6 is applied.

Whenever the GP process finds an algorithm with a higher fitness than those found in previous generations, the algorithm is pushed to femtocells in the simulation scenario. While the GP process is running, the model building process continues to refine itself. Additionally, in order to evaluate the quality of the solutions that were evolved using the equivalent fitness from the generated model, the evolution is additionally performed using the same population size and mutation rates, but instead of the generated model, the actual simulation scenario is used to drive the selection and mutation process. In addition, a brute force search approach, which tries combinations of femtocell coverage in order to obtain the combination that has the highest fitness was applied. As a complete global search of all the possible power level combinations would not have been practically possible due to the very large number of combinations, the brute force search is constrained to the region of the solution space where the highest fitness values were obtained.

In Fig. 3.10, the performance of these schemes is compared over 100 generations. It is shown that both of the GP schemes provide a near-optimal solution. In

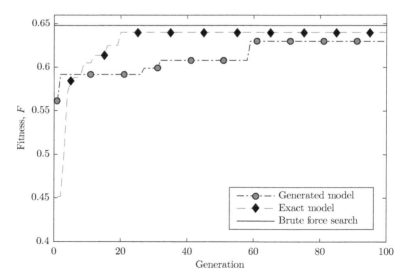

Figure 3.10. Comparison of fitness obtained from the exact model and the generated model.

addition, the model building procedure seems to be effective and is able to produce algorithms which have a level of performance close to evolution using perfect environment information. For the first time, this capability enables a BS to rewrite its own optimization code while operating without disrupting the network operation.

3.5 SUMMARY AND CONCLUSIONS

In this chapter, the motivations behind introducing SON functions were discussed. Given the current trends of traffic demand growth and the increasing complexity of the future HetNet architecture, intelligent mechanisms for self-configuration, self-optimization and self-healing functions of the network elements are critically important. Then, the main 3GPP SON use cases where introduced and briefly explained. These uses cases aim at addressing areas where deploying intelligent SON techniques can be substantially beneficial both in the deployment and operation phases. Self-configuration of BSs, cell ID planning, and configuration of neighbor relations are examples of SON functions, which are designed to facilitate the deployment of cells. Moreover, optimization of capacity and coverage, handling mobility and load balancing as well as issues such as inter-cell interference coordination are important SON functions that are related to self-optimization mechanisms. Finally, cell outage detection and compensation are main examples of self-healing functions.

In addition to introducing major 3GPP SON use cases, key requirements for SON functions are introduced and three categories of techniques used to implement intelligence in the network, namely rule-based expert systems, machine learning, and evolutionary computing, were briefly described. Rule-based expert systems are considered as powerful tools to integrate the available domain expert knowledge into a system. Machine learning methods can be broadly classified into three categories of supervised learning, unsupervised learning and reinforcement learning methods. Genetic algorithms and genetic programming were introduced as key examples of evolutionary computing methods.

Finally, two case studies were presented in this chapter. The first case study considered the problem of coverage and capacity optimization through adjustment of BSs antenna tilt angle. It was shown that a combination of fuzzy logic and reinforcement learning is a suitable candidate for such distributed optimization tasks. Subsequently, a solution for the credit distribution problem was introduced and discussed. Simulation results confirm the feasibility of this framework and show potential performance gains. The second case study considered the issue of coverage optimization for a group of femtocells in a live network. The optimization is achieved by using a genetic programming approach and by automatically building a model of the environment such that individual algorithms can be evaluated in an offline manner without disrupting the network operation. The simulation results show that a near optimal performance can be achieved using this scheme. For the first time, this enables a BS to rewrite its own optimization code during operation.

REFERENCES

1. "Self-optimizing networks: The benefits of SON in LTE," 4G America, White Paper, Oct. 2013.
2. S. Hämäläinen, H. Sanneck, and C. Sartori, *LTE Self-Organising Networks (SON): Network Management Automation for Operational Efficiency.* John Wiley & Sons, 2012.
3. R. Razavi, D. López-Pérez, and H. Claussen, "Neighbour cell list management in wireless heterogeneous networks," in *Proc. IEEE Wireless Commun. Netw. Conf. (WCNC)*, Apr. 2013, pp. 1220–1225.
4. "Evolved Universal Terrestrial Radio Access (E-UTRA); Physical channels and modulation," 3GPP TS 36.211 v14.2.0, Mar. 2017. [Online]. Available: https://portal.3gpp.org/desktopmodules/Specifications/SpecificationDetails.aspx?specificationId=2425
5. A. Zakrzewska, D. López-Pérez, S. Kucera, and H. Claussen, "Dual connectivity in LTE HetNets with split control- and user-plane," in *Proc. IEEE Global Telecommun. Conf. (GLOBECOM)*, Dec. 2013, pp. 390–395.
6. A. Checco, R. Razavi, D. Leith, and H. Claussen, "Self-configuration of scrambling codes for WCDMA small cell networks," in *Proc. 23rd IEEE Pers. Indoor Mobile Radio Commun. (PIMRC)*, Sep. 2012, pp. 149–154.
7. Evolved Universal Terrestrial Radio Access Network (E-UTRAN); Self-configuring and self-optimizing network (SON) use cases and solutions, 3GPP TR 36.902 v9.3.1, Apr. 2011. [Online]. Available: https://portal.3gpp.org/desktopmodules/Specifications/SpecificationDetails.aspx?specificationId=2581
8. Telecommunication management; Self-organizing networks (SON) policy network resource model (NRM) integration reference point (IRP); Requirements, 3GPP TS 32.521 v11.1.0, Dec. 2012. [Online]. Available: https://portal.3gpp.org/desktopmodules/Specifications/SpecificationDetails.aspx?specificationId=2041
9. R. Razavi, S. Klein, and H. Claussen, "Self-optimization of capacity and coverage in LTE networks using a fuzzy reinforcement learning approach," in *Proc. 21st IEEE Pers. Indoor Mobile Radio Commun. (PIMRC)*, Sep. 2010, pp. 1865–1870.
10. R. Razavi, S. Klein, and H. Claussen, "A fuzzy reinforcement learning approach for self-optimization of coverage in LTE networks," *Bell Labs Tech. J.*, vol. 15, no. 3, pp. 153–175, 2010. [Online]. Available: http://dx.doi.org/10.1002/bltj.20463
11. M. Amirijoo, P. Frenger, F. Gunnarsson, J. Moe, and K. Zetterberg, "On self-optimization of the random access procedure in 3G long term evolution," in *Proc. IFIP/IEEE Integr. Netw. Manag.-Workshops (IM)*, Jun. 2009, pp. 177–184.
12. R. Razavi and H. Claussen, "Urban small cell deployments: Impact on the network energy consumption," in *Proc. IEEE Wireless Commun. Netw. Conf. Workshops (WCNCW)*, Apr. 2012, pp. 47–52.
13. W. Wang, J. Zhang, and Q. Zhang, "Cooperative cell outage detection in self-organizing femtocell networks," in *Proc. Int. Conf. Comput. Commun. (INFOCOM)*, Apr. 2013, pp. 782–790.
14. M. Amirijoo, L. Jorguseski, R. Litjens, and R. Nascimento, "Effectiveness of cell outage compensation in LTE networks," in *Proc. IEEE Consum. Commun. Netw. Conf. (CCNC)*, Jan. 2011, pp. 642–647.

15. Evolved Universal Terrestrial Radio Access (E-UTRA); Study on minimization of drive-tests in next generation networks, 3GPP TS 36.805 v9.0.0, May 2010. [Online]. Available: https://portal.3gpp.org/desktopmodules/Specifications/SpecificationDetails.aspx?specificationId=2484

16. O. Aliu, A. Imran, M. Imran, and B. Evans, "A survey of self organisation in future cellular networks," *IEEE Commun. Surveys Tuts.*, vol. 15, no. 1, pp. 336–361, 2013.

17. E. H. Shortliffe and B. G. Buchanan, "A model of inexact reasoning in medicine," *Math. Biosci.*, vol. 23, no. 3–4, pp. 351–379, 1975.

18. F. Hayes-Roth, D. A. Waterman, and D. B. Lenat, *Building Expert Systems*. Boston, MA: Addison-Wesley Longman Publishing Co., Inc., 1983.

19. L. Zadeh, "Fuzzy sets," *Inform. Control*, vol. 8, no. 3, pp. 338–353, 1965.

20. A. A. Atayero, M. K. Luka, and A. A. Alatishe, "Article: Neural-encoded fuzzy models for load balancing in 3GPP LTE," *Int. J. Appl. Inform. Syst.*, vol. 4, no. 1, pp. 34–40, Sep. 2012, published by Foundation of Computer Science, New York.

21. P. d'Orey, M. Garcia-Lozano, and M. Ferreira, "Automatic link balancing using fuzzy logic control of handover parameter," in *Proc. 21st IEEE Pers. Indoor Mobile Radio Commun. (PIMRC)*, Sep. 2010, pp. 2168–2173.

22. P. M. Lee, *Bayesian Statistics: An Introduction*. John Wiley & Sons, 2012.

23. Z. Tu, "Learning generative models via discriminative approaches," in *Proc. IEEE Comput. Vis. Pattern Recognition (CVPR)*, Jun. 2007, pp. 1–8.

24. R. Khanafer, B. Solana, J. Triola, R. Barco, L. Moltsen, Z. Altman, and P. Lazaro, "Automated diagnosis for UMTS networks using bayesian network approach," *IEEE Trans. Veh. Technol.*, vol. 57, no. 4, pp. 2451–2461, Jul. 2008.

25. A. Fujino, N. Ueda, K. Saito, "A hybrid generative/discriminative approach to semi-supervised classifier design," in: *Proceedings of the Proc. 20th National Conference on Artificial Intelligence* – Volume 2, AAAI'05, AAAI Press, pp. 764–769, 2005.

26. R. Daniels, C. Caramanis, and R. Heath, "A supervised learning approach to adaptation in practical MIMO-OFDM wireless systems," in *Proc. IEEE Global Telecommun. Conf. (GLOBECOM)*, Nov. 2008, pp. 1–5.

27. A. Farooqi and A. Munir, "Intrusion detection system for IP multimedia subsystem using k-nearest neighbor classifier," in *Proc. IEEE Multitopic Conf. (INMIC)*, Dec. 2008, pp. 423–428.

28. A. K. Jain and R. C. Dubes, *Algorithms for Clustering Data*. Prentice-Hall, Inc., 1988.

29. V. L. Brailovsky, "A probabilistic approach to clustering," *Pattern Recognit. Lett.*, vol. 12, no. 4, pp. 193–198, 1991.

30. C. Zahn, "Graph-theoretical methods for detecting and describing gestalt clusters," *IEEE Trans. Comput.*, vol. C-20, no. 1, pp. 68–86, Jan. 1971.

31. K. Haldar, H. Li, and D. Agrawal, "A cluster-aware soft frequency reuse scheme for inter-cell interference mitigation in LTE based femtocell networks," in *Proc. 14th IEEE World Wireless, Mobile Multimedia Netw. (WoWMoM)*, Jun. 2013, pp. 1–6.

32. A. G. Barto, *Reinforcement Learning: An Introduction*. MIT Press, 1998.

33. L. P. Kaelbling, M. L. Littman, and A. W. Moore, "Reinforcement learning: A survey," *J. Artif. Intell. Res.*, vol. 4, pp. 237–285, 1996.

34. C. Gaskett, D. Wettergreen, and A. Zelinsky, "Q-learning in continuous state and action spaces," in *Advanced Topics in Artificial Intelligence*. Springer, 1999, pp. 417–428.

35. A. E. Eiben and J. E. Smith, *Introduction to Evolutionary Computing*. Springer, 2003.

36. M. Mitchell, *An Introduction to Genetic Algorithms*. Cambridge, MA: MIT Press, 1996.

37. L. Ho, H. Claussen, and D. Cherubini, "Online evolution of femtocell coverage algorithms using genetic programming," in *Proc. 24th IEEE Pers. Indoor Mobile Radio Commun. (PIMRC)*, Sep. 2013, pp. 3033–3038.

38. J. Niemela, T. Isotalo, J. Borkowski, and J. Lempiainen, "Sensitivity of optimum downtilt angle for geographical traffic load distribution in WCDMA," in *Proc. 62nd IEEE Veh. Technol. Conf. (VTC-2005-Fall)*, vol. 2, Sep. 2005, pp. 1202–1206.

39. A. Bonarini, " Evolutionary learning of fuzzy rules: Competition and cooperation," in *Fuzzy Modelling*, ser. International Series in Intelligent Technologies, vol. 7, W. Pedrycz, Ed. Springer US, 1996, pp. 265–283.

40. M. Wang, "Ultra mobile broadband technology overview," in *Proc. Commun. Netw. Services Res. Conf. (CNSR)*, May 2008, pp. 8–9.

PART II

COVERAGE AND CAPACITY OPTIMIZATION

4

FREQUENCY ASSIGNMENT AND ACCESS METHODS

4.1 INTRODUCTION

When deploying small cells, one of the first choices a network operator needs to make is the configuration of the frequency spectrum and the access model for user equipments (UEs). This choice has an important impact on the interaction with an existing macrocellular network and the resulting network coverage, capacity, and therefore costs for achieving a desired network performance. As a result, it is important to understand the trade-offs of the possible choices.

For frequency allocation, the main options are using new frequency bands or reusing frequencies that are currently in use by existing macrocells, known as co-channel operation. The benefits of using a separate carrier is that there is no interference between the small cells and any existing macrocellular network. While this avoids any potential issues caused by interference, for example mobility problems, the disadvantage is the need for more spectrum resources, which may be more expensive. Alternatively, the spectrum of existing macrocells can be reused, which leads to a higher spectral efficiency, but limits the flexibility of the access model for small cells, and requires more careful optimization. A further challenge of co-channel operation is achieving sufficient coverage, in particular when small cells are deployed close to a high-power macrocell. Partial frequency reuse can be used to overcome some of

Small Cell Networks: Deployment, Management, and Optimization, First Edition. Holger Claussen,
David López-Pérez, Lester Ho, Rouzbeh Razavi, and Stepan Kucera.
© 2017 by The Institute of Electrical and Electronic Engineers, Inc. Published 2017 by John Wiley & Sons, Inc.

these issues and can be used to provide more flexibility in terms of access models or some range extension to small cells.

The second choice is whether small cells should be openly accessible for all customers to complement the macrocellular network or restricted to a closed subscriber group. A closed-access model can be used, for example, to provide access only to customers in a home or in an office. This model has been widely deployed for residential femtocells, which are installed by the users to enhance cellular service in the home and they pay for backhaul and power. Disadvantages of a closed-access model are a reduced offload from the macrocell, and that it cannot be used in combination with reusing the macrocell frequencies since preventing access to the best serving cell would lead to dropped calls.

These problems can be addressed by selecting an open-access model. This is the most beneficial option in terms of network capacity, since all UEs can connect to and benefit from small cells, which reduces interference, and also maximizes offload from the macrocell. For user deployed cells, it is possible to give the primary user preferential access to achieve a similar user experience as in the closed-access model while retaining the benefits of open access.

In this chapter, these different frequency assignment approaches and access methods are discussed.

Section 4.2 focuses on the different frequency assignment approaches such as separate frequency deployments, co-channel deployments and partial frequency reuse. Their advantages and disadvantages are discussed, and their mapping to different radio access technologies is presented.

Then, possible UE access schemes such as closed access and open access are presented. Their applicability for different frequency assignments, and opportunities for the operator are discussed in Section 4.3. These concepts are then illustrated further in two examples in the following sections.

In Section 4.4, the feasibility of co-channel operation of a large number of residential femtocells with an existing macrocellular network is investigated using an open-access model. It is shown that in combination with power self-optimization, femtocells can achieve high performance indoors while keeping the interference impact on the macrocell very low.

In Section 4.5, a co-channel scheme where macrocells reduce their transmit power in part of the spectrum to provide range extension for outdoor small cells is explored. This allows small cells to increase the coverage in this spectrum part resulting in significantly increased network capacity when the powers are correctly configured.

4.2 FREQUENCY ASSIGNMENT APPROACHES

For frequency assignment, no single approach is optimal for all deployments. Several factors play a role in determining the best assignment including the cost and availability of wireless spectrum, and the desired choice of access methods described in Section 4.3.

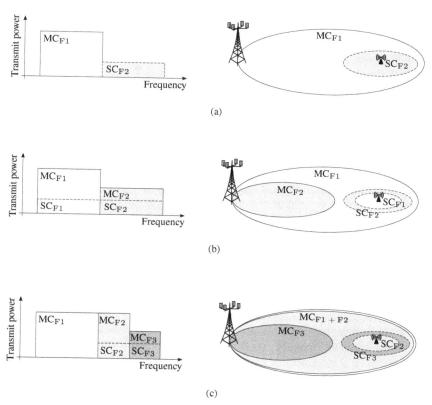

Figure 4.1. Frequency allocation options for Macrocells (MC) and Small Cells (SC). (a) Separate frequency bands, (b) Shared frequency bands and (c) Partially shared frequency bands.

Figure 4.1 illustrates the different frequency assignment approaches, which are described in more detail below.

4.2.1 Separate Frequency Bands

The simplest approach of spectrum allocation is to assign separate frequency bands F1 to macrocells and F2 to small cells as shown in Fig. 4.1a.

This way small cells can operate independently of the underlying macrocellular network without causing interference and potential mobility issues to existing networks. As a result, this assignment also provides the most freedom of choice for access methods including closed-access for a restricted UE group (see Section 4.3). For these reasons, this frequency deployment approach is currently most widely used for residential femtocells.

However, these advantages also come with the drawback of reduced spectral efficiency since both the macrocell network and the small cell network only have access to a fraction of the total available frequency bandwidth. Therefore, this approach is not suitable where the available frequency spectrum is limited or expensive.

When using a closed-access model, interference between small cells can become a problem in dense deployments, where UEs that are close to a neighbor's femtocell cannot connect to their own femtocell due to high interference of the other femtocell to which they cannot connect.

As a result, this approach is most suitable if sufficient frequency spectrum is available. For example, this can be the case in rural areas where the demand is not high enough that the full spectrum needs to be used for macrocells, or if access to spectrum is cheap. Alternatively, this assignment is also attractive if part of the available spectrum is in higher frequency bands, for example, exceeding 5 GHz, that are more suitable for small cells due to their high attenuation. In upcoming 5G networks, mm-wave frequencies would be another good example for small cells on a separate frequency.

4.2.2 Shared Frequency Bands

Alternatively, small cells can fully reuse the frequency bands F1 + F2 of the existing macrocellular networks as shown in Fig. 4.1b. This frequency allocation is only compatible with an open-access model where UEs can connect to all cells of an operator.

Co-channel deployment of small cells and macrocells provides the highest spatial reuse of the available frequency resources, and provides the full available bandwidth to both network tiers. Since the capacity scales linearly with the bandwidth [1], this leads to a high network capacity.

However, co-channel deployment also has a few limitations, which can be addressed by selecting the appropriate UE access scheme and through careful network optimization.

First, since the coverage is defined by the transmit power of the base stations (BSs), the power imbalance between macrocells and small cells can limit the coverage range of small cells, particularly if deployed closely to a macrocell. This can lead to an imbalance of UE throughput due to the limited ability to offload macrocell UEs. To resolve this problem, part of the macrocell spectrum F2 can be used at reduced power. In long-term evolution (LTE), this can be achieved by using almost blank subframe (ABS) (see Chapter 8) or, alternatively, via multi-carrier soft reuse (MCSR) as explained in this chapter. MCSR allows significant range expansion for small cells within this part of the frequency spectrum. For optimal performance, both the amount of spectrum and the power needs to be carefully optimized across multiple macrocells. In Section 4.5, this optimization is discussed in more detail for MCSR.

Second, the resulting interference between macrocells and small cells requires more careful optimization for mobility to avoid handover problems since the overlapping coverage areas between cells are small. Finally, the interference between network tiers limits the feasible access methods. Since UEs have to be able to hand over to the cell with the highest received power to avoid a dropped call, a closed-access model where the access to certain cells is restricted is only feasible if fast handovers to alternative frequencies or access technologies are possible.

The scenarios that benefit the most from co-channel deployment are indoor deployments, which are shielded from the macrocells, such as enterprises, shopping malls, airports and other open venues, as well as sparse outdoor deployments where coverage is not the limiting factor or can be increased by reducing power in part of the macrocell spectrum. Residential deployments are also a good candidate as shown in Section 4.4, but they require a change from today's prevalent closed-access model to open access as discussed in more detail in Section 4.3 below.

4.2.3 Partially Shared Frequency Bands

In order to benefit from increased spectral efficiency of frequency reuse, but retain the flexibility for allowing a closed-access model where only registered UEs can access the small cell, partial frequency reuse can be used. Here, macrocells use the full available spectrum F1 + F2 + F3 and small cells reuse a fraction of that frequency spectrum F2 + F3 as shown in Fig. 4.1(c).

This frequency allocation can improve spectral efficiency over separate frequency bands by allowing the macrocells to use the whole available bandwidth, and provides full flexibility for using all access methods including closed access. This is achieved by retaining a "clean" part of frequency spectrum for macrocells that can be used by mobile UEs, which pass through small cells for which they do not have access rights. This way, UEs can pass through those cells on this interference-free part of the frequency spectrum without dropping a call or data session. As described previously, when using a closed-access model, interference between small cells operating on the same frequency band can become a problem in dense deployments.

Since only part of the frequency spectrum is used by small cells, the overall spectral efficiency is reduced compared to full co-channel operation. This is the price to pay for enabling the closed-access model. As small cells operate in the spectrum used by the macrocell, partial frequency reuse suffers from the same limitation of limited coverage due to the power imbalance between macrocell and small cells. As described above, this can be prevented by using ABS or alternatively via MCSR, that is, reducing the transmit power of the macrocell in part of the shared bandwidth F3.

Partial frequency reuse is mainly useful for residential femtocell deployments if a closed-access model is required.

4.3 ACCESS METHODS

While traditionally, cellular networks operate an open-access model where all customers of an operator can connect to each cell, a closed-access model was introduced with femtocells to cater for the residential deployment in customer's, homes. With this model, only a few registered UEs can access the cell, similar to those wireless fidelity (Wi-Fi) access points in the home. However, transitioning residential cells to an open-access model provides a significant opportunity for operators, as these cells can then effectively complement existing macrocellular cellular networks. These options are described and discussed in more detail in this section.

4.3.1 Closed Access

The closed-access model restricts the access to a closed subscriber group. It was introduced for femtocells, which are deployed in the user's home and used the broadband Internet connection as backhaul for cellular communication.

Advantages of the closed-access model are that users are already comfortable with this as it is equivalent to what is currently in use by Wi-Fi routers that are widely deployed in homes for providing wireless access to the Internet. In addition, closed access is compliant with the terms and conditions of Internet access providers, which often do not allow to make the service publicly available. This is of particular importance if the Internet access provider is different from the mobile provider, which is often the case. Finally, closed access provides all radio resources for the customers that deployed the femtocell, which provides a high and consistent quality of experience.

The main disadvantages of closed access are a reduced network efficiency since less UEs can be offloaded to small cells. For an operator, this is a significant lost opportunity for increasing network capacity. Second, closed access is incompatible with co-channel deployment, the most spectral efficient frequency allocation approach. This is illustrated in Fig. 4.2a showing the coverage probability of a macrocell UE when coming close to a co-channel femtocell, which the UE

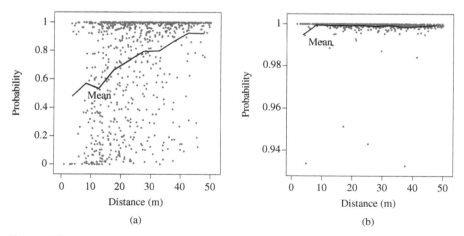

(a)   (b)

Figure 4.2. Coverage probability for un-registered macrocell UEs as a function of the distance from a small cell with closed and open access [2]. (a) Private access and (b) Public access.

cannot access. It is shown that the coverage probability becomes close to zero when approaching the small cell, which results in service outage and dropped calls for the UE. Similarly, since small cells typically share the same spectrum to maximize the spatial reuse of the frequency spectrum, interference between small cells can cause problems for the UE in dense deployments, such as apartment blocks. Since UEs can only connect to their own cell, it needs to be placed such that it provides the best signal in the complete intended coverage. This may not be possible if the neighbors deploy sub-optimally. Therefore, while each UE has access to the full radio resources of their cell, this can still lead to poor performance when the UE is located close by a neighboring small cell. However, in this case, the macrocell can be used as fallback, so this does not result in a coverage hole when good macrocell coverage exists.

Closed access is currently used extensively for residential femtocells, but due to the disadvantages to both operator and user, a transition to an open-access model should be considered where possible. In areas where the fixed backhaul connection is provided by the same operator, this is a relatively straightforward change.

4.3.2 Open Access

Open access is the standard access method in macrocellular networks where each customer of an operator can access all cells in the network and is handed over the one which provides the best service. This is usually the cell with the highest received signal strength, but small cells can be prioritized since they share their bandwidth with a much lower number of UEs.

Open access solves the interference problem both between macro- and small cells, and between small cells as it allows all UEs always to connect to the best cell. As a result, the network efficiency is significantly improved. This is illustrated in Fig. 4.2b, showing that with open access, the coverage holes around co-channel small cells that exist for closed access completely disappear.

Moreover, open access enables a much greater traffic offload to small cells. UEs that now can connect to a small cell can achieve much higher data rates since the bandwidth per UEs is typically much higher compared to a macrocell. Due to the higher offload, macrocell UEs benefit as well since the radio resources do not need to be shared amongst as many UEs. This increases the network capacity significantly and provides improved quality of experience for UEs.

In addition, open access in combination with customer deployment also presents a very attractive small cell deployment strategy for an operator, since end-user deployment is much more cost-effective than operator-deployed small cells. It has been shown in [3] that in urban areas, an end-user–deployed small cell network with open access can quickly serve the majority of traffic, and macrocells are only required for providing the remaining coverage, as illustrated in Fig. 4.3. For example, for an operator with 40% market share, if only 20% of customers have small cells deployed, they can already serve 80% of the total traffic demand in an urban area. As a result, the total network cost is reduced significantly.

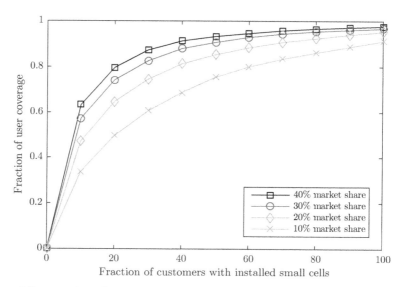

Figure 4.3. Fraction of total user demand covered by small cells with open access in an urban environment for operators with different market share [3].

The main challenges with applying the open-access model to small cells are business challenges. First, if fixed broadband is provided by a different operator than the mobile broadband, an agreement between the operators may be required to allow opening up the UEs broadband connection to serve other UEs. Second, as the end user provides the site, backhaul and power, it would be reasonable to compensate the him in some way. Examples for this could be earning free services via serving others, or getting a revenue share. This would provide the end user an incentive to deploy small cells where they are needed and offload UEs from the macrocellular network.

A minor issue when small cells are deployed by the end user in homes or in enterprises is that the owners would often like to have priority. This can be achieved by introducing a preferential access for registered UEs, which prioritizes registered UEs in case the small cell is also serving other UEs at the same time. Despite the prioritization, other UEs typically still benefit from connecting to the small cell compared to being served by the macrocell.

In summary, due to the many advantages, open access should be the preferred access scheme for small cell deployments.

4.4 CO-CHANNEL OPERATION OF RESIDENTIAL FEMTOCELLS

As a first example, the technical feasibility of co-channel operation of residential femtocells with an existing macrocellular network is investigated based on [4–6].

The achievable capacities and the impact on the performance of the macrocellular network are investigated for both downlink and uplink.

4.4.1 Scenario

Figure 4.4 illustrates the considered scenario with seven macrocells, each with three sectors, and 100 femtocells per macrocell sector. Femtocells, which appear as solid dots, are deployed randomly within the coverage area, and reuse the same frequency as the macrocells, shown as squares, in a hierarchical cell structure. Figure 4.5 shows the floor plan for the explicit indoor wall model for the modeled house, and an example of the resulting path gains for a femtocell located in the center of the house.

Key simulation parameters including the propagation models for path loss, shadow fading, and the antenna gain for the macrocell sectors assumed are shown in Table 4.1.

The macrocell downlink transmit power is set such that the received signal-to-noise ratio inside of a house at the cell edge is 10 dB, assuming an additional wall loss of 15 dB. As a result, the obtainable throughputs from the macrocell are interference limited. The femtocell power is controlled to achieve a target cell radius of 10 m for line-of-sight propagation subject to a maximum value of 125 mW. The

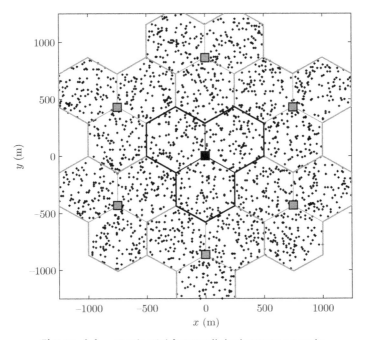

Figure 4.4. Residential femtocell deployment scenario.

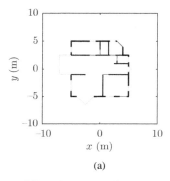 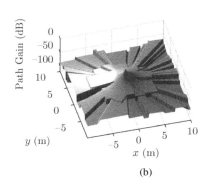

(a) (b)

Figure 4.5. Floor plan of modeled house (a) and an example of the resulting path gain (b) located at the center of the house.

calculation of the required pilot power is described in more detail in Section 5.2.1, and the pilot accounts of 1/10 of the total transmit power. A target small cell radius of 10 m is considered to cover only the users' homes, in order to minimize handovers from passing UEs outside. The maximum power for UEs in the uplink is 125 mW. The power of UEs connected to femtocells is additionally controlled to limit the total interference received at the closest macrocell to a value 3 dB above the thermal noise level. This way it can be guaranteed that the uplink of the macrocell is not degraded significantly. Here, a simple approach is considered where the total interference

TABLE 4.1 Simulation Scenario Parameters

Parameters	Value
Macro BS placement	7 macrocell BSs, 800 m inter-site distance
Small cell BS placement	0, 10, 100 femto BSs per macrocell BS sector
Scenario resolution	4 m (macro), 0.6 m (femto)
Carrier frequency	2000 MHz
Bandwidth	3.84 MHz (HSPA)
Height	25 m (macro), 1.5 m (femto)
BS transmit power	21.6 W (macro), 125 mW (femto)
SNR target at cell edge	10 dB
Macro BS antenna	See Section A.3.1
Path loss	$28 + 35 \log_{10}(d[\text{m}])$ outdoors
	$38.5 + 20 \log_{10}(d) + L_{\text{walls}}$ indoors
Indoor wall loss L_{walls}	15/10/7 dB for external/internal/light internal walls, 3 dB for doors, 1 dB for windows
Shadow fading (SF)	8 dB outdoor, 4 dB indoor (std dev.)
Macro SF correlation	$R = e^{-1/20d}$, 50% inter-site
Noise figure	7 dB (UE, femtocell); 4 dB (macrocell)

allowance is shared equally among all femtocells in each macrocell sector irrespective of how many are actively receiving data from UEs. Interference generated by UEs to neighboring sectors or cells is not taken into account [6].

System level simulations were performed to identify the possible downlink and uplink throughputs at any location of the scenario for both the macrocell and for femtocells operating simultaneously in the same frequency band. Throughput statistics are obtained for femtocells located in the three sectors covered by the central macrocell, so that interference from the adjacent cells can be taken into account. It is assumed that each UE connects to the BS (femto or macro) with the best downlink signal-to-interference ratio (i.e., open access). Transmissions to multiple UEs are scheduled in time as in high-speed downlink packet access (HSDPA). The obtainable throughput can be calculated as described in more detail in Section A.9 based on the Shannon–Hartley theorem [1]. Here an operation point 3 dB from the capacity limit is assumed.

In the downlink, interference resulting from all macrocell sectors and all femtocells is taken into account. In the uplink, interference results from all currently active UEs. Macrocell UEs are located randomly in the coverage area of each sector (one active UE per sector at each time instant). Femtocell UEs are located randomly in each house (one active UE per femtocell at each time instant).

The simulated resolution is 40 m for the macrocell throughput. Areas where femtocells are deployed were simulated with a higher resolution of 0.6 m. Hundred simulation iterations were performed with different shadow fading values, house locations, and femtocell locations for each scenario to collect throughput statistics from which cumulative distribution functions (CDFs) of the possible throughputs were derived. They describe the probability distribution of the possible macro- and femtocell throughputs over the area covered by the three sectors of the central macrocell, or the area covered by femtocells, respectively. In order to evaluate the impact of different numbers of deployed femtocells on the macrocell performance, the scenario was also simulated with 0, 10, and 100 femtocells per sector.

The simulation results show that co-channel deployment of femtocells on the same frequency as an existing macrocellular network is feasible with only minor impact on macrocell performance when using the proposed femtocell power auto-configuration. In addition, due to the small distance between transmitter and receiver, and the additional shielding due to walls in typical situations, the femtocells can provide very high throughputs. A more detailed description of the results is discussed below.

4.4.2 Downlink Results

The throughput CDFs for the downlink of both macrocells and femtocells with different numbers of N active femtocells, and examples of downlink throughput distributions around one macrocell and femtocell are shown in Figs. 4.6 and 4.7.

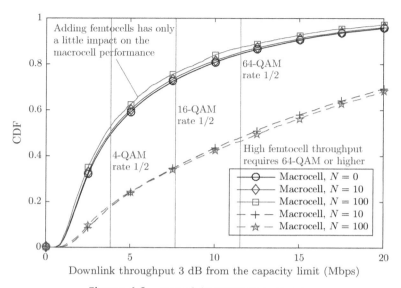

Figure 4.6. CDF of downlink throughputs.

As Fig. 4.6 shows, the drop in macrocell throughput as a result of the additional interference caused by 10 or 100 deployed femtocells is only minimal. This is essential for a successful co-channel femtocell deployment since any significant degradation in coverage or capacity of the existing macrocellular network would be unacceptable. In the downlink, this results from the low femtocell transmit power, the strong signal falloff at the femtocell boundary due to the path loss at such short ranges,

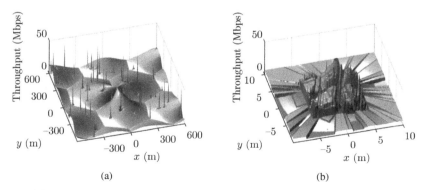

Figure 4.7. Distributions of downlink throughputs. (a) Example downlink throughput distribution in the simulated area and (b) Example downlink throughput distribution in the area around one femtocell.

and in most cases, an additional wall separation. It is also shown that due to high indoor signal-to-interference-plus-noise ratios (SINRs), the theoretically achievable femtocell throughput is very high, so that in most of the cases, 64-quadrature amplitude modulation (QAM) or higher order modulation would be required to achieve it, assuming 1/2 rate coding. The high indoor SINRs are a result of the small distance to the femtocell and the wall separation, which shields the signal from interference. It is also shown that increasing the number of femtocells from 10 to 100 does not significantly affect the downlink femtocell throughput due to the small cell size and a typically strong wall separation between houses. Therefore, large numbers of femtocells can be deployed in the same frequency band of existing macrocell networks without a significant downlink performance degradation of the macrocells, when open access to the femtocells is granted.

Figure 4.7a shows one example of a corresponding downlink throughput distribution around the central macrocell with 10 deployed femtocells per macrocell sector, for clarity without shadow fading. As expected, the macrocell throughput is highest close to the BS and in the directions of the main lobe of the directional antenna of each sector, and falls off toward the edge of each sector due to the lower signal levels and increased interference from neighboring sectors. The spikes of higher throughput shown are in locations where femtocells are deployed. Figure 4.7b shows one example of a downlink throughput distribution around one of the femtocells located at the coordinates [100 m, 300 m] within the central macrocell, for clarity without shadow fading. As a result of the short distances to the femtocell and the walls that shield the house from interference, the achievable indoor throughputs are very high. Outside of the house, throughput is reduced since those areas are served primarily by the macrocell, which provides the strongest signal, and because the interference caused by other macrocells and femtocells is higher than indoors.

4.4.3 Uplink Results

Throughput and examples of throughput distributions for the uplink are shown in Figs. 4.8 and 4.9.

Similar to that of the downlink, the drop in macrocell uplink performance resulting from the addition of 10 or 100 active femtocells is minimal, as shown in Fig. 4.8. This is a result of the fast maximum power control in the uplink for femtocell UEs that limits the interference caused to macrocell UEs to a predefined level. Despite this maximum uplink power limitation, it is shown that the possible throughput of femtocell UEs is very high as a result of high signal-to-interference ratios due to the small distance to the femtocell and the wall separation to the macrocell. When the number of femtocell UE transmissions is increased from 10 to 100 per macrocell sector, the maximum allowed interference per UE is reduced by a factor of 10. Therefore, the SINR can drop by up to 10 dB, resulting in a reduction in uplink throughput. However, since the uplink throughput performance is very high, this is not problematic. As for the downlink, the theoretically achievable uplink throughput is so high

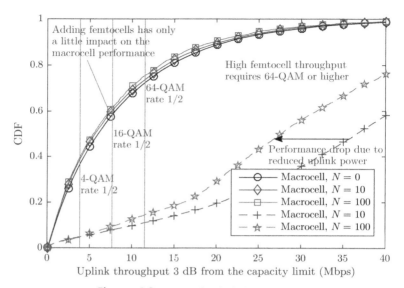

Figure 4.8. CDF of uplink throughputs.

that in most of the cases 64-QAM or higher order modulation would be required to achieve it.

Figure 4.9a shows one example of a corresponding uplink throughput distribution for 10 deployed femtocells per macrocell sector, for clarity without shadow fading. As expected, the achievable uplink throughput is high in the direction of the main lobe of the directional antenna of each sector close to the BS, and falls off with increasing distance. The levels of achievable throughput are different for each sector, since

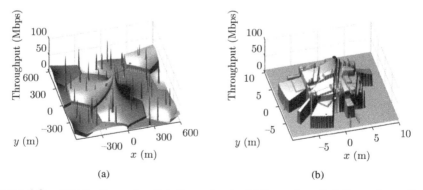

Figure 4.9. Distributions of uplink throughputs. (a) Example uplink throughput distribution in the simulated area and (b) Example uplink throughput distribution in the area around one femtocell.

they are impacted by the interference from currently transmitting mobile devices in the neighboring sectors, which are randomly located within each of the sectors. The spikes of high throughput are in locations with femtocell coverage.

Figure 4.9b shows one example of an uplink throughput distribution around one of the femtocells located at the coordinates [100 m, 300 m] within the central macrocell, for clarity without shadow fading. It is shown that the achievable indoor throughput is very high. This is also the case for areas outside with good channels to the femtocell, such as in front of windows within the range where the received downlink signal from the femtocell is still the strongest. This is due to the assumption that in the uplink, the mobile device is connected to the same BS as in the downlink (i.e., the BS from which it receives the strongest signal power).

4.5 MULTI-CARRIER SOFT-REUSE

If a larger coverage is intended for small cells (e.g., outdoors), reducing the macrocell transmit power in the shared band is required to meet the coverage target without using excessive transmit power at the small cell. This concept is known as soft-reuse, and requires optimization of the carrier transmit power. In this section, an example of a MCSR deployment and its optimization is discussed.

As discussed in Section 4.2, with MCSR, small cells use two carriers F1 + F2 with full transmit power. The macrocells use the same carriers, but F2 at a reduced transmit power. The effect of the reduced macrocell transmit power at F2 is that it increases the coverage area of the small cell on F2, resulting in range extension for small cells. This creates a frequency reuse pattern illustrated in Fig. 4.1b. The use of MCSR can increase the overall system performance by offering a higher UEs offload of macrocell UEs to small cells.

However, in order to maximize the gains from using MCSR, the correct macrocell F2 power reduction has to be chosen. If too little reduction is used, this causes a degradation in macrocell SINR, without the benefits of small cell offloading. Conversely, if the power of F2 is reduced too much, this causes F2 on the macrocell to be under-utilized, while F2 on the small cell becomes over-utilized. These issues can cause MCSR to actually reduce the overall system performance, and there is therefore a need to set the power reduction level of F2 on the macrocell correctly.

The optimization of the F2 power reduction on the macrocell is dependent on many different aspects, including the location and number of small cells, the UE traffic distribution, the macrocell and small cell maximum transmit powers and the key performance indicators (KPIs) to improve, such as total system throughput, cell-edge throughput and so on. The optimization of parameters in MCSR is a topic that has been widely studied [7–10]. However, the techniques that have been proposed rely on the availability of SINR information of the UEs in order to optimize the radio resource allocations. These measurements are typically not available in deployed networks, and estimations of SINR tend to be inaccurate. In this section, a simulation study of a practical MCSR optimization that uses standardized measurements available in the X2 interface is presented.

4.5.1 Scenario

A simulated heterogeneous network scenario is used to examine the use of MCSR for an outdoor small cell deployment, and the main simulation parameters used are described in Table 4.2. The two carrier frequencies for F1 and F2 are 740 MHz and 2.135 GHz, respectively.

There are two alternatives in which the load balancing between carriers in multi-carrier BS deployments can be handled: with and without carrier aggregation, which is a feature introduced in the LTE Release 10 specification.

Without carrier aggregation, UEs are strictly assigned to a single carrier (F1 or F2) and scheduled on a single carrier. The load balancing between the two carriers can be managed with inter-frequency handovers to transfer a UE from one carrier to another.

With carrier aggregation, a UE can be served simultaneously by multiple carriers, called component carriers. Each UE is assigned to one primary cell (PCell) and at least one secondary cell (SCell), which correspond to different component carriers. The physical downlink control channel (PDCCH) is carried by the PCell that the UE is connected to. At a single transmission time interval (TTI), each UE may be scheduled

TABLE 4.2 Simulation Scenario Parameters

Parameters	Value
Macrocell network topology	7×3 macrocells, 500 m inter-site distance
Carrier frequencies	F1 = 740 MHz, F2 = 2.135 GHz, 10 MHz bandwidth
F2 path loss	Macrocell to UE =15.9 + 39 $\log_{10}(d)$
	Small cell to UE = 31.3 + 25 $\log_{10}(d)$
F1 path loss	Macrocell to UE = 25 + 39 $\log_{10}(d)$
	Small cell to UE = 31.3 + 25 $\log_{10}(d)$
Traffic model	Full buffer
Scheduler	Proportional fair
Shadow fading standard deviation	6 dB for macrocell, 4 dB for small cell
Maximum macrocell BS transmit power	30 W for each carrier
Small cell BS maximum transmit power	0.5 W for each carrier
UE placement	60 UEs per macrocell sector, 33% of UEs placed within 40 m of small cell BS
Small cell BS placement	Placed with a minimum distance of 100 m between Macrocell and 40 m from other small cell BSs

in resources from a single or both carriers based on channel quality indicator (CQI) feedback, which is available from both carriers, and the assignment can change from TTI to TTI. Therefore, the resource allocation across the carriers is more efficient and load balancing between the F1 and F2 carriers is achieved as a side effect of the scheduling. In the simulated scenario, carrier aggregation functionality is assumed, where the small cell PCell is configured in the F2 carrier and the macrocell PCell is configured in the F1 carrier.

4.5.2 Effect of Power Reduction

In order to understand the effect of MCSR on the performance of the network, a simple scenario with a single small cell is used. The small cell is placed 140 m away from the center of a macrocell sector, as shown in Fig. 4.10. Simulations were performed with the transmit powers of the F2 carrier of the macrocells, $P_{MacroF2}$, reduced in the range of 0 dB to 20 dB, and the UE throughputs and the fraction of physical resource blocks (PRBs) available for the small cell and its underlay macrocell is calculated. The transmit powers of the F1 carrier on the macrocell, and the F1 and F2 carriers of the small cell are set to the maximum. The UEs are scheduled resources in order to maintain a target throughput of 500 kbps per UE.

Figure 4.11 shows the fraction of free PRBs available, and Fig. 4.12 shows the number of UEs unable to achieve the 500 Kbps target throughput (i.e., in outage conditions) at the macrocell and small cell with increasing reductions in the transmit power of F2 at the macrocell. When there is no power reduction, the macrocell is

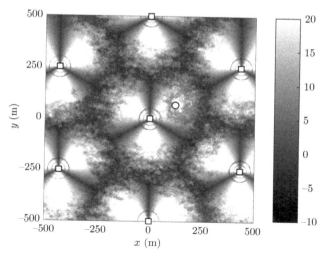

Figure 4.10. Simulation scenario with a single small cell, showing the locations of the macrocell and small cell BSs and the maximum SINR.

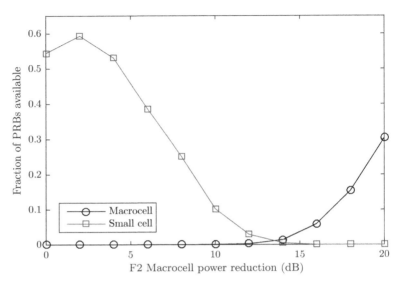

Figure 4.11. Fraction of free physical resource blocks in a single small cell scenario for different $P_{MacroF2}$ reduction.

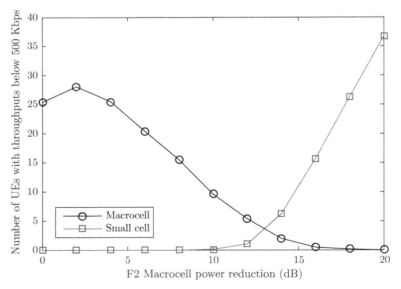

Figure 4.12. Number of UEs unable to achieve the 500 Kbps throughput target for different $P_{MacroF2}$ reduction.

overloaded, with all PRBs utilized and a high number of UEs in outage conditions. As the macrocell F2 transmit power is gradually reduced by up to 14 dB, more UEs are offloaded from the macrocell to the small cell due to the resulting range expansion in the small cell. This results in the reduction in PRB utilization at the macrocell, and a corresponding increase of PRB utilization at the small cell. The total number of UEs in outage conditions also reduces. However, when the power reductions are higher than 14 dB, the small cell starts to become overloaded as it has to serve more UEs on its F2 carrier. At this stage, the total number of UEs in outage conditions starts to increase.

The results of this simulation serve to illustrate the importance for configuring the F2 power reduction properly in order to optimize the gains offered by the use of MCSR. One observation that can be made is that the best configuration occurs when the load between the macrocell and the small cell is balanced, which corresponds in this example to a power reduction of roughly 12–14 dB. Achieving this load balancing between the macrocell and small cell BSs will therefore be used as the objective of the MCSR optimization algorithm.

In the results shown in Figs. 4.10 and 4.11, P_{MacroF2} of all the macrocells in the scenario were reduced by the same amount. If the power reduction is applied only to the macrocell in which the small cell is present (i.e., the local macrocell), this results in a degradation of SINR on the F2 carrier of the local macrocell due to interference from neighboring macrocells. Figure 4.13 shows the impact of this interference when the power reduction is local, compared with a global power reduction.

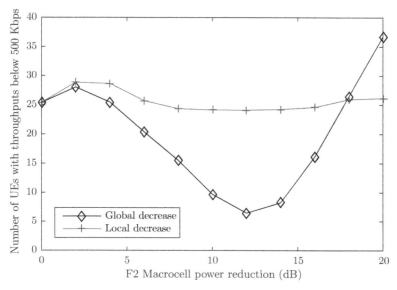

Figure 4.13.   Comparison between global and local power reduction.

The results show that little to no gains in performance can be achieved if MCSR is applied locally. The power reduction of P_{MacroF2}, therefore, has to be applied globally to all macrocells.

The proposed MCSR optimization algorithm uses mean PRB utilization (i.e., the load) of the macrocell and small cell BSs as its input. The basic principle of the algorithm is to monitor the load of the BSs and adjust P_{MacroF2} in order to equalize the load between the macrocell and small cell BSs. With a single small cell, this process is straightforward. Once the mean PRB utilization measurements have been collected over a period of time, if the macrocell's PRB utilization is higher than the small cell, decrease P_{MacroF2}, and if the small cell's PRB utilization is higher than the macrocell, increase P_{MacroF2}. This gradual increase or decrease in P_{MacroF2} is performed until the load between the macrocell and small cell is balanced.

However, when there are multiple small cells deployed within the coverage area of several macrocells, this will result in some macrocells wanting to increase P_{MacroF2}, while others will want to decrease P_{MacroF2}. In order to overcome this conflict, the decision on power adjustment is given to the most highly loaded macrocell/small cell grouping following the steps below:

1. Identification of macrocell and small cell group. The small cells measures the macrocell with the highest reference signal received power (RSRP) as the macrocell it "belongs" to. A small cell can belong to more than one macrocell group if it is located at the macrocell edge and measures two or more macrocell RSRPs that are very similar (e.g., within 4 dB of each other).
2. Each macrocell and small cell group determines its action. Either to increase P_{MacroF2} to offload more UEs to the small cells, or increase P_{MacroF2} to offload more UEs to the macrocell.
3. Calculation of the load metric of the most highly loaded cell within each macrocell and small cell group. The load metric used here is the total average PRB utilization of all the cells in the group, normalized to 1.
4. Perform the action of the macrocell and metrocell group with the highest load metric (either through central controller or X2 coordination).

Steps 2 to 4 are repeated, and the algorithm will eventually converge to an oscillating point, whereupon the optimization is stopped.

Figure 4.14 shows an example scenario with 9 small cells deployed randomly, and the resulting macrocell/small cell grouping from Step 1 described above. P_{MacroF2} is initialized to the maximum of 30 W, and the algorithm is iterated. Figure 4.15 shows the normalized load metric of the macrocell and small cell groups over the algorithm iterations. Between iterations 1 and 6, Group 3 wants to decrease P_{MacroF2} to offload UEs to the small cells, and has the highest priority as it has the highest load, so the power is reduced at each iteration by 1 dB. At iteration 7 and 8, Group 2 now has the highest load and is given priority. Group 2 further decreases P_{MacroF2} by 1 dB at

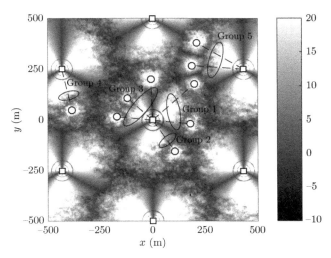

Figure 4.14. Example scenario with small cells grouped to their respective macrocells.

each iteration. From iteration 9 onward, the group with the highest priority oscillates between Group 4, who wants to decrease P_{MacroF2} to offload UEs to the macrocell, and Group 2 who wants to increase P_{MacroF2}. Once this oscillation is detected, the optimization stops. In this example, the algorithm has reduced P_{MacroF2} by a total of

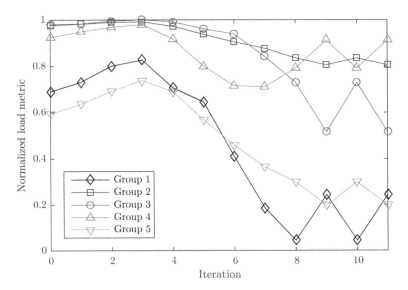

Figure 4.15. Load metrics of macrocell/small cell groups during optimization.

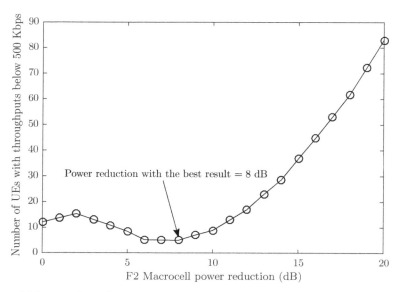

Figure 4.16. Number of UEs unable to achieve the 500 Kbps in multiple small cell scenario for different $P_{MacroF2}$ reduction.

8 dB. This is the optimal configuration, as shown by the result of a global search of all power reductions of the scenario in Fig. 4.16, which achieves an improvement of nearly 60% compared with the baseline of no power reductions.

4.6 SUMMARY AND CONCLUSIONS

In this chapter, frequency assignment and access methods, which are some of the first configuration choices an operator has to make when deploying small cells, have been discussed. These choices have a significant impact on the performance and capabilities of the overall network.

For the frequency assignment, tree different options were presented. Deployment of small cells in separate frequency bands is the simplest approach which avoids interference between macrocells and small cells, and provides full flexibility in terms of access methods, but comes with the drawback of reduced spectral efficiency. As a result, this allocation is most suitable when sufficient spectrum is available, or when deploying small cells in higher frequency bands that are less suitable for macrocells such as mm-wave. Alternatively, the deployment of small cells in shared frequency bands with macrocells results in higher spectral efficiency but due to the resulting interference between macrocells and small cells, a more careful network optimization, and an open-access model are required. MCSR can be employed to increase the coverage range of small cells and increase the offload of users to small cells, which improves the overall network performance. Co-channel deployment is particularly suitable for indoor deployments that are shielded from the macrocells such as enterprises, airports, shopping malls, and residential femtocell deployments with open access. As a third option, the deployment of small cells in partially shared frequency bands with macrocells improves spectral efficiency compared to separate frequency deployment while retaining the full flexibility in terms of access models. Partial frequency reuse is mainly useful for residential femtocell deployments if a closed-access model is required.

Depending on the frequency assignment selected, the different access methods can be chosen. Closed access was introduced for residential femtocells to restrict the access to a predefined subscriber group. This approach comes with significant disadvantages for both operator and user, resulting in a lower network efficiency since less UEs are offloaded to small cells, and in some cases highly increased interference when users cannot connect to the best cell. While closed access is currently used extensively for residential femtocells, a transition to an open-access model should be considered to address the disadvantages where possible. Open access is the standard access method for cellular networks where all subscribers of an operator can connect to all cells. It significantly increases the offload to small cells and solves the interference problem, resulting in improved network and user performance. Open access is also a key enabler for allowing user-deployed small cells to effectively complement macrocellular coverage and capacity. To guarantee high performance of user deployed cells for the owners, user priorities can be configured. Due to many advantages, open access should be the preferred access scheme for small cell deployments.

In a first example, it has been shown that co-channel operation of residential femtocells with existing macrocellular networks is feasible with only minor impact on the macrocellular network when an open-access model is adopted or if other networks are available to cover unregistered UEs. This allows efficient spatial frequency reuse, resulting in a significantly higher spectral efficiency per area. In the downlink, the transmit power optimization is necessary to achieve a roughly constant coverage of the residential homes due to the varying power of the macrocellular network. In the uplink, a power control method was used that ensures a low predefinable impact on the uplink performance of the macrocell network. Despite the low resulting femtocell transmit powers for downlink and uplink, the short distance between femtocell BS and UE and the wall separation to other interference sources result in very high achievable theoretical femtocell throughputs for both downlink and uplink in most of the covered femtocell area. In reality, those data rates can be achieved only by using 64-QAM or higher-order modulation schemes.

In a second example, it has been shown that MCSR can be used for range extension of small cells to allow small cells to offload more traffic from macrocells, resulting in significant gains in capacity when configured properly. This range extension can be achieved through the adjustment of the macrocell transmit power on one of the carriers, and does not require the close coordination required to implement ABS in enhanced inter-cell interference coordination (eICIC). However, the adjustment of the macrocell transmit powers needs to be optimized to balance the load between the different carriers of the macrocell and small cells. In addition, the effect of inter-cell interference between macrocells needs to be considered when applying MCSR. Here, a reduction of transmit power at one macrocell carrier should be matched by reductions of the neighboring cells on the same carrier to maintain the same level of SINR. A self-organizing network (SON) algorithm has been presented that can optimize the transmit powers automatically.

REFERENCES

1. C. E. Shannon, "A mathematical theory of communication," *The Bell Syst. Tech. J.*, vol. 27, no. 3, pp. 379–423, Jul. 1948.

2. J. D. Hobby and H. Claussen, "Deployment options for femtocells and their impact on existing macrocellular networks," *Bell Labs Tech. J.*, vol. 13, no. 4, pp. 145–160, Feb. 2009.

3. H. Claussen, L. T. W. Ho, and L. G. Samuel, "Financial analysis of a pico-cellular home network deployment," in *Proc. IEEE Int. Conf. Commun. (ICC)*, Glasgow, UK, Jun. 2007, pp. 5604–5609.

4. H. Claussen, L. T. W. Ho, and L. G. Samuel, "An overview of the femtocell concept," *Bell Labs Tech. J.*, vol. 13, no. 1, pp. 221–245, May 2008.

5. H. Claussen, "Performance of macro- and co-channel femtocells in a hierarchical cell structure," in *Proc. IEEE Int. Symp. Pers., Indoor Mobile Radio Commun.*, Athens, Greece, Sep. 2007, pp. 1–5.

6. H. Claussen, "Co-channel operation of macro- and femtocells in a hierarchical cell structure," *Int. J. Wireless Inform. Netw.*, vol. 15, no. 3, pp. 137–147, Dec. 2008.

7. M. Qian, W. Hardjawana, Y. Li, B. Vucetic, X. Yang, and J. Shi, "Adaptive soft frequency reuse scheme for wireless cellular networks," *IEEE Trans. Veh. Technol.*, vol. 64, no. 1, pp. 118–131, Jan. 2015.

8. Y. Yu, E. Dutkiewicz, X. Huang, and M. Mueck, "Adaptive power allocation for soft frequency reuse in multi-cell LTE networks," in *Proc. Commun. Inform. Technol. (ISCIT)*, Oct. 2012, pp. 991–996.

9. L. Chen and D. Yuan, "Soft frequency reuse in large networks with irregular cell pattern: How much gain to expect?" in *Proc. IEEE Int. Symp. Pers., Indoor Mobile Radio Commun.*, Sep. 2009, pp. 1467–1471.

10. X. Mao, A. Maaref, and K. H. Teo, "Adaptive soft frequency reuse for inter-cell interference coordination in SC-FDMA based 3GPP LTE uplinks," in *Proc. IEEE Global Telecommun. Conf. (GLOBECOM)*, Nov. 2008, pp. 1–6.

5

COVERAGE AND CAPACITY OPTIMIZATION FOR INDOOR CELLS

5.1 INTRODUCTION

When deploying a large number of small cells as overlay to an existing macrocellular network, coverage optimization is an important problem that needs to be addressed to avoid mobility and interference problems, provide full coverage of the intended area and maximize user equipment (UE) performance.

The coverage of a cell is typically controlled by its pilot signal power, or more specifically, its received pilot signal strength compared to neighboring pilot signals and noise. In addition, it is usually possible to define an offset to associate a UE to a cell even if its received pilot signal is not the strongest. In typical configurations, the pilot power is set to around 10% of the total base station (BS) transmit power. To keep inter-cell interference at the cell boundary balanced, the total transmit power for the BS is typically adapted with the pilot power when changing the coverage.

For coverage optimization, one needs to distinguish between residential femtocell deployments, which require optimization of individual cells, and enterprise deployments in which groups of cells need to be optimized jointly.

For residential femtocells, the deployment differs significantly from the carefully planned deployment of the conventional underlay macrocells. Femtocells are deployed by users, and consequently little to no cell planning is carried out. The location is often determined based on where the broadband cable connection is available,

Small Cell Networks: Deployment, Management, and Optimization, First Edition. Holger Claussen, David López-Pérez, Lester Ho, Rouzbeh Razavi, and Stepan Kucera.
© 2017 by The Institute of Electrical and Electronic Engineers, Inc. Published 2017 by John Wiley & Sons, Inc.

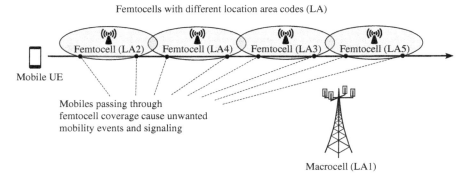

Figure 5.1.   Residential femtocell deployment scenario and resulting mobility problem.

and not optimally from a radio perspective. Therefore, a femtocell BS can often be deployed in an unsuitable location where its pilot signal radiates to areas outside of its intended area of coverage. This has the undesirable effect of triggering handovers from unwanted transient UEs who happen to be passing by the femtocell.

In dense urban areas where many femtocells are deployed, a very large number of mobility events can consequently occur, which has an impact both on the network due to an increased signaling load, and on battery life of UEs. In some cases, the handover frequency can be also exacerbated by the need for closed-access operation (see Section 4.3), which requires the location area code used by a femtocell to be different from the one used by the macrocell underlay, and for closed access even to be different from those used by neighboring femtocells. The effect of such disparate allocation of location area codes is that the rise in unwanted mobility events not only includes handover requests when the UE passes through the cell in active mode during a call, but also cell reselection requests when the UE is in idle mode.

This problem is illustrated in Fig. 5.1 showing a UE moving along a street where femtocells are deployed, into and outside of the coverage areas of multiple femtocells, which triggers messages for mobility management each time, even when the UE is not active. The handover problem can be alleviated by optimizing the femtocell coverage such that it matches as good as possible the shape of the house, covering everything inside, but avoids coverage of busy areas outside.

For enterprise femtocell deployments such as for office buildings, shopping malls and conference centers, the problem is slightly different. The area is typically covered by multiple cells, which need to collaborate to jointly provide service coverage. Cells are deployed in a more planned manner using guidelines for distances but without detailed radio planning to minimize costs. A typical enterprise femtocell deployment scenario is shown in Fig. 5.2. The objectives for coverage optimization for enterprise femtocell deployments are to provide contiguous coverage, balance load to maximize UE capacity, and to a lesser extent minimize leakage outside of building. Therefore, the requirements for coverage optimization for the deployment of femtocell groups

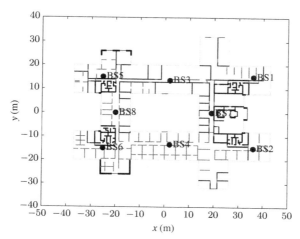

Figure 5.2. Deployment scenario with eight enterprise femtocell BSs covering an office building.

differ significantly from residential femtocell deployments, and the related solutions are not transferable.

In this chapter, different approaches are presented for initial configuration of coverage and self-optimization of coverage for both residential femtocells and enterprise femtocells deployments.

First, the initialization of coverage is discussed for both co-channel operation with macrocells and dedicated channel deployment. For co-channel femtocells, two methods are presented: distance-based (from macrocell) and measurement-based [1]. Both perform similarly, but the measurement-based approach is preferable since it requires much less information about the macrocell network. For dedicated channel deployment, a simpler approach based on the noise power and handover trigger threshold is presented as well.

For residential femtocells, three coverage self-optimization approaches are discussed assuming a single antenna, a switched multi-element antenna system, or a tuneable multi-element antenna system. All the three approaches use information on mobility events and UE measurement reports. Depending on the scenario (e.g., open or closed access to femtocells), such events can be handovers or cell reselection events, both successful or rejected (e.g., a mobile might be rejected by a femtocell using a closed-access mode). All self-optimization methods are able to adapt to different types, shapes, and sizes of indoor environments.

The single-antenna coverage self-optimization solution can reduce resulting mobility events significantly by up to 80% in the considered scenarios compared to using coverage auto-configuration methods only [1]. The method has a relatively short convergence time and is easily applicable in existing femtocells since it can be implemented via software and does not add any hardware costs.

The switched multi-element antenna coverage self-optimization solution uses multiple antennas connected to a switch. This allows the selection of one or two antennas with different patterns to provide additional gain pattern flexibility to achieve better adaptation of the femtocell coverage to the shape of the indoor environment compared to that of a single-antenna solution [2]. The additional flexibility gained by the switched multi-element antenna solution results in further improvement, on average a 20% decrease in resulting mobility events and a 20% increase in indoor coverage compared to the best single-antenna solution in the considered scenario. Moreover, due to the simplicity of the approach, the additional hardware costs for antennas and switch are relatively low. Further, the small footprint of the antennas means that they can easily be integrated into existing femtocell designs.

The tuneable multi-element antenna coverage self-optimization solution uses multiple patch antennas connected with attenuators, which allow the gain of each lobe to be individually adjusted [3]. This allows a further flexibility increase in terms of achievable antenna gain patterns. The proposed method achieves a performance similar to the switched multi-element antenna solution, which suggests that the flexibility of the switched solution is usually sufficient. However, the proposed tuneable multi-element antenna solution has the advantage of faster convergence compared to the switched multi-element antenna solution, since each antenna lobe can be adjusted individually in each optimization step. A disadvantage of this solution is its increased size due to the use of four patch antennas.

For enterprise femtocell deployments, a decentralized joint coverage optimization algorithm [4] is presented that solves a multi-objective problem entailing partially conflicting objectives of providing joint coverage, balancing load, and minimizing the necessary transmit power. By acquiring the relevant information from its neighbors, the algorithm periodically adjusts the transmitted pilot power for each femtocell to achieve these objectives. Compared to the fixed pilot power allocation, a performance improvement of 18% in terms of femtocell-supported traffic and 80% in terms of pilot transmit power reduction has been shown.

An alternative approach for addressing the joint coverage optimization problem was also presented in Chapter 3, where Genetic Programming, a machine learning technique which allows to automatically evolve self-optimization algorithms, is applied to this problem.

Since the approaches discussed in this chapter are based on standards independent measurements (e.g., received power, mobility events, load), and the objectives are broadly applicable, they can be applied independently of the air interface standard used (e.g., universal mobile telecommunication system (UMTS), long-term evolution (LTE), etc.).

5.2 INITIAL CONFIGURATION OF COVERAGE

When a femtocell is first deployed, it needs to initialize its coverage. Here, we can differentiate between two deployment scenarios: co-channel and dedicated channel deployment. While the concept is fairly similar in both scenarios, the difference is that for co-channel operation, the cell boundary is typically at the locations where the femtocell pilot power is equal to the macrocell pilot power. Therefore, the achieved coverage is dependent on the femtocell pilot power, the macrocell pilot power and the location of the femtocell within the macrocell. For dedicated channel deployments, the cell boundary is at locations where the received pilot signal-to-noise ratio (SNR) exceeds the camping or handover threshold. Since the noise level is more or less constant, the coverage is only dependent on the femtocell pilot power and the camping or handover threshold, but independent of the femtocell location.

5.2.1 Co-channel Operation

For co-channel operation of femtocells in the same frequency band as existing macro-cellular networks, the following pilot power auto-configuration schemes can be considered: distance-based and measurement-based.

5.2.1.1 Distance-Based Pilot Power Configuration

The femtocell pilot power is configured such that it is received on average with equal strength as the pilot power received from the strongest macrocell at a defined target cell radius of d_c, subject to its maximum power $P_{max}^{tx,sc,pilot}$ [1]. The initial femtocell pilot power can be calculated as

$$P^{tx,sc,pilot} = \min \left(\frac{P^{tx,m,pilot} G^{a,m} G^{p,m}}{G^{p,sc}(d_c)}, P_{max}^{tx,sc,pilot} \right), \tag{5.1}$$

where $P^{tx,m,pilot}$ is the pilot power transmitted by the macrocell, $G^{a,m}$ is the macro-cell antenna gain in the direction of the femtocell, $G^{p,m}$ is the path loss between the macrocell and the femtocell and $G^{p,sc}(d_c)$ is the estimated path loss from the femto-cell to a UE at the target femtocell radius d_c. The antenna gain can be modeled as described in Section A.3, and the path loss between a transmitter and a receiver can be calculated based on a path-loss model as described in Section A.4.

This achieves a roughly constant cell range independent of the distance to the macrocell in the co-channel hierarchical cell structure considered. Note that variations caused by the shadow fading losses or additional environment losses are unpredictable and cannot easily be taken into account using the path-loss model. While this approach is suitable for the initial power configuration for femtocell deployments, it has the disadvantage that a large amount of information on the macrocellular network, such as cell locations, power levels, antenna orientation, and gain, is required. The approach also relies on a reliable path-loss model for L_{mBS}. An example of the power configuration is shown in Fig. 5.3, which shows the configured femtocell transmit power values (f) and (g), the resulting received maximum power and pilot power values from the macrocell (a) and (b) and from the femtocell (d) and (e), and the femtocell coverage range as function of the distance from the macrocell.

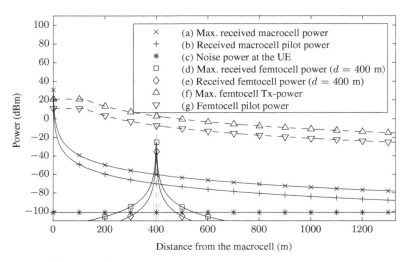

×	(a) Max. received macrocell power
+	(b) Received macrocell pilot power
＊	(c) Noise power at the UE
□	(d) Max. received femtocell power ($d = 400$ m)
◊	(e) Received femtocell power ($d = 400$ m)
△	(f) Max. femtocell Tx-power
▽	(g) Femtocell pilot power

Figure 5.3. Distance-based power configuration example.

5.2.1.2 Measurement-Based Pilot Power Configuration

The femtocell pilot power is configured using the same principle as above, but with the difference that the received power $P^{\mathrm{rx,m,pilot}}$ is not estimated using a path-loss model, but measured at the femtocell. Then the initial femtocell pilot power can be calculated as

$$P^{\mathrm{tx,sc,pilot}} = \min\left(\frac{P^{\mathrm{rx,m,pilot}}}{G^{\mathrm{p,sc}}(d_{\mathrm{c}})}, P^{\mathrm{tx,sc,pilot}}_{\max} \right). \tag{5.2}$$

As with the distance-based auto-configuration, this approach achieves a roughly constant cell range independent of the distance to the macrocell in a co-channel hierarchical cell structure. Measurement-based auto-configuration has the advantage that no information on the macrocell network is required. However, the received signal from the macrocell $P^{\mathrm{rx,m,pilot}}$ depends highly on the deployment location and the resulting indoor wall losses, which causes some variability in the achieved cell range.

5.2.1.3 Effects of Maximum Transmit Power on the Coverage

For co-channel operation, the maximum femtocell power is important since it has an effect on the size of the area within the macrocell where the femtocell can achieve its target cell coverage. This effect is also shown in Fig. 5.3, curves (f) and (g), where the maximum configured power and pilot power remains flat at distances up to approximately 100 m from the macrocell. In this area, the femtocell cannot achieve its target cell radius due to the power limit of 20 dBm in this example. The area in which the coverage is not achievable is only a very small fraction of the overall coverage area of the macrocell, and shielding effects of walls usually help as well. The dimensioning of the maximum femtocell power should not be overlooked.

5.2.2 Dedicated Channel Operation

When femtocells are deployed on a dedicated carrier, the initial configuration of coverage becomes very simple since it is independent of the location of the femtocell with respect to the macrocell. The required pilot power for a target cell radius of d_c can be calculated as

$$P^{tx,sc,pilot} = \min\left(\frac{P^{rx,sc,pilot}_{HO}}{G^{p,sc}(d_c)}, P^{tx,sc,pilot}_{max}\right), \tag{5.3}$$

where $P^{rx,sc,pilot}_{HO}$ is the femtocell reference signal received power (RSRP) which is required to trigger a handover or camping event (i.e., idle mode cell reselection).

5.3 COVERAGE OPTIMIZATION OF RESIDENTIAL FEMTOCELLS

While the auto-configuration approaches in the previous section provide a useful starting point, for residential femtocells, it is necessary to fine-tune the coverage to the deployment location within a building (e.g., large house or small house) and the traffic (e.g., if deployed next to a busy street or in the countryside). The objective of the self-optimization process is to both minimize the resulting mobility events of transient UEs and to maximize the indoor coverage.

In this section, different mobility event-based approaches for the self-optimization of coverage are presented to address this problem for residential femtocells with different antenna configurations. The benefits in coverage flexibility obtained by multi-element antenna solutions are illustrated in Fig. 5.4 in a conceptual way (without wall effects) when mobility events of passing UEs are prevented. Figure 5.4a shows a conventional single-antenna solution, Fig. 5.4b shows additional flexibility gained through a switched multi-element antenna solution and Fig. 5.4c shows a solution with individually tuneable antennas.

5.3.1 Coverage Optimization for Single-Antenna Femtocells

In most current designs of femtocell BSs, the antenna system is of a static nature with only one single antenna. Usually, a simple dipole or printed circuit-board antenna with low gain and fixed pattern is used. For such BSs, the only way to increase coverage or reduce mobility events is by increasing or decreasing the pilot power. This case is illustrated in Fig. 5.4a. A mobility event-based coverage self-optimization method for femtocells with a single antenna is described below.

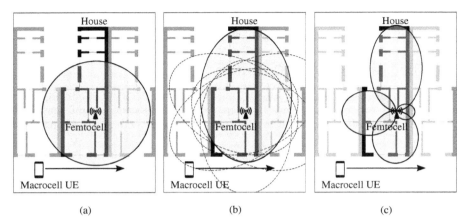

(a)   (b)   (c)

Figure 5.4. Coverage optimization approaches with different antenna configurations [3].

5.3.1.1 Minimization of Mobility Events of Passing Users

This optimization method takes only unwanted mobility events from transient UEs into account that briefly hand over to the femtocell and hand back immediately after passing the house. The objective is to maximize the indoor coverage under the constraint of limiting the mobility events from such passing UEs. Starting from a measurement-based pilot auto-configuration, the femtocell counts mobility events and classifies them into wanted and unwanted events over time, dependent on whether the mobile is registered at the femtocell and the time a UE spends in the cell before moving back to the macrocell. If the number of unwanted mobility events of passing UEs n_{t1} exceeds a predefined value of n_1 events per timeframe t_1, the femtocell reduces its pilot power by a step ΔP_{dec}. The femtocell then starts a new mobility event count for the updated configuration. If the number of unwanted events n_{t2} is smaller or equal to a predefined acceptable value of n_2 events for a time frame t_2, the femtocell increases its pilot power by a step ΔP_{inc} to provide improved indoor coverage and starts a new mobility event count for the updated configuration.

The measurement intervals and the optimization process are illustrated in Fig. 5.5. For the evaluation in Section 5.3.4, it is assumed that all mobility events of passing UEs shall be prevented, therefore $n_1 = n_2 = 0$. The measurement interval t_1 is equal to the optimization iteration time, and for any occurring event the femtocell pilot power is reduced by $\Delta P_{dec} = 3$ dB. In order to speed up the cell radius increase after the initial auto-configuration, t_2 is set to a low value of 120 seconds until a first unwanted mobility event is detected. Then t_2 is increased significantly to 6 hours so that the coverage is dominated by the decreases caused by unwanted mobility events. The step size ΔP_{inc} for the pilot power increase was selected to be 0.3 dB. While the selected parameters perform well in the considered scenario, automatic optimization

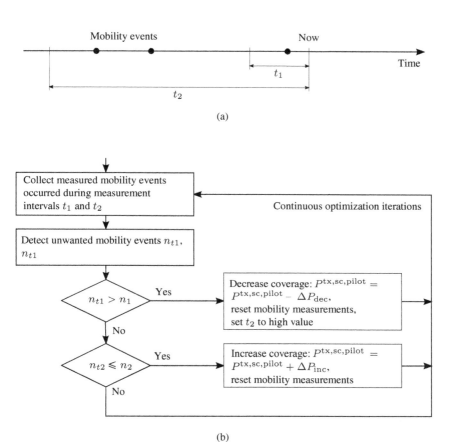

Figure 5.5. Overview of measurement intervals (a) and flowchart of mobility event-based coverage optimization (b) that minimizes mobility events of passing users [1].

of these parameters is also a possibility. Such optimized parameters would have to result in a fast convergence to the point with the best trade-off between femtocell coverage and increase in core network signaling for the operator.

5.3.2 Coverage Optimization Using Switched Multi-element Antennas

A significant drawback of single antenna femtocell BSs is that when deployed in an unsuitable location (e.g., near a window close to a busy street in the front of the house), the BS needs to reduce its output power to such an extent that the coverage inside the building is compromized in order to limit the number of mobility events caused by outdoor UEs [1]. This can be particularly problematic if the antenna pattern is erratic (which is the case for most printed circuit board (PCB)- and dipole-antennas in consumer grade equipment) and if, for example, a stronger beam of the pattern points into the direction of the bypassing UEs and a rather low-gain section of the pattern points toward the inside of the home. In this case, in order to reduce the outdoor coverage, it might happen that the output power of the femtocell would have to be reduced to such an extent that the signal strength inside the building may get too weak to provide sufficient coverage. It is obvious that with a single antenna, it is difficult to adapt the femtocell coverage to its environment and only a limited reduction of mobility events can be achieved without compromising the coverage within the femtocell.

However, if the femtocell could choose from a certain number of diverse antenna patterns, then a significant improvement would be possible: if the patterns show enough spatial diversity, a pattern with a rather lower gain pointing to the outside and a rather higher gain pointing to the inside of the building could be chosen (see Fig. 5.4b). This makes it possible to reduce the exposure outside of the building, and at the same time maintain sufficient coverage inside.

5.3.2.1 Multi-element Antenna Concept   Since for residential femtocell BSs, a low-cost implementation for such a multi-antenna system is essential, a simple switching network can be used that allows switching between multiple antennas to choose between several fixed antenna patterns. For this scheme to work, it is required that the patterns of the antennas are diverse and show a reasonable directivity, so that either good coverage or low interference in each horizontal direction can be achieved. The antennas then should be arranged in a way that their patterns do not overlap too much, so that not too many antennas are needed to cover the complete horizontal direction. The number of antennas needs to be chosen such that a reasonable large variety of pattern and coverage scenarios can be realized. Figure 5.6 shows a simplified illustration of a switching network with four antennas and four varying patterns.

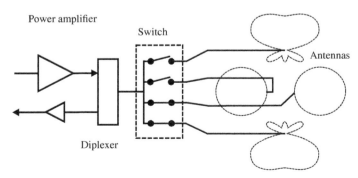

Figure 5.6.   Antenna circuitry for switched multi-element antenna solution [2].

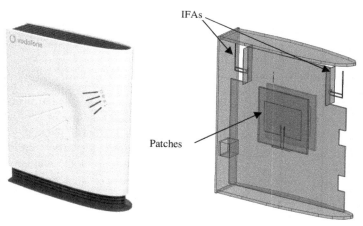

Figure 5.7. Physical design study of a Femtocell and schematic of patch antenna and IFA location within the housing [3].

By choosing the right pattern or pattern combination, it is possible to provide coverage within the femtocell where it is required as well as reducing the number of mobility events and interference. In order to choose the right pattern or pattern combination, an optimization process has to be performed. This process needs to be executable by the femtocell BS itself without further knowledge or coordination with the macrocell environment.

5.3.2.2 Antenna Design Since the available space inside the volume of a femtocell BS is limited, only a few types of antennas are suited to be used. In Fig. 5.7, an example for a consumer-grade femtocell BS is shown; the picture shows the upright positioned housing and the placements of two patch- and two Inverted-F antennas (IFAs) within the femtocell BS. The size of the BS is approximately 170 mm × 150 mm × 35 mm ($h \times d \times w$). It contains a single main circuit board (MCB), which is located approximately in the center of the housing. The used antennas have to be of low cost and must be easy to integrate into the available volume of the housing.

The patch antennas were chosen for their directional pattern, flat form factor and the moderate implementation cost. They can be easily integrated into the existing BS by placing them parallel to the MCB. The IFAs were also chosen for their pattern, low cost, and ease of integration into the circuit board as a printed structure. With these four antennas, the above stated requirements can be fulfilled: the antennas offer good diversity, in the shown orientation they cover the complete azimuth, and they offer good directivity at the same time. The pattern of the patch antenna shows a very good front-to-back ratio, such that two back-to-back placed antennas, one on either side of the MCB, provide two very diverse antenna patterns. The IFA also shows a reasonable front-to-back ratio, and if it is placed in the upper corners of the MCB, it generates a pattern, which complements the pattern of the patches. In addition, a combination of both patch antenna and IFAs can be chosen.

5.3.2.3 Switch Design   To make use of the different patterns, it is required to be able to switch between those antennas. In order to keep the cost and complexity of the switching circuitry as low as possible, a simple arrangement of four parallel switches is considered as shown in Fig. 5.6. This allows the use of either one of the four antennas at a time, or any antenna combination thereof in parallel. This means no beamforming in the classical sense (meaning the use of an array-configuration of several elements with amplitude- and phase control) is used. Of course, it has to be taken into account that if several antennas are operated in parallel, a certain impedance mismatch will occur, resulting in a lower efficiency. In order to keep the power losses resulting from impedance mismatch acceptable, no more than two antennas are combined at any time.

5.3.2.4 Prototype   A prototype of a femtocell BS with integrated patch antenna and IFA according to Fig. 5.7 was built and measured. Figure 5.8 illustrates the measured antenna patterns.

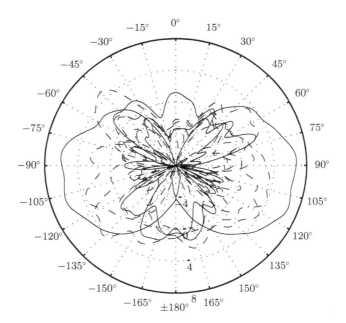

Figure 5.8.   Resulting measured antenna patterns achievable by selecting either one single antenna or any combination of two antennas [2].

5.3.2.5 Optimization Approach The optimum pattern and pilot power configuration can be identified by a global search over all possibilities. However, this would lead to a long convergence time, which is not well suited for real-life systems. Therefore, a different solution is required that achieves the same objectives and comes as close as possible to the optimal performance, but is better suited for implementation. One possible solution is described in the following.

First, the femtocell pilot power is initialized for all antenna patterns using the measurement-based auto-configuration method described above, and an initial antenna pattern is selected that provides similar gain in all directions. In this example, the pattern resulting from a combination of the two inverted-F antennas would be suitable. Note that each pattern is associated with its own individual pilot power value.

During operation, the femtocell collects the information on mobility events such as handovers and idle mode mobility. Mobility events are classified as wanted or unwanted events over time, dependent on whether the mobile is registered at the femtocell and on the time a UE spends in the cell before moving back to the macrocell. In addition, the femtocell periodically collects and stores the coverage information for different antenna patterns through path-loss measurements. These measurements can be performed by the UE using standardized methods and by the femtocell itself using its own measurement functionality and information on transmit powers and antenna gains, or by a combination of both.

The objective for the ongoing optimization is to maximize the indoor coverage under the constraint of limiting the mobility events from passing UEs. Figure 5.9 shows a flowchart of the method.

If the number of unwanted mobility events of passing UEs n_{t1} exceeds a predefined value of n_1 events per timeframe t_1, the femtocell reduces its pilot power for the currently selected antenna pattern by a step ΔP_{dec}. The femtocell then starts a new mobility event count for the updated configuration.

If the number of unwanted events n_{t2} is smaller or equal to a predefined acceptable value of n_2 events for a timeframe t_2, the femtocell increases its pilot power for the currently selected antenna pattern by a step ΔP_{inc} to provide improved indoor coverage and starts a new mobility event count for the updated configuration.

If the pilot power for the current antenna pattern was changed, the pattern selection is reevaluated, and the best pattern and pilot power combination is selected. For this, the estimated received powers at all measured points for the current power values for each antenna pattern are calculated. Then, the pattern and power combination that maximizes the mean of the worst 10% of the estimated received powers is selected as best new pattern and pilot power.

The parameters chosen for n_1, n_2, t_1, t_2, ΔP_{dec}, ΔP_{inc} are equivalent to the ones used for the femtocell coverage optimization for a BS with a single antenna, as described in Section 5.3.1.

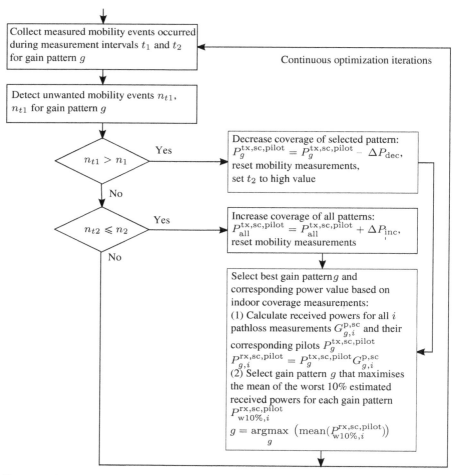

Figure 5.9. Flowchart of the switched multi-element antenna configuration process [2].

5.3.3   Coverage Optimization Using Multi-element Antennas with Individually Tuneable Attenuation

When using a switched multi-element antenna solution as described above, it is possible to use more than one pattern at a time. However, since only one transmitter is available, the power that is radiated by the two antennas cannot be individually controlled. In order to do so, each antenna could be connected to a separate transceiver, which would significantly increase the cost. Alternatively, the power level at each antenna can be tuned by individually adjustable attenuators, which is much more

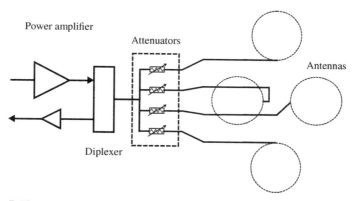

Figure 5.10. Antenna circuitry for a tuneable multi-element antenna solution [3].

cost-effective. Using attenuators results in losses for both downlink and uplink when adjusting the antenna configuration, but the transmit power for low-power residential femtocell BSs only accounts for a very small portion of the total energy consumption, so that these inefficiencies do not have a significant impact on the energy consumption. This increases the flexibility for the coverage optimization even further compared to the previously discussed approaches as illustrated in Fig. 5.4c. A schematic of the antenna circuitry with tuneable attenuators is shown in Fig. 5.10. The proposed self-optimization method for femtocells with gain-tuneable multi-element antennas is described below.

5.3.3.1 Concept   An antenna configuration with four patch antennas, which are placed in a 90 degree orientation is considered as shown in Fig. 5.10. This antenna configuration will further improve the coverage versatility, since the maximum beam directions of each antenna falls into the same direction of the nulls of the remaining antennas. Therefore each direction can be either covered by an adjustable main beam or by a null.

5.3.3.2 Attenuator Design   The use of attenuators reduces the efficiency of the transceiver, but this is acceptable, considering that in most cases only a fraction of the overall power needs to be absorbed. For example, in a situation in which the coverage of one of these antennas causes the vast number of unwanted mobility events, and if this sector is turned off by absorbing all transmitted energy in that branch, then the overall power loss is only 25%. The impact on the receiver noise due to the additional attenuation is also negligible since the path loss between the UE and the BS is low compared to the available power levels from the UE and considering that the sectors in which a strong attenuation will be applied to the transmitted signal are exactly those sectors from which unwanted traffic is to be avoided.

5.3.3.3 *Optimization Approach*   The optimum antenna and power configuration that minimizes the total number of mobility events, while maximizing indoor coverage can be identified via a global search over all possibilities. Due to the high number of patterns resulting from the possibility to fine-tune the attenuation for each lobe individually and the time-consuming evaluation of each potential pattern, this approach is impractical.

Alternatively, instead of jointly optimizing the coverage for all adjustable lobes, they can be optimized individually with the aim to prevent unwanted mobility events on each lobe. This is performed in a similar fashion as for the mobility event optimization for the single antenna femtocell case in Section 5.3.1, with the difference that instead of increasing or decreasing the pilot power, the attenuation for the serving lobe is modified. The serving lobe is defined as the one from which the mobile receives the strongest signal. As a result, one benefit over the switched pattern approach described above is a faster convergence of the solution, since only four attenuations need to be configured compared to 10 different pilot power levels (one for each pattern). The identification of the serving lobe can be performed by periodic attenuation modifications for each lobe to create a unique signature. This signature can be detected based on the standardized received power reporting capability of the UE. The strongest detected signature determines best-serving antenna.

5.3.4 Performance Comparison

The performance of the different auto-configuration and self-optimization methods for residential femtocells was analyzed using system-level simulations. All methods are compared in terms of the resulting mobility events, indoor coverage and pilot powers. In addition, the usage of the different gain patterns for the switched multi-element antenna solution is discussed. Further details and results can be found in [1–3].

UMTS system-level simulations were performed to derive both the number of mobility events for passing and indoor UEs and indoor coverage as a function of the distance from the femtocell BS. A scenario with seven macrocells with three sectors each is considered (cell of interest with Tier 1 neighbors), as described in Section A.2.1. Femtocells are deployed randomly in houses within the coverage area and they reuse the same frequency as the macrocells in a hierarchical cell structure.

Figure 5.11 illustrates the simulated area around a typical London terraced house, as considered for this investigation, and shows one example of the received pilot power levels within the coverage areas for both macrocell and femtocell. This house type was chosen as a worst-case example due to its narrow shape and its proximity and lack of wall separation to the footpath in front of the house. An overview of the simulation assumptions and models is described in Appendix A.

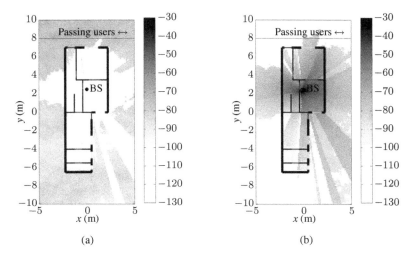

(a)   (b)

Figure 5.11. Coverage example for macrocell (a) and femtocell (b). The color-bar on the right shows the received pilot powers for covered areas in dBm. White areas are not covered by the corresponding BS [1].

To simulate the behavior of the femtocell UE, a waypoint-based indoor UE mobility model was used. Underlay macrocell UEs were simulated using an outdoor UE model, which models UEs passing by on the sidewalk outside of the house with an exponentially distributed inter-arrival time. The models are described in Section A.11.2, and illustrated in Figs. A.18 and A.19.

5.3.4.1 Mobility Events

Figure 5.12 shows a comparison of the resulting average number of mobility events per hour as a function of the femtocell distance from the footpath. The single-antenna, distance-based, auto-configuration method achieves a performance similar to the simpler single-antenna measurement-based auto-configuration method. The mobility event-based, single-antenna, self-optimization method is able to significantly outperform both auto-configuration methods that do not take mobility events into account, and reduces the number of mobility events by up to 90%. In [3], it is shown that the performance of the mobility event-based self-optimization method approaches the optimum performance for this antenna configuration.

The proposed switched multi-element antenna solution can reduce the number of mobility events further, by up to 44% compared to a single-element antenna solution with a similarly sophisticated coverage self-optimization. The improvement depends on where the femtocell is deployed in the house. In [3], it is shown that the performance of the implementable optimization approaches the optimal performance

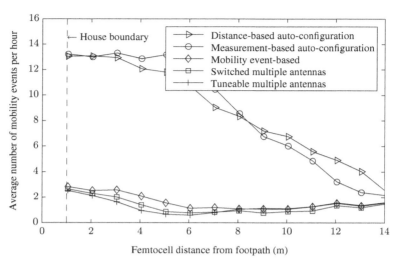

Figure 5.12. Average number of mobility events per hour for different femtocell locations [1–3].

for this antenna configuration. The tuneable multi-element solution with four patch antennas achieves an additionally performance improvement of 31% for short distances from the footpath, but loses this advantage at larger distances where the performance is reduced due to its inability to focus the transmit power toward one direction without losses in the attenuators.

5.3.4.2 Indoor Coverage Figure 5.13 shows a comparison of results of indoor coverage, defined as the fraction of the area inside of the house where the pilot power of the femtocell is received stronger than the pilot power of the strongest macrocell, for the single-antenna versus multiple-antenna solutions. The single-antenna auto-configuration methods considered achieve very similar coverage. With a single antenna, the mobility event-based self-optimization method achieves a high rate of coverage when femtocells are deployed in good locations, but a lower rate of coverage when deployed in unsuitable locations close to the footpath. This is a trade-off resulting from the reduction in core network signaling which is of higher priority. Note that the lower performance in unsuitable locations is not necessarily a disadvantage, since it would encourage the user to redeploy the femtocell (e.g., via short message service (SMS) notification), which will result in both better coverage for the UE and a lower number of mobility events.

For the multi-element antenna solutions, it is shown that in addition to the reduced number of mobility events discussed above, the indoor coverage can also be improved compared to a single-antenna solution with a similar sophisticated coverage self-optimization. The most significant improvements in coverage are achieved when

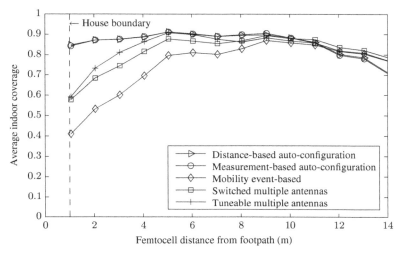

Figure 5.13.   Average resulting indoor coverage for different femtocell locations [1–3].

the femtocell is deployed in unsuitable locations close to the footpath, where the flexibility in terms of gain patterns allows better coverage of the rest of the house without increasing the number of unwanted mobility events and the associated core network signaling. In [3], it is shown that the performance of the implementable optimization for the switched multi-element antenna solution approaches the optimal performance. The tuneable multi-element antenna solution results in slightly higher coverage for femtocells deployed close to the footpath due to the higher flexibility of the gain pattern compared with the switched pattern approach. However, with this solution, the coverage is slightly worse than that of the switched pattern solution for femtocell locations in the back of the house due to its inability to focus the transmit power toward one direction without losses in the attenuators. Note that since the model only considers transmission loss through walls, but neglects reflections on walls, floor and ceiling, the coverage results are relatively conservative and slightly higher coverage can be expected in real deployments.

5.3.4.3 Pilot Power   Figure 5.14 shows the cumulative distribution functions (CDFs) of the resulting pilot power for both scenarios. In general, all methods result in a large variation of pilot powers depending on the distance from the macrocell and the deployment location within the house. The mobility event-based self-optimization methods show the largest variation in power. While in some cases they use a comparable power as the other auto-configuration methods in order to maximize the indoor coverage for the given environment, outdoor mobility, and deployment location, they also reduce the pilot power in some cases to −50 dBm. This happens to prevent excessive mobility events in cases where the user deploys the femtocell

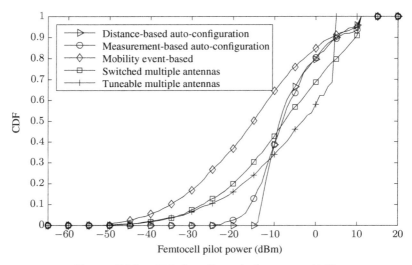

Figure 5.14.   CDF of configured pilot powers [1–3].

in an unsuitable location, for example, close to the window near the sidewalk. Some cellular standards such as UMTS have not been designed for femtocells and therefore specify much higher values for the minimum pilot power.

For the multi-element antenna solutions, the typically radiated pilot power can be around 10 dB higher than that of the single-antenna solution for minimizing the total number of mobility events. This is enabled by the flexibility in terms of antenna gain of the proposed multi-element antenna solution and contributes to the better indoor coverage. The radiated power of the tuneable antenna solution can be even higher, but the power per antenna is limited due to the fact that the total power is split among the antennas.

5.3.4.4 Pattern Selection   Table 5.1 shows the probability of being optimal for each pattern in the proposed switched multi-element antenna solution. The simulations show that the patch antennas, with their clean directional gain patterns

TABLE 5.1   Probability of Antenna Pattern Use

Pattern	1	2	3	4	5	6	7	8	9	10
IFA 1	•				•	•			•	
IFA 2		•					•	•	•	
Patch 1			•		•			•		•
Patch 2				•		•	•			•
Probability of use (%)	10.2	5.7	21.6	18.7	9.0	8.6	6.1	7.6	6.9	4.9

* indicates activated antennas.

and high spatial diversity, seem to be most useful (patterns 3 and 4), since they are chosen far more often than the other patterns. Therefore, it can be given as a design rule that it is beneficial to achieve as diverse patterns as possible for switched multi-element antenna systems.

5.4 COVERAGE OPTIMIZATION OF FEMTOCELL GROUPS

For enterprise femtocell deployments, the requirements for coverage optimization are very different compared to residential deployments. In this section, the problem of joint coverage optimization is discussed, where a group consisting of N femtocell BSs is deployed to jointly provide user services, and a distributed coverage optimization approach [4] is described.

Here, joint coverage optimization is performed by means of updating the pilot power only, to satisfy the following three objectives:

1. To balance the UE load among the N collocated BSs for efficient radio resource utilization (prevent overloading or under-utilization).
2. To minimize radio coverage holes or gaps.
3. To achieve objectives 1 and 2 with the minimum pilot power possible, that is, to reduce the coverage overlap with the neighboring BSs. Some degree of coverage overlap can be advantageous in terms of smooth UE handovers. However, excessive overlap may translate to greater interference to neighboring cells [5].

This constitutes a multi-objective optimization problem with partially conflicting objectives. For example, increasing the coverage of a BS reduces the amount of coverage holes, but doing so may increase the UE load of the BS, and/or increase the coverage overlap with its neighbors. Therefore, there is a need to balance the requirements of all three objectives.

5.4.1 Distributed Algorithm for Joint Coverage Optimization

To facilitate that neighboring BSs can effectively collaborate, it is assumed that each BS has the ability to communicate with its neighboring BSs in the group via the backhaul connection. The membership of the neighbor set depends on how a BS discovers its neighbors. For example, each BS can perform pilot signal measurements to detect BSs that are located nearby. Alternatively, BSs can also perform a query-and-respond procedure over the backhaul network. In any case, the BSs in the neighbor set of each BS should be those whose pilot power updates can have a significant impact on its performance.

　　The approach described in this section makes use of the local status information of the BS. This information can be obtained from measurement reports communicated

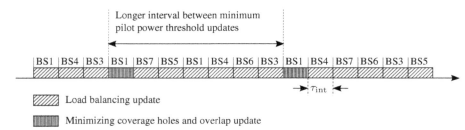

Load balancing update

Minimizing coverage holes and overlap update

Figure 5.15.   Snapshot of the update timeline illustrating the length of intervals for the two updates.

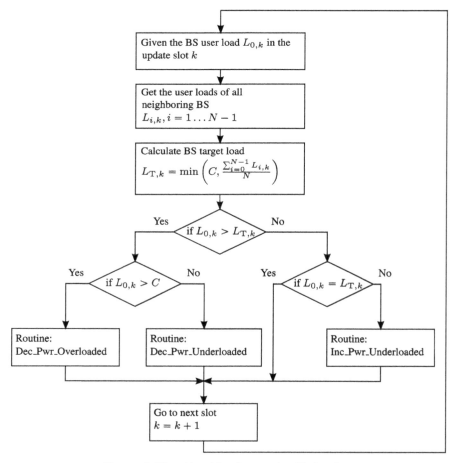

Figure 5.16.   Algorithm for user load balancing.

back by the connected UEs, or from inputs from the neighboring BSs. The BS utilizes this information and makes local decisions based on the defined algorithm routines. Each BS stores thresholds for maximum and minimum pilot transmit power. The maximum pilot power threshold $P_{max}^{tx,sc,pilot}$ is fixed and depends on how much of the total BS transmit power is allocated to the pilot channel by the network designer. The minimum pilot power threshold $P_{min}^{tx,sc,pilot}$ is set adaptively by the algorithm. The algorithm, which runs individually at each BS, consists of two update cycles as shown in Fig. 5.15): a more frequent $P^{tx,sc,pilot}$ update to achieve UE load balancing and an infrequent $P_{min}^{tx,sc,pilot}$ update to handle coverage holes and overlap minimization. The reason for more frequent load balancing updates is to cater for the mobility of the UEs and to converge to a balanced load before the positions of the UEs change.

5.4.1.1 Pilot Power Updates for Load Balancing   Figure 5.16 shows the algorithm procedure that each BS employs to adjust its UE load to that of the calculated mean of all N neighbor BSs. Given the current UE load L_0 of the BS, it collects via backhaul the UE loads of its $N - 1$ neighbors, L_i, where $i = 1 \ldots N - 1$. It then computes the average load L_T of the neighbor set (including its own load L_0) given as $L_T = \min(C, \sum_{i=0}^{N-1} L_i/N)$. Next, the BS compares L_0 with L_T, and decreases or increases its pilot transmit power to drop or acquire the desired number of UEs, respectively. When overloaded ($L_0 > C$), it employs Routine 5.1: *Dec_Pwr_Overloaded* which decreases the pilot transmit power of the BS to drop the ($L_0 - C$) number of extra UEs. For the case ($C \geq L_0 > L_T$), the BS uses Routine 5.2: *Dec_Pwr_Underloaded*, which drops a UE only if it can be picked up by another femtocell. If ($L_0 < L_T$) (underloaded case), it employs Routine 5.3: *Inc_Pwr_Underloaded*, which increases the pilot transmit power of the BS to acquire the required number of UEs ($L_T - L_0$). Here, η is the receiver sensitivity, $P^{rx,pilot}$ is the RSRP, $P_L^{rx,pilot}$ is the lowest RSRP, $P_B^{rx,pilot}$ is the RSRP from best BS, $P_{NB}^{rx,pilot}$ is the RSRP from next best BS, $P_{diff}^{rx,pilot} = P_B^{rx,pilot} - P_{NB}^{rx,pilot}$, and $P_{diff,L}^{rx,pilot}$ is the lowest $P_{diff}^{rx,pilot}$.

Routine 5. 1 *Dec_Pwr_Overloaded*

1: For each connected mobile receiver, calculate its $P_{diff}^{rx,pilot}$.

2: Sort, in ascending order, the calculated $P_{diff}^{rx,pilot}$ of all mobile receivers.

3: Store the sorted $P_{diff}^{rx,pilot}$ in an array $\mathfrak{R}$.

4: Decrease BS pilot transmit power

$$P^{tx,sc,pilot} = P^{tx,sc,pilot} - \text{MEAN}[\mathfrak{R}(L_0 - C), \mathfrak{R}(L_0 - C + 1)].$$

5: **if** $\left(P^{tx,sc,pilot} < P_{min}^{tx,sc,pilot} \right)$ **then**

6:    $P_{min}^{tx,sc,pilot} = P^{tx,sc,pilot}$.

7: **end if**

8: Go to Routine: *Dec_Pwr_Underloaded*.

Routine 5. 2 *Dec_Pwr_Underloaded*

1: **for** $i = 1$ to $(L_0 - L_T)$ **do**
2:    For each connected mobile receiver, calculate its $P_{\mathrm{diff}}^{\mathrm{rx,pilot}}$.
3:    Sort, in ascending order, the calculated RSRP differences of all mobile receivers.
4:    Store the sorted $P_{\mathrm{diff}}^{\mathrm{rx,pilot}}$ in an array $\mathfrak{R}$.
5:    Pick the mobile receiver R_1 with $P_L^{\mathrm{rx,pilot}}$.
6:    Pick the mobile receiver R_2 with $P_{\mathrm{diff},L}^{\mathrm{rx,pilot}}$.
7:    Calculate pilot power update as

$$\Delta P = \min\left(P_{\min}^{\mathrm{tx,sc,pilot}}, \mathrm{MEAN}[\mathfrak{R}(1), \mathfrak{R}(2)]\right).$$

8:    **if** (R_1 is the same as R_2) **then**
9:        Pick the mobile receiver R_3 with the second lowest $P^{\mathrm{rx,pilot}}$, i.e. $\hat{P}_L^{\mathrm{rx}}$.
10:       **if** $\left((\hat{P}_L^{\mathrm{rx}} - \Delta P) > \eta\right)$ **then**
11:          For R_2, check the following:
12:           **if** (next best BS is not a Macro) AND (next best BS is not fully loaded) AND $\left(P_{\mathrm{NB}}^{\mathrm{rx,pilot}} > \eta\right)$
              then
13:              $P^{\mathrm{tx,sc,pilot}} = P^{\mathrm{tx,sc,pilot}} - \Delta P$.
14:           **else**
15:              Break.
16:           **end if**
17:       **else**
18:          Break.
19:       **end if**
20:    **else**
21:       **if** $\left((P_L^{\mathrm{rx,pilot}} - \Delta P) > \eta\right)$ **then**
22:          For R_2, check the following:
23:           **if** (next best BS is not a Macro) AND (next best BS is not fully loaded) AND $\left(P_{\mathrm{NB}}^{\mathrm{rx,pilot}} > \eta\right)$
              then
24:               $P^{\mathrm{tx,sc,pilot}} = P^{\mathrm{tx,sc,pilot}} - \Delta P$.
25:           **else**
26:              Break.
27:           **end if**
28:       **else**
29:          Break.
30:       **end if**
31:    **end if**
32: **end for**

Routine 5. 3 *Inc_Pwr_Underloaded*

1: For each neighbor-connected mobile receiver, calculate its RSRP difference between the own pilot signal and the neighbor BS's pilot signal.
2: Sort, in ascending order, the above-calculated RSRP differences, and store in an array $\mathfrak{R}$.
3: Increase BS pilot transmit power

$$P^{\mathrm{tx,sc,pilot}} = \min\left(P_{\max}^{\mathrm{tx,sc,pilot}}, P^{\mathrm{tx,sc,pilot}} + \mathrm{MEAN}[\mathfrak{R}(L_T - L_0), \mathfrak{R}(L_T - L_0 + 1)]\right).$$

5.4.1.2 Minimum Pilot Power Updates for Reducing Coverage Holes and Overlap

Figure 5.17 illustrates the algorithmic procedure for adapting the minimum pilot transmit power threshold $P_{\min}^{tx,sc,pilot}$ for each BS in accordance with the measured coverage metric ξ (e.g., a metric incorporating the UEs call drop probability and coverage overlap coefficient) over a certain specified amount of time. The BS compares the measured metric against the corresponding threshold and decides to increase (by ΔP_{inc}) or decrease (by ΔP_{dec}) the threshold $P_{\min}^{tx,sc,pilot}$ accordingly. The threshold increment is for removal of coverage holes while the threshold decrement serves to minimize the coverage overlap. For instance, if it is desired to give more preference to coverage holes minimization, ΔP_{dec} should be set much lower than ΔP_{inc} to ensure that the BS coverage decreases gradually, and the coverage adjustment is dominated by the increment step.

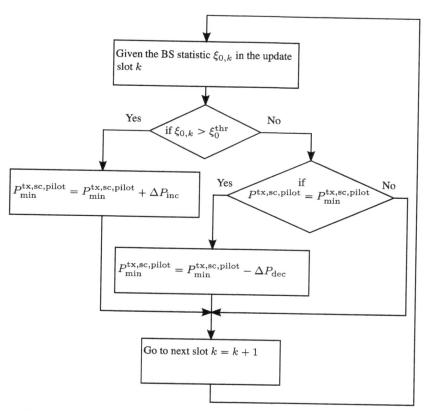

__Figure 5.17.__   The algorithm for minimizing coverage holes and overlap.

5.4.2 Performance Evaluation

The performance of the proposed algorithm was evaluated in a typical office environment shown in Fig. 5.2. The simulated office building contains open plan offices in partitioned cubicles, closed meeting rooms, toilets, and kitchen facilities. It is assumed that the femtocells are deployed in the building with some rudimentary planning, but without performing a detailed cell planning survey of the building. Since the distribution of UEs throughout the building is relatively even, the femtocells are deployed roughly equally apart. This depicts a realistic plug-and-play femtocell deployment, where the placement of femtocells is done fairly intelligently, but can be sub-optimal due to the lack of cell planning.

Each femtocell has a maximum capacity of $C = 8$ active UEs, and an underlay macrocell coverage is assumed. A path-loss map-based simulation model is used, as described in Appendix A. A waypoint-based UE mobility model is used, as described in Appendix Section A.11.2 and illustrated in Fig. A.20. A total of $U = 48$ UEs are modeled, and each user produces 1 Erlang of traffic (always active).

For the generation of results presented hereafter, the values of η, $P_{max}^{tx,sc,pilot}$, and ΔP_{inc} are set to -100 dBm, 11 dBm, and 3 dB, respectively. The parameters $P^{tx,sc,pilot}$ and $P_{min}^{tx,sc,pilot}$ are initialized to 11 dBm and -10 dBm, respectively. The coverage metric ξ for each BS is represented by the number of handovers to and from the macrocell, with ξ^{thr} set to 1. For each femtocell, the $P_{min}^{tx,sc,pilot}$ update is performed after every 10 $P^{tx,sc,pilot}$ updates. Finally, in each update slot, a randomly chosen single femtocell BS runs the algorithm. Simultaneous updates are not modeled to ensure that the latest neighbor information is available to a BS before it makes its local decision.

5.4.2.1 Load Balancing Performance
Let $E[L_i]$ be the average UE load supported by the ith femtocell, where $i = 1 \ldots N$, and L_M denotes the target average load for each femtocell, such that $L_M = \min(C, \lfloor \frac{U}{N} \rfloor)$. Then, the load balancing performance of the algorithm can be characterized using the following error function:

$$\varepsilon = \frac{1}{N} \sum_{i=1}^{N} (| E[L_i] - L_M |). \tag{5.4}$$

Figure 5.18 highlights the performance of the algorithm for UE load balancing between the femtocells, where the evolution of the load difference ε is illustrated over time. The plot shows that ε is reduced with more frequent updates (shorter τ_{int}) because each femtocell can then keep up with the mobility of the UEs. However, more frequent updates consequently generate large amounts of backhaul traffic, which is examined below.

5.4.2.2 Coverage Optimization Performance
Next the performance of the proposed algorithm is evaluated in terms of its ability to reduce coverage holes and

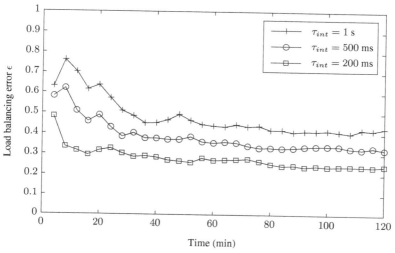

Figure 5.18. Load balancing performance of the proposed algorithm as a function of different update intervals. $\Delta P_{inc} = \Delta P_{dec} = 3$ dB.

minimize the pilot transmit power, and thus the coverage overlap between femtocells. For the ith femtocell BS, $i = 1 \ldots N$, the corresponding average pilot transmit power is expressed as a percentage of the maximum allowable pilot transmit power, such that

$$\rho_i = \frac{E[P_{tx,i}]}{P_{max}^{tx,sc,pilot}} \times 100 \qquad (5.5)$$

with $E[P_{tx,i}]$ and $P_{max}^{tx,sc,pilot}$ expressed in linear units. Also, $E[\rho] = \bar{\rho} = \sum_{i=1}^{N} \rho_i / N$ represents the average pilot transmit power of the whole femtocell network.

Figure 5.19 illustrates the performance of the proposed algorithm in terms of the average pilot transmit power reduction for the overall network, as a function of varying amounts of power decrement steps ΔP_{dec}. The plot shows that the algorithm readily converges to the reduced average pilot transmit power of the network, with $\Delta P_{dec} = 20$ dB providing the most reduction. From the same simulation run, the corresponding values of the average supported traffic ϕ are given in Table 5.2. The results show that, compared to the $\Delta P_{dec} = 3$ dB case, an aggressive decrease ($\Delta P_{dec} = 20$ dB) of $P_{min}^{tx,sc,pilot}$ causes only a 2.59% reduction in the supported network traffic ϕ, but provides a reduction of approximately 33.25% in average pilot transmit power. Moreover, when simulated with a fixed pilot transmit power of 11 dBm, the system supported an average of $\phi = 39.83$ Erlangs UE traffic. Therefore, the proposed algorithm's performance gain, with ΔP_{inc} and ΔP_{dec} both set to 3 dB, is approximately 18% relative to the fixed power allocation.

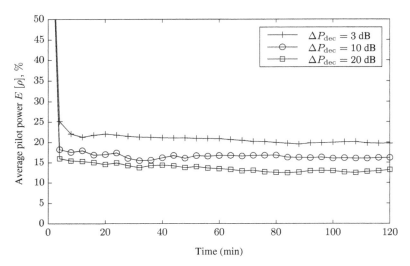

Figure 5.19. Average pilot transmit power of the whole femtocell network, as a percentage of the maximum allowable pilot transmit power $P_{max}^{tx,sc,pilot}$ = 11 dBm. τ_{int} is set to 500 ms.

Figure 5.20 shows the coverage of the femtocells at the end of the simulation run, with white regions denoting the areas without femtocell coverage. The pilot transmit powers for the femtocells 1 and 5 is adjusted to the maximum value (11 dBm) by the algorithm because they are located next to the concrete stairwells that significantly block the propagation of their pilot channel broadcasts to many UE waypoints. This leads to an initial imbalance of load with the femtocells placed in the corridors. That imbalance is addressed when the algorithm increases the pilot powers of the corner femtocells and decreases those of the corridor femtocells.

5.4.2.3 Backhaul Signaling Overhead In this section, the amount of information that has to be exchanged within the femtocell group as a result of the proposed algorithm is examined. It is assumed that the femtocell group is inter-connected via an Ethernet switch and each femtocell BS broadcasts its load information to its M number of neighbors at the beginning of the update slot. As a result, each femtocell

TABLE 5.2 Values for $E[\rho]$ and ϕ from the proposed algorithm

ΔP_{dec} (dB)	$E[\rho]$ (%)	ϕ (Erlangs)
3	19.76	46.95
10	16.19	46.56
20	13.19	45.73

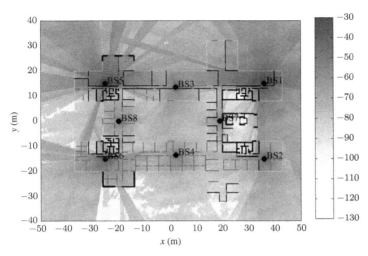

Figure 5.20. A snapshot of the radio coverage using the proposed algorithm, with the color-bar on the right showing received pilot channel powers in dBm. White areas are not covered by the small cells.

generates $(M + 1)$ messages for every update. Moreover, a relatively worst-case scenario is considered where the information is exchanged irrespective of a change in the corresponding femtocell's load conditions. Therefore, the total number of messages exchanged on the Ethernet backhaul is $O(N \cdot (M + 1))$ within a window of length τ_{int}. By recalling that the backhaul signaling packet length (bits) is denoted by λ, the backhaul signaling data rate Ω (kbps) required for load information transportation is given as

$$\Omega = \frac{\lambda \cdot N \cdot (M + 1)}{\tau_{int}}. \tag{5.6}$$

From [6], the minimum value of λ equates to 72 bytes, which contains a minimum necessary payload of 46 bytes. Using (5.6) and setting $M = N - 1$, Table 5.3

TABLE 5.3 Backhaul Data Rate Requirements of the Proposed Algorithm

	Number of femtocells, N	Ω (kbps)
$\tau_{int} = 200$ ms	8	184.32
	16	737.28
$\tau_{int} = 1$ s	8	36.864
	16	147.46

contains the backhaul signaling data rates required for different update intervals τ_{int} and varying number of femtocells. It shows that the proposed algorithm's backhaul requirements are not significant and can be met easily with the Ethernet data rate capacity.

5.5 SUMMARY AND CONCLUSIONS

In this chapter, the optimization of coverage for both residential and enterprise femtocells was discussed. First, coverage initialization methods were presented that aim at achieving a given target cell radius based on estimates of measurements of the macrocell power or noise level. While these are a good starting point, the cell coverage needs to be further optimized depending on the deployment scenario. For self-optimization of coverage, one needs to distinguish between residential and enterprise deployments.

For residential femtocells, the objective is to ideally cover as much of the building in which the femtocell BS is deployed, while avoiding to cover neighboring houses or outside areas frequented by other UEs. Different coverage optimization methods were investigated that use information on mobility events of passing outdoor and indoor UEs to optimize the femtocell coverage in order to minimize the increase of core network mobility signaling. It was shown that for femtocells with a single antenna, mobility event-based self-optimization of coverage can significantly reduce the total number of mobility events caused by residential femtocell deployments. In addition, the indoor coverage for femtocells deployed in suitable locations could be improved compared to simpler methods that aim to achieve a constant cell radius.

In addition, two low-cost multi-element antenna solutions were presented to increase the flexibility in terms of good deployment locations within the home where both good indoor coverage and a low number of resulting mobility events can be achieved. First, a switched multi-element antenna solution and the corresponding self-optimization approach were presented, that jointly optimizes both power and pattern selection. It was shown that the flexibility gained due to the ability to select from multiple antenna patterns allows shaping the femtocell coverage such that both the number of mobility events and the indoor coverage can be improved considerably. The second multi-element antenna solution uses four individually adjustable patch antennas, which further increases the flexibility of the achievable gain patterns. While the increased flexibility did not result in significant performance improvements compared to the switched multi-element antenna solution in the investigated scenario, it has some advantages in terms of convergence speed.

For enterprise femtocell deployments, the objectives are to jointly cover an area with multiple cells, maximize the capacity by load balancing, and achieve this with a minimum necessary transmit power. A decentralized joint coverage optimization algorithm was presented to solve this multi-objective problem entailing conflicting objectives. By acquiring the relevant information from its neighbors, the algorithm periodically adjusts the pilot transmit power of the femtocell BS to share the system

UE load evenly across the network, and minimize coverage holes and overlap. Compared to the fixed pilot power allocation, a performance improvement of 18% in terms of femtocell supported traffic and 80% in terms of pilot transmit power reduction has been shown.

REFERENCES

1. H. Claussen, L. T. W. Ho, and L. G. Samuel, "Self-optimization of coverage for femtocell deployments," in *Proc. Wireless Telecommun. Symp. (WTS)*, Los Angeles, USA, Apr. 2008, pp. 278–285.

2. H. Claussen and F. Pivit, "Femtocell coverage optimization using switched multi-element antennas," in *Proc. IEEE Int. Conf. Commun. (ICC)*, Dresden, Germany, Jun. 2009, pp. 1–6.

3. H. Claussen, L. T. W. Ho, and F. Pivit, "Self-optimization of femtocell coverage to minimize the increase in core network mobility signalling," *Bell Labs Tech. J.*, vol. 14, no. 2, pp. 155–184, Aug. 2009.

4. I. Ashraf, H. Claussen, and L. T. W. Ho, "Distributed radio coverage optimization in enterprise femtocell networks," in *Proc. IEEE Int. Conf. Commun. (ICC)*, Cape Town, South Africa, May 2010.

5. D. Fagen, P. A. Vicharelli, J. Weitzen, "Automated wireless coverage optimization with controlled overlap," *IEEE Trans. Veh. Technol.*, vol. 57, no. 4, pp. 2395–2403, Jul. 2008.

6. IEEE Std 802.3-2008: Carrier Sense Multiple Access with Collision Detection (CMSA/CD) Access Method and Physical Layer Specifications. Available: http://standards.ieee.org/getieee802/802.3.html

6

COVERAGE AND CAPACITY OPTIMIZATION FOR OUTDOOR CELLS

6.1 INTRODUCTION

Cell selection in cellular networks is usually driven by downlink (DL) pilot reference signal strength (RSS) or reference signal quality (RSQ) measurements, evaluated at the user equipment (UE) and fed back to the network, which finally decides UE cell association (see Chapter 12). Cell selection mechanisms (e.g., handover (HO) mechanisms) are then usually designed with the objective that the most favorable cell in the DL should serve the UE. This does not always mean that the UE is served by the most favourable cell in the uplink (UL), resulting in an *UL/DL coverage imbalance*. For example, different base stations (BSs) may have receiver chains with different sensitivity, low noise amplifier configuration, and may be subject to different UL interference levels [1, 2].

This UL/DL coverage imbalance problem is aggravated in heterogeneous networks (HetNets) due to the large transmission power difference between macrocell and small cell BSs. This difference implies that a UE connected to a macrocell BS may have a more favourable path gain to/from a non-serving small cell BS than to/from the serving macrocell BS because cell selection is not evaluated based on

Small Cell Networks: Deployment, Management, and Optimization, First Edition. Holger Claussen,
David López-Pérez, Lester Ho, Rouzbeh Razavi, and Stepan Kucera.
© 2017 by The Institute of Electrical and Electronic Engineers, Inc. Published 2017 by John Wiley & Sons, Inc.

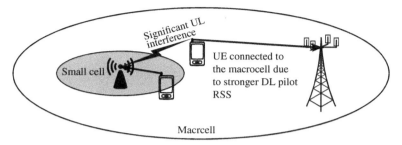

Figure 6.1.   Macrocell UE jamming the UL of a nearby small cell.

path gain but on DL pilot RSS or RSQ. In other words, a UE may not connect to the BS with the shortest path loss.

As a consequence, due to the indicated DL-based cell selection procedure, UEs connected to a macrocell BS may cause severe UL interference to a small cell BS in their vicinity due to the lower path loss. Figure 6.1 illustrates a macrocell UE, served by a macrocell, which induces significant UL interference to a nearby small cell. Note also that if this macrocell UE would be connected to the small cell, it would use much less UL transmit power and create much less UL interference, due to the lower path loss to the small cell.

Another drawback of HetNets using a DL-based cell selection procedure is poor offloading. Due to their smaller coverage, small cells may attract very little traffic and thereby macrocell traffic may not be offloaded as desired. The closer the small cell BS is to the macrocell BS, the smaller the small cell coverage is. Figure 6.2 shows how the coverage of a small cell BS nearby the macrocell BS is smaller than that of a small cell BS far away from the macrocell BS, which leads to poor offloading.

In order to address these issues, UL interference, and poor offloading, an artificially biased serving cell selection can be considered, where the DL pilot RSS is evaluated considering a bias in favor of the small cell to increase its DL coverage footprint. This is sometimes referred to as cell range expansion (CRE) or just as range

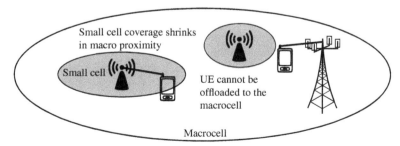

Figure 6.2.   Small cell coverage shrinks in the proximity of a macrocell BS leading to poor offloading.

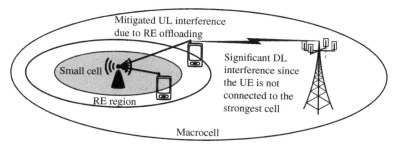

Figure 6.3. Range expansion mitigates UL interference and facilitates offloading at the expense of increasing DL interference.

expansion (RE), the bias is referred to as range expansion bias (REB) or cell selection offset (CSO), and the coverage extension region is referred to as *RE region*. Figure 6.3 illustrates the concept of CRE.

CRE can thus be used to make the serving cell selection more UL relevant, and enhance the coverage and capacity of outdoor small cells through an optimized offloading. However, it is important to note that UEs in the RE region have a more favourable DL from the non-serving aggressor macrocell BS than from the serving small cell BS. This creates significant DL interference for those victim UEs in the RE region, as also illustrated in Figure 6.3.

Due to its significance, CRE and its impact on HetNets will be discussed in more detail in this chapter. The remainder of this chapter is organized as follows: In Section 6.2, the concept of CRE to aid UL interference mitigation and macrocell offloading in HetNets is formally introduced. Moreover, the benefits and drawbacks of CRE in both the DL and the UL are discussed, and simulation results are shown to exemplify such benefits and disadvantages. In Section 6.3, the different CRE strategies that an operator may follow to set up the REB of a particular small cell BS located nearby the center of the hotspot it is covering are presented. In addition, closed-form expressions are derived to calculate such appropriate REBs, according to the different CRE range expansion strategies. In Section 6.4, it is shown how multi-antenna techniques and machine learning procedures can help to optimize the small cell BS coverage by using beamforming techniques. These schemes are particularly useful when the small cell BS is not located at the center of the hotspot it is covering or the UE distribution around the small cell BS is not uniform. Finally, conclusions are drawn in Section 6.5.

6.2 CELL RANGE EXPANSION

In the Third-Generation Partnership Project (3GPP), CRE has been recently investigated for increasing the DL coverage footprint of small cells by adding a positive cell individual offset to the DL pilot RSS of the small cell during the serving cell selection procedure [3–9]. With a larger REB, more UEs are offloaded from the macrocell BSs to the small cell BSs, at the cost of increased co-channel DL interference for range-expanded UEs. Without interference coordination, CRE has been shown to degrade the throughput of the overall network due to this co-channel DL interference, but improve the sum capacity of macrocell UEs due to offloading.

The idea of a biased serving cell selection is not new. It has been discussed since long as a means to share and balance the load between adjacent cells. For example, in [10], the authors discussed a HO criteria for global system for mobile communication (GSM) networks with cell individual offsets that in turn are functions of the respective cell load. A similar approach is presented in [11], where the HO margin between two adjacent cells is adjusted based on cell loads.

Due to the interest from standards organizations for investigating the merits of CRE, there are also several recent CRE-related works available in the academic literature. In [12], closed-form analytical expressions of outage probability with CRE in HetNets have been provided, which verifies that CRE without inter-cell interference coordination (ICIC) degrades the outage probability of the overall network. In [13], joint cell selection and scheduling for picocells with RE have been discussed for the DL. However, UL aspects of CRE have not been covered. The DL/UL imbalance problem and trade-offs for cell selection have been presented and investigated in [14–16]. Moreover, in [17–19], DL performance with different REB values has also been studied for small cell deployments, and resource partitioning is proposed to combat the detrimental effects of increased co-channel DL interference for range-expanded UEs. In this chapter, the impact of range expansion is analyzed, while ICIC mechanisms are presented and studied in Chapters 7–10.

6.2.1 Downlink Aspects of Range Expansion

The HO mechanism is typically supported by event-triggered measurement reporting from the UE to its serving cell (see Chapter 12). The event triggering condition is based upon DL pilot measurements,[1] M_i (in dBm or dB), possibly considering a configurable cell individual offset (CIO) CIO_i for each cell i in dB. If the event triggering condition is configured to trigger a report when a candidate cell is evaluated stronger than the serving cell considering the CIO, and the serving cell is changed accordingly (e.g., through a HO), then the serving cell selection can be described as

$$\hat{i} = \arg\max_{i \in C}\{M_i + CIO_i\}, \tag{6.1}$$

where C is the set of BSs comprised by the currently serving cell as well as the candidate cells.

---

[1] Such as reference signal received power (RSRP) and reference signal received quality (RSRQ) in long-term evolution (LTE), and pilot RSS and signal-to-interference-plus-noise ratio (SINR) in more general terms.

With CRE, a UE adds a positive REB to the small cell measurements, such that

$$\begin{cases} \text{CIO}_i = 0, & \text{if } i \text{ is a macrocell,} \\ \text{CIO}_i > 0, & \text{if } i \text{ is a small cell.} \end{cases} \qquad (6.2)$$

The REB may be UE-specific, reflecting the UE capabilities and the transmit power difference between macrocell BSs and small cell BSs, as well as the benefits of offloading a particular UE. Typical values of REB $\lambda_i^{(dB)}$ may range from few dBs to close to 10 dB.

In LTE, the event-triggering condition for intra-frequency measurements is usually based on the event A3 condition [20], which is triggered when

$$M_{n_j} > M_s + \text{Hyst}_s - \text{CIO}_{s,n_j}, \qquad (6.3)$$

where M_s is the L3 RSRP sample of the serving cell s, not taking into account any offset, M_{n_j} is the L3 RSRP sample of neighboring cell n_j, not taking into account any offset, Hyst_s is the hysteresis parameter, and CIO_{s,n_j} is the mentioned CIO of the neighbor cell n_j with respect to the serving cell s [20]. Refer to Section 12.2 for more details on the HO process.

The actual measurement report is triggered when the criterion has been fulfilled during a configurable time-to-trigger (TTT). A zero hysteresis would imply HO decisions taken directly based on (6.1).

Figure 6.4 illustrates that CRE, as in (6.1), artificially increases the DL coverage footprint of small cells, and thus has the potential to offload more UEs from macrocell BSs.

6.2.2 Uplink Aspects of Range Expansion

The natural cell boundaries between a macrocell BS and a small cell BS are *different in DL and UL* as opposed to between macrocells in a homogeneous network deployment.

Figure 6.4a shows that the DL SINRs observed from a macrocell BS and a small cell BS are equivalent at a location that is closer to the small cell BS than to the macrocell BS, which defines the natural DL cell boundary. This is due to the difference in transmit power between both cells types. In contrast, Fig. 6.4b shows that, since the UE transmits with the same transmit power to all cells, the location of the natural UL cell boundary is where the path losses to the macrocell BS and the small cell BS are equivalent.

In practice, with traditional deployments, the UE serving cell selection is based on DL measurements and it is typically not feasible to associate a UE to different

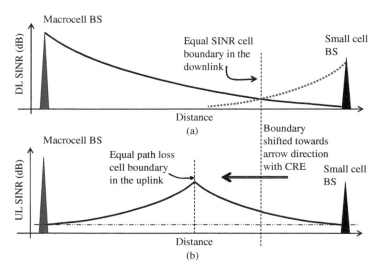

Figure 6.4. Natural cell boundaries during (a) DL and (b) UL in a scenario with heterogeneous DL transmission powers.

cells in the DL and the UL.[2] Moreover, fractional UL power control in LTE is based on path loss compensation [21, 22]:

$$P = \min\{P_{\max}, P_0 + 10\log_{10} M + \alpha L_{\mathrm{DL}}\} \qquad (6.4)$$

where $P_{\max}$ is the maximum UE transmit power, P_0 is a UE-specific (optionally cell-specific) target received power, M is the number of assigned physical resource blocks (PRBs) to the UE, α is the cell-specific path-loss compensation factor (equal to 1 for full compensation), and L_{DL} is the estimated DL path loss. In addition, there can be additional corrections based on decoding performance [21, 23]. As a consequence, the UL/DL imbalance may cause interference problems in the UL, since a UE that has a lower path loss to a small cell BS may be forced to connect to the umbrella macrocell BS, where the path loss is higher due to the large difference in transmit power. This means that the UE (aggressor UE) has to transmit with a high transmit power due to the fractional UL power control, thus causing a high UL interference to the nearby small cell BS (victim BS). In contrast, with CRE, UEs in similar situations may be offloaded to the small cell BS, where the UL transmit powers of UEs and hence the interference level present in the network, will be reduced due to decreased path loss to the small cell BS.

---

[2] However, in more coordinated deployments, this can be feasible. Systems with soft HO can also be seen to enable this to some extent.

Therefore, another important merit of CRE is that due to UL power control CRE reduces the total UL interference observed in the network. The drawback is that, as depicted in Fig. 6.3, these offloaded UEs become victim UEs and observe significant DL interference from the macrocell BS, the aggressor BS.

6.2.3 Behavior of Range Expansion

In order to demonstrate the different trade-offs associated with CRE in the DL and the UL, system-level simulation results are presented in this section. Simulation parameters are mostly based on [24], which are specified by the 3GPP for evaluating picocell scenarios. A seven eNodeB deployment is considered, where each eNodeB is further divided into three sectors. Two co-channel picocell eNodeBs are positioned uniformly at the cell edge in all eNodeB sectors. UEs use fractional UL power control, where the UEs' UL transmit power is given by (6.4) with $P_{max} = 23$ dBm, $P_0 = -78$ dBm and $\alpha = 0.8$. The system throughput is evaluated at the center eNodeB, and ICIC is not implemented.

Simulation results in Fig. 6.5a show that the UL sum throughputs of both macrocell and picocells are improved with increasing REB values. The reason for the improvement is that, with increased REB values, the picocell coverage area in Fig. 6.4a is shifted closer to its ideal UL cell-boundary in Fig. 6.4b. As a consequence,

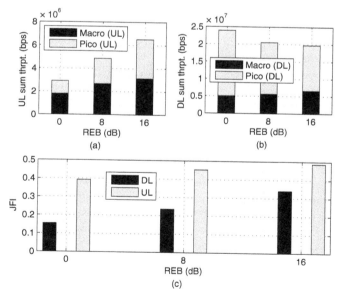

Figure 6.5. (a) Average UL UE throughput; (b) average DL UE throughput; and (c) JFI values in the DL/UL with different cell range expansion bias values.

since UEs get connected to closer BSs, they decrease their UL transmit powers, and thus cause lower UL interference to other BSs. Moreover, low-SINR UEs that were originally connected to the macrocell are now offloaded to the cell-edge picocells, which further improves the aggregate macrocell UL throughput.

In contrast, simulation results in Fig. 6.5b show that the DL sum throughputs decrease with increasing REB values. While CRE improves the macrocell UE DL throughput due to offloading of low-SINR macrocell UEs to the cell-edge picocells, it degrades the aggregate small cell UE DL throughput. This degradation dominates the overall sum throughput, and occurs because the small cell resources have to be shared among more small cell UEs and also because range-expanded small cell UEs observe high interference. In order to optimize the performance of CRE, this DL performance degradation has to be addressed, which calls for enhanced inter-cell interference coordination (eICIC) techniques to mitigate the DL interference in the RE region. eICIC techniques will be presented in Chapters 7–10.

Finally, Fig. 6.5c also shows Jain's fairness index (JFI) for different REB values. JFI jointly considers all macrocell UEs and small cell UEs, and is defined as

$$\text{JFI}(R_1, R_2, \ldots, R_{N_{\text{tot}}}) = \frac{\left(\sum R_j\right)^2}{N_{\text{tot}} \sum R_j^2}, \tag{6.5}$$

with j being the UE index, R_j being the throughput of UE-j, and N_{tot} being the total number of UEs in the system. Figure 6.5c shows that JFI increases with REB values for both the UL and the DL, since larger number of UEs are able to benefit from the under-utilized spectrum resources of the picocell eNodeBs. Moreover, due to the implementation of fractional UL power control, throughputs are more fairly distributed among UEs in the UL, when compared with throughputs in the DL.

6.3 CELL RANGE EXPANSION BIAS OPTIMIZATION

In this section, CRE is analyzed in more detail together with the computation of REB values, through theoretical analysis. Closed-form expressions to calculate appropriate REBs are presented, considering different range expansion strategies [25]. Note that these expressions are not targeted at finding an accurate REB for every small cell BS, since it will vary according to the different scenario conditions (e.g., scenario, deployment, channel conditions), but will help to better understand the different trade-offs that should be considered when optimizing REBs.

First of all, analytical expressions for the boundaries of the coverage area of an small cell BS with and without range expansion are derived. These small cell coverage area boundaries will help to compute the REB, and are defined as follows:

- The equal downlink receive signal strength boundary (ESB) of a small cell is comprised of the 2D plane points at which the DL RSSs from the umbrella macrocell and the small cell are the same. Conventionally, the ESB determines the small cell DL coverage boundary.

- The equal path-loss boundary (EPB) of a small cell is comprised of the 2D plane points at which the path loss from the umbrella macrocell and the small cell are the same.

- The hotspot boundary (HSB) encloses the area wherein the spatial density of UEs is high. As an example, the radius of a small cell hotspot is specified to be 40 m in [24] (Appendix A.2.1.1.2, pp. 65).

Figure 6.6 shows both ESB and EPB boundaries.

Accordingly, two different small-range expansion strategies are considered [25]:

- In the first strategy, the small coverage is expanded from the area enclosed by the ESB to the whole area within the EPB, so that UEs can always connect to the cell with the least path loss. However, this strategy may result in a large REB and thus an aggressive range expansion, which may overwhelm the small cell and cause mobility problems among others.

- In the second strategy, the small cell coverage is expanded from the ESB to the HSB, so that the whole hotspot area is covered.

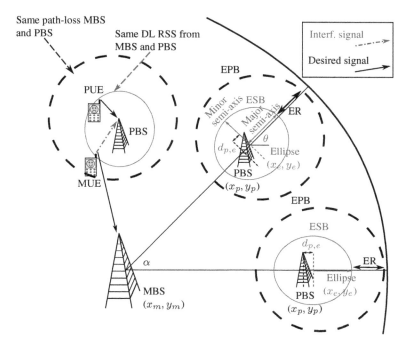

Figure 6.6. Small cell equal DL RSS boundary (ESB) and equal path-loss boundary (EPB).

In the following, closed-form expressions for calculating the REB ΔER_p of small cell BS C_p^P will be provided for the second proposed range expansion strategy, taking 3GPP modeling assumptions into account.

6.3.1 Equal DL RSS Boundary

Without CRE, the DL coverage area of a small cell is delimited by the equal downlink receive signal strength boundary (ESB).

Theorem 1 *The ESB of a small cell can be approximated by an ellipse (given by (6.A.14) in Appendix 6.A, with the center $\mathbf{x}_e = (x_e, y_e)$ and the two semi-axes s_1 and s_2 given as follows*

$$x_e = \frac{cd - bf}{b^2 - ac}, \quad y_e = \frac{af - bd}{b^2 - ac}, \tag{6.6}$$

$$s_1 = \sqrt{\frac{2(af^2 + cd^2 + gb^2 - 2bdf - acg)}{(b^2 - ac) \cdot \left[\sqrt{(a - c)^2 + 4b^2} - (a + c)\right]}},$$

$$s_2 = \sqrt{\frac{2(af^2 + cd^2 + gb^2 - 2bdf - acg)}{(b^2 - ac) \cdot \left[-\sqrt{(a - c)^2 + 4b^2} - (a + c)\right]}}, \tag{6.7}$$

where a, b, c, d, f, g are defined in (6.A.15)–(6.A.20) in Appendix 6.A, and this ellipse is depicted in Fig. 6.6.

Proof: See Appendix 6.A.   □

Using this formulation, it is easy to show that the rotation angle of a small cell ESB (i.e., the counter-clockwise angle Θ from the positive x-axis to the major-axis of the ESB ellipse [26]) is given by

$$\Theta = \arctan\left(\frac{y_p - y_e}{x_p - x_e}\right) = \begin{cases} 0, & \text{if } b = 0 \text{ and } a < c \\ \frac{\pi}{2}, & \text{if } b = 0 \text{ and } a > c \\ \frac{1}{2}\cot^{-1}\left(\frac{a-c}{2b}\right) & \text{if } b \neq 0 \text{ and } a < c \\ \frac{\pi}{2} + \frac{1}{2}\cot^{-1}\left(\frac{a-c}{2b}\right) & \text{if } b \neq 0 \text{ and } a > c \end{cases}, \tag{6.8}$$

where $\mathbf{x}_p = (x_p, y_p)$ is the position of small cell BS C_p^P. The rotation angle Θ is illustrated in Fig. 6.6.

From (6.6), it is also important to note that the center $\mathbf{x}_e$ of the ESB ellipse does not overlap with the small cell BS position $\mathbf{x}_p$, as one may think. Instead, $\mathbf{x}_e$ is on the straight line defined by the macrocell BS location $\mathbf{x}_m = (x_m, y_m)$ and $\mathbf{x}_p$, and in between $\mathbf{x}_m$ and $\mathbf{x}_e$, as shown in Fig. 6.6 for $P_m > P_p$. Note that $P_{m(p)} = G_{m(p)} p_{m(p)}^{\text{pilot}} / (\varphi_{m(p)} \, \xi_{m(p),u})$. As a consequence,

- the rotation angle Θ of the ESB ellipse is equal to the counter-clockwise angle $\alpha = \arctan(\frac{y_p - y_m}{x_p - x_m})$ from the positive x-axis to the straight line defined by $\mathbf{x}_m$ and $\mathbf{x}_p$.
- the major axis of the ESB ellipse (6.A.14) always overlaps with the straight line connecting $\mathbf{x}_m$ and $\mathbf{x}_p$, regardless of the transmission power difference between the macrocell and the small cell BS.

6.3.2 Equal Path-Loss Boundary

The coverage area of the equal path-loss boundary (EPB) can be derived in a similar way to the derivation of the coverage area of the ESB, as shown in the following theorem.

Theorem 2 *The EPB of a small cell is an ellipse given by (6.B.2) in Appendix 6.B. The center $\mathbf{x}'_e = (x'_e, y'_e)$, semi-axes s'_1 and s'_2, and rotation angle Θ' of the EPB ellipse can be computed by replacing a, b, c, d, f, g with a', b', c', d', f', g' in (6.6), (6.7), and (6.8), respectively. a', b', c', d', f', g' are defined in (6.B.3)–(6.B.9) in Appendix 6.B, and this ellipse is depicted in Fig. 6.6.*

Proof: See Appendix 6.B. ☐

Similar to the ESB ellipse, the center $\mathbf{x}'_e$ of the EPB ellipse does not overlap with the small cell BS position $\mathbf{x}_p$. Instead, it is located on the straight line defined by $\mathbf{x}_m$ and $\mathbf{x}_p$, and in between $\mathbf{x}'_e$ and $\mathbf{x}_m$.

6.3.3 Shifts of ESB and EPB Centers from the Small Cell BS Location

It is interesting to note that, the distance between the ESB ellipse center and the small cell BS location is given by $d_{p,e} = |\mathbf{x}_e - \mathbf{x}_p|$, while the distance between the EPB ellipse center and the small cell BS location is given by $d'_{p,e} = |\mathbf{x}'_e - \mathbf{x}_p|$.

Theorem 3 *Both $d_{p,e}$ and $d'_{p,e}$ can be well approximated as linear functions of the distance $d_{m,p}$ between the macrocell and the small cell BSs, that is, $d_{p,e} \approx \dfrac{d_{m,p}}{P_m/P_p - 1}$*

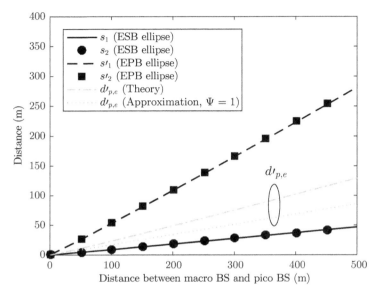

Figure 6.7. Semi-axis lengths (s_1 and s_2) of the ESB and (s'_1 and s'_2) of the EPB ellipses, Also shown is the distance $d'_{p,e}$ between the small cell BS and its EPB ellipse center using a typical HetNet scenario.

and $d'_{p,e} \approx \dfrac{d_{m,p}}{G_m/G_p - 1}$, where $G_m = \dfrac{1}{\varphi_m \, \xi_{m,u}}$, and $G_p = \dfrac{1}{\varphi_p \, \xi_{p,u}}$. Moreover, $d_{p,e} < d'_{p,e}$, and thus the EPB always contains the ESB.

Proof: See Appendix 6.C.   □

For illustration purposes, in Fig. 6.7, $d'_{p,e}$ is plotted versus $d_{m,p}$, using the theoretical value $|\mathbf{x}_p - \mathbf{x}'_e|$ based on (6.6) and the linear approximation in Theorem 3, respectively. Typical 3GPP simulation parameters are adopted [25]. Both the theoretical value and the linear approximation show that $d'_{p,e}$ increases with $d_{m,p}$, indicating that for a small cell BS located at a larger distance from the macrocell BS, the EPB ellipse center $\mathbf{x}'_e$ shifts further away from the small cell BS location $\mathbf{x}_p$ (and the MBS location $\mathbf{x}_m$) along the straight line defined by $\mathbf{x}_m$ and $\mathbf{x}_p$. Note that there is a gap between the theoretical value and the linear approximation. The gap exists because $\psi = 3.76/3.67$ is used[3] in the numerical evaluation of $|\mathbf{x}_p - \mathbf{x}'_e|$, while $\psi = 1$ is used to obtain the linear approximation in Theorem 3.

---

[3] Since $\psi \approx 1$, the theoretical value also shows a linear behavior, but the linear behavior is lost as ψ diverges from 1. See the path-loss models for macrocells and small cells in Section A.4.

Figure 6.7 also shows the semi-axis lengths of the ESB and EPB ellipses versus $d_{m,p}$. It can be seen that the semi-axes s_1' and s_2' of the EPB ellipse (6.B.2) are always larger than the semi-axes s_1 and s_2 of the ESB ellipse (6.A.14), with s_1' and s_2' increasing with $d_{m,p}$ faster than s_1 and s_2. This indicates that the EPB ellipse always contains the ESB ellipse. Moreover, for this typical network setup, it can be seen that $a \approx c$ and $a' \approx c'$, and consequently, $s_1 \approx s_2$ and $s_1' \approx s_2'$, which implies that both the ESB and EPB ellipses can be well approximated by circles.

As follows from Theorem 3, it is important to note that since the distance $d_{p,e}$ (or $d_{p,e}'$) between the ESB (or EPB) ellipse centre and the small cell BS location grows with the distance between the macrocell and the small cell BSs $d_{m,p}$, a side of the ESB (or EPB) ellipse gets closer to the small cell BS location as $d_{m,p}$ grows. Consequently, some macrocell UEs may get closer to the small cell BS than some small cell UEs in the expanded region. This causes significant UL interference to the small cell UEs in the expanded region.

In order to illustrate this UL interference issue, Fig. 6.8 depicts a HetNet, where a macrocell BS is located at $(0, 0)$ m and a small cell BS is located at $(202, 0)$ m. Here,

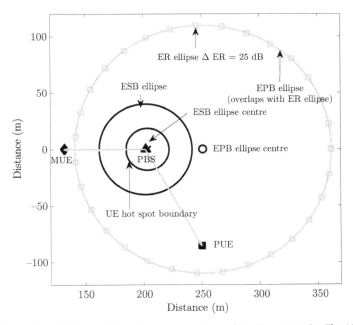

Figure 6.8. Plot of ESB and EPB ellipses using a typical HetNet scenario. The UE hotspot boundary indicates the area where small cell UEs are distributed, and it is discussed in Section 6.3.4. Even if we expand the small cell coverage from ESB to EPB, due to the shift of the EPB ellipse center from the small cell BS, an macrocell UE may still be closer to the small cell BS than a small cell UE (i.e., $d_1 > d_2$), causing severe UL interference to the small cell BS.

the 3GPP simulation parameters presented in [25] are also used. The ESB (6.A.14) and EPB (6.B.2) of the small cell are also plotted. It can be observed that the ESB center shifts only slightly away from the small cell BS, while the EPB center shift from the small cell BS is considerable. Figure 6.8 also illustrates that the distance d_1 between a small cell UEs in the expanded region and the small cell BS may be larger than the distance d_2 between macrocell UE and the small cell BS. If these two UEs are assigned with the same resource block (RB), the expanded region UE may suffer from strong UL interference caused by the macrocell UE. Therefore, expanding the small cell coverage area from the ESB to the EPB requires coordinated radio resource allocation between macrocell and small cell BSs to mitigate not only the DL interference but also the UL interference from the macrocell to expanded region UEs due to the presented effect.[4] Due to these complexity issues, these may not be the right strategy.

6.3.4 Range Expansion Bias Calculation

In order to avoid aggressive range expansion and the previously presented problems, the small cell coverage area can be expanded to just cover the targeted hotspot. The coverage area required for a hotspot could be learned by operators using UE localization and hotspot identification techniques, and can be used for optimizing the network deployment. Without loss of generality, we define the HSB as a circle centered at $\mathbf{x}_h = (x_h, y_h)$ with a radius r_h, and assume that the ESB ellipse center $\mathbf{x}_e$ (which is near the small cell BS location as shown in Fig. 6.8) overlaps with the HSB center $\mathbf{x}_h$.

Since the ESB ellipse can be well approximated by a circle (as shown in Section 6.3.3), the expansion of the small cell coverage area from the ESB to the HSB can be realized by the UEs increasing the DL RSSs of the small cell pilot signals by an REB $(\Delta \mathrm{ER}_p^{\mathrm{HSB}})_{\mathrm{dB}}$, which can be computed based on (6.7) and the given HSB radius r_h.

For $\psi = 1$, $\Delta \mathrm{ER}_p^{\mathrm{ESB}}$ can be approximated by the following closed-form expression

$$\Delta \mathrm{ER}_p^{\mathrm{HSB}} = \frac{\left(2r_h^2 + d_{m,p}^2\right) \mathcal{P}_m - \mathcal{P}_m d_{m,p} \sqrt{d_{m,p}^2 + 4r_h^2}}{2r_h^2 \mathcal{P}_p}, \tag{6.9}$$

which is derived in Appendix 6.D.

Figure 6.9 plots the REB $(\Delta \mathrm{ER}_p^{\mathrm{HSB}})_{\mathrm{dB}}$ required for providing a targeted small cell coverage versus the targeted hotspot radius, that is, r_h for several values of the macro to small cell BS distance $d_{m,p}$. Both the theoretical value of $(\Delta \mathrm{ER}_p^{\mathrm{HSB}})_{\mathrm{dB}}$ calculated using (6.7) and the approximation in (6.9) are plotted. From Fig. 6.9, it can

---

[4] See, for example, [27–29] for some recent discussions in the 3GPP for UL interference mitigation in HetNets through power control.

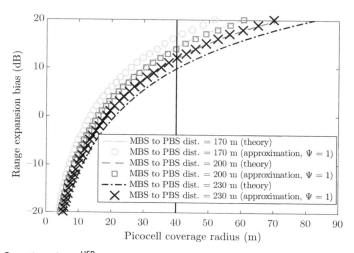

Figure 6.9. Plot of ΔER_p^{HSB} required for a targeted small cell DL coverage radius at several macro to small cell BS distances. The vertical green line denotes the 40 m hotspot radius specified in [24].

be seen that the REB increases with the increase of the targeted hotspot radius r_h at a given $d_{m,p}$, but decreases with the increase of the macro to small cell BS distance $d_{m,p}$ for a given hotspot radius r_h. The approximated ΔER_p^{HSB} in (6.9) is about 2 dB larger than the theoretical value. This is because $\psi = 3.76/3.67$ is used in the numerical evaluation based on (6.7), while $\psi = 1$ is assumed for (6.9) in Appendix 6.D. Negative values of $(\Delta ER_p^{HSB})_{dB}$ indicate that the ESB without range expansion already contains the targeted HSB. Range expansion is not necessary in this case.

6.4 COVERAGE OPTIMIZATION BY USING SWITCHED MULTI-ELEMENT ANTENNA SYSTEMS

This section introduces a solution to further enhance the performance of outdoor small cells, which relies on deploying switched multi-element antennas (MEAs) at the small cell BSs, and its objective is to improve the small cell coverage or the UE capacity through SINR maximization.

As shown in Chapter 5, the combination of the simple antenna patterns of the different antenna elements in a switched MEAs system can create a rich diversity of antenna patterns with different gains in different directions in a cost-effective manner. Given the availability of different antenna pattern choices, the small cell BS may follow different strategies. It may decide to select the antenna pattern that maximizes its coverage toward a given region, which could be quite useful to optimize coverage when the small cell BS cannot be deployed at the center of the hotspot it is covering, or may select a given antenna pattern for each UE in order to maximize the signal quality of its data channels, a sort of data channel beamforming. In this section, the focus is on this switched MEAs system and the selection of a given antenna pattern for each UE. In more detail, the criteria for antenna pattern selection is based on UEs radio frequency (RF) fingerprints, and a supervised machine learning classification method is proposed to select the optimal antenna pattern.

6.4.1 Switched Multi-element Antenna System

With an omni-directional antenna (ODA), the only way a BS can interact with its environment is through resource allocation and power control. A simple alternative solution to further benefit from beam directionality, while keeping the complexity and cost at minimum, is to equip a BS with multiple antenna elements with distinctive directional antenna patterns. A switch can then be used to select a single or a combination of antenna elements. In its simplest form, this can be realized using a number of patch antennas. Using this switched MEA system, the BS performance would be heavily dependent on the accuracy of the antenna pattern selection algorithm, which selects the appropriate antenna pattern for each scenario. In the following, we embrace this architecture for small cell BSs, and propose an efficient antenna pattern selection algorithm.

In this section, the proposed switched MEA system is not applied to pilot signals but only to data channels, that is, the focus is on selecting a given antenna pattern for each UE. Therefore, it is proposed that the pilot channel is always transmitted on a separate fixed antenna (perhaps an ODA) and the switched multi-element antenna system is only used for data channels. The same principles could also be used to beam-form pilot channels, thus shaping the small cell BS coverage toward a given region.

6.4.2 Antenna Pattern Selection Algorithm

Consider a small cell BS equipped with a set of antenna patterns, $A = \{a_1, a_2, ..., a_K\}$. For the sake of simplicity, we first consider a case where the BS serves no more than a single UE. The multi-UE case is discussed separately in following subsections.

In its simplest form, the antenna selection decisions can be made in a distributed and uncoordinated fashion by each BS. The objective is to select an antenna pattern that provides the maximum directional gain toward UE U_u, which is being served by the small cell BS. Therefore, the optimal antenna pattern for UE U_u, represented by $a_u^* \in A$, maximizes the received power at UE U_u, that is,

$$a_u^* = \arg\max_{a \in A}(\mathcal{R}_u^a), \qquad (6.10)$$

where $\mathcal{R}_u^a$ is the received power from the small cell BS at the intended UE U_u when the former selects antenna pattern a. Moreover, in dense deployments, it is possible to consider coordination amongst the small cell BSs to select antenna patterns that minimize inter-cell interference. However, this approach is not considered in this chapter, due to the resulting increased signaling overhead and complexity.

In order to assist the antenna pattern selection, RF fingerprints can be used as an indicator to decide the optimal antenna pattern for UE U_u. Radio fingerprinting based techniques have been extensively studied for localization. Generally, this procedure can be decomposed into two steps. First, the RF fingerprints with appropriate features are collected and associated to known geographical coordinates, and thereafter a machine learning-based classifier estimates the coordinates of other measurement points.

A training set $\mathcal{T}$ can be defined as a collection of individual fingerprint samples f for which the optimal antenna patterns are known. Let F be the set of fingerprint samples in the training set and A^* be the corresponding vector of optimal antenna patterns (i.e., $\mathcal{T} := (F, A^*)$). Then, the task of the classifier C is to take the training set $\mathcal{T}$ and predict the optimal antenna pattern for UE U_u, which reports the fingerprint sample f_u. In other words,

$$\hat{a}_u \leftarrow C(\mathcal{T}, f_u), \qquad (6.11)$$

where $\hat{a}_u$ is the predicted optimal antenna pattern for UE U_u.

It is important to note that unless $\hat{a}_u = a_u^*$, the prediction is considered as misclassification. In fact, the expected misclassification error rate $\zeta_{\mathcal{T}}$ is one of the common measures to benchmark the performance of a classifier for a given training set $\mathcal{T}$. This refers to the number of incorrect classification decisions divided by the total number of decisions made. In this context, a classifier may also consider the *prior* misclassification probabilities of each class. In most cases, this can be empirically derived from A^*. However, the performance of the classification would be further enhanced if the classifier is provided with as accurate as possible prior misclassification probabilities. This is especially useful when the size of the training set is restricted.

Most classifiers can also provide a measure of certainty behind their predictions. This metric can be different from one classifier to another. In this light, (6.11) can be extended as

$$[\hat{a}_u, Cr_u] \leftarrow C(\mathcal{T}, f_u, \Pi_A), \qquad (6.12)$$

where Π_A is the vector of prior probabilities of available antenna patterns and Cr_u is the prediction certainty associated with UE u.

6.4.3   Use of RF Fingerprints

RSS is considered as the most common feature for RF fingerprints and is subsequently used here. As a routine part of their operation, UEs periodically send measurement reports to their serving BSs indicating the RSS of their serving and neighboring cells. In addition, a BS can explicitly request its UEs to perform and report measurements where it can additionally specify the measurements' characteristics. These measurement reports are critical for making handover decisions (see Chapter 12). This simply implies that collecting RF-fingerprints from UEs does not require additional or nonstandard operations and overheads. Therefore, the final step toward constructing $\mathcal{T}$ would be to find the optimal antenna patterns A^* associated with the set of fingerprint observations F.

Since the antenna patterns in set $A = \{a_1, a_2, ..., a_K\}$ are rather distinct and because small cell BSs are deployed densely in urban areas, a small training set is expected to be sufficient. Therefore, the small cell BS can quickly scan through the available antenna patterns, and learn which antenna pattern yields the highest received power at the UE u location. In that mode, the small cell BS can change the antenna patterns and request initial UEs to perform measurements. It should be noted that the absolute values of these measurements are not relevant for the purpose of antenna selection as the BS is only required to identify the best pattern amongst all. Subsequently, depending on the radio access technology in use, different measurements shall be performed and different indicative parameters may be reported back. Once the optimal antenna pattern is decided, it will be mapped to the RF fingerprint and the training set is constructed subsequently. We refer to this procedure as the exploration phase. Evidently, changing antenna patterns and immediately performing such measurements may be acceptable for few initial UEs, but due to implications on UE battery life and because of the time and complexity of such measurements, it may not be considered as a routine practice to be carried out for all UEs at all times. For this reason, a classification method should be deployed.

Another challenging issue is to decide when to stop collecting training samples and to start the classification process. Too small training sets result in inaccurate predictions, whereas constructing unnecessary large training sets would lead to increased overhead and potential implications for storage at the small cell BS. In practice, a classifier may seek to minimize the size of the training set, whilst assuring that $\zeta_{\mathcal{T}}$ is lower than a certain threshold. In order to estimate $\zeta_{\mathcal{T}}$ for future and unseen fingerprints, the k-fold cross-validation (CV) test can be applied where the original training set is partitioned into k equal size subsamples uniformly at random. One of the subsamples is then retained as the validation data for testing the classifier, while the remaining $k-1$ subsamples are used as the training set. The process is repeated until all subsamples are tested against the rest.

The average classification error rate of CV tests $\zeta_{\mathcal{T}}^{cv}$ can be used as an indicator of the future performance (i.e., $\zeta_{\mathcal{T}}^{cv} \approx \zeta_{\mathcal{T}}$). It should be noted that correct classification decisions are known when considering CV tests, hence calculating the misclassification rate is straightforward. Moreover, if Cr_u falls below a certain threshold T_{Cr}^-, the small cell BS can simply discard the classification outcome, and instead identify a set of potentially appropriate antenna patterns $A_{\text{pot}} \subset A$ and explore them individually to find the best one. Identification of such candidates depends on the classification method used. For example, in case of probabilistic methods, in which antenna patterns are associated with a probability of being the optimal choice, any antenna pattern with its probability exceeding a certain threshold can be added to the set A_{pot}. This exploration is important for scenarios where relevant and similar samples are non-existent in the training set. An example of this case is when the small cell BS is covering both outdoor and indoor but the training set contains samples of only outdoor UEs (or very few indoor samples) and an antenna pattern is to be selected for an indoor UE, which has a distinctly different RF fingerprint. If $\mathcal{T}$ has not yet reached its maximum size limit l_{max}, the findings of recent explorations can be incorporated into $\mathcal{T}$. In other words,

if $Cr_u < T_{Cr}^-$ **then**
    $\hat{a}_u \leftarrow$ explore(A_{pot}) **if** $\mid \mathcal{T} \mid < l_{max}$ **then**
        add f_u to F
        add $\hat{a}_u$ to A^*
    end
end

where explore(A_{pot}) represents the exploration operation in that the small cell BS evaluates all candidate antenna patterns $a \in A_{\text{pot}}$ and selects the best one $\hat{a}_u$. Moreover, as a result of bias sampling, the prior probabilities might be misleading. Observations over long periods and adjusting the training set may discount the issue, but this might have implications on the size of the training set. As a solution, the prior probabilities can be updated separately and fed into the classifier. With this objective in mind, the outcome of any prediction exceeding a specific certainty threshold T_{Cr}^+ can be used to update the prior probability vector. For this, the small cell BS would need to associate the number of total observations N to the existing prior probabilities P_A. If P_A is initially constructed from the training set, then $N = \mid \mathcal{T} \mid$. Generally speaking,

if $Cr_u > T_{Cr}^+$ **then**

$$P_A = \begin{cases} \frac{N P_A(a)+1}{N+1} & \text{if } a = \hat{a}_u; \\ \frac{N P_A(a)}{N+1} & \text{Otherwise;} \end{cases}$$

    N$\leftarrow$N+1;
end

6.4.4 Multi-user Scenarios

In order to serve multiple active UEs that require different antenna patterns, the strategy for selecting antenna patterns can be combined with the scheduling decisions. In

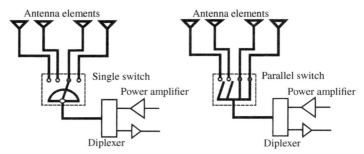

Figure 6.10. Two design schematics for the switched multi-element antenna system.

other words, UEs with an identical optimal antenna pattern can be scheduled in the same time slot, and UEs with different optimal antenna patterns can be multiplexed in the time domain. However, given that scheduling decisions are rather complex, taking into account such additional requirements may further restrict the scheduler. Moreover, scheduling in time domain may not be possible for some radio access technologies. A simple and conservative alternative would be to immediately switch back to the ODA mode when the optimal antenna pattern is not the same for all UEs that are to be served in a given time slot. However, this solution may impact the performance gains when the number of UEs starts to increase.

Figure 6.10 compares two design schematics, one with a single switch and the other one with a parallel switch, capable of selecting more than one antenna pattern subset at the time. Using a parallel switch allows for support of more UEs in the MEA mode, at the cost of scarifying antenna gain and directionality. This is because when multiple antenna elements are simultaneously selected, the input signal is split between the selected antenna patterns, hence the effective directionality of the combined pattern is reduced.

The impedance mismatch is another challenging issue that should be taken into account when selecting more than one antenna pattern at a time. The resulting reflection loss coefficient Γ is defined as the ratio of the amplitude of the reflected wave to the amplitude of the incident wave, and is expressed as

$$\Gamma = \frac{Z_L - Z_S}{Z_L + Z_S},$$ (6.13)

where Z_S and Z_L are the impedances toward the source and load, respectively. In case of a single switch, where only one antenna element is selected at a time, impedance matching can be applied to minimize the reflection loss. However, in case of the parallel switch, the source impedance will be dependent on the number of selected antenna elements. More specifically, the source impedance when selecting n antenna elements Z_S^n can be viewed as the combined parallel impedances of the individual antenna elements (i.e., $Z_S^n = Z_S^1/n$). Therefore, selecting the value of Z_S^l close to Z_L results in minimizing the reflection loss for the case when a single antenna element is selected, but is suboptimal when multiple antenna elements are combined. Combining too many antenna elements may additionally reduce directionality gains, and thus it makes sense to restrict the number of antenna patterns selected at a time to a certain maximum value. If UEs are distributed such that the combination of a large number of antenna elements is required, the small cell BS switches to the ODA mode.

6.4.5 Classification Methods

Broadly speaking, supervised learning classification techniques can be categorized into two groups of deterministic and probabilistic methods [30]. Deterministic methods try to estimate the class of an observation by considering observation values, whereas probabilistic methods consider observations as part of a random process. In the following, the performance of four well-known classification algorithms is investigated for the above antenna pattern selection problem. For more information about these classifiers, readers can consult [30].

The considered classification algorithms are as follows:

- k-NN classifier is the first classification method considered, which is a non-parametric method, and is based on closest training samples in the feature space. As the name implies, an observation is classified by a majority vote of the k-nearest neighbors in the training set. It is additionally a common practice to apply weights to the vote contributions of the neighbors. This study considers such weights to be proportional to the inverse of the Euclidean distance of the neighbors to the given observation sample. This classifier is computationally very simple making it suitable for applications where real-time classification decisions are required.

- Naive Bayes (NB) classifier is a simple probabilistic classifier based on applying Bayes' theorem with the strong assumptions that classification features are independent (i.e., presence, absence, or value of a specific feature is unrelated to those of other features). Despite this assumption, this classifier has been shown to perform well in many applications.

- Linear discriminant analysis (LDA) is a method based upon the concept of searching for a linear combination of variables that can best separate two classes. This classifier assumes that the conditional class densities are Gaussian.

- Decision tree is a simple and widely used method in that the classifier organizes a series of consecutive evaluations based on some specific features of the variables and makes the classification decisions.

6.4.6 Simulation Scenarios and Results

System-level simulations were performed to confirm the feasibility of the proposed scheme and get an understanding of the trade-offs involved in terms of accuracy of the antenna pattern selection with respect to relevant parameters, including the size of the training set, the number of visible neighboring cells and the type of classification method used. Moreover, it was also of interest to understand and visualize parts of the coverage area where misclassification of antenna patterns occurs more frequently. For this reason, the scenarios considered for the first set of experiments were rather small and specific to allow traceability of results.

Due to the sensitivity of the results to the RF propagation model, the wireless system engineering (WiSE) [31] package was used, which is a comprehensive 3D ray-tracing based simulation tool. The simulation results from the WiSE simulator have been validated by comparison against measurement data in a number of indoor and outdoor environments.

In order to examine whether RF fingerprint measurements can provide sufficient information for selecting the optimal antenna pattern, a real deployment scenario was considered. More specifically, the simulation study considers the Stachus square, which is a large crowded shopping area in central Munich, Germany. Figure 6.11a shows the simulation scenario, which is 300 m × 300 m large and consists of three small cell BSs deployed around the square. Additionally, macrocell BSs of a top-tier operator in Germany were considered. Figure 6.11b shows the predicted strongest received pilot power from the three small cell BS using the WiSE simulator. Within the simulation scenario, it was considered that a UE could detect and measure up to 12 different neighboring macrocells.

Calculation of UE throughput follows a similar approach presented in Appendix A. As mentioned earlier, a simple solution to realize the proposed MEA system is to deploy a number of patch antennas with different azimuth orientations. Figure 6.12 shows the antenna gain patterns of individual antenna subsets used in the simulations where the horizontal and vertical gain patterns are similar.

Figure 6.13 shows the expected antenna pattern misclassification rate ζ_T for different classification methods when the size of the training sets $|\ T\ |$ increases. In this case, UE measurement reports are assumed to be perfectly accurate. As expected, ζ_T starts to decrease with an increase in the number of training samples. Figure 6.13 also shows that k-NN and decision tree (DT) perform better than the other two classification schemes. ζ_T stands just below 10 % for a training set consisting of 300 samples when using k-NN. Moreover, Fig. 6.13 shows the average classification error rate of the 10-fold cross-validation tests for the k-NN method with $k = 1$. It is important to note that, considering only the closest neighbor ($k = 1$) was shown to be a superior strategy in the k-NN method.

As mentioned earlier, the results shown in Fig. 6.13 are for the scenario where UE measurement reports are perfectly accurate. In practice, inaccuracies in measurement reports are very common. According to the 3GPP model in [32], the

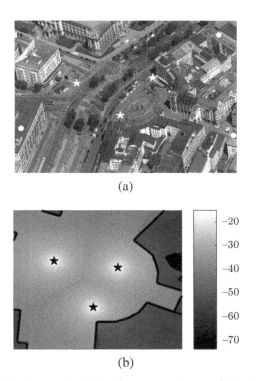

(a)

(b)

Figure 6.11.   Simulated scenario. (a) Stachus square in central Munich, Germany where the location of small cell BSs and some of the neighboring macrocell BSs are represented by star and circle markers, respectively and (b) The WiSE simulator prediction representing the strongest received pilot power (in dBm) from small cell BSs.

measurement errors can be modeled as a zero-mean Gaussian distribution with a standard deviation set to 1.21 dB. Following this recommendation, the results presented in Fig. 6.14 compares the performances of the different classification methods when measurement errors were additionally considered for both training and test samples. The results confirm that the performance of k-NN is not significantly impacted as a result of measurement errors, and the misclassification rate is just slightly above 10% with 300 training samples. In other words, such measurement errors were not significant enough to be able to shift the decision boundaries of the classifier, and the proposed method is robust enough. It is also interesting to notice that, with sufficient training samples, considering more than a single neighbor ($k > 1$) can better compensate the effect of measurement errors. In contrast, DT is shown to be very sensitive to measurement errors. Moreover, construction of optimal

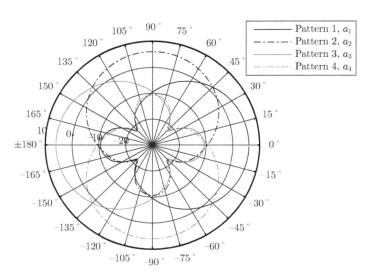

Figure 6.12.   Gain antenna pattern of the antenna elements used in small cell BSs.

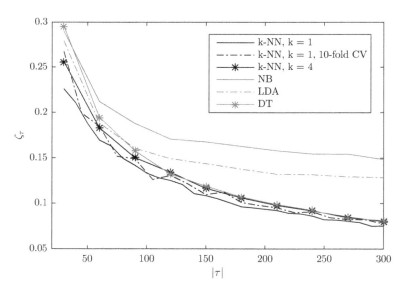

Figure 6.13.   Misclassification rate versus size of the training set for four different classification methods.

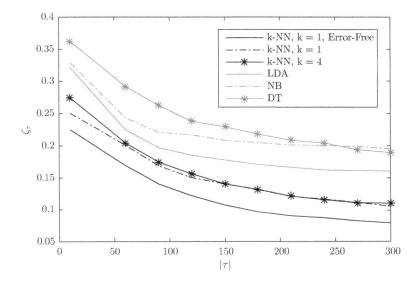

igure 6.14. Misclassification rate versus size of the training set for four different clasfication methods when measurement errors are additionally considered.

ecision trees is known to be a NP-complete problem [30] making it unsuitable for he antenna pattern selection task. As a result, k-NN is considered as the classification nethod throughout the rest of this study.

Figure 6.15 shows the variation of ζ_T against the number of cells measured in ach RF fingerprint report. Considering more neighbors can improve the classification accuracy, and fortunately the number of visible neighbors is sufficiently high in rban areas. However, the classification performance is shown to be acceptable even vith a moderate number of neighbors.

In order to define a certainty measure Cr_u for a k-NN classifier, which identifies vhich antenna element $\hat{a}_u$ is the optimal choice for UE u, let $A_k^* = \{a_1^*, a_2^*, ..., a_k^*\}$ e the optimal antenna patterns associated to the k-nearest fingerprint matches in the raining set (i.e., $A_k^* \subset A^*$). Moreover, let $W = \{w_1, w_2, ...w_k\}$ be the corresponding veights. Then, Cr_u is defined as

$$Cr_u = \frac{\sum_{\forall n: a_n^* = \hat{a}_u} w_n}{\sum_{n=1}^k w_n} . \tag{6.14}$$

As described earlier, a simple way to improve the misclassification rate is to discard predictions which are associated with low certainty (i.e., $Cr_u < T_{Cr}^-$). The trade-off vould be the additional overhead for exploring candidate patterns.

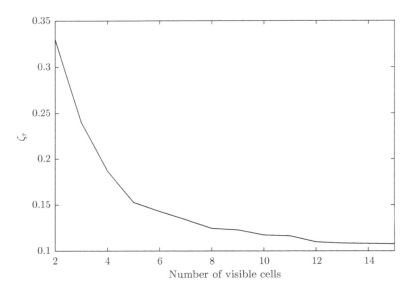

Figure 6.15.   Misclassification rate versus the number of visible cells.

Figure 6.16a confirms how classification accuracy is enhanced by rejecting predictions where Cr_u is below 0.9 and 0.7, respectively. The results are presented for a k-NN classifier where k was set to 4. Moreover, Fig. 6.16b shows the ratio of those rejected predictions for which the small cell BS needs to explore at least two candidate antenna patterns. The overhead starts to decline with the increase in the number of training samples.

To what extent a small cell BS needs to be conservative when making decisions in relation to antenna pattern selection significantly depends on the level of performance degradation as a consequence of selecting a sub-optimal antenna pattern. For example, if the small cell BS selects the second best antenna, the overall performance might not be significantly impacted. Nevertheless, the case is considered as a misclassification.

Figure 6.17a shows the cumulative distribution function (CDF) of UE SINRs with ODA and MEA systems. While the average SINR is improved by 5.1 dB when using the proposed MEA scheme, the effect of imperfect antenna pattern selection is shown to be insignificant in that the optimal antenna pattern selection could merely improve the average SINR by 0.7 dB with respect to the proposed MEA scheme. In addition to the mean, the worst 5th percentile of UE performance is of great importance for the operator. According to Fig. 6.17a, the worst 5th percentile of UE SINR is improved by 4.3 dB when using the proposed MEA scheme. In addition, the

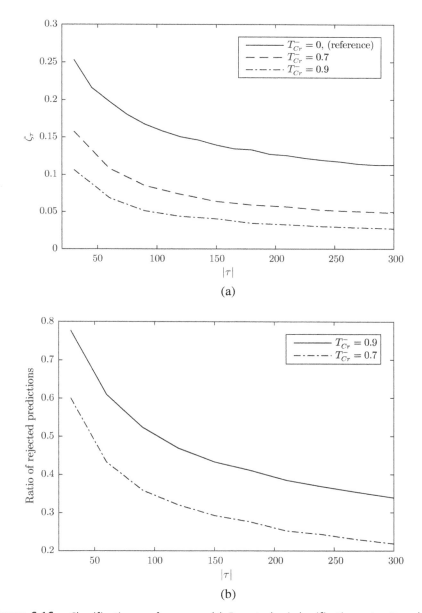

Figure 6.16. Classification performance. (a) Expected misclassification rate, ζ_T, when rejecting classifications with Cr_u smaller than T'_{Cr} and (b) Ratio of rejected classifications.

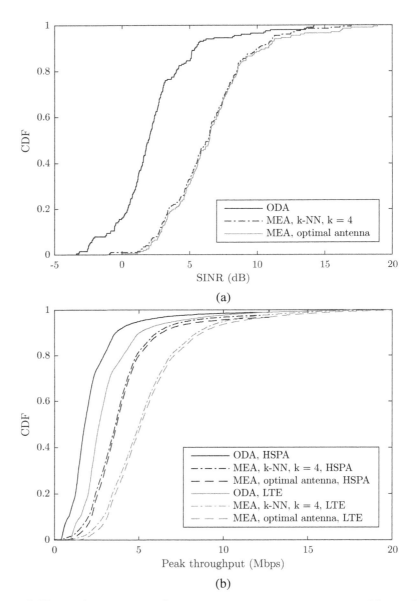

Figure 6.17.   Performance gain of MEA scheme over existing ODA systems. (a) CDF of UE SINRs and (b) CDF of UE throughputs.

difference between the SINR performances when always selecting the antenna patterns optimally and the proposed scheme is more noticeable for higher SINR values. This is mainly because misclassification of patterns occurs more frequently when UEs are closer to the BS where small UE movements may correspond to notable changes in the angular direction of a UE relative to small cell BS. According to Fig. 6.17a, the 95th percentile of UE SINR is 12.5 dB for the optimal antenna selection strategy while it is 11.2 dB and 8.1 dB for the proposed scheme and the ODA scheme, respectively. However, due to the limitations imposed by the modulation schemes, very high SINR values may not be directly translated into throughput gains. Figure 6.17b shows the corresponding gains in UE peak throughput for a 5 MHz high-speed packet access (HSPA) and LTE network. The term "peak" refers to the fact that the BS serves a single UE at the time. It can be seen that using the MEA scheme can provide up to 72 % improvement in UE peak throughput compared with the existing ODA systems.

6.5 SUMMARY AND CONCLUSIONS

This chapter focused on two important aspects for the optimization of coverage and capacity for outdoor small cells.

First, CRE and its impact on HetNets have been discussed in detail. CRE has been introduced as a method to provide UL interference mitigation and macrocell offloading in HetNets and its benefits and drawbacks in both the DL and the UL have been evaluated using simulation results. Different CRE strategies that an operator may follow to set up the REB of a particular small cell BS located nearby the center of the hotspot it is covering has been presented, and closed-form expressions have been derived to calculate such appropriate REBs.

Moreover, a solution for enhancing small-cell throughput by deploying simple switched antenna beamforming techniques using a MEA system has been introduced. These schemes are particularly useful when the small cell BS is not located at the center of the hotspot it is covering or the UE distribution around the small cell BS is not uniform. The antenna pattern selection was considered as a supervised machine-learning classification problem where the small-cell BS decides the optimal antenna pattern based on the RF fingerprint of the UEs and previous observations. Comparing various classification methods, k-NN is shown as an appropriate choice. Simulations were carried out using a 3-D ray tracing model of a real urban deployment in central Munich, Germany. The results confirm the feasibility of the proposed scheme, even with relatively limited training samples and few neighboring BSs. An average UE SINR improvement of 5.1 dB was shown in the investigated scenario. In addition, the results indicate a clear superiority of the scheme over existing ODA BSs, with an improvement of up to 72% of user's mean throughput.

APPENDIX 6.A: PROOF OF THEOREM 1

For a given UE U_u located on the ESB of picocell C_p^P, the DL RSSs from the umbrella macrocell C_m^M and picocell C_p^P are equal, that is,

$$p_m^{\text{pilot}} \cdot \frac{G_m}{\xi_{m,u}\, \delta_{m,u}} = p_p^{\text{pilot}} \cdot \frac{G_p}{\xi_{p,u}\, \delta_{p,u}}, \tag{6.A.1}$$

where pilot signal transmit power p_m^{pilot} and p_p^{pilot} are used since the HO process is typically driven by pilot signals.

Plugging path-loss expressions into (6.A.1), we get

$$\frac{\mathcal{P}_m}{(d_{m,u})^{\alpha_m}} = \frac{\mathcal{P}_p}{(d_{p,u})^{\alpha_p}}, \tag{6.A.2}$$

where $\mathcal{P}_m = G_m\, p_m^{\text{pilot}} / (\varphi_m\, \xi_{m,u})$ and $\mathcal{P}_p = G_p\, p_p^{\text{pilot}} / (\varphi_p\, \xi_{p,u})$.

Taking the $(2/\alpha_p)$th root of both sides in (6.A.2), we obtain

$$\frac{\mathcal{P}_m^{2/\alpha_p}}{(d_{m,u})^{2\psi}} = \frac{\mathcal{P}_p^{2/\alpha_p}}{(d_{p,u})^2}, \tag{6.A.3}$$

where $\psi = \frac{\alpha_m}{\alpha_p}$. Since outdoor path-loss exponents typically range from 2 to 4, while indoor path-loss exponents can vary from 4 to 6 [33], it is likely that $\alpha_m \leq \alpha_p$ and thus $0 < \psi \leq 1$.

Assuming that UE U_u is located at the point (x, y), its distances to the BSs of macrocell C_m^M and picocell C_p^P are given, respectively, by

$$d_{m,u} = \sqrt{(x - x_m)^2 + (y - y_m)^2}, \tag{6.A.4}$$

$$d_{p,u} = \sqrt{(x - x_p)^2 + (y - y_p)^2}, \tag{6.A.5}$$

where (x_m, y_m) and (x_p, y_p) are the BS locations of macrocell C_m^M and picocell C_p^P, respectively.

Plugging (6.A.4) and (6.A.5) into (6.A.3) and performing some manipulations, we obtain

$$\mathcal{P}_p^{2/\alpha_p}(x^2 + x_m^2 - 2x_m x + y^2 + y_m^2 - 2y_m y)^\psi$$
$$- \mathcal{P}_m^{2/\alpha_p}(x^2 + x_p^2 - 2x_p x + y^2 + y_p^2 - 2y_p y) = 0. \tag{6.A.6}$$

Because $x^2 + x_m^2 - 2x_m x + y^2 + y_m^2 - 2y_m y$ is a polynomial of order 2 and since $0 < \psi \leq 1$, the order of (6.A.6) is at most 2. We thus approximate $(x^2 + x_m^2 - 2xx_m + y^2 + y_m^2 - 2yy_m)^\psi$ using the following second-order Taylor series evaluated at point $i, j)$

$$T(i,j) = f(i,j) + (x-i)f_x(i,j) + (y-j)f_y(i,j) + \frac{1}{2}(x-i)^2 f_{xx}(i,j)$$

$$+\frac{1}{2}(y-j)^2 f_{yy}(i,j) + (x-i)(y-j)f_{xy}(i,j), \qquad (6.A.7)$$

where

$$f(i,j) = (i^2 + j^2 - 2ix_m - 2jy_m + x_m^2 + y_m^2)^\psi \qquad (6.A.8)$$

$$f_x(i,j) = 2\psi[f(i,j)]^{\frac{\psi-1}{\psi}}(i - x_m) \qquad (6.A.9)$$

$$f_y(i,j) = 2\psi[f(i,j)]^{\frac{\psi-1}{\psi}}(j - y_m) \qquad (6.A.10)$$

$$f_{xx}(i,j) = 4\psi(\psi-1)[f(i,j)]^{\frac{\psi-2}{\psi}}(i - x_m)^2 + 2\psi[f(i,j)]^{\frac{\psi-1}{\psi}} \qquad (6.A.11)$$

$$f_{yy}(i,j) = 4\psi(\psi-1)[f(i,j)]^{\frac{\psi-2}{\psi}}(j - y_m)^2 + 2\psi[f(i,j)]^{\frac{\psi-1}{\psi}} \qquad (6.A.12)$$

$$f_{xy}(i,j) = 4\psi(\psi-1)[f(i,j)]^{\frac{\psi-2}{\psi}}(i - x_m)(j - y_m). \qquad (6.A.13)$$

Plugging (6.A.7) into (6.A.6) and carrying out some manipulations, we get

$$ax^2 + 2bxy + cy^2 + 2dx + 2fy + g = 0, \qquad (6.A.14)$$

where

$$a = \frac{1}{2}P_p^{2/\alpha_p} f_{xx}(i,j) - P_m^{2/\alpha_p}, \qquad (6.A.15)$$

$$b = \frac{P_p^{2/\alpha_p} f_{xy}(i,j)}{2}, \qquad (6.A.16)$$

$$c = \frac{1}{2}P_p^{2/\alpha_p} f_{yy}(i,j) - P_m^{2/\alpha_p}, \qquad (6.A.17)$$

$$d = \frac{P_p^{2/\alpha_p}[f_x(i,j) - if_{xx}(i,j) - jf_{xy}(i,j)] + 2P_m^{2/\alpha_p} x_p}{2}, \qquad (6.A.18)$$

$$f = \frac{P_p^{2/\alpha_p}[f_y(i,j) - jf_{yy}(i,j) - if_{xy}(i,j)] + 2P_m^{2/\alpha_p} y_p}{2}, \qquad (6.A.19)$$

$$g = P_p^{2/\alpha_p}[f(i,j) - if_x(i,j) - jf_y(i,j) + \frac{i^2}{2}f_{xx}(i,j),$$

$$+ \frac{j^2}{2}f_{yy}(i,j) + ijf_{xy}(i,j)] - P_m^{2/\alpha_p}(x_p^2 + y_p^2). \tag{6.A.20}$$

If $a \neq c$ and $b^2 < 4ac$, (6.A.14) is a non-degenerated ellipse, which may rotate depending on the term $2bxy$. The center and the two semi-axes s_1 and s_2 of the ellipse in (6.A.14) are given in (6.6) and (6.7), respectively [26].

APPENDIX 6.B: PROOF OF THEOREM 2

For a UE U_u located at point (x, y) on the EPB of picocell C_p^P, the path losses from the umbrella macrocell C_m^M and picocell C_p^P are equal, that is,

$$\frac{\mathcal{G}_m}{(d_{m,u})^{\alpha_m}} = \frac{\mathcal{G}_p}{(d_{p,u})^{\alpha_p}}, \tag{6.B.1}$$

where $\mathcal{G}_m = 1/(\varphi_m \xi_{m,u})$ and $\mathcal{G}_p = 1/(\varphi_p \xi_{p,u})$. Following similar steps as in Appendix 6.A, we find that the EPB can also be approximated by an ellipse, which is given by

$$a'x^2 + 2b'xy + c'y^2 + 2d'x + 2f'y + g' = 0, \tag{6.B.2}$$

where

$$a' = \frac{1}{2}\mathcal{G}_p^{2/\alpha_p}f_{xx}(i,j) - \mathcal{G}_m^{2/\alpha_p}, \tag{6.B.3}$$

$$b' = \frac{\mathcal{G}_p^{2/\alpha_p}f_{xy}(i,j)}{2}, \tag{6.B.4}$$

$$c' = \frac{1}{2}\mathcal{G}_p^{2/\alpha_p}f_{yy}(i,j) - \mathcal{G}_m^{2/\alpha_p}, \tag{6.B.5}$$

$$d' = \frac{\mathcal{G}_p^{2/\alpha_p}[f_x(i,j) - if_{xx}(i,j) - jf_{xy}(i,j)] + 2\mathcal{G}_m^{2/\alpha_p}x_p}{2}, \tag{6.B.6}$$

$$f' = \frac{\mathcal{G}_p^{2/\alpha_p}[f_y(i,j) - jf_{yy}(i,j) - if_{xy}(i,j)] + 2\mathcal{G}_m^{2/\alpha_p}y_p}{2}, \tag{6.B.7}$$

$$g' = \mathcal{G}_p^{2/\alpha_p}[f(i,j) - if_x(i,j) - jf_y(i,j) + \frac{i^2}{2}f_{xx}(i,j),$$

$$+ \frac{j^2}{2}f_{yy}(i,j) + ijf_{xy}(i,j)] - \mathcal{G}_m^{2/\alpha_p}(x_p^2 + y_p^2). \tag{6.B.9}$$

Based on (6.B.2), the center (x'_e, y'_e), semi-axes s'_1 and s'_2 and rotation angle Θ' of the EPB ellipse can be formulated in a similar way to (6.6), (6.7), and (6.8), respectively.

APPENDIX 6.C: PROOF OF THEOREM 3

Without loss of generality, we set $x_m = 0$, $y_m = 0$, $x_p = d_{m,p}$, $y_p = 0$, $\alpha_p = 2$ and $\psi = 1$. For $i = x_p$ and $j = 0$, plugging all these values into (6.A.8)–(6.A.13) yields

$$f(i,j) = d_{m,p}^2, f_x(i,j) = 2d_{m,p}, f_y(i,j) = 0,$$
$$f_{xx}(i,j) = 2, f_{yy}(i,j) = 2, f_{xy}(i,j) = 0. \tag{6.C.1}$$

Substituting (6.C.1) into (6.A.15)–(6.A.20), we have

$$a = P_p - P_m, b = 0, c = P_p - P_m = a,$$
$$d = P_m d_{m,p}, f = 0, g = -P_m d_{m,p}^2. \tag{6.C.2}$$

Based on (6.6) and (6.C.2), we calculate $d_{p,e}$ as follows

$$d_{p,e} \approx |\mathbf{x}_e - \mathbf{x}_p| = \left| \frac{cd - bf}{b^2 - ac} - x_p \right| = \left| -\frac{d}{a} - x_p \right|$$
$$= \left| \frac{P_m d_{m,p}}{P_p - P_m} + d_{m,p} \right| = \frac{d_{m,p}}{P_m/P_p - 1}, \tag{6.C.3}$$

which proves the linear relationship between $d_{p,e}$ and $d_{m,p}$. Note that for ψ different than 1, (6.C.3) would be a non-linear function of $d_{m,p}$. As $\frac{P_m}{P_p} \to \infty$, $d_{p,e} \to 0$, and $\mathbf{x}_e \to \mathbf{x}_p$, which is aligned with the observation in [34].

In a similar way, the distance between the EPB ellipse center and the PBS is calculated as

$$d'_{p,e} \approx |\mathbf{x}'_e - \mathbf{x}_p| = \left| \frac{G_m d_{m,p}}{G_p - G_m} + d_{m,p} \right| = \frac{d_{m,p}}{G_m/G_p - 1}, \tag{6.C.4}$$

which is also linearly dependent on $d_{m,p}$. Note that as $G_m \to G_p$, we have $d_{m,p} \to \infty$, and the EPB converges to a line which is equidistant to the PBS and the MBS.

Comparing (6.C.4) with (6.C.3), we can show that $d'_{p,e} > d_{p,e}$, because in a HetNet, we typically have $P_m \gg P_p$ and hence $P_m/P_p \gg G_m/G_p$.

Moreover, since we have shown in Sections 6.3.1 and 6.3.2 that the major axis of the EPB ellipse overlaps with that of the ESB ellipse, it is sufficient to show that the EPB always contains the ESB if the following condition is satisfied

$$|x_e' - x_e| + s_1 \leq s_1'. \tag{6.C.5}$$

Plugging in the values of x_e', x_e, s_1, s_1' (see also Appendix 6.D) into (6.C.5), we have

$$\left| \frac{\mathcal{G}_m d_{m,p}}{\mathcal{G}_m - \mathcal{G}_p} - \frac{\mathcal{P}_m d_{m,p}}{\mathcal{P}_m - \mathcal{P}_p} \right| + \sqrt{\frac{\mathcal{P}_p \mathcal{P}_m}{(\mathcal{P}_p - \mathcal{P}_m)^2}} d_{m,p} \leq \sqrt{\frac{\mathcal{G}_p \mathcal{G}_m}{(\mathcal{G}_p - \mathcal{G}_m)^2}} d_{m,p}. \tag{6.C.6}$$

Upon some manipulation, (6.C.6) simplifies to

$$\mathcal{G}_m \mathcal{P}_p + \mathcal{P}_m \mathcal{G}_p \geq 2\sqrt{\mathcal{G}_m \mathcal{G}_p \mathcal{P}_m \mathcal{P}_p}, \tag{6.C.7}$$

which is equivalent to

$$\left(\sqrt{\mathcal{G}_m \mathcal{P}_p} - \sqrt{\mathcal{P}_m \mathcal{G}_p} \right)^2 \geq 0. \tag{6.C.8}$$

Since (6.C.8) is always satisfied, (6.C.5) is also satisfied, and hence the EPB contains the ESB.

APPENDIX 6.D: DERIVATION OF REB FOR A DESIRED COVERAGE RADIUS

Similar to Appendix 6.C, let $x_m = 0$, $y_m = 0$, $x_p = d_{m,p}$, $y_p = 0$, $\psi = 1$, $i = x_p$, $j = 0$, and $\alpha_p = 2$, which yields (6.C.1) and (6.C.2). We also define the DL RSS enlarged by the REB $\Delta \text{ER}_p^{\text{HSB}}$ as

$$\mathcal{P}_p' = \Delta \text{ER}_p^{\text{HSB}} \mathcal{P}_p = \Delta \text{ER}_p^{\text{HSB}} G_p p_p^{\text{pilot}} / (\varphi_p \xi_{p,u}). \tag{6.D.1}$$

Then, based on (6.7) and (6.C.2), we have

$$r_{\text{h}} = s_1 = \sqrt{\frac{f^2 + d^2 - ag}{a^2}} = d_{m,p} \sqrt{\frac{\mathcal{P}_p' \mathcal{P}_m}{(\mathcal{P}_p' - \mathcal{P}_m)^2}}, \tag{6.D.2}$$

where $r_{\text{h}} = s_1$ is used to realize the expansion from the ESB to the HSB, under the assumption that the PBS is placed at the HSB center, which should typically be known

y operators, and based on the observation that the ESB center is close to the PBS
ocation.

Solving (6.D.2) for P'_p, we obtain

$$P'_p = \frac{(2r_h^2 + d_{m,p}^2)P_m \pm P_m d_{m,p}\sqrt{d_{m,p}^2 + 4r_h^2}}{2r_h^2}. \tag{6.D.3}$$

Note that the two values of P'_p in (6.D.3) correspond to two different circles of
adius r_h. The smaller value of P'_p results in the HSB circle around the PBS, while
he larger value of P'_p leads to a circle surrounding the MBS instead, because the
orresponding value of REB is so large that the picocell DL coverage overwhelms
hat of the macrocell. Substituting the smaller value of (6.D.3) into (6.D.1) and after
ome manipulation, we get the approximated expression of ΔER_p^{HSB} in (6.9).

REFERENCES

1. A. Lozano, D. Cox, and T. Bourk, "Uplink-downlink imbalance in TDMA personal communication systems," in *Proc. IEEE Int. Conf. Universal Pers. Commun. (ICUPC)*, vol. 1, Florence, Italy, Oct. 1998, pp. 151–155.

2. W. Lee and D. Lee, "The impact of front end LNA on cellular system," in *Proc. IEEE Veh. Technol. Conf. (VTC)*, vol. 5, Boston, MA, 2000, pp. 2180–2184.

3. Qualcomm, "Importance of serving cell selection in heterogeneous networks," 3GPP Standard Contribution (R1-100701), Valencia, Spain, Jan. 2010.

4. NTT DOCOMO, "Performance of eICIC with control channel coverage limitation (R1-103264)," 3GPP TSG RAN WG1 Meeting-61, Montreal, Canada, May 2010.

5. Kyocera, "Potential performance of range expansion in macro-pico deployment (R1-104355)," 3GPP TSG RAN WG1 Meeting-62, Madrid, Spain, Aug. 2010.

6. Motorola, "On range extension in open-access heterogeneous networks (R1-103181)," 3GPP TSG RAN WG1 Meeting-61, Montreal, Canada, May 2010.

7. Huawei, "Evaluation of Rel-8/9 techniques and range expansion for macro and outdoor hotzone (R1-103125)," 3GPP TSG RAN WG1 Meeting-61, Montreal, Canada, May 2010.

8. Alcatel-Lucent, "DL pico/macro Het Net performance: Cell selection (R1-102808)," 3GPP TSG RAN WG1 Meeting-61, Montreal, Canada, May 2010.

9. Ericsson, ST-Ericsson, "System performance evaluations on feICIC," 3GPP Standard Contribution (R1-113482),Zhuhai, China, Oct. 2011.

10. J. Wigard, T. Nielsen, P. Michaelsen, and P. Morgensen, "On a handover algorithm in a PCS1900/GSM/DCS1800 network," in *Proc. IEEE Veh. Technol. Conf. (VTC)*, vol. 3, Houston, TX, Jul. 1999, pp. 2510–2514.

11. A. Tolli, I. Barbancho, J. Gomez, and P. Hakalin, "Intra-system load balancing between adjacent GSM cells," in *Proc. IEEE Veh. Technol. Conf. (VTC)*, vol. 1, Orlando, FL, Apr. 2003, pp. 393–397.

12. H.-S. Jo, Y. J. Sang, P. Xia, and J. G. Andrews, "Outage probability for heterogeneous cellular networks with biased cell association," in *Proc. IEEE Global Telecommun. Conf. (GLOBECOM)*, Houston, TX, Dec. 2011.

13. R. Madan, J. Borran, A. Sampath, N. Bhushan, A. Khandekar, and T. Ji, "Cell association and interference coordination in heterogeneous LTE-A cellular networks," *IEEE J. Sel. Areas Commun. (JSAC) – Spec. Issue Coop. Commun. MIMO Cell. Netw.*, vol. 28, no. 9, pp. 1479–1489, Oct. 2010.

14. D. Ghosh and C. Lott, "Uplink-downlink imbalance in wireless cellular networks," in *Proc. IEEE Int. Conf. Commun. (ICC)*, Glasgow, Scotland, Jun. 2007, pp. 4275–4280.

15. K. Azarian, C. Lott, D. Ghosh, and R. A. Attar, "Imbalance issues in heterogeneous DO networks," in *Proc. IEEE Int. Workshop Femtocell Netw. (FEMnet)*, Miami, FL, Dec. 2010.

16. S. Landstrom, H. Murai, and A. Simonsson, "Deployment aspects of LTE pico nodes," in *Proc. IEEE Int. Workshop Heterogen. Netw. (HETnet)*, Kyoto, Japan, Jun. 2011.

17. K. Balachandran, J. H. Kang, K. Karakayali, and K. Rege, "Cell selection with downlink resource partitioning in heterogeneous networks," in *Proc. IEEE Int. Workshop Heterogen. Netw. (HETnet)*, Kyoto, Japan, Jun. 2011.

18. K. Okino, T. Nakayama, C. Yamazaki, H. Sato, and Y. Kusano, "Pico cell range expansion with interference mitigation toward LTE-Advanced heterogeneous networks," in *Proc. IEEE Int. Workshop Heterogen. Netw. (HETnet)*, Kyoto, Japan, Jun. 2011.

19. M. Vajapeyam, A. Damnjanovic, J. Montojo, T. Ji, Y. Wei, and D. Malladi, "Downlink FTP performance of heterogeneous networks for LTE-Advanced," in *Proc. IEEE Int. Workshop Heterogen. Netw. (HETnet)*, Kyoto, Japan, Jun. 2011.

20. "Radio Resource Control," 3GPP Protocol Specification, Tech. Rep., TS 36.331 v.10.4.0, Mar. 2011.

21. "Physical layer procedures," 3GPP Technical Specification, Evolved Universal Terrestrial Radio Access (E-UTRA), Tech. Rep. TS 36.213 v.12.1.0, Dec. 2014.

22. A. Simonsson and A. Furuskar, "Uplink power control in LTE - overview and performance, subtitle: Principles and benefits of utilizing rather than compensating for SINR variations," in *Proc. IEEE Veh. Technol. Conf. (VTC)*, Sep. 2008, pp. 1–5.

23. C. U. Castellanos, D. L. Villa, C. Rosa, K. I. Pedersen, F. D. Calabrese, P. H. Michaelsen, and J. Michel, "Performance of uplink fractional power control in UTRAN LTE," in *Proc. IEEE Veh. Technol. Conf. (VTC)*, Singapore, May 2008, pp. 2517–2521.

24. "Further advancements for E-UTRA physical layer aspects," 3GPP Technical Specification, Evolved Universal Terrestrial Radio Access (E-UTRA) , Tech. Rep., TR 36.814 v.9.0.0, Mar. 2010.

25. D. López-Pérez, X. Chu, and I. Güvenç, "On the expanded region of picocells in heterogeneous networks," *IEEE J. Sel. Topics Signal Process. (J-STSP)*, vol. 6, no. 3, pp. 281–294, 2012.

26. J. D. Gersting, *Technical Calculus with Analytic Geometry*. Dover Publications, Oct. 2010.

27. Intel Corporation, "Performance impact of open loop power control parameter settings in HetNet," 3GPP Standard Contribution (R1-113938), San Francisco, CA, Nov. 2011.

8. Intel Corporation, "Discussion of uplink open loop power control algorithm in heterogeneous networks," 3GPP Standard Contribution (R1-113666), San Francisco, CA, Nov. 2011.

9. NTT DOCOMO, "Uplink transmission power control scheme in HetNet scenario," 3GPP Standard Contribution (R1-114078), San Francisco, CA, Nov. 2011.

0. C. M. Bishop and N. M. Nasrabadi, *Pattern Recognition and Machine Learning*. Springer-Verlag, 2006.

1. S. J. Fortune, D. M. Gay, B. W. Kernighan, O. Landron, R. A. Valenzuela, and M. H. Wright, "Wise design of indoor wireless systems: practical computation and optimization," *IEEE Comput. Sci. Eng.*, vol. 2, no. 1, pp. 58–68, 1995.

2. "Requirements for support of radio resource management," 3GPP Technical Specification, 'Evolved Universal Terrestrial Radio Access (E-UTRA), Tech. Rep., TS 36.133 v11.2.0, Nov. 2012.

3. Z. Rong and T. S. Rappaport, *Wireless Communications: Principles & Practice*, 2nd ed. Prentice Hall, Jan. 2002.

4. İsmail Güvenç, M.-R. Jeong, F. Watanabe, and H. Inamura, "A hybrid frequency assignment for femtocells and coverage area analysis for co-channel operation," *IEEE Commun. Lett.*, vol. 12, pp. 880–882, Dec. 2008.

PART III

INTERFERENCE MANAGEMENT

7

FREQUENCY-DOMAIN INTER-CELL INTERFERENCE COORDINATION

7.1 INTRODUCTION

Conventional cellular networks for mobile wireless communications are character-ized by homogeneous layouts. Adjacent macrocells cover comparable areas and expe-ience a similar service demand. Careful frequency planning helps in maximizing the reuse of licensed frequency spectrum and the overall network performance. Yet, costly drive testing and configuration parameter fine-tuning are still unavoidable in ensuring a predefined level of service quality and availability.

In contrast, small cells deployments in heterogeneous networks, that is the (pos-sibly unplanned) addition or expansion of a secondary small cell tier complement-ing the primary macrocell tier, necessarily creates new unaccounted interference. To avoid the consequent degradation of network performance, configuration parameters should be optimized, ideally in both network tiers, to reflect the change in networking conditions after the small cell deployment.

Each of the four dimensions of the wireless resource space, namely frequency, time, power, and coding, represents a degree of freedom can be used for orthogonal-izing small cell and macrocell transmissions. The coding dimension is available in

Small Cell Networks: Deployment, Management, and Optimization, First Edition. Holger Claussen, David López-Pérez, Lester Ho, Rouzbeh Razavi, and Stepan Kucera.
© 2017 by The Institute of Electrical and Electronic Engineers, Inc. Published 2017 by John Wiley & Sons, Inc.

networks employing code division multiple access (CDMA). Our focus in the next three Chapters 7, 8, and 10 is on the Third-Generation Partnership Project (3GPP) universal mobile telecommunication system (UMTS) and long-term evolution (LTE) networks using orthogonal frequency division multiple access (OFDMA) so attention will be paid only to the frequency, time, and power dimensions of interference coordination.

This chapter explores the possibility of avoiding inter-tier interference by using frequency-domain multiplexing. Herein, potentially interfering simultaneous transmission are accommodated in different frequency bands. By assigning complementary bands to small cells, the macrocell configuration can be left unmodified even after small cells are deployed and activated, that is, the operator investment into macrocell reliability and performance is protected. Yet, the operator must bear substantial costs related to licensing scarce and fragmented electromagnetic spectrum from national regulators.

The following Chapter 8 summarizes the time-domain multiplexing techniques for sharing a given band between the small cell and macrocell tiers, that is scheduling transmissions in mutually disjoint time intervals. Although spectrally more efficient, the periods of alternating cell activity and inactivity fragment ongoing transmissions and may affect the efficiency of radio resource allocation.

A combination of both frequency-domain and time-domain multiplexing is described in Chapter 10 in the context of interference coordination tailored to LTE control channels.

In general, different approaches can be taken to implement frequency-domain multiplexing of network transmissions. The technique of fractional frequency reuse takes the division of a single communication band (carrier) into logical sub-bands as a basis for frequency-domain multiplexing [1, 2]. A generalization of this approach, the technique of shared/dedicated carriers (see Chapter 4), implements the same multiplexing functionality assuming the availability of multiple carriers and their shared/dedicated assignment to individual network cells.

Although similar conceptually, both techniques differ substantially in performance and complexity due to their technical implications. The former single-carrier technique is straightforward to implement by software-level updates of scheduling policies, but it is less efficient performance-wise than the latter multi-carrier technique. Yet, the performance gains of the multi-carrier technique may not be sufficient to justify its deployment in view of the fact that inter-band scheduling and load balancing implies relatively frequent and potentially unreliable inter-band handovers as well as the related measurement and signaling overhead [3, 4].

Noticing the potential throughput improvements of multi-carrier transmissions for both interference coordination, the 3GPP has recently defined within the LTE standard native support for multi-carrier transmissions in general, and efficient frequency-domain multiplexing in heterogeneous networks in particular [5,6]. More specifically, tailored signaling has been introduced that significantly simplifies the handover-based approaches without sacrificing the achievable performance.

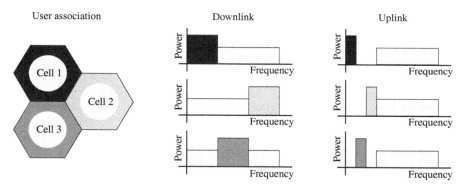

Figure 7.1. Concept of fractional frequency reuse. Soft reuse is employed for the downlink while hard reuse is employed for the uplink.

The description of the technical nature of the so-called carrier aggregation framework [7,8] in LTE and its applications to frequency-domain interference coordination in heterogeneous networks are the main topic of this chapter.

Fractional frequency reuse was the first technique adopted in the early LTE Releases as part of the framework for so-called inter-cell interference coordination (ICIC) [9]. The concept is based on (a) dividing a *single* carrier into multiple logical fragments, and (b) creating different reuse patterns of these carrier fragments to reflect varying interference conditions among network user equipments (UEs).

As illustrated in Fig. 7.1, typically two reuse patterns are configured to accommodate the needs of cell-edge and cell-center UEs [10, 11]. The first reuse pattern is achieved by dividing the available LTE subcarriers into several disjoint subsets, and assigning each subset to one evolved NodeB (eNodeB) such that strongly interfering (neighboring) cells do not share the same subcarrier subset. Transmissions associated with cell-edge UEs are then accommodated over these reserved cell-specific subcarriers to minimize the inter-cell interference. However, the spectral reuse is reduced by a factor equal to the number of distinct subcarrier subsets.

The second reuse pattern is created by letting each eNodeB use another (generally unrelated) spectrum fragment for communications with UEs experiencing low interference such as the cell-center UEs. For downlink communications, the entire system bandwidth is typically reused by all cells. However, hard orthogonality for allocated spectrum fragments may be more beneficial for uplink communications (see also Fig. 7.1) to reflect the geometrical asymmetry between macrocells and small cells, as well as the fact that all UEs have the same power limit, that is, that macrocell UEs can cause very strong interference to small cell UEs as their transmission power must be sufficient to overcome a distance (path loss) that is typically much longer than the small cell diameter.

Spectrum partitioning can be accompanied by the configuration of non-uniform power spectral density. Yet, power control techniques including power boosting and beamforming are usually not very effective as they can create only soft orthogonality, while requiring good coordination in practice [12].

In terms of 3GPP support, coordinated fractional frequency reuse and adaptive scheduling of frequency-partitioned data channels (namely, the physical downlink shared channel (PDSCH)) have been enabled in the LTE Releases 8 and 9 through the addition of dedicated signaling, namely the relative narrowband transmit power (RNTP) indicator, the interference overload indicator (IOI), and the high-interference indicator (HII) in the *LOAD INFORMATION* message to the X2 interface [13].

The RNTP indicator is used to inform the neighboring eNodeBs that the downlink transmission power of indicated physical resource blocks (PRBs) will be greater than a given threshold in the near future. The HII is used in a similar manner for uplink transmissions to effectively indicate PRBs associated with cell-edge UEs. In addition to the proactive messages, the IOI messages are reactively triggered when high uplink interference is detected to help identify potential sources.

According to these messages, the neighbor eNodeBs can proactively anticipate interference and update their scheduling policies rather than react to the UE channel quality indicator (CQI) reports reflecting past transmission collisions. These indicators can also be readily used to identify the bands used in a frequency partitioning scheme.

Unfortunately, the indicator-based approach cannot be applied to interference coordination in LTE control channels because the UE-specific control channels (namely, the physical downlink control channel (PDCCH)) are pseudo-randomly spread across the entire system bandwidth [14]. In addition, cell-specific signals such as the common PDCCHs and the cell-specific reference symbol (CRS) cannot be transmitted at reduced powers, even partially, otherwise their detectability throughout the entire cell is jeopardized.

Moreover, the standardized ICIC proposals do not vary reference signal power based on frequency bands, giving rise to demodulation issues of quadrature amplitude modulation (QAM). It may also be difficult to find a sufficient number of reuse patterns in a dense heterogeneous network (HetNet) deployment as efficient search algorithms are typically characterized by a non-negligible communication overhead [2].

In conclusion, it can be stated that the applicability of the fractional frequency reuse is only limited to homogeneous networks characterized by a single macrocell tier.

Whenever *multiple* carriers are available to the operator, inter-tier interference can be mitigated by adaptively dividing active network UEs among the available carriers. As described in more detail in Chapter 4, these can be of dedicated or shared nature. Within the shared carriers, the macrocell uses either low or even zero transmit power to enable small cell transmissions. UEs are assigned to the desired carriers by means of (forced) inter-frequency handovers.

The shared or dedicated carrier approach is compliant with the capabilities of any legacy UE but must be supported by (a) adaptive coverage optimization or load balancing and (b) timely handovers to ensure an efficient and fair carrier occupancy and usage, as well as to prevent radio link failures.[1]

While it is relatively straightforward to execute adaptive load balancing, mobility control is typically difficult considering the individual UE mobility patterns, the relatively long duration of inter-frequency handovers, and the relatively small cell diameter. The necessity of using the so-called *Compressed mode* in 3G systems [15] and *Measurement GAPs* in 4G systems [3] for multi-frequency measurements, a prerequisite of inter-frequency handovers, further reduces the handover flexibility in time as well as implies a capacity loss due to the inter-frequency measurements.

To resolve the signaling-related issues of multi-carrier transmissions without sacrificing their advantages, the carrier aggregation (CAG) framework has been introduced in LTE Release 10 (so-called long-term evolution advanced (LTE-A)) [7]. Conceived originally as a scalable means for boosting network capacity, the CAG feature of LTE allows natively aggregating multiple carrier components (CaCos) of smaller bandwidth into one logical block, characterized by the combined effective bandwidth. Up to five carriers can currently be aggregated, while the support for a higher CaCo number is being discussed [16].

The aggregated CaCos can be discontiguous and even located in different frequency bands. Thanks to this measure, maximum flexibility in utilizing the available radio spectrum is provided to operators as contiguous transmission bandwidth is rare and its size may vary depending on the geographical area and national regulations [17].

From the interference mitigation point of view, the most notable feature of CAG is that control channels do not need to be broadcasted over all CaCos. Instead, one selected CaCo can carry control channels containing information such as scheduling grants and power control commands applicable to all the aggregated CaCos. Hence, adaptive cross-scheduling can be natively used to flexibly orthogonalize both control and data transmissions of potentially interfering CaCos with a subframe resolution.

The benefits of CAG go beyond providing better throughput due to frequency-domain interference coordination. Operators can recombine fragmented spectrum into one usable resource and complement it even by unlicensed (wireless fidelity (Wi-Fi)) bands as discussed subsequently. The possibility of dynamic traffic distribution among multiple carriers brings the advantage of statistical multiplexing, resulting in gains associated with congestion reduction stemming from a tight adaptation of used bandwidth to rate variations of application data [17]. In the frequency division duplexing (FDD) transmission mode, any misbalance between the downlink and uplink data traffic can be also addressed directly by allocating different numbers of CaCos for downlink and uplink.

---

[1] Control channel reliability and capacity is degraded by co-channel interference in the shared band, especially in the case of non-zero macrocell transmit power. This problem is explicitly addressed in Chapter 10.

The remainder of this chapter is organized as follows. The concept of CAG is introduced on a high level in Section 10.2, describing targeted deployment scenarios and operational modes. A more detailed view on the CAG protocol architecture, standardized signaling, and implementation aspects is offered in Section 7.2. Application of CAG to frequency-domain interference coordination in heterogeneous networks follows in Section 7.4, together with performance evaluation of discussed strategies. Conclusions are drawn in Section 7.5.

7.2 CARRIER AGGREGATION IN LTE-ADVANCED

The concept of CAG has been first introduced for the high-speed packet access (HSPA) standards for wireless networking under the name of dual-carrier high-speed packet access (DC-HSPA). The HSPA technology allows aggregation of up to four downlink and at most two uplink carrier components of equal (5 MHz) bandwidth [18]. Moreover, both carriers must be located within the same band in a contiguous fashion.

Subject to spectrum availability and the UE capability, LTE standards currently envisage the aggregation of up to five CaCos without any restrictions on the bandwidth or band of individual CaCo [8]. The aggregation of a larger CaCo number is under discussion [16].

In order to facilitate commercial CAG rollout and maximize the reuse of legacy infrastructure and design, the LTE standard requires that individual CaCos are backward compatible with the LTE Release 8/9 requirements in terms of both frequency band location and channel structure. CAG allows the aggregation of up to five component carriers with bandwidths of 1.4, 3, 5, 10, 15, and 20 MHz, resulting in an allocation of up to 100 MHz of bandwidth (5×20 MHz). The transmission of all downlink physical channels and signals are supported as envisaged by the LTE Releases 8 and 9 [19].

The carrier aggregation functionality can be used in the classical FDD and time division duplexing (TDD) transmission modes, but FDD and TDD carriers can be jointly combined as well [20]. The combining of operation modes allows the operator to reuse spectrum more efficiently and flexibly adapt to the variations of uplink and downlink traffic requirements. In this context, CAG allows the allocation of downlink bands independently from how the uplink bands are configured [21, 22]. The only restriction is that the number of uplink carrier components cannot exceed the number of downlink carrier components.

7.2.1 Band Allocation Types

Depending on the spacing of aggregated bands, both contiguous and non-contiguous CaCo allocation are supported as illustrated by Fig. 7.2.

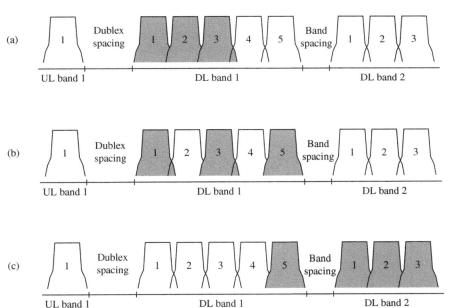

Figure 7.2. Carrier aggregation arrangements for (a) intra-band contiguous allocation, (b) intra-band non-contiguous allocation, and (c) inter-band non-contiguous allocation.

Under the contiguous CAG arrangement, contiguous CaCos within the same frequency band as defined by LTE Releases 8 and 9 are aggregated. The document [23] lists the supported combinations of LTE bands. This option primarily targets new carrier frequencies with sufficient bandwidth (e.g., at 3.5 GHz) since usable bandwidths larger than 20 MHz are currently rare in the radio frequency plans of most countries, making its adoption in near future less likely.

Advantageously, the implementation complexity of the transceiver chain for contiguous intra-band CAG can be reduced by employing a single inverse discrete Fourier transform (IDFT) processor [24]. In general, large guard bands between adjacent carriers of the same eNodeBs are not necessary which makes the use of the available spectrum more efficient.

The actual band spacing is standardized based on the notion of the widest configured CaCo [23]. More specifically, the spacing between the CaCo center frequencies must be a multiple of 300 kHz (the least common multiple of the 15 kHz subcarrier spacing and the 100 kHz frequency raster of the LTE Releases 8 and 9) to ensure backward compatibility. Consequently, the guard-band effect can also be achieved by any unused subcarriers in the frequency raster. For example, if two bands of 20 MHz are aggregated, then a "guard band" of 100 kHz is created (2×10 MHz half-bands — 67×0.3 MHz raster step = 0.1 MHZ band gap).

Under the non-contiguous CAG arrangement, CaCos belonging to the same frequency band (intra-band allocation) or different frequency bands (inter-band allocation) are aggregated. The document [23] lists the supported combinations of LTE bands. The main use case for intra-band allocation is the efficient reuse of fragmented frequency bands as well as band sharing by multiple operators. The ideas of inter-band allocation is to enhance the capacity of legacy CaCos, for example, in hotspot areas, by adding new high-bandwidth CaCos on higher carrier frequencies.

Each CaCo is treated as an independent entity: it must be surrounded by guard bands and its power density profile must meet the existing LTE requirements on out-of-band and spurious emissions as well as adjacent channel leakage. Inter-band aggregation logically adds to the complexity of the UE radio frequency front end by requiring multiple independent transceivers.

7.2.2 Deployment Scenarios

The concept of inter-band CAG permits the use of frequency bands that have potentially very different radio propagation characteristics such as path loss and Doppler shift experienced by mobile UEs.

Acknowledging that CAG is possible only where the coverage of the CaCos of the same eNodeB overlap, the 3GPP expects four typical deployment scenarios of the CAG feature [7] as visualized by Fig. 7.3. For clarity purposes, the figure illustrations are limited to only two CaCos.

The first and possibly most likely deployment scenario (a) is characterized by co-located eNodeB antennas having the same beam orientation for all CaCos. Any type of intra-band allocation, or inter-band allocation of bands with little frequency separation would ensure similar radio propagation characteristics in all bands, that is, the same coverage for each CaCo.

The second scenario (b) differs only in that it assumes the aggregated spectrum to be dispersed across different bands. The larger the inter-band frequency separation, the smaller the coverage of higher-frequency CaCos compared to the lower frequency CaCos. Consequently, lower-frequency bands would serve for coverage provisioning and mobility management, while the higher frequency bands would be added for throughput improvement. In inter-band scenarios, the scaling of the coverage footprint can be achieved by adjusting the CaCo power output.

The third scenario (c) leverages the observation from Chapter 4 that service continuity can be efficiently ensured by combining CaCos with partially disjoint coverage areas. Herein, the CaCos are transmitted by eNodeB antennas, whose antenna radiation patterns are either partitioned (e.g., sectorization schemes based on three or six sectors), or intentionally offset for selected CaCos to improve the throughput at sector boundaries (see Chapter 9). Clearly, mobility must be based on the CaCos with better coverage, implying the preference of CaCos having lower frequencies and higher powers.

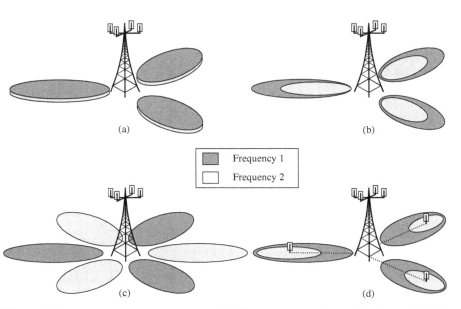

Figure 7.3. Carrier aggregation use cases for (a) *Scenario 1*—co-located and overlaid cells with nearly identical coverage, (b) *Scenario 2*—co-located and overlaid cells with different coverage, (c) *Scenario 3*—co-located cells with different spatial antenna orientation, and (d) *Scenario 4*—dislocated cells (e.g., remote radio heads).

The fourth scenario (d) assumes that one CaCo is used to provide wide-range (macrocell) coverage and mobility management, whereas remote radio heads (RRHs) extend the capacity in local hotspot areas by using another CaCo.

7.2.3 Operation Modes

Simple voice services generate symmetric downlink and uplink traffic having a ratio of downlink and uplink traffic share of approximately 1:1. However, as reported in [25], broadband data services may be characterized by skewed downlink-to-uplink ratios of up to 10:1. A combined stream of voice and data as seen from the backhaul perspective yields typically a traffic share ratio of 3:1.

To reflect the downlink dominance in mobile data networks, the LTE Release 11 enables the aggregation of so-called supplemental downlink bands [26] that are not associated with complementary uplink bands (see Fig. 7.4a). This additional downlink bandwidth enables faster downloads for mobile broadband and generally increases the number of servable UEs. The FDD Band 29 at 700 MHz has been designated as a supplemental downlink band [26].

Two other LTE features help promoting a better reuse of licensed spectrum in general and addressing traffic asymmetry in particular.

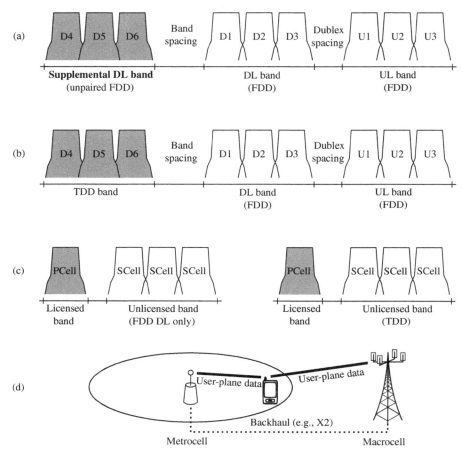

Figure 7.4. Recent advances in LTE carrier aggregation: (a) Aggregation of supplemental downlink bands, (b) joint aggregation of FDD-TDD carrier components, (c) carrier aggregation of secondary cells (SCell) with licensed assisted access from a primary cell (PCells), and (d) inter-site carrier aggregation (dual connectivity).

The first feature is the possibility of jointly aggregating FDD and TDD CaCos (see Fig. 7.4b). This so-called mixed mode of operation has been introduced in LTE Release 12 [27], whereas in the first CAG-capable LTE Releases 10 and 11, FDD only bands or TDD only bands could be aggregated to enhance the UE experience under same mode of operation.

From the CaCo management point of view, FDD and TDD CaCos may differ in their coverage. The coverage of TDD CaCos is limited not only by the radio propagation loss like in the FDD case, but also by the length of the guard interval specified for switching between downlink and uplink transmissions.

The second feature consists of the so-called licensed assisted access introduced first in LTE Release 12 [28–30]. As illustrated in Fig. 7.4c, the dedicated spectrum licensed by operators is further leveraged by the possibility of accessing the unlicensed spectrum, for example, around 3.5 GHz and 5 GHz [31].

Given the transmit power restrictions imposed by regulators on unlicensed bands, licensed assisted access will be primarily used for small cell deployments. The most likely configuration assumes that a primary CaCo in the licensed spectrum is complemented by secondary CaCos in the unlicensed spectrum, whereby all control signaling is carried out over the primary licensed CaCo. Secondary unlicensed CaCos can be used as the supplemental downlink or for both downlink and uplink transmissions.

The support of up to 32 bands is currently being discussed [16]. To this end, standardization work items have been created for LTE Release 13 [16] to define the management of scaling CAG, mobility, QoS provisioning, and security. Means for establishing the coexistence of unlicensed LTE and Wi-Fi must also be established and followed by a study of how both technologies can complement each other.

The above discussion so far assumed that CAG is carried out by a single eNodeB. The LTE Release 12 introduces a new operation mode referred to as dual connectivity or inter-site CAG (see Fig. 7.4d) that lets UEs consume radio resources provided by at least two different network points connected with a non-ideal backhaul, typically a macrocell and a small cell [32–34].

The idea of dual connectivity operation is to combine the benefits of macrocell coverage and small cell capacity. The radio bearers of a dual-connected UE are split between the Master eNodeB (MeNodeB) [35], typically a macrocell, and the secondary eNodeB (SeNodeB), typically a small cell. A UE is configured with one MeNodeB and at least one SeNodeB. Importantly, the SeNodeB owns its radio resources, and is responsible for their allocation.

Clearly, a certain level of direct coordination between the MeNodeB and the SeNodeB(s) is needed to which end the LTE Xn interface signaling will be used (currently Xn = X2). The same transport layer protocol as for S1/X2 could be also assumed for Xn, that is, stream control transmission protocol (SCTP) over Internet protocol (IP) for the control plane and GPRS tunnel protocol—user plane (GTP-U) over user datagram protocol (UDP)/IP for the user plane [13]. Dependent on the actual connection configuration, signaling overhead toward the core network can potentially be saved by keeping the mobility anchor in the macrocell, while aggregating small cells provides extra user plane capacity, increasing the throughput.

Multiple options have been evaluated in order to specify which is the best way of splitting the control and data plane between the MeNodeB and the SeNodeB(s).

In the control plane, a lower complexity centralized solution has been adopted as a baseline solution [35] where the MeNodeB generates the final radio resource control (RRC) messages to be sent toward the UE after the coordination of radio resource management (RRM) functions between MeNodeB and SeNodeB. Hence,

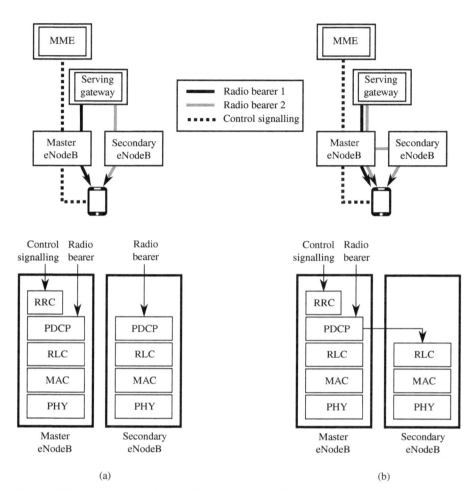

Figure 7.5. Dual connectivity architectures where (a) radio bearers terminate individually in MeNodeB and SeNodeB, (b) radio bearers terminate in MeNodeB only and data are forwarded over an RLC bearer to SeNodeB.

the UE RRC entity communicates only with the MeNodeB RRC entity. This fact is illustrated in Fig. 7.5 by a single MeNodeB instance of the RRC layer.

More specifically, the MeNodeB provides the input parameters such as UE capabilities and radio resource configuration to the SeNodeB. The SeNodeB decides the parameters relevant for it (e.g., physical uplink control channel (PUCCH) configuration), and signals these to the MeNodeB. Using the input from the SeNodeB, the MeNodeB generates the final RRC message, and signals this message to the UE. The signaling and processing overhead over the Xn is considered to be small.

In the user plane, two architectural options have been selected for further conideration [35].

The so-called option 1A (see Fig. 7.5a) terminates one evolved packet system EPS) bearer for each eNodeB, that is, there is no bearer split between the two nodes. The bearer transmissions may happen simultaneously from the MeNodeB and the SeNodeB. The benefits of this alternative are as follows:

- MeNodeB does not need to route or process packets for the EPS bearer transmitted by the SeNodeB.
- Backhaul requirements are low.
- Required modifications of the packet data convergence protocol (PDCP)/radio link control (RLC) and GTP-U/UDP/IP layers are negligible.

The disadvantages of this alternative include:

- SeNodeB mobility events are visible to the core network.
- Offloading to SeNodeB must be performed by the mobility management entity (MME), that is, cannot be very dynamic.
- Security is impacted due to ciphering being required in both MeNodeB and SeNodeB.
- Sharing of radio resources across MeNodeB and SeNodeB for the same bearer is not possible.
- SeNodeB bearers suffer handover-like interruption during SeNodeB change.

The other option referred to as 3C (see Fig. 7.5b) terminates the EPS bearers only in the MeNodeB. The PDCP layer always resides in the MeNodeB. Data are forwarded between the MeNodeB and the SeNodeB over a separate and independent RLC bearer. The benefits of this alternative are as follows:

- SeNodeB mobility is hidden to the core network.
- Security is not impacted as ciphering is required in MeNodeB only.
- No data forwarding between SeNodeB is required at SeNodeB change.
- RLC processing of SeNodeB traffic is offloaded from MeNodeB to SeNodeB.
- Impact to RLC is negligible.
- MeNodeB and SeNodeB can share radio resources for the same bearer.
- Requirements for SeNodeB mobility management are relaxed.

The corresponding disadvantages include:

- Routing, processing, and flow control of all data need to be done at MeNodeB.
- More complex PDCP layer.

7.3 PROTOCOL ARCHITECTURE AND IMPLEMENTATION ASPECTS

7.3.1 Support Evolution

The migration to fully CAG-capable networks is planned to occur in several stages, differentiated by UE capabilities [23], and reflecting progressive commercial penetration as well as technological maturity.

The CAG framework started in LTE Release 10 with a basic support of CAG features, generally enabling aggregation of up to five carriers of the same frame structure [36]. LTE Release 11 extended the supported CAG combinations and defined CAG with multiple uplink timing advances as well as inter-band TDD CAG with different uplink-downlink configurations [37]. LTE Release 12 added the support for the aggregation of carriers with different frame structures through FDD-TDD CAG [27]. The support of 32 CaCos is being discussed for LTE Release 13 [16].

More specifically, in terms of band support, LTE Release 10 defines the support for intra-band contiguous CAG (FDD Band 1 at 2 GHz and TDD Band 40 at 2.3 GHz (see [23] for detailed band specifications), and inter-band CAG (between FDD Band 1 at 2 GHz and FDD Band 5 at 0.9 GHz).

The LTE Release 11 [37] offers more CAG configurations including non-contiguous intra-band combinations. A novel feature is the addition of downlink-only inter-band carrier aggregation that is not associated with an uplink complementary band (supplemental downlink). Additional downlink bandwidth enables faster downloads for mobile broadband and generally increases the number of servable small cells. For example, FDD Band 29 at 0.7 GHz has been designated as a supplemental downlink band.

Within the LTE Release 12 framework [27], the aggregation of two uplink CaCos and three downlink CaCos is possible as well as CAG configurations for uplink inter-band and intra-band non-contiguous CAG. Similar to the Release 11, downlink-only FDD Band 32 at 1.5 GHz has been added. An important feature of the LTE Release 12 is dual connectivity.

The support of 32 CaCos is being discussed for LTE Release 13 [16], targeting primarily the 5 GHz spectrum under the licensed assisted access in combination with Wave 2 deployment of Institute of Electrical and Electronics Engineers (IEEE) 802.11ac (Wi-Fi) networks.

LTE releases differ not only in band combination support, but also in the way uplink transmissions are synchronized—a feature that may have a secondary impact on band selection.

The LTE Release 10 assumes time-synchronized uplink transmissions with co-located receiver antennas, that is, the propagation delays for all CaCos are equal. Accordingly, a single timing advance is issued based on the measurements of the primary CaCo. The initial timing correction for an uplink transmission over the random access channel is determined based on the downlink reference timing.

To enable more advanced CAG configurations, the LTE Release 11 defines separate timing and synchronization management for uplink transmissions [38,39]. The motivation is to reflect possible differences in CaCo propagation delays that occur when frequency-selective repeaters or relays are used to boost the poor radio propagation of higher-frequency CaCos. A propagation delay difference can occur when using non-co-located antennas, for example, as part of a coordinated multi-point (CoMP) scheme.

It is also noteworthy that the LTE Release 10 restricts the TDD transmission mode by requiring all CaCos to apply the same downlink and uplink transmission pattern. This restriction has been removed in the LTE Release 11. Herein, the pattern configuration must be identical across component carriers in the same band, but may differ across component carriers in different bands. Two additional configurations have been added, one for the case of normal cyclic prefix and one for the case of extended cyclic prefix. Moreover, the LTE Release 12 supports joint carrier aggregation of FDD and TDD frequency bands under the mixed mode of operation.

7.3.2 Configuration Options

Both FDD and TDD CaCos can be aggregated as well as their combinations [27].

In the TDD transmission mode, the resource allocation relationship between the uplink and downlink is simplified because the adjustment of the ratio of the uplink and downlink time slots can additionally be used to reflect any imbalance between the uplink and downlink traffic.

In the FDD transmission mode, LTE-A UE can be configured with UE-specific configurations of uplink and downlink CaCos [21], which greatly simplifies the implementation of load balancing, interference coordination, or power control. The number of uplink CaCos cannot exceed the number of the downlink CaCos. An unequal allocation of downlink and uplink CaCos is referred to as an asymmetric configuration.

In general, any CaCo is treated by higher layers as a serving cell and is also referred to as such. When an LTE-A UE (re)-establishes an RRC connection, the serving eNodeB always configures exactly one so-called primary cell (PCell) [40] that is associated with one primary carrier component (PCC) for downlink communications and one PCC for uplink communications, identically to a Release-8/9 serving cell. In this manner, compliance with requirements of legacy Release-8/9 UEs is ensured. Accordingly, a PCell also provides nonaccess stratum (NAS) mobility information and maintains the connection to the EPS. A PCell can be changed only by performing a full handover. Both TDD and FDD cells can serve as primary cells.

After establishing the PCell connection, the serving eNodeB must configure the Release 10 UE with other additional cells to operate in the CAG mode. These cells are referred to as secondary cells (SCells), and can be associated with up to four secondary CaCos [40]. Up to 32 SCells may be supported from LTE Release 13 [16]. The SCell set is configured using the *RRC Connection Reconfiguration* procedure (also

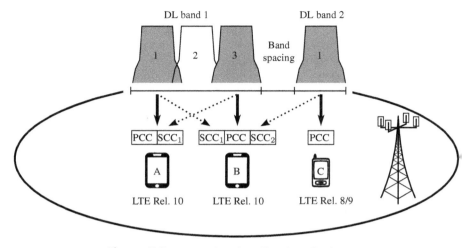

Figure 7.6. Principles of PCell and SCell selection.

used as part of the LTE handover procedure), depending on the UE requirements and capabilities as well as channel conditions. A cell index is used to reference individual SCells during all subsequent signaling.

PCells and SCells as well as the total number of configured uplink/downlink CaCos may be different for different UEs. For example, a cell at an eNodeB can be configured as the PCell for one UE, and as an SCell for another UE (see Fig. 7.6).

In terms of synchronization, if the PCell and all the SCell(s) are associated by co-located antennas of the same eNodeB, only one (grouped) uplink timing advance is issued for all the configured uplink CaCos. Otherwise, multiple timing advances are necessary, but as mentioned above, this functionality is supported only from LTE Release 11. Once the UE is synchronized with the PCell, synchronization with the dislocated SCells is obtained by means of random access channel (RACH) request following the SCell activation.

To avoid ambiguity between random access preambles sent by different UEs on different uplink CaCos, a dedicated system information block (SIB) is also used to indicate on a cell level the linkage between individual downlink and uplink CaCos.[2] Explicit uplink–downlink linkage is necessary to indicate which downlink CaCo should be expected to convey the random access response, as well as for relating downlink path-loss measurements with uplink power control.

- A PCell cannot be deactivated. PCC allocation can, however, be modified by the network during handover procedures.
- An SCell never has a PUCCH.

---

[2] In LTE Releases 8 and 9, the downlink carrier is explicitly associated with a given uplink carrier. Their linkage is described in the SIB.

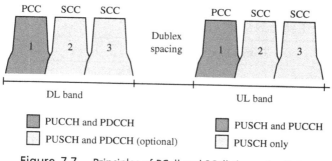

Figure 7.7. Principles of PCell and SCell channel activity.

- Radio link monitoring and random access are based on the PCell CaCo. Unlike for PCells, an RRC reestablishment is not carried out when an SCell experiences a radio link failure (RLF).
- PCells provide the Release 8/9-compliant non-access stratum (NAS) mobility information (e.g., the tracking area identifier).
- Semi-persistent scheduling is limited to PCells only as it is typically used for low-throughput voice over IP applications that do not require CAG.
- UE PUCCH transmissions can be accommodated only in the PCell CaCo.

Therefore, cells with robust and wide coverage should be preferably chosen as PCells. Their selection should be also updated to reflect the small movements across the network and variations of the cell signal quality. Other considerations include load balancing. Practically, an acceptable balance between fast CaCo assignment adaptation and communication overhead must be found as PCell updates are performed on the basis of the standard handover procedure, a relatively lengthy process involving random access, security key change, and RLC/PDCP reestablishment.

The unique role of PCells also implies that not all aspects of CAG scale directly with the increasing number of CAG-capable UEs or the aggregated CaCos. For example, the PCells would become disproportionally loaded given their key features such as the PUCCH transmission, unavailable for SCells. In addition, the PCell PUCCH payload sizes per CAG UE grows with the number of supported CaCos, which creates even more PCell uplink overload affecting primarily non-CAG UEs. To this end, the addition of PUCCH to CAG SCells is being discussed for the LTE Release 13 [16]. The PCell and SCell channel structure based on Releases 10 and 11 is visualized in Fig. 7.7).

7.3.3 Data-Plane Signaling

The overall protocol structure for CAG-capable LTE is the same as the protocol structure of the original LTE Releases 8 and 9, but several enhancements are made to support CAG. Their summary can be found in Table 7.1.

TABLE 7.1   Summary of Carrier Aggregation Impact on
LTE Protocol Stack

Layer	Carrier Aggregation Impact
RRC	Configuration and updates of SCell
PDCP	None
RLC	Larger buffer
MAC	Multi-CC scheduling,
	HAQR processes for each CC
PHY	PDCCH, ACK/NACK, CSI reporting for each CC

In terms of the data (user) plane, the multi-carrier nature of the physical (PHY) layer is only exposed to the medium access control (MAC) layer but remains transparent to the upper unmodified RLC layers and PDCP. The CAG protocol stack of the LTE Layer 2 [7] is summarized in Fig. 7.8.

The MAC layer performs UE scheduling and priority handling across all the active UE cells, and delivers a separate data stream to each CaCo. Clearly, the eNodeB

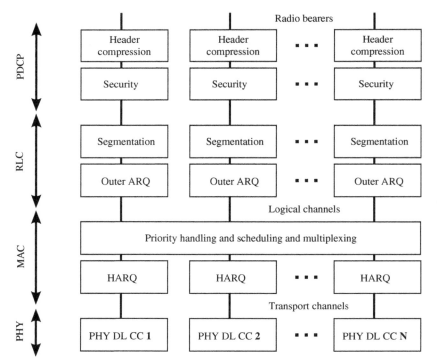

Figure 7.8.   LTE protocol stack with configured carrier aggregation.

cheduler must consider the downlink and uplink channel conditions across the entire aggregated bandwidth in order to take advantage of the aggregated bandwidth. In the absence of multiple-input multiple-output (MIMO) transmission, a CAG-enabled scheduler allocates, at most, one transport block per shared data channel per transmission time interval (TTI). To this end, an independent hybrid automatic repeat request (HARQ) process is managed for each downlink and uplink Caco to support per-Caco link adaptation, resulting in a minor signaling overhead.

The notion of CAG is defined only for the *RRC_CONNECTED* state. In the *RRC_IDLE* state, the CAG-capable UE camps on a single cell and follows the same protocol for connection (re)-establishment and paging as a UE compliant with LTE Releases 8 and 9.

7.3.4 Control-Plane Signaling

As for the control plane, physical channel and signals such as the primary synchronization channel (PSS), secondary synchronization channel (SSS), and physical broadcast channel (PBCH) are transmitted in accordance with the Releases 8 and as they serve to establish the synchronization of an active UE. The physical control format indicator channel (PCFICH), physical HARQ indicator channel (PHICH), PDCCH, and PDSCH channels are Caco-specific.

Minimum impact of the CAG enhancement on implementation and backward compatibility is ensured in that there is only one RRC connection with the network per UE, which is (re)-established with a single serving cell and over which additional serving cells can be (re)-configured [4]. The UE uses a single cell radio network temporary identifier (C-RNTI) to uniquely identify its RRC connection and scramble the PDCCH transmissions. Naturally, the UE RRM measurements are performed for each Caco to assist the eNodeB in CAG management.

The evolved packet core (EPC) is informed at the RRC level on the UE support for CAG by using the so-called *UE Capability Transfer* procedure during the EPS bearer establishment [4]. The CAG-related information sent by the UE includes the UE category and the supported band combinations and configurations. In addition, the UE indicates the support of subsequently discussed features such as cross-carrier scheduling, simultaneous PUCCH and physical uplink shared channel (PUSCH) transmission, multi-cluster PUSCH and non-contiguous uplink resource allocation within a Caco, event A6 support, SCell addition during handover to evolved UTRA (E-UTRA) and periodic sounding reference signals (SRS) transmission on all Cacos.

7.3.4.1 Cell Activation
Dynamic management of serving cells can be used at the MAC layer in full transparency from the upper layers to reflect variations in data activity. Two approaches can be used to extend the UE battery life time.

The first approach consists of the discontinuous reception (DRX) control as known from the LTE Releases 8 and 9, but enhanced to support multi-cell configurations. For example, the UE can switch to the DRX mode when no data are scheduled for the duration of a timer normally used for specifying the inactivity period for PDCCH monitoring. The eNodeB can also force the UE into an immediate transition to DRX operation.

Alternatively, the downlink bands of the configured SCells can be activated and deactivated depending on the UE buffer status, achieved QoS and CaCo load-balancing strategy. Dedicated MAC control elements containing a (de)-activation bit map and implicit deactivation timers are to be used for this purpose. In this way, redundant PDCCH monitoring is avoided. However, RRM is still required. Importantly, the uplink band of any SCell should not be deactivated to allow the UE access their PUSCH if an appropriate scheduling grant is issued.

7.3.4.2 Uplink Control Information The uplink control information (UCI) includes HARQ feedback, channel state information, and scheduling request. As in LTE Releases 8 and 9, the UCI can be transmitted over the PCell regardless of the number of aggregated CaCos, either in the PUSCH or in the PUCCH, if there is no PUSCH transmission in the given subframe.[3] To improve resource usage efficiency, LTE-A also allows the network to configure simultaneous PUCCH and PUSCH transmissions within one subframe. In order to distinguish which UCI belongs to a given CaCo, the header of the UCI contains a carrier indicator field (CIF).

In addition to the HARQ feedback on transport block reception on a per-CaCo basis as known from the LTE Releases 8 and 9 characterized by one-to-one association between the uplink and downlink CaCos, the network can also configure HARQ feedback that is associated with multiple PDSCHs [21]. This is necessary in case of an asymmetric allocation of downlink and uplink CaCos. Several acknowledgments are bundled together as known from the TDD transmission mode to form single-carrier acknowledgments. HARQ feedback compression schemes are designed for the TDD transmission mode, in which a larger number of HARQ bits must be transmitted in an uplink subframe.

Independent PHY management incurs a significant uplink signaling overhead. Mitigation techniques are discussed in [41].

7.3.4.3 Uplink Power Control The maximum uplink transmit power of a UE is a constraint, which limits the maximum number of active CaCos [42,43]. The reason is that simultaneous transmissions over multiple CaCos increase the overall peak-to-average power ratio (PAPR) of the orthogonal frequency division multiplexing (OFDM) waveform [44], that is, the overall waveform power must be reduced to fit within the linear range of the UE power amplifier. Inter-modulation effects may

---

[3] However, CQI can also be reported over the PUSCH. So if a PUSCH transmission is scheduled at the time in which a (periodic) CQI reporting occurs, the CQI feedback could be transmitted over an SCell PUSCH if only an SCell uplink transmission is scheduled without any simultaneous uplink PCell transmission.

rce a further reduction of the effective transmit power in order to satisfy the speci-
ed out-of-band emissions.

Any CAG-induced power back-off directly reduces the UE throughput. An
nsuitable Caco configuration for power-limited UEs may even turn the gains stem-
ing from multiple-Caco transmissions into a performance loss compared to the
on-CAG scenario [43].

In order to maintain the PUSCH/PUCCH reception reliability, LTE Releases 8
nd 9 use uplink power control consisting of an open loop adjustment of the transmit
ower to compensate for the path loss between the UE and the serving eNodeB, and of
closed loop transmit power adjustment following the PDCCH transmission power
ontrol commands of the eNodeB [40].

Similar to the LTE Releases 8 and 9, uplink channel quality is estimated at the
ase station via SRS transmitted by the UE. LTE Release 10 supports CAG by per-
nitting the eNodeB to request periodic SRS transmission on SCells in addition to
Cells, although this function is optional at the UE.

The uplink power control processes are cell-specific, that is, they are indepen-
lent among active cells. Depending on the PUCCH/PUSCH timing configuration and
he UCI transmission type, a UE may transmit in a given subframe one PUCCH and
ne or more PUSCHs with or without UCIs[40]. When the sum of the UE's nominal
ransmit powers exceeds the maximum output power configured for a given subframe,
hen the UE transmit power is distributed in the priority order PUCCH, PUSCH with
JCI, PUSCH without UCI as control information is more important than data infor-
nation and does not benefit from the use of HARQ. Multiple PUSCH transmissions
without UCI are scaled by the same factor to achieve the same quality of service on
ll CaCos.

Power headroom reporting provides the network with an accurate information on
UE transmission power [40] since the exact UE transmit power may not be known,
for example, due to missed transmit power control (TPC) commands and UE-specific
implementation of maximum or additional power reduction imposed to meet regulatory requirements.

7.3.5 Transceiver Design

As depicted in Fig. 7.9a, contiguous CAG allows for efficient signal processing thanks
to the fact that a single IDFT block can be shared for modulating data to all CaCos
[24], but its main challenge remains the radio frequency (RF) design. Current power-
amplifier technology can ensure the required linearity and energy-conversion efficiency over the bandwidth of several tens of MHz. Yet the CAG functionality stretches
this requirement up to 100 MHz, making the trade-off between cost, complexity, and
performance challenging. Inter-modulation and cross-modulation can be also an issue
in a wideband design.

In case of intra-band non-contiguous and inter-band allocation of transmission
bandwidth, the downlink eNodeB transmitter comprises an independent transmitter

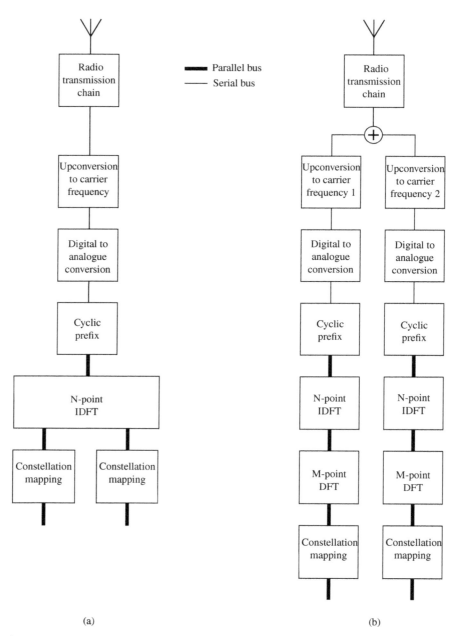

(a)   (b)

Figure 7.9. Exemplary implementation architectures: (a) eNodeB downlink transmitter for contiguous carrier aggregation, (b) UE uplink transmitter for non-contiguous carrier aggregation.

chain for each CaCo. These chains must be isolated to such an extent that any unde-sirable signal mixing and spurious emissions are prevented [23].

For the uplink UE transmitter, a new waveform referred to as N×discrete Fourier transform spread (DFTS)-OFDM has been introduced to support multiple uplink transmission over CaCos [45]. Herein, DFTS signals are divided into several fre-quency components and then mapped non-contiguously to the frequency resources, as shown in Fig. 7.9b). This extension of the DFTS-OFDM approach used for single-carrier uplink transmissions in the LTE Releases 8 and 9 is backward compatible and more efficient than downlink OFDM in terms of PAPR [46]. Moreover, it also permits per-CaCo HARQ and link adaptation as well as a similar hardware implementation for both contiguous and non-contiguous CAG [43].

On the other hand, the N×DFTS-OFDM waveform imposes more stringent lin-earity requirements on the power amplifier compared to the LTE Releases 8 and 9 [23]. Moreover, transmitting over multiple carriers does not preserve the single car-rier property of uplink transmissions (the edges of each CaCo are usually reserved for uplink control channels), implying a higher PAPR (cubic metric) [47]. Hence, a larger power back-off is needed to preserve the power amplifier linearity. Additional power back-offs effectively reduce the range of the UE as well as power efficiency. However, as noted in [48], cell-edge small cells will most likely transmit only in one CaCo, such that usually no coverage reduction compared to the Releases 8 and 9 occurs.

Multiple variants of possible UE RF chain architectures can be found in [8]. In general, the fundamental choices for CAG architectures consist of either a wideband or a narrowband approach.

A single wideband transceiver chain that cover all aggregated bands is practi-cal only for intra-band aggregation of contiguous CaCos. Wideband transceivers rely on costly wideband RF components, high-power amplifiers and high-performance analog-to-digital converters (ADCs), and digital-to-analog converters (DACs). In addition to the fact that additive noise and interference likelihood increases with band-width, path loss, and Doppler effects are not constant in wideband applications. They vary as a non-linear function of frequency.

CAG architectures for non-contiguous and inter-band aggregation types must be implemented on the basis of multiple narrowband transceivers that operate in individ-ual aggregated bands. The number of transceivers increases also with the number of active MIMO antenna elements. The overall hardware complexity and battery drain are higher, but the specification requirements for individual chains are milder than in the above wideband case.

For conformance verification purposes, the 3GPP defines requirements for trans-mitters in terms of peak and dynamic output power, output signal quality, adjacent channel leakage, spurious emissions, and inter-modulation, as well as for receivers in terms of sensitivity, selectivity, blocking, spurious response, inter-modulation, and spurious emissions [23, 49].

7.4 Hetnet DEPLOYMENTS

7.4.1 Configuration Approaches

In accordance with the CAG concept, a Release-10 UE may be scheduled simultaneously on multiple CaCos. To this end, a new type of scheduling grant has been defined that allows to indicate the associated CaCo in a new three-bit field (so-called CFI).

More specifically, the eNodeB can signal the PDSCH allocations either collectively over the PDCCH of a selected CaCo (see UE A in Fig. 7.10, whose data allocation in CaCo 1 and CaCo 2 is signaled only over CaCo 1), or in each CaCo separately over the corresponding PDCCH (see UEs B and C in Fig. 7.10, whose data allocation in CaCo 1 and CaCo 2 is signaled separately over the respective PDCCHs) [50].

The former per-CaCo scheduling is oblivious to the notion of PCell and SCells, and reuses the PDCCH structure and downlink control information (DCI) formats as defined in the LTE Releases 8 and 9. The latter cross-carrier scheduling is a key feature for interference coordination in heterogeneous networks.

When using cross-scheduling, PDCCH grants for all CaCos must be transmitted by the PCell, that is, PCells cannot be cross-scheduled. The potential PDCCH absence makes PCFICH decoding redundant, so LTE-A allows directly announcing the control format indicator (CFI) parameter, that is, the PDCCH region size in the time domain, by using a dedicated mechanism referred to as PDSCH-Start.

The availability of multiple CaCos as well as the flexibility of PDCCH control signaling allow efficient coordination of interference in HetNets. In general,

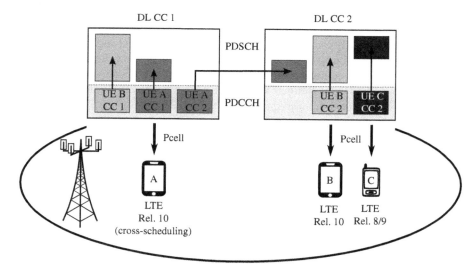

Figure 7.10. Scheduling options for carrier aggregation: (a) cross-scheduling by UE A, and (b) per-CC scheduling by UEs B and C.

Figure 7.11. Principle of carrier aggregation usage in heterogeneous networks. Significant path loss and reduced macrocell transmit power can be used to protect small cell transmissions in shared carrier components.

victim-cell transmissions can be protected by configuring at least one CaCo of the aggressor cell—be it a macrocell or a small cell with closed subscriber group—to use lower output power or a band with higher frequency characterized by more significant attenuation. The former case is visualized in Fig. 7.11.

The more CaCos are available, the more configuration combinations are possible, and the more diverse the PDCCH distribution among the available CaCos. Advantageously, PDCCH load balancing and interference coordination can be carried out on a subframe time scale.

Figure 7.12 shows four configuration examples assuming the availability of two CaCos, whereby the macrocell and the small cell are the aggressor cell and the victim cell, respectively.

The first CaCo configuration (a) is characterized by both the macrocell and the small cell operating on dedicated CaCos causing thereby no mutual interference. Zero

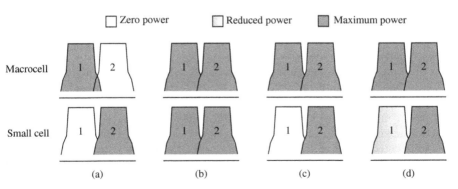

Figure 7.12. Carrier aggregation deployment strategies for HetNets availing of two carrier components: (a) Dedicated CCs, (b) full CC reuse, (c) asymmetric CC allocation, and (d) soft CC reuse.

interference translates into the possibility of transmitting at maximum powers and configuring high bias values for small cell selection, that is, efficient offload. Assuming two CaCos for simplicity and in accordance with Fig. 7.12, the resulting frequency reuse factor is 2, if both CaCos have the same bandwidth.

In fact, this first configuration (a) is not a pure CAG deployment. It is effectively just a multi-carrier extension of LTE Releases 8 and 9 networks, for which no UE CAG functionality is required. Accordingly, inter-tier handovers between the macrocell and a small cell are inter-frequency handovers, while the macrocell-to-macrocell and small cell-to-small cell handovers are intra-frequency handovers (see Chapter 4 for more details).

The second CaCo configuration (b) is based on equal sharing of all CaCos by both cell tiers, leading to a frequency reuse factor 1. Similar to the first configuration with dedicated CaCos, the transmit powers for each CaCo are set to the maximum value. All handovers may be either intra-frequency or inter-frequency handovers. CAG-incapable UEs are assigned to only one of the available CaCos. Yet, given the unmitigated inter-tier interference, the potential of bias-based small cell range extension is limited. Depending on the scenario, around 10% of macrocell traffic is offloaded by one small cell.

The third CaCo configuration (c) alleviates the interference problem, impacting especially the PDCCH, by combining the above two strategies. The small cell eNodeB operates in both available CaCos, which is a measure conforming to the philosophy of deploying small cells as a means of providing for a localized capacity boost. However, the macrocell eNodeB is designated to operate only in a single CaCo. Given the inactivity of the macrocell on the other CaCo, this CaCo configuration is effectively characterized by a CaCo dedicated to the small cell tier.

As a result of the macrocell interference contained in only one CaCo, higher cell-selection biases can be used for small cell range expansion without sacrificing the CAG functionality. Clearly, the communication with UEs located at the edge of the small cell should happen over the dedicated small cell CaCo. In particular, all control PDCCH transmissions should be accommodated in the dedicated CaCo, that is, the dedicated CaCo should be configured as the PCell and cross-scheduling should be used to access the non-dedicated CaCo.

The fourth CaCo configuration (d) is based on the principle of soft-frequency reuse. Assuming again that all CaCos are shared by both network tiers as in the second CaCo configuration example, the small cell eNodeB uses its maximum transmit power in all CaCos. Yet, to protect the small cell transmissions, the macrocell eNodeB applies a reduced transmit power in one of its CaCo. The protected CaCo can be located in a high frequency band characterized by a more severe propagation path loss to additionally or on its own contribute to the mitigation of macrocell interference.

The advantage of this configuration compared to the second reuse-1 example is that the macrocell has access to more resources, that is, the macrocell can offer better service to its cell-center users, and its small cells benefit from the reduced power transmissions. Similar to the above third configuration case, cell-edge UEs of the small cell are scheduled in the protected CaCo managed as their PCell.

7.4.2 Mobility Management

Handovers in LTE Releases 10–12 are performed in essentially the same manner as in LTE Releases 8 and 9 (see Chapter 12 for more details on the handover process). Importantly, measurement-related RRC signaling messages refer only to the PCell.

To support the addition and removal of SCells as shown in Fig. 7.13, the LTE Release 10 introduces a new measurement event [4], Event A6, for detecting and reporting neighbor cells on the same frequency as the SCell that becomes better than the SCell. Legacy Events A1-A5 do not allow efficiently selecting the best SCell for a given PCell. The addition of a new Event A6 targets primarily inter-band CAG and its monitoring can also be used for the addition of a new SCell or the removal of an already configured SCell.

The support of CAG generally implies RRC signaling overhead stemming from mobility management, since the PCell/SCell signal quality of a mobile UE is controlled only by means of dynamic inter/intra-frequency handovers and CaCo selection. The signaling overhead is exacerbated by several factors including varying CaCo coverage [51], UE uplink power limits [41], etc. Different CaCo management policies and their effect on RRC signaling overhead reduction are investigated in [51–53].

Typical inputs for CaCo management consist not only of RRM measurements, but also include load and interference conditions, as well as cell quality. To support CaCo management, the eNodeB can configure RRM measurements on selected or all configured CaCos, whereby it is also possible to perform RRM measurements on non-configured CaCos.

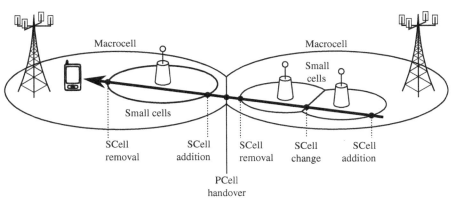

Figure 7.13. Mobility management under carrier aggregation.

7.4.3 Performance Comparison

In this section, the four CAG deployment strategies described above and depicted in Fig. 7.12—(a) dedicated CaCos, (b) full CaCo reuse, (c) asymmetric CaCo allocation, and (d) soft CaCo reuse—are evaluated.

A network with two CaCos of 5 MHz bandwidth is assumed [50]. Other simulation parameters are summarized in Table 7.2. Most notably, for the soft reuse strategy (Fig. 7.12d), it is assumed that the macrocell eNodeBs reduce the transmit power of the second CaCo from their nominal power of 46 dBm to the transmission power applied by the small cell eNodeBs, of 30 dBm.

Figure 7.14 shows the cell border throughput and the mean UE throughput for the four deployment strategies and small cell range expansion bias (REB) values varying between 0 dB and 15 dB [50].

TABLE 7.2 Simulation Parameters

Parameters	Value
System Parameters	
Scenario	Outdoor RRH/hotzone [54]
Bandwidth	2×5 MHz
Carrier frequency	2 GHz
Cell Topology	
Macrocells	19 sites with 3 sectors (57 totally)
	500 m inter-site distance
Small cells	2 per macrocell (randomly placed)
User Distribution	
Configuration 4b [54]	2/3 near small cells
	1/3 uniformly randomly distributed
Antenna Models	
Macrocells	3D (15 degrees downtilt)
Small cells	2D (omni-directional)
Antenna elements	$1/2\lambda$ spacing
Transmit Powers	
Macrocells	46 dBm
Small cells	30 dBm
Path Loss (Model 1 [54])	
Macrocells	$PL = 128.1 + 37.6 \log_{10}(R)$ [R in km]
Small cells	$PL = 140.7 + 36.7 \log_{10}(R)$ [R in km]
Shadow fading	variance 10 dB
Site-to-site correlation	0.5
Other Parameters	
Small cell bias	0–15 dB
Transmission mode	Closed-loop spatial multiplexing, single-user MIMO
Scheduler	Proportional fair with exponent $\alpha = 1$ [55]
PDCCH error rate	$\leq 1\%$
Traffic model	Full buffer

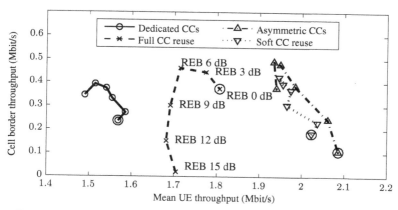

Figure 7.14. Performance evaluation of carrier aggregation deployment strategies in HetNets [50].

We observe that the full reuse strategy (Fig. 7.12b) only performs well for small REBs (less than approximately 6 dB). Higher REB values result in strong unmitigated wideband interference, which significantly degrades the cell border throughput.

The other three deployment strategies provide some level of channel protection for small cell UEs, which allows using higher REB values. In particular, the usage of dedicated CaCos (Fig. 7.12a) yields slightly higher mean UE throughput than the full reuse strategy thanks to significantly lower interference even though network cells are restricted to operating only in a single CaCo.

In terms of the mean UE throughput, the poor spectral efficiency of the dedicated CaCo approach is somewhat mitigated by the approach of asymmetric CaCo allocation (Fig. 7.12c). Yet it is the strategy of soft CaCo reuse (Fig. 7.12d) that results in the highest mean UE throughput in the studied scenario. Macrocell power reduction on the second CaCo is the best compromise between increasing the number of available resources and reducing the interference to small cell UEs.

7.5 SUMMARY AND CONCLUSIONS

Following the recent spread of smart phones and tablets, mobile UEs now have higher expectations in terms of service quality and will continue to demand faster guaranteed data rates. Carrier aggregation is a feature of LTE-A that allows operators to efficiently address the demands of their subscribers, as can be seen from its fast commercial rollout.

The single most important reason for the smooth commercial adoption of LTE carrier aggregation techniques is their native support for the aggregation of fragmented spectrum resources. This is a key feature enabling the multiplication of

UE-available bandwidth as operators acquire carriers in a somewhat uncoordinated fashion via auctions as well as through mergers or acquisitions of other operators. Since joint aggregation of FDD and TDD bands is possible as well, the capacity of existing LTE-FDD networks can be augmented also by the TDD spectrum, which is relatively large in some markets. By bonding unpaired non-contiguous spectrum into a single logical channel, operators can also efficiently address the typical dominance of the downlink traffic over uplink data flows. International roaming is also easier when mobile UEs support many different bands.

From the HetNet perspective, carrier aggregation is a potent tool, opening a new frequency dimension for efficient interference management and inter-tier coexistence provisioning. The cross-scheduling of control transmissions as well as the orthogonalization of data transmissions by frequency-domain multiplexing and power control enhances the overall service quality through interference reduction. To enhance the user experience at the cell edge, operators can also supplement their primary transmission bands by low-frequency carriers, characterized by better propagation and coverage. By enabling reliable deployments of small cells, carrier aggregation opens up another possibility for operators to address the capacity crunch—spatial reuse.

Carrier diversity comes hand in hand with better use of network resources through load balancing. Traffic can be dynamically and more evenly distributed across multiple carriers, improving the overall resource utilization as well as user experience through congestion reduction.

Successful commercialization of carrier aggregation in LTE consequently sparked interest in aggregating unlicensed (Wi-Fi) bands, triggering a novel line of research focusing on establishing coexistence between LTE and Wi-Fi. The emerging discussion on fifth-generation (5G) standardization takes carrier aggregation as a granted feature for future networks.

In the context of carrier aggregation evolution, the current focus on backward-compatible designs brings the advantage of simplified system design and development, but incurs an irreducible overhead on each component carrier, stemming from the transmissions of reference, synchronization, and control signals. Such an overhead is an issue especially in component carriers with small bandwidth. Future definitions of backward-incompatible carriers, for example, within the 5G framework, could further improve the energy and spectrum efficiency, especially when fragmented or narrowband bands are aggregated.

REFERENCES

1. T. Novlan, J. G. Andrews, S. Illsoo, R. K. Ganti, and A. Ghosh, "Comparison of fractional frequency reuse approaches in the OFDMA cellular downlink," in *Proc. IEEE GLOBECOM*, Miami, USA, Dec. 6–10, 2010, pp. 1–5.
2. A. L. Stolyar and H. Viswanathan, "Self-organizing dynamic fractional frequency reuse in OFDMA systems," in *Proc. IEEE/ACM INFOCOM*, Phoenix, USA, Apr. 2008.

3. "Requirements for support of radio resource management," 3GPP Technical Specification Group Radio Access Network, Evolved Universal Terrestrial Radio Access (E-UTRA), Tech. Rep. TS 36.133 v12.6.0, Jan. 2015. [Online]. Available: http://www.3gpp.org

4. "Radio Resource Control (RRC) protocol specification," 3GPP Technical Specification Group Radio Access Network, Evolved Universal Terrestrial Radio Access (E-UTRA), Tech. Rep. TS 36.331 v12.4.1, Sep. 2014. [Online]. Available: http://www.3gpp.org

5. "Carrier aggregation: Base station (BS) radio transmission and reception," 3GPP Technical Specification Group Radio Access Network, Evolved Universal Terrestrial Radio Access (E-UTRA), Tech. Rep. TR 36.808 v10.1.0, Jul. 2013. [Online]. Available: http://www.3gpp.org

6. "Carrier aggregation enhancements: User equipment (UE) and base station (BS) radio transmission and reception," 3GPP Technical Specification Group Radio Access Network, Evolved Universal Terrestrial Radio Access (E-UTRA), Tech. Rep. TR 36.823 v11.0.1, Nov. 2013. [Online]. Available: http://www.3gpp.org

7. "E-UTRA and E-UTRAN: Overall description, stage 2," 3GPP Technical Specification Group Radio Access Network, Evolved Universal Terrestrial Radio Access (E-UTRA), Tech. Rep. TS 36.300 v12.4.0, Jan. 2015. [Online]. Available: http://www.3gpp.org

8. "Feasibility study for further advancements for E-UTRA (LTE-Advanced) (Release 12)," 3GPP Technical Specification Group Radio Access Network, Evolved Universal Terrestrial Radio Access (E-UTRA), Tech. Rep. TR 36.912 v12.0.0, Sep. 2014. [Online]. Available: http://www.3gpp.org

9. "Overview of 3GPP Release 9," 3GPP Technical Specification Group Radio Access Network, Evolved Universal Terrestrial Radio Access (E-UTRA), Tech. Rep. v0.3.4, Sep. 2014. [Online]. Available: http://www.3gpp.org

10. M. Rahman, H. Yanikomeroglu, and W. Wong, "Enhancing cell-edge performance: A downlink dynamic interference avoidance scheme with inter-cell coordination," *IEEE Trans. Commun.*, vol. 9, no. 4, pp. 1414–1425, Apr. 2010.

11. R. Madan, J. Borran, A. Sampath, N. Bhushan, A. Khandekar, and T. Ji, "Cell association and interference coordination in heterogeneous LTE-A cellular networks," *IEEE J. Sel. Areas Commun.*, vol. 28, no. 9, pp. 1479–1489, Dec. 2010.

12. "Interference conditions in heterogeneous networks," 3GPP Technical Specification Group Radio Access Network, Evolved Universal Terrestrial Radio Access (E-UTRA), Tech. Rep. TSG-RAN Meeting 59b, R1-100700, Jan. 2010. [Online]. Available: http://www.3gpp.org

13. "X2 application protocol (X2AP)," 3GPP Technical Specification Group Radio Access Network, Evolved Universal Terrestrial Rad io Access (E-UTRA), Tech. Rep. TS 36.423 v12.4.2, Jan. 2015. [Online]. Available: http://www.3gpp.org

14. "Physical channels and modulation," 3GPP Technical Specification Group Radio Access Network, Evolved Universal Terrestrial Radio Access (E-UTRA), Tech. Rep. TS 36.211 v12.4.0, Jan. 2015. [Online]. Available: http://www.3gpp.org

15. "Multiplexing and channel coding," 3GPP Technical Specification Group Radio Access Network, Evolved Universal Terrestrial Radio Access (E-UTRA), Tech. Rep. TS 36.212 v12.1.0, Sep. 2014. [Online]. Available: http://www.3gpp.org

16. "Work plan for LTE carrier aggregation enhancement beyond 5 carriers," 3GPP Technical Specification Group Radio Access Network, Evolved Universal Terrestrial Radio Access (E-UTRA), Tech. Rep. TSG-RAN Meeting 66, RP-141997, Dec. 2014. [Online]. Available: http://www.3gpp.org

17. 4G Americas. (2014). LTE Carrier Aggregation - Technology Development and Deployment Worldwide. [Online]. Available: http://www.4gamericas.org/documents/

18. "Dual-cell high speed downlink packet access (HSDPA) operation," 3GPP Technical Specification Group Radio Access Network, Evolved Universal Terrestrial Radio Access (E-UTRA), Tech. Rep. TR 25.825 v1.0.0, Jun. 2008. [Online]. Available: http://www.3gpp.org

19. "Further advancements of E-UTRA physical layer aspects," 3GPP Technical Specification Group Radio Access Network, Evolved Universal Terrestrial Radio Access (E-UTRA), Tech. Rep. TR 36.814 v1.0.0, Feb. 2009. [Online]. Available: http://www.3gpp.org

20. "LTE TDD-FDD joint operation including carrier aggregation," 3GPP Technical Specification Group Radio Access Network, Evolved Universal Terrestrial Radio Access (E-UTRA), Tech. Rep. TSG-RAN Meeting 63, RP-140465, Mar. 2014. [Online]. Available: http://www.3gpp.org

21. "Medium access control (MAC) protocol specification," 3GPP Technical Specification Group Radio Access Network, Evolved Universal Terrestrial Radio Access (E-UTRA), Tech. Rep. TS 36.321 v12.4.0, Jan. 2015. [Online]. Available: http://www.3gpp.org

22. "User equipment (UE) radio access capabilities," 3GPP Technical Specification Group Radio Access Network, Evolved Universal Terrestrial Radio Access (E-UTRA), Tech. Rep. TS 36.306 v12.3.0, Jan. 2015. [Online]. Available: http://www.3gpp.org

23. "User equipment (UE) radio transmission and reception," 3GPP Technical Specification Group Radio Access Network, Evolved Universal Terrestrial Radio Access (E-UTRA), Tech. Rep. TS 36.101 v12.6.0, 2015. [Online]. Available: http://www.3gpp.org

24. R. Ratasuk, D. Tolli, and A. Ghosh, "Carrier aggregation in LTE-Advanced," in *Proc. IEEE VTC Spring 2010*, Taipei, Taiwan, May 16–19, 2010, pp. 1–5.

25. Ceragon. (2011). Broadband Backhaul - Asymmetric Wireless Transmission. [Online]. Available: http://www.ceragon.com/images/

26. "L-band for supplemental downlink," 3GPP Technical Specification Group Radio Access Network, Evolved Universal Terrestrial Radio Access (E-UTRA), Tech. Rep. TR 37.814 v12.0.0, Oct. 2014. [Online]. Available: http://www.3gpp.org

27. "Overview of 3GPP Release 12," 3GPP Technical Specification Group Radio Access Network, Evolved Universal Terrestrial Radio Access (E-UTRA), Tech. Rep. v0.1.4, Sep. 2014. [Online]. Available: http://www.3gpp.org

28. "LTE in unlicensed spectrum: European regulation and co-existence considerations," 3GPP Technical Specification Group Radio Access Network, Evolved Universal Terrestrial Radio Access (E-UTRA), Tech. Rep. TSG-RAN Workshop on LTE in Unlicensed Spectrum, RWS140002, Jun. 2014. [Online]. Available: http://www.3gpp.org

29. "Use cases & scenarios for licensed assisted access," 3GPP Technical Specification Group Radio Access Network, Evolved Universal Terrestrial Radio Access (E-UTRA), Tech. Rep. TSG-RAN Workshop on LTE in Unlicensed Spectrum, RWS140020, Jun. 2014. [Online]. Available: http://www.3gpp.org

30. "Views on LAA for unlicensed spectrum—Scenarios and initial evaluation results," 3GPP Technical Specification Group Radio Access Network, Evolved Universal Terrestrial Radio Access (E-UTRA), Tech. Rep. TSG-RAN Workshop on LTE in Unlicensed Spectrum, RWS140026, Jun. 2014. [Online]. Available: http://www.3gpp.org

31. "In the matter of amendment of the Commission's rules with regard to commercial operations in the 3550-3650 MHz band, further notice of proposed rulemaking," Federal Communications Commission, Tech. Rep. GN Docket No. 12-354, Apr. 2013. [Online]. Available: http://www.fcc.gov

32. "Discussions on some issues of dual connectivity," 3GPP Technical Specification Group Radio Access Network, Evolved Universal Terrestrial Radio Access (E-UTRA), Tech. Rep. TSG-RAN Meeting WG2 81, R2-130055, Feb. 2013. [Online]. Available: http://www.3gpp.org

33. "Discussion on small cell enhancement and dual connectivity," 3GPP Technical Specification Group Radio Access Network, Evolved Universal Terrestrial Radio Access (E-UTRA), Tech. Rep. TSG-RAN Meeting WG2 81, R2-130100, Feb. 2013. [Online]. Available: http://www.3gpp.org

34. "What is dual connectivity?" 3GPP Technical Specification Group Radio Access Network, Evolved Universal Terrestrial Radio Access (E-UTRA), Tech. Rep. TSG-RAN Meeting WG2 81, R2-130117, Feb. 2013. [Online]. Available: http://www.3gpp.org

35. "Study on small cell enhancements for E-UTRA and E-UTRAN: Higher layer aspects," 3GPP Technical Specification Group Radio Access Network, Evolved Universal Terrestrial Radio Access (E-UTRA), Tech. Rep. TR 36.842 v12.0.0, Jan. 2014. [Online]. Available: http://www.3gpp.org

36. "Overview of 3GPP Release 10," 3GPP Technical Specification Group Radio Access Network, Evolved Universal Terrestrial Radio Access (E-UTRA), Tech. Rep. v0.2.1, Jun. 2014. [Online]. Available: http://www.3gpp.org

37. "Overview of 3GPP Release 11," 3GPP Technical Specification Group Radio Access Network, Evolved Universal Terrestrial Radio Access (E-UTRA), Tech. Rep. v0.2.0, Sep. 2014. [Online]. Available: http://www.3gpp.org

38. "E-UTRA and E-UTRAN: Overall description, stage 2," 3GPP Technical Specification Group Radio Access Network, Evolved Universal Terrestrial Radio Access (E-UTRA), Tech. Rep. TS 36.300 v11.12.0, Jan. 2015. [Online]. Available: http://www.3gpp.org

39. "Physical layer procedures," 3GPP Technical Specification Group Radio Access Network, Evolved Universal Terrestrial Radio Access (E-UTRA), Tech. Rep. TS 36.213 v11.9.0, Jan. 2015. [Online]. Available: http://www.3gpp.org

40. "Physical layer procedures," 3GPP Technical Specification Group Radio Access Network, Evolved Universal Terrestrial Radio Access (E-UTRA), Tech. Rep. TS 36.213 v12.4.0, Jan. 2015. [Online]. Available: http://www.3gpp.org

41. Y. Wang, K. Pedersen, M. Navarro, P. Mogensen, and T. Srensen, "Uplink overhead analysis and outage protection for multi-carrier LTE-Advanced systems," in *Proc. IEEE PIMRC*, Tokyo, Japan, Sep. 13–16, 2009, pp. 11–17.

42. H. Wang, C. Rosa, and K. Pedersen, "Uplink component carrier selection for LTE-Advanced systems with carrier aggregation," in *Proc. IEEE ICC*, Kyoto, Japan, Jun. 5–9, 2011, pp. 1–6.

43. H. Wang, C. Rosa, and K. Pedersen, "Performance of uplink carrier aggregation in LTE-Advanced systems," in *Proc. IEEE VTC Fall*, Ottawa, Canada, Sep. 6–9, 2010, pp. 1–5.

44. "LTE-A MC RF requirements for contiguous carriers," 3GPP Technical Specification Group Radio Access Network, Evolved Universal Terrestrial Radio Access (E-UTRA), Tech. Rep. TSG-RAN 4 Meeting-51, R4-091910, Apr. 2009. [Online]. Available: http://www.3gpp.org

45. "DFTS-OFDM extension for LTE-A," 3GPP Technical Specification Group Radio Access Network, Evolved Universal Terrestrial Radio Access (E-UTRA), Tech. Rep. TSG-RAN 1 Meeting-55, R1-084422, Nov. 2008. [Online]. Available: http://www.3gpp.org

46. T. Iwai, A. Matsumoto, and S. Takaoka, "System performance of clustered DFT-S-OFDM considering maximum allowable transmit power," in *Proc. IEEE ICC*, Kyoto, Japan, Jun. 5–9, 2011, pp. 1–6.

47. "User equipment (UE) conformance specification, radio transmission and reception, Part 1: Conformance testing," 3GPP Technical Specification Group Radio Access Network, Evolved Universal Terrestrial Radio Access (E-UTRA), Tech. Rep. TR 36.521 v12.4.0, Sep. 2014. [Online]. Available: http://www.3gpp.org

48. "Uplink access for LTE-A—Non-aggregated and aggregated scenarios," 3GPP Technical Specification Group Radio Access Network, Evolved Universal Terrestrial Radio Access (E-UTRA), Tech. Rep. TSG-RAN 1 Meeting-54b, R1-083820, Sep. 2008. [Online]. Available: http://www.3gpp.org

49. "Base station (BS) radio transmission and reception," 3GPP Technical Specification Group Radio Access Network, Evolved Universal Terrestrial Radio Access (E-UTRA), Tech. Rep. TS 36.104.v12.6.0, Sep. 2014. [Online]. Available: http://www.3gpp.org

50. O. Stanze and A. Weber, "Heterogeneous networks with LTE-Advanced technologies," *Bell Labs Tech.*, vol. 18, no. 1, pp. 41–58, May 2013.

51. K. Yagyu, T. Nakamori, H. Ishii, M. Iwamura, N. Miki, T. Asai, and J. Hagiwara, "Investigation on mobility management for carrier aggregation in LTE-Advanced," in *Proc. IEEE VTC Fall*, San Francisco, USA, Sep. 5–8, 2011, pp. 1–5.

52. L. Liu, M. Li, J. Zhou, X. She, L. Chen, Y. Sagae, and M. Iwamura, "Component carrier management for carrier aggregation in LTE-Advanced systems," in *Proc. IEEE VTC Spring*, Budapest, Hungary, May 15–18, 2011, pp. 1–6.

53. F. Wu, Y. Mao, S. Leng, and X. Huang, "A carrier aggregation based resource allocation scheme for pervasive wireless networks," in *Proc. IEEE DASC*, Sydney, Australia, Dec. 12–14, 2011, pp. 196–201.

54. "Further advancements of E-UTRA physical layer aspects," 3GPP Technical Specification Group Radio Access Network, Evolved Universal Terrestrial Radio Access (E-UTRA), Tech. Rep. TR 36.814 v9.0.0, Mar. 2010. [Online]. Available: http://www.3gpp.org

55. "Frequency domain scheduling for E-UTRA," 3GPP Technical Specification Group Radio Access Network, Evolved Universal Terrestrial Radio Access (E-UTRA), Tech. Rep. TSG-RAN 1 Meeting -44b, R4-060877, Mar. 2006. [Online]. Available: http://www.3gpp.org

8

TIME-DOMAIN INTER-CELL INTERFERENCE COORDINATION

8.1 INTRODUCTION

The deployment of small cells in areas with poor macrocell coverage and in traffic hotspots can enhance the overall network performance, but the potential cell-splitting gains can be negated by inter-cell interference. The intra-cell inter-carrier/symbol interference is negligible in modern orthogonal frequency division multiplexing (OFDM) systems such as the fourth-generation (4G) long-term evolution (LTE), but inter-cell interference is not.

The impact of inter-cell interference on the network performance is particularly adverse in co-channel deployments. Two typical scenarios can be distinguished [1–3]: the deployment of outdoor metrocells where the aim is to maximize the offload of user equipments (UEs) from macrocells, and the deployment of femtocells that are typically deployed to serve UEs within a particular targeted area. As detailed in Section 4.3, femtocells can be set up either with a closed-access model, where only subscribers that have been registered in a white list are allowed to connect, or an open-access model, where all subscribers can connect. While open access is preferred in co-channel deployments to avoid dropped calls to UEs not in the white list

Small Cell Networks: Deployment, Management, and Optimization, First Edition. Holger Claussen, David López-Pérez, Lester Ho, Rouzbeh Razavi, and Stepan Kucera.
© 2017 by The Institute of Electrical and Electronic Engineers, Inc. Published 2017 by John Wiley & Sons, Inc.

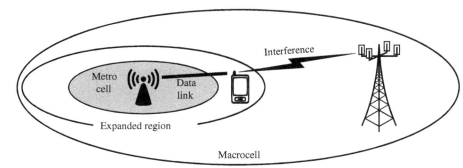

Figure 8.1. Metrocell UEs in the expanded region are the victims of macrocell interference.

that are within the femtocell's coverage area, a closed-access models are also possible if interference from the femtocell can be managed.

In the first metrocell example (see Fig. 8.1), the network operator selects the metrocell locations in a planning phase where the radio-frequency and medium access conditions of the network are carefully evaluated and optimized. The metrocell evolved NodeBs (eNodeBs) are open-access nodes with a protocol stack and backhaul identical to the macrocell eNodeBs. However, compared to macrocells, their lower antenna gain and output power results in a smaller coverage range, typically at the order of tens of meters.

Because of this, several approaches have been developed to artificially expand the small cell range to ensure macrocell traffic offload to a level desired by the network operator. A positive range expansion bias (REB) achieves this goal by making the metrocell eNodeB appear to be the strongest transmitter in a larger area but makes the metrocell UEs in the expanded region more vulnerable to co-channel interference [4].

One can therefore see co-channel metrocells as the victims of interference originating at the macro (aggressor) cell. The coexistence problem is exacerbated by the fact that larger REBs are required for metrocells that are located either closer to the macrocell or further from a hotspot center. For more details on REB configuration for metrocells, please also refer to Chapter 6.

On the other hand, in the femtocell deployment scenario (see Fig. 8.2), it is assumed that the placement of the femtocell eNodeB is determined by the femtocell owner rather than the operator. Importantly, the owner is usually independent of the macrocell network operator, for example, a small business, so the femtocell deployment can be considered as unplanned and uncoordinated with respect to the macrocell tier [5]. Apart from a lower power output, femtocell eNodeBs differ from metrocell eNodeBs mainly in their deployment, targeting a specific coverage area and hence a particular subset of UEs. In the absence of a dedicated X2 link to the hosting macrocell, handovers are carried out over the S1 interface [6, 7].

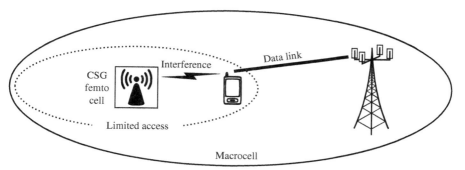

Figure 8.2. Mobile macrocell UEs are the victims of interference from a femtocell restricting its service to a closed subscriber group.

Developing further the comparison of metrocells and femtocells, it is noteworthy that the aggressor and victim roles are reversed in the femtocell case—the mobile UEs of the macrocell are the victims of the interference stemming from the aggressor femtocell. The co-channel interference is also less manageable in the case of closed-access femtocells because macrocell UEs affected by the activity of a nearby closed-access femtocell cannot be freely handed over to the interfering femtocell to resolve the interference problem as in the metrocell scenario.

It becomes clear from the above observations that inter-cell interference coordination must be implemented to support co-channel deployments of small cells. In general, such coordination can take place in the time, frequency, and power domain.

The main focus of this chapter is on time-domain related techniques, together with the review of applicable power-domain approaches. Frequency-domain methods were examined in the previous Chapter 7.

The idea of time-domain interference coordination is simple: the transmissions of cooperating cells are multiplexed in time, that is, in certain time intervals, only a subset of selected cells can engage in active transmissions to its UEs. For example, the aggressor (macro) cell engages in transmission activity only in preselected subframes, while the victim (small) cells transmit in the complementary subframes. Intuitively, the ratio of subframes within a transmission frame that are associated with the aggressor and victim cell should reflect the relative load of each cell for optimized performance.

The remaining part of this chapter discusses the technical implications of time-domain inter-cell interference coordination in 4G LTE systems, notably the required modifications on the UE-to-base-station feedback, the basis of adaptive network operation. Efficient algorithms for optimizing the performance of cell multiplexing in the time domain as well as their computationally fast implementations are offered subsequently.

Before proceeding, it is worth mentioning the concept of coordinated multipoint (CoMP) transmission, an extension of the multi-user MIMO (MU-MIMO) techniques, that has been introduced by the LTE Release 11 [8] as an alternative way of enhancing cell-edge rates [9]. Its principle is to combine signals from multiple cells, which is practically possible only on the basis of high-speed inter-cell communication and massive data exchange, as well as extremely tight time-phase frequency synchronization among collaborating cells [10]. More costly requirements of antenna properties such as gain and steerability are also likely to promote the effects of constructive and destructive signal interference in a wide spatial area. So even though the CoMP-related overhead could be compressed to certain extent [11], the software-configurable enhanced inter-cell interference coordination (eICIC) approach is currently more likely to be adopted faster.

8.2 ALMOST BLANK SUBFRAMES

8.2.1 3GPP Evolution

Anticipating the demand for efficient spectrum reuse, the Third-Generation Partnership Project (3GPP) continues evolving techniques for time-domain coordination and cancellation of inter-cell interference in heterogeneous networks that orthogonalize the small cell and macrocell transmissions.

Fractional frequency reuse was the first technique adopted in the early LTE Releases 8 and 9 [12] as part of the framework for so-called inter-cell interference coordination (ICIC). The concept is based on dividing the spectrum into fragments and creating different reuse patterns to reflect varying interference conditions among network UEs. A detailed description of this concept can be found in Chapter 7 where it is also mentioned that pseudo-random spreading of control channels across the system bandwidth and other technical issues limit the applicability of fractional frequency reuse to homogeneous macrocell networks.

To resolve these issues, the support for explicit time-domain multiplexing of macrocell and small cell transmissions has been defined in the LTE Release 10 as part of the so-called eICIC framework [13, 14]. Namely, the LTE standard has been expanded by the capability of configuring so-called almost blank subframes (ABSs), a concept compatible with ICIC. As illustrated by Fig. 8.3, these are subframes in which certain selected physical channels are not transmitted, that is, "blanked," by the aggressor (macro) cell to protect the victim (small cell) transmissions from its interference, in particular in the small cell expanded region.

The blanking is applied to the physical downlink shared channel (PDSCH) and the majority of UE-specific physical downlink control channel (PDCCH) signals. These two channels consume the majority of wireless resources. However, in order to support legacy LTE Release-8/9 UE, the cell-specific reference symbol (CRS) as well as the common PDCCH are active to maintain a minimum control traffic by using few wireless resources. Hence, the protective macrocell subframes are not entirely devoid of any transmission—they are "almost blank."

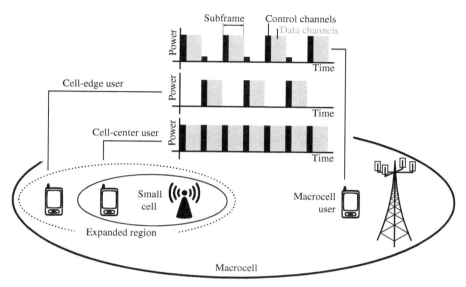

Figure 8.3. Concept of almost blank subframes.

The LTE Release 11 further develops the concept of time-domain eICIC by enabling UE-based cancellation of aggressor CRS interference to help victim-cell UEs perceive the almost blank subframes at a higher level of blanking [8]. In order to reconstruct the aggressor CRS for interference cancellation purposes, the more advanced eICIC protocol provides the UE with information on the aggressor physical layer cell identity (PCI) and the cycle prefix (CP) type, as these parameters determine the CRS waveform. Otherwise, the only way to reduce the CRS pollution is to configure multicast-broadcast single-frequency network (MBSFN) subframes that are characterized by much lower CRS density (present solely in the PDCCH region).

In addition, to reduce the impact of the reoccurring inactivity periods on the capacity of cooperating cells, the LTE Release 11 also introduces the possibility of low power transmissions to cell-center UEs within almost-blank subframes since such transmissions have a low potential to harm the victim-cell UEs.

8.2.2 Concept Review

Adaptive subframe blanking requires symbol and subframe-level synchronization between the cooperating eNodeBs. eNodeBs compliant with LTE Release 10 can configure ABSs either as normal ABSs or as MBSFN ABSs.

In normal ABSs, control signals such as primary synchronization channel (PSS), secondary synchronization channel (SSS) and CRS are transmitted whenever

TABLE 8.1   Relative Strength (in dB) of LTE Physical Channels and Signals with Respect to Cell-Reference Symbols in Non-ABS and ABS Subframes

		Non-MBSFN Subframes				MBSFN Subframes			
		1 × 2 Antennas		2 × 2 Antennas		1 × 2 Antennas		2 × 2 Antennas	
		(with PBCH)		(without PBCH)		(no PBCH)		(no PBCH)	
		Non-ABS	ABS	Non-ABS	ABS	Non-ABS	ABS	Non-ABS	ABS
PSS	S^{other}	0	0	−3	−3	0	N/A	−3	N/A
SSS	S^{other}	0	0	−3	−3	0	N/A	−3	N/A
PBCH	S^{other}	0	0	−3	−∞	0	N/A	−3	N/A
	S^{CRS}	0	0	−3	−∞	0	N/A	−3	N/A
PCFICH	S^{CRS}	0	0/−∞	1	−∞	0	−∞	1	−∞
PHICH	S^{other}	0	−∞	−3	−∞	0	−∞	−3	−∞
	S^{CRS}	0	−∞	−3	−∞	0	−∞	−3	−∞
PDCCH	S^{other}	0	0	1	−∞	0	−∞	1	−∞
	S^{CRS}	0	0	1	−∞	0	−∞	1	−∞
PDSCH	S^{other}	0	0	−3	−∞	0	−∞	−3	−∞
	S^{CRS}	0	0	−3	−∞	0	−∞	−3	−∞

S^{other} — Transmitted OFDM symbols *not* containing CRS resource elements.
S^{CRS} — Transmitted OFDM symbols containing CRS resource elements.

applicable, but nothing else, that is, the physical control format indicator channel (PCFICH), physical HARQ indicator channel (PHICH) and PDCCHs transmissions do not take place, although uplink grants can still be sent over the PDCCHs in an ABSs. The physical broadcast channel (PBCH) may also be switched off, as shown by the configuration examples in Table 8.1 [15]. If the ABS coincides with an MBSFN subframe, the CRS transmission in the data region is muted as well.

   Since timely acknowledgments (ACKs) and negative acknowledgments (NACKs) over the PHICH are the basis of the hybrid automatic repeat request (HARQ) processes that largely improve the LTE data rate performance, configured ABS patterns should be aligned with the uplink HARQ timing to avoid the ABS-induced muting of the PHICH carrying ACKs and ACKs datagrams.

   Keeping the HARQ functionality in mind, the 3GPP have specified multiple basic ABS patterns with the period identical to the period of the HARQ processes to simplify the ABS and HARQ alignment [15]. Eight basic ABS patterns having the same period and spanning over four frames (40 subframes) have been specified for the frequency division duplexing (FDD) transmission mode as shown in Table 8.2. Generally, two or more basic patterns can be combined to create a more complex pattern having the desired ABSs over non-ABSs ratio. This property is useful in the time division duplexing (TDD) mode, in which the uplink/downlink ratio defines the number of supported HARQ processes.

TABLE 8.2   ABS Patterns in LTE FDD

	Frame 1	Frame 2	Frame 3	Frame 4
ABS pattern 1	1000000010	0000001000	0000100000	0010000000
ABS pattern 2	0100000001	0000000100	0000010000	0001000000
ABS pattern 3	0010000000	1000000010	0000001000	0000100000
ABS pattern 4	0001000000	0100000001	0000000100	0000010000
ABS pattern 5	0000100000	0010000000	1000000010	0000001000
ABS pattern 6	0000010000	0001000000	0100000001	0000000100
ABS pattern 7	0000001000	0000100000	0010000000	1000000010
ABS pattern 8	0000000100	0000010000	0001000000	0100000001
	1 = Almost-blank subframes		0 = Active subframes	

It is also noteworthy that the MBSFN subframes are transmitted in subframes 1, 2, 3, 6, 7, and 8 (assuming in accordance with the 3GPP subframe indexing from 0). Hence, an MBSFN subframes alone cannot support the 8 ms periodicity of synchronous uplink HARQ processes, that is, non-MBSFN ABS must be used as well.

In the LTE TDD mode, the subframes of different cells must be synchronized so that PSS/SSS/PBCH signals required for cell acquisition and (re)selection overlap. To protect the PSS/SSS and PBCH from the collision with the aggressor PDSCH, ABSs 0 and 5 should be prioritized for use.

In the LTE FDD mode, an additional level of interference reduction to PSS/SSS/PBCH can be achieved by offsetting the subframe alignment between the two cells. Another constraint for ABS pattern selection consists of the protection of the paging messages and periodically issued grants to system information block (SIB) information that are allocated to the common PDCCHs search spaces.

8.2.3 eNodeB Configuration Support

Metrocells typically use the X2 interface and protocol stack and can therefore dynamically negotiate ABS patterns with their neighbors, reflecting variations in networking conditions such as the relative cell/tier load.

In accordance with [16], the ABS configuration parameters can be requested over the X2 interface by using the *Invoke* message, or announced by using the *Status* message. These messages have the form of indicator bitmaps and are included in the *Load Indication* report and *Resource Status Report Indication* report. In general, the update frequency should not exceed the exchange rate of the Release-8/9 indicators such as relative narrowband transmit power (RNTP), high-interference indicator (HII), and interference overload indicator (IOI). More frequent changes of the ABS pattern are not only undesired in view of the associated overhead, but also because they can destabilize the network by creating "ping pong" requests for ABS pattern (re)configuration.

The following information element (IE) of the X2 protocol [16] are to be used primarily:

- *Invoke Indication IE* in the *Load Information* message: The sender indicates a request for information on non-zero ABS patterns.
- *ABS Information IE* in the *Load Information* message: The sender indicates the currently used ABS pattern. The receiving eNodeB may use the indicated *Measurement Subset IE* for the configuration of specific measurement sets as discussed in the next subsection. The uplink/downlink configuration to be used by the indicated cell can be included in the *Load Information* and can be employed to define the scheduling policies of the receiving eNodeB.
- *ABS status IE* in the *Resource Status Request* message: The sender requests an update on the percentage of used ABS resources and the employed ABS pattern, typically for the purpose of evaluating the need for modifying current ABS patterns.

In the case of femtocells, the X2 connection is typically not assumed and a semi-static configuration is instantiated by a centralized operation, administration, and maintenance (OAM) service. Pattern adjustments are triggered by processing the collected network statistics. The OAM service must be able to ensure the dissemination of the ABS information associated with the closed subscriber group (CSG) femto-cells to the hosting (victim) macrocells to enable optimized scheduling of macrocell edge UEs.

8.2.4 UE Measurement Support

In order to support eNodeB decision-making with UE feedback on its actual local conditions, UEs conduct measurements of specific reference signals such as CRS. Yet subframe blanking directly affects the signal-to-interference-plus-noise ratio (SINR) in ABSs compared to the SINR in non-ABSs. Consequently, all subframes are not anymore equally representative for measurement purposes as inherently assumed in the LTE Releases 8 and 9 [12].

The SINR disparity between signal measurements in ABSs and non-ABSs primarily affects three operational and optimization processes of the eNodeB that use UE measurements as their input:

1. Cell (re)selection and handovers, whose execution is governed by the radio resource management (RRM)-based evaluation of the reference signal received power (RSRP) and reference signal received quality (RSRQ) feedback [15],
2. Monitoring of the UE link quality and synchronization to the serving cell which is a part of the radio link monitoring (RLM) functionality [10], and
3. Channel state information reporting that drives optimized scheduling, link adaptation, as well as coding and modulation scheme selection [15].

More specifically, regarding the first-mentioned process, the RSRP measurements on the CRS are not affected by the configuration of non-MBSFN ABSs as these do not mute the CRS for backward compatibility reasons. However, the partial CRS muting in the MBSFN ABSs requires so-called restricted measurements in case such subframes are in use.

Restricted measurements, supported from the LTE Release 10 [17], are measurements that are selectively conducted in restricted subframe subsets. The idea is to provide for differentiated reports in the protected ABSs and unprotected non-blanked subframes. By default, the LTE Release-8/9 UEs report the perceived channel quality to the service eNodeB based on the measurements of all subframes.

As for the RSRQ metric, it must be always measured on the basis of restricted measurements because the RSRQ is defined as the ratio of the CRS strength and the total carrier signal strength, the latter being affected by ABS muting.

Both the radio link failure evaluation and channel state information—mentioned as the second and the third processes, respectively—must also reflect the ABS patterns to provide dedicated evaluation of protected and unprotected subframes to prevent unnecessary link outages and unrealistic channel state estimates.

Ideally, UEs should measure only subframes relevant to their networking condition. For example, the UEs of the aggressor cell, or the UEs in the center of the victim cell, can be configured to provide measurement reports averaged over all the subframes, but the UEs at the edge of the victim cell must report measurements conducted over the protected blanked subframes only, in which their transmission would be scheduled.

For implementation, the LTE standard categorizes the subframes into two groups for which independent channel state reporting is carried out by the UE: normal subframes and coordination subframes that are associated with non-ABSs and ABSs, respectively. Importantly, restricted measurements are limited to a single carrier only [17]. This limitation implies that carrier aggregation, that is, frequency domain ICIC as described in Chapter 7, cannot be practically combined with the time-domain ABS-based approach to ICIC. This problem could be alleviated by future LTE Releases.

In terms of the protocol, UE-specific radio resource control (RRC) signaling [17] is used to configure dedicated measurement patterns that reflect the ABS patterns in use. The serving cell specifies subframe subsets for RRM and RLM measurements separately, as well as measurement patterns for specific subsets of neighboring cell PCIs. Signals from cells with unspecified PCI are measured in all subframes as defined in LTE Releases 8 and 9.

8.3 CONFIGURATION AND OPTIMIZATION

Intuitively, the configuration of ABS patterns is a trade-off in which the aggressor interference to victim cells can be reduced at the expense of restricting the aggressor-cell capacity. However, finding a "fair" throughput share for each cell or cell tier depends on the cell coverage, that is, the setting of transmit powers and REBs is closely coupled with optimizing the transmit silence patterns [4, 18–22]. In a multi-channel scenario, another dimension of the problem consists of distributing the UE transmissions among the available communication channels.

This section shows that finding an optimum solution to the joint problem of setting ABSs, REBs, and transmit powers is NP-hard when multiple small cells are present. Hence, a reasonable degree of problem relaxation is necessary to ensure computational efficiency when configuring dense/large networks. To this end, approximate reasonably-well performing algorithms are discussed in Sections 8.4–8.5.

8.3.1 System Model

The system model consists of a network of macrocell and small cell eNodeBs. $\mathcal{M}$ denotes the set of macrocells and $\mathcal{P}$ denotes the set of small cells. The indices m and p are used to denote macrocells and small cells, respectively.

The set of macrocells that interfere with a small cell typically depends on the coverage area of the small cell, which in turn depends on the REB of the small cell. Specifically, if $RSRP_1$ is the RSRP (in dBm) of the best macrocell m as measured by a UE u, and $RSRP_2$ is the UE-measured RSRP from small cell p, then the UE associates with the small cell p if

$$RSRP_1 < RSRP_2 + b \qquad (8.1)$$

where b is the small cell REB in dB. Thus, for a higher value of b, the small cell coverage area increases, and so does the number of interfering macrocells.

To define specifically the notion of macrocell-to-small cell interference, the main point of concern of eICIC, the set of macrocells that interfere with a small cell p is denoted by $\mathcal{I}_p \subseteq \mathcal{M}$.

Intuitively speaking, a macrocell m is called an interfering macrocell of a small cell p if the downlink transmission of macrocell m causes interference to the transmissions of small cell p, that is, the interference to the downlink transmission of small cell p is above a threshold in most of the coverage area of p.

The macrocells in the set $\mathcal{I}_p$ are silent during any ABSs, causing no interference to the small cell p. Thus, the UEs of small cell p can only be interfered by $m \in \mathcal{I}_p$ during regular subframes (non-ABS for short). Under frequency reuse one, macrocells can interfere with each other and small cells can interfere with each other; that is, a macrocell UE can receive interference from another macrocell only during non-ABSs, while small cell UE can receive interference from another small cell both during ABS and non-ABSs.

The downlink SINR of a macrocell UE u can be written as

$$SINR(u) = \frac{P_{Rx}(u)}{P_{Int}^{SC}(u) + P_{Int}^{MC}(u) + n_u} \qquad (8.2)$$

where $P_{Int}^{SC}(u)$ and $P_{Int}^{MC}(u)$ denote the interference from interfering small cells and macrocells, respectively. n_u is the additive noise power for UE u.

However, the muting of macrocell transmission during ABS affects the small cell UE SINR. Suppose that the total interference powers a small cell UE u receives from all other interfering small cells and all interfering macrocells (those in the set $\mathcal{I}_p$) are denoted as $P_{\text{Int}}^{\text{SC}}(u)$ and $P_{\text{Int}}^{\text{MC}}(u)$, respectively. Denoting by $P_{Rx}(u)$ the received downlink power of UE u from its serving cell, the downlink SINR of small cell UE $u \in \mathcal{U}_p$ can be modeled as

$$
\text{SINR}(u) = \begin{cases} \dfrac{P_{Rx}(u)}{P_{\text{Int}}^{\text{SC}}(u)+n_u} & \text{for ABSs,} \\[3mm] \dfrac{P_{Rx}(u)}{P_{\text{Int}}^{\text{SC}}(u)+P_{\text{Int}}^{\text{MC}}(u)+n_u} & \text{for non-ABSs.} \end{cases} \tag{8.3}
$$

During ABSs, the only accounted interference is from the interfering small cells of p as all interfering macrocells of small cell p remain silent. In contrast, during non-ABS, there is interference from both small cells and macrocells.

As for the UE model, a set $\mathcal{U}$ of static UEs is considered. For each UE u, one knows (i) the best candidate macrocell in terms of RSRP together with the average physical-layer data rate r_u^{MC} from that macrocell, (ii) the best candidate small cell, if any, and the average data rate in both ABSs and non-ABS, denoted $r_u^{\text{SC,ABS}}$ and r_u^{SC}, respectively. The data rate r_u can be obtained from the SINR expressions (8.2) and (8.3) using LTE look-up tables. Alternatively, one can use the Shannon capacity formula [23] as described in Appendix Section A.9.

The main parameters and some of the optimization variables (introduced later) are summarized in Table 8.3.

8.3.2 Problem Formulation and Hardness Statement

The goal of the eICIC configuration and optimization is to compute the optimal association rules for UEs (i.e., small cell REBs), as well as the way how macrocells and small cells share radio resources in the time domain (i.e., the ABS pattern configuration).

To proceed with the problem formulation, let N_{sf} denote the total number of subframes over which ABSs are reserved ($N_{sf} = 40$ in practice, see Table 8.2). The quantity N_{sf} is referred to as the ABS period. Let also N_m be the number of subframes for which macrocell m can transmit during each ABS period (clearly, $N_{sf} - N_m$ ABSs are offered by macrocell m in each ABS period).

Furthermore, let x_u be the time-average air-time[1] in subframes per ABS period that UE u gets from m_u (the best candidate macrocell for UE u). Note that x_u

---

[1] This time-average airtime can be achieved through medium access control (MAC) scheduling, particularly weighted proportional-fair scheduling. Two time scales are implicitly assumed: the ABS selection happens at a slow time scale, and the MAC scheduling happens at a fast time scale resulting in a time-average airtime for each UE.

TABLE 8.3   List of Parameters and Key Optimization Variables

Notation	Description
$\mathcal{U}$, u, N	Set of UEs, index for a typical
$(u \in \mathcal{U})$	UE, number of UEs, respectively
$\mathcal{M}$, m, M	Set of macrocells, index for a typical
$(m \in \mathcal{M})$	macrocell, number of macrocells, respectively
$\mathcal{P}$, p, P	Set of small cells, index for a typical
$(p \in \mathcal{P})$	small cell, number of small cells, respectively
m_u	The macrocell that is *best* for UE u
r_u^m	Data rate achievable by UE u from m_u (in bits/subframe)
p_u	The small cell that is *best* for UE u
$r_u^{p,ABS}$	Data rate achievable by UE u from p_u when all interfering macrocells are muted (in bits/subframe)
r_u^p	Data rate achievable by UE u from p_u when all/some interfering macrocells are transmitting (in bits/subframe)
$\mathcal{I}_p$	Set of macrocell eNBs that interfere with small cell p
$\mathcal{U}_m$	Set of UEs for whom macrocell m is the best macrocell eNB
$\mathcal{U}_p$	Set of UEs for whom small cell p is the best small cell eNB
A_p	Variable denoting ABS subframes used/received by small cell p
N_m	Variable denoting non-ABS subframes used by macrocell m
x_u	Variable denoting UE u's air-time from macrocell
y_u^A, y_u^{nA}	Variables denoting UE u's air-time from small cell over ABS and non-ABS subframes, respectively
R_u	Variable denote UE u's average throughput
z, **p**	Vector of all primal and dual variables, respectively

need not be an integer. The airtime that UE u gets from a small cell can be dur-ing ABSs or regular (non-ABS) subframes because small cell eNodeBs can trans-mit during ABSs and non-ABSs overlapping subframes. To this end, let y_u^A and y_u^{nA} denote the time-average airtime in subframes per ABS period that UE u gets from small cell p_u (the best candidate small cell of UE u) in ABSs and non-ABSs overlapping subframes, respectively. Also, let R_u be the average throughput UE u achieves.

Then, the eICIC optimization problem, referred to as the *OPT-ABS Problem* in the next, can be stated as follows.

Problem 8.1   *(OPT-ABS)*
Given

- *a set of UEs $\mathcal{U}$,*
- *a set of macrocells $\mathcal{M}$,*

- *a set of small cells* $\mathcal{P}$, *and*
- *total number* N_{sf} *of subframes,*
- *for each UE* $u \in \mathcal{U}$,
  - *the best candidate parent macrocell* m_u *along with data rate* r_u^{MC} *to* m_u,
  - *the best candidate parent small cell* p_u *along with ABS data rate* $r_u^{SC,ABS}$ *and non-ABS data rate* r_u^{SC} *to* p_u,
- *for each small cell* p, *the set* $\mathcal{I}_p$ *of interfering macrocells,*

solve the optimization problem

$$maximize_{\{x_u, y_u^A, y_u^{nA}, A_p, N_m, R_u\}} \sum_u w_u \ln R_u = Util(\mathbf{R}) \qquad (8.4)$$

by computing

- *the number* N_m *of non-ABSs left for macrocell* m,
- *the number* A_p *of ABSs for each small cell* p,
- *a binary decision on whether each UE* u *associates with its best candidate macrocell or best candidate small cell,*
- *throughput* R_u *of each UE* u,

subject to

1. *Association constraint: A UE can associate with either the macrocell or a small cell but not both.*

$$\forall\, u \in \mathcal{U}\ : x_u(y_u^A + y_u^{nA}) = 0 \qquad (8.5)$$
$$x_u \geq 0,\ y_u^A \geq 0,\ y_u^{nA} \geq 0. \qquad (8.6)$$

2. *Throughput constraint: The average throughput* R_u *for a UE* u *cannot be more than what is available based on the air-times from associated macrocell / small cell.*

$$\forall\, u \in \mathcal{U} : R_u \leq r_u^{MC} x_u + r_u^{SC,ABS} y_u^A + r_u^{SC} y_u^{nA}. \qquad (8.7)$$

3. *Interference constraint: The ABSs used by a small cell* p *are offered by all macrocells in the set* $\mathcal{I}_p$ *that interfere with the small cell.*

$$\forall\, (p, m \in \mathcal{I}_p) : A_p + N_m \leq N_{sf}. \qquad (8.8)$$

4. *Total airtime constraint: The total time-average airtime allocated to a UE from a macrocell or a small cell is less than the total usable subframes.*

$$\forall\, m \in \mathcal{M} \;:\; \sum_{u \in \mathcal{U}_m} x_u \leq N_m \tag{8.9}$$

$$\forall\, p \in \mathcal{P} \;:\; \sum_{u \in \mathcal{U}_p} y_u^A \leq A_p \tag{8.10}$$

$$\forall\, p \in \mathcal{P} \;:\; \sum_{u \in \mathcal{U}_p} \left(y_u^A + y_u^{nA} \right) \leq N_{sf} \tag{8.11}$$

for all $p, m \in \mathcal{I}_p$ and A_p, $N_m \in \mathbb{Z}^+$ where $\mathbb{Z}^+$ denotes the space of non-negative integers.

The weighted proportional-fair optimization objective which maximizes $\sum_u w_u \ln R_u$, where w_u represents a weight associated with UE u, is well known to strike a very good balance between system throughput and UE-throughput fairness [24]. This bodes well with the goal of improving cell-edge throughput using eICIC. The weights w_u provides a means for service-differentiation [25]. As seen in the following, other utility functions can be used as well, with no effect on the overall algorithmic solution.

Unfortunately, the ABS-optimization problem as defined above is NP-hard unless $P = NP$ even with a single macrocell but multiple small cells [26]. This can be shown by reducing the *SUBSET-SUM Problem* [27] to an instance of the *OPT-ABS Problem* [26]. Hence, only approximate solutions can be obtained in polynomial time.

To proceed, empirically motivated heuristics approaches are first reviewed, consisting typically in load balancing techniques. Then, in the following Sections 8.4 and 8.5, more rigorous analytic approaches with polynomially complexity that solve relaxed instances of the *OPT-ABS Problem* with close-to-optimum performance are discussed. The idea in Section 8.4 is to relax the integer constraints and then round the obtained solution to ensure feasibility. Alternating maximization of parameter subspaces is discussed in Section 8.5.

8.3.3 Empirical Solution Approaches

Computationally light heuristics typically divide the *ABS-OPT Problem* (8.1) into three independent (consecutive) steps:

- the selection for small cell REBs,
- the selection of the macrocell ABS patterns, and
- the assignment of small cell UEs to the available ABSs.

In the simplest setting, the first step of REB selection can be based on ensuring a fair balance of UE load in the macrocell and small cell tiers, which may also have a positive effect on the uplink interference distribution [28, 29]. Assuming the knowledge of UE/demand distribution[2] and the flexibility in eNodeBs deployment[3] (e.g., close to the hotspot centers), the small cell REBs can also be set to offload a certain fraction of macrocell UEs or demand by small cells.

In the second step consisting of ABS configuration, the ratio of UE load in the small cell and macrocell tiers can be used as in [28] to directly define the macrocell ABS duty cycle, that is, the ratio of ABSs and non-ABSs in an unspecified ABS pattern. A less coarse but still scalable strategy [30] is to set the ABS duty cycle D_m of the macrocell m as the ratio

$$D_m = \alpha_m \frac{\max_{\forall s \in m}(N_s)}{N_m} \tag{8.12}$$

between (i) the maximum number N_s of interference-limited UEs in the expanded region of the small cell s, considering all small cells hosted by the macrocell m, and (ii) the total number of UEs at the macrocell m. The multiplicative factor α_m is used to prioritize the small cell tier ($\alpha_m < 1$) or the macrocell tier ($\alpha_m > 1$) in terms of subframe allocation to reflect the fact that balancing UE numbers is not equivalent to balancing the actual data traffic or demand [29]. For example, α_m can be set proportionally to the ratio of the small- and macrocell mean or aggregate UE demand.

The ABS duty cycle can also be set by comparing the conditions before and after applying small cell REBs, as the motivation for configuring ABSs is to protect small cell UEs in the expanded region. For example, [29] sets the macrocell ABS duty cycle D_m to

$$D_m = 1 - \alpha_m^{-1} \frac{N_m^{REB>0}}{N_m^{REB=0}}, \tag{8.13}$$

where $N_m^{REB=0}$ and $N_m^{REB>0}$ are the number of UEs of the macrocell m before and after the small cell offload (zero and non-zero small cell REBs), respectively. Eq. (8.13) is equivalent to configuring S^{ABS} subframes in a block of S subframes, that is, a duty cycle $D_m = S^{ABS}/S$, such that

$$\frac{S}{N_m^{REB=0}} = \frac{S - S^{ABS}}{N_m^{REB>0}} \alpha_m. \tag{8.14}$$

---

[2] A difficult task given the poor spatial resolution of localization methods that is even lower in 4G LTE networks compared to its third-generation (3G) predecessor due to the lack of soft handover.
[3] Another potentially difficult task given the practical constraints on small cell placement in terms of site rental availability, costs, and backhaul options, as well as in view of the physical mobility and time-varying demand of network UEs.

The idea here is to maintain the subrame-per-macrocell-UE ratio before and after the small cell range expansion.

The third phase of UE-to-subframe mapping within each small cell can then be used to maximize the performance of the worst nth percentile of small cell UEs [29]. This effectively corresponds to defining the cell-edge and cell-center UEs.

8.4 NON-LINEAR RELAXATION APPROACH

This section discusses approximate algorithms to find reasonably well-performing solutions to the joint problem of setting ABSs, REBs, and transmit powers.

8.4.1 Formulation

On an analytical level, the NP-hardness of the *OPT-ABS Problem* (8.1) can be also approached by

1. first solving a non-linear program obtained by ignoring integrality constraints on A_p and N_m, as well as the constraint that a UE can receive data either from small cell or macrocell but not both, followed by

2. appropriate rounding of the output of the non-linear optimization to obtain a feasible solution to the original problem.

More specifically on the first step, the term Util(**R**) is maximized subject to the constraints (8.7)–(8.11) whereby A_p and N_m are allowed to be non-integer values. Note that ignoring the constraint (8.5) implies that UEs can receive radio resources from both macrocell and small cell.

By denoting for notational convenience the vector of constraints (8.7)–(8.11) in a compact form as $\mathbf{g}_R(.) \leq 0$, the relaxed *OPT-ABS Problem* (8.1), referred to as the *RELAXED-ABS Problem*, can be formulated as follows.

Problem 8.2 *(RELAXED-ABS).*

$$Maximize_{\{x_u, y_u^A, y_u^{nA}, A_p, N_m\}} \sum_u w_u \ln R_u \tag{8.15}$$

$$subject\ to\ \mathbf{g}_R(.) \leq 0 \tag{8.16}$$

$$\forall p, m \in \mathcal{I}_p\ such\ that\ A_p,\ N_m \in \mathbb{R}^+, \tag{8.17}$$

where $\mathbb{R}^+$ *denotes the space of non-negative real numbers, and* $\mathbf{g}_R(.)$ *denotes the vector of constraints (8.7)–(8.11).*

There are certain challenges in resolving the above steps. The usual approach in solving such non-linear program is through a dual or a primal-dual approach. The dual-based approach [31, 32] typically greatly reduces the algorithmic complexity and leads to distributed solutions, while retaining the core essence [33, 34].

However, the dual of the problem is not differentiable everywhere and this necessitates the use of subgradients instead of gradients. The key challenge is to find a suitable subgradient ensuring the converge and scalability of the algorithmic optimization search. The second challenge lies in rounding the non-linear program solution to obtain a feasible solution to the original ABS optimization problem so that the final solution is a good approximation to the optimal solution.

The following description summarizes the approach of [19].

8.4.2 Duality Optimization

Let the subspace Π be defined as

$$\Pi = \left\{ \mathbf{x}, \mathbf{y}, \mathbf{A}, \mathbf{N} : A_p \le N_{sf}, N_m \le N_{sf}, \sum_{u \in \mathcal{U}_m} x_u \le N_{sf}, \right. \tag{8.18}$$

$$\left. \sum_{u \in \mathcal{U}_p} y_u^A \le N_{sf}, \sum_{u \in \mathcal{U}_p} y_u^{nA} \le N_{sf}, \; \forall \, m, p \right\} \tag{8.19}$$

Bold-face notations denote vectors of variables. For example x denotes the vector of values x_u. Clearly, any solution that satisfies the constraints described in the previous section lies within Π. In the following discussion, even without explicit mention, it is understood that optimization variables always lie in Π.

By treating A_p and N_m as real numbers, the Lagrangian of the relaxed non-linear program can be expressed as

$$\mathcal{L}(\mathbf{x}, \mathbf{y}, \mathbf{A}, \mathbf{N}, \lambda, \mu, \beta, \alpha) = \sum_u w_u \ln R_u \tag{8.20}$$

$$- \sum_u \lambda_u (R_u - r_u^{MC} x_u - r_u^{SC,ABS} y_u^A - r_u^{SC} y_u^{nA}) \tag{8.21}$$

$$- \sum_{p,m \in I_p} \mu_{p,m} (A_p + N_m - N_{sf}) \tag{8.22}$$

$$- \sum_m \beta_m \left(\sum_{u \in \mathcal{U}_m} x_u - N_m \right) - \sum_p \beta_p \left(\sum_{u \in \mathcal{U}_p} y_u^A - A_p \right) \tag{8.23}$$

$$- \sum_p \alpha_p \left(\sum_{u \in \mathcal{U}_p} (y_u^A + y_u^{nA}) - N_{sf} \right), \tag{8.24}$$

where the variables $\lambda, \mu, \beta, \alpha$ are dual variables and so-called Lagrange multipliers, which also have a price interpretation. The term $\mathbf{p}$ denotes the vector of all dual variables, that is, $\mathbf{p} = (\lambda, \mu, \beta, \alpha)$.[4] Similarly, the variables $\mathbf{x}, \mathbf{y}, \mathbf{A}, \mathbf{N}$ are referred to as primal variables. The vector $\mathbf{z}$ denotes the vector of all primal variables, that is, $\mathbf{z} = (\mathbf{x}, \mathbf{y}, \mathbf{A}, \mathbf{N})$.

Thus, the Lagrangian $\mathcal{L}(\mathbf{z}, \mathbf{p})$ can be expressed as

$$\mathcal{L}(\mathbf{z}, \mathbf{p}) = \text{Util}(\mathbf{R}) - \mathbf{p}' \mathbf{g}_R(\mathbf{z}). \tag{8.25}$$

Accordingly, the dual *RELAXED-ABS Problem* (8.2) can be reformulated as

$$\min_{\mathbf{p} \geq 0} \mathcal{D}(\mathbf{p}), \tag{8.26}$$

where

$$\mathcal{D}(\mathbf{p}) = \max_{\mathbf{z} \in \Pi} \mathcal{L}(\mathbf{z}, \mathbf{p}). \tag{8.27}$$

The dual problem has now the form of a maximization problem with a concave objective and convex feasible region. Therefore, it follows that there is no duality gap [32], and thus

$$[\text{RELAXED-ABS}] - [\text{OPT-ABS}] = \min_{\mathbf{p}} \mathcal{D}(\mathbf{p}). \tag{8.28}$$

The following rules are used for iteratively solving the reformulated dual problem, having initialized the primal variables to any value in Π and the dual variables are to zero:

1. *Greedy primal update:* The primal variables $\mathbf{z}_t$ in iteration $(t + 1)$ are set as

$$\mathbf{z}_{t+1} = \arg\max_{\mathbf{z} \in \Pi} \mathcal{L}(\mathbf{z}, \mathbf{p}_t). \tag{8.29}$$

2. *Subgradient descent based dual update:* The dual variables are updated in a gradient descent-like manner as

$$\mathbf{p}_{t+1} = [\mathbf{p}_t + \gamma \mathbf{g}_R(\mathbf{z}_t)]^+, \tag{8.30}$$

where $\mathbf{p}_t$ is the dual variable at iteration t, γ is the step size and $[.]^+$ is the component-wise projection into the space of non-negative real numbers.

---

[4] The dual variables are often denoted by $\mathbf{p}$ because they have the interpretation of prices.

The above steps are continued for sufficiently large number of iterations T, and the optimal solution to *RELAXED-ABS Problem* is produced as

$$\hat{\mathbf{z}}_T = \frac{1}{T} \sum_{t=1}^{T} \mathbf{z}_t. \tag{8.31}$$

The computation of the greedy primal update is not immediate at a first glance. The document [19] shows how the primal update can be efficiently decomposed into optimization actions carried out independently of the users, macrocells, and small cells. It also concretely specifies how to set the step size γ and the number of iterations T.

8.4.3 Solution Configuration

In general, the solution to the *RELAXED-ABS Problem* (8.2) must be converted to a feasible solution for the original *OPT-ABS Problem*. There are two challenges in performing this step.

First, unlike in the *RELAXED-ABS Problem* (8.2), each UE can receive in the *OPT-ABS Problem* (8.1) resources either from a macrocell or a small cell but not both. Second, as with all dual-based subgradient algorithms, the solution may violate feasibility after T iterations running *RELAXED-ABS Problem* albeit by a small margin [31].

Thus, each UE must be associated with a macrocell or a small cell, and the values of N_m and A_p must be rounded so that the overall solution is feasible and has provable performance guarantee.

To this end, [19] introduces three steps for converting *RELAXED-ABS Problem* solutions to configurable values, using a rounding function

$$\mathrm{Rnd}_{N_{sf}}(x) = \begin{cases} \lfloor x \rfloor & \text{for } x \geq \frac{N_{sf}}{2}, \\ \lceil x \rceil & \text{for } x < \frac{N_{sf}}{2}. \end{cases} \tag{8.32}$$

1. *User Association*: In the first step, each UE, which gets higher throughput from a macrocell in the solution of *RELAXED-ABS Problem* (8.2) is associated with a macrocell, and each UE, which gets higher throughout from a small cell gets associated with a small cell. More specifically, for all $u \in \mathcal{U}$,

   a. the throughput that UE u gets from the macrocell and the small cell in *RELAXED-ABS Problem* (8.2) solution is computed as

   $$R_u^{\mathrm{MC}} = r_u^{\mathrm{MC}} \hat{x}_u \tag{8.33}$$
   $$R_u^{\mathrm{SC}} = r_u^{\mathrm{SC,ABS}} \hat{y}_u^A + r_u^{\mathrm{SC}} \hat{y}_u^{nA}, \tag{8.34}$$

   where $\hat{x}_u, \hat{y}_u^A, \hat{y}_u^{nA}$ are the obtained optimization solutions, and

b. if $R_u^{MC} > R_u^{SC}$, UE u associates with the macrocell; otherwise, with the small cell.

Lastly, the set $\mathcal{U}_m^*$ ($\mathcal{U}_p^*$) of UEs associated with every macrocell m (small cell p) after the *UE association* step is computed.

2. *ABS rounding*: In the second step, the UE association decisions are used to obtain the ABSs and non-ABSs by computing integral N_m^* and A_p^* as

$$N_m^* = \text{Rnd}_{N_{sf}}(\hat{N}_m) \forall m \in \mathcal{M} \tag{8.35}$$

$$A_p^* = \text{Rnd}_{N_{sf}}(\hat{A}_p) \forall p \in \mathcal{P}, \tag{8.36}$$

where $\hat{N}_m$ and $\hat{A}_p$ are the obtained optimization solutions.

3. *Throughput computation*: In the third step, each UE's available average airtime x_u^* is scaled to fill up the available subframes, allowing us to compute the final throughput R_u^* of each UE. In particular, for each macrocell m and for all $u \in \mathcal{U}_m^*$,

$$x_u^* = \frac{\hat{x}_u N_m^*}{X_m} \tag{8.37}$$

$$R_u^* = r_u^{MC} x_u^*, \tag{8.38}$$

where $X_m = \sum_{u \in \mathcal{U}_m^*} \hat{x}_u$. The ABS-utilization Y_p^A and non-ABS-utilization Y_p^{nA} is computed for every small cell as

$$Y_p^A = \sum_{u \in \mathcal{U}_p^*} \hat{y}_u^A, \tag{8.39}$$

$$Y_p^{nA} = \sum_{u \in \mathcal{U}_p^*} \hat{y}_u^{nA}, \tag{8.40}$$

while the final values of $y_u^A *, y_u^{nA} *$ and R_u^* are computed for each small cell p and for all $u \in \mathcal{U}_m^*$ as

$$y_u^{A*} = \frac{\hat{y}_u^A A_p^*}{Y_p^A} \tag{8.41}$$

$$y_u^{nA*} = \frac{\hat{y}_u^{nA}(N_{sf} - A_p^*)}{Y_p^{nA}} \tag{8.42}$$

$$R_u^* = r_u^{SC,ABS} y_u^{A*} + r_u^{SC} y_u^{nA*} \tag{8.43}$$

The system utility is computed as $\text{Util}(\mathbf{R}^*) = \sum_u w_u \ln R_u^*$.

The worst-case performance guarantee of the output produced by the above three steps (1)–(3) depends on the number of iterations and step size used for running

RELAXED-ABS Problem (8.2). Detailed recommendations of the step size and number of iterations are offered in [19].

Denoting the throughput computed in *RELAXED-ABS Problem* by R_u^* and the throughput computed by *OPT-ABS Problem* (8.1) by R_u^{OPT}, and using mild technical assumptions, [19] shows that for any given $\delta > 0$, there exists T and γ such that if the steps (1)–(3) are applied to the output of the *RELAXED-ABS Problem* algorithm with this T and γ, then by [26],

$$\text{Util}(2(1 + \delta)\mathbf{R}^*) \geq \text{Util}(\mathbf{R}^{OPT}). \tag{8.44}$$

It follows that for sufficiently large but polynomial number of iterations, the worst-case approximation factor is close to 2. It is important to realize that, as with all NP-hard problems, this is simply a worst case result. The subsequent evaluations of several real-life network topologies suggest that the performance of the described algorithm is typically within 90% of the optimal.

It is intuitively understandable that converting an algorithmic solution to cell-specific biases that precisely achieve the desired UE association may not always be feasible. To this end, [19] proposes to compute biases using several steps aiming at minimizing the "association error" of an actual UE association as compared to the computed optimal association.

1. Set all biases of macrocells to zero.
2. Let $C_{p,m}$ be the set of UEs, which have small cell p as the best candidate small cell as well as macrocell m as the best candidate macrocell, that is, $C_{p,m} = \mathcal{U}_p \cap \mathcal{U}_m$. From the UEs in the set $C_{p,m}$, let $W_{p,m}^*$ be the total weight of UE's associated to small cell p under the UE association produced by optimization algorithm. Also from UEs in $C_{p,m}$, as a function of bias b, let $W_{p,m}(b)$ be the total weight of UEs that would associate with small cell p if the bias of small cell p were set to b. In other words, $W_{p,m} = \sum_{u \in D_{p,m}(b)} w_u$ where

$$D_{p,m}(b) = \{u \in C_{p,m} : \text{RSRP}_{u,p} + b \geq \text{RSRP}_{u,m}\}, \tag{8.45}$$

where $\text{RSRP}_{u,m}$ and $\text{RSRP}_{u,p}$ are the received power (in dBm) of the reference signal at UE u from the best candidate macrocell and best candidate small cell, respectively. Then compute $W_{p,m}^*$ and $W_{p,m}(b)$ for every interfering small cell/macrocell pair (p, m) and every permissible bias value.
3. For every small cell p, cell selection bias b_p is set as

$$b_p = \arg\min_b \left[\sum_{m \in \mathcal{I}_p} |W_{p,m}(b) - W_{p,m}^*|^2 \right]. \tag{8.46}$$

Thus, the bias values are chosen as the ones that minimize the mean square error of the association vector of number of UEs to different small cells.

In many scenarios, operators that deploy small cells may desire to have a maximum or minimum bias for a small cell p (say, $b_{p,\max}$ and $b_{p,\min}$). For example, if a small cell is deployed to fill a coverage hole or high-demand area, then the value $b_{\min}$ should be set such that UEs around the coverage hole get associated with the small cell. This can be handled using the following steps:

1. First, run the joint UE-association and ABS-determination algorithm by setting $r_u^{MC} = 0$ for all UEs that get associated with the small cell even with minimum bias, and $r_u^{SC} = 0$ for all UEs that do not get associated with the small cell even with maximum bias.

2. Next, execute the three above steps for cell bias determination but with a minor modification in (8.46) so that the arg min operation is restricted to $b \in [b_{p,\min}, b_{p,\max}]$.

A last remark is that the described techniques allow computing the number of ABSs for every N_{sf} subframes (i.e., ABSs per ABS period). In practice, one must specify the exact subframes in an ABS period that are used as ABSs. The ABS number can be converted into a pattern as follows.

1. Index the subframes in an ABS period in a *consistent* manner across all macrocells and small cells.

2. Suppose a macrocell m leaves out k out of N_{sf} subframes as ABSs. Then macrocell m offers the *first* k subframes as ABSs where the *first* k relates to the indexing in the previous step.

This simple scheme works provided that all macrocells have the same set of permissible subframes (i.e., there is no restriction on a certain macrocell that it cannot offer certain subframes as ABS). Since a small cell can effectively use the least number of ABSs offered by interfering macrocells, this scheme would naturally ensure a provably correct mapping between the number of ABSs and ABS pattern.

8.4.4 Performance Evaluation

Using signal propagation maps generated by a commercially available radio network planning tool using the parameters and drive-test data of a real operational LTE network in a major US metropolitan area, the algorithm for solving the *RELAXED-ABS Problem* (8.2) discussed in this section has been evaluated [19].

Since the solution to *RELAXED-ABS Problem* (8.2) is an upper bound to the optimal solution of *OPT-ABS Problem* (8.1), the optimality gap g is obtained by comparing the solution to the *RELAXED-ABS Problem* with the solution to the *OPT-ABS Problem*.

TABLE 8.4 Optimality Gap g for Different Macrocell UE Density and Small Cell Powers

Macro-Cell UE Density	DU	DU	DU	U	SU
Small cell power	4 W	1 W	0.5 W	4 W	4 W
% of OPT-ABS	93.77%	95.64%	95.86%	92.98%	97.03%

DU, U, SU stand for dense urban, urban, and sub-urban UE density.

In particular, suppose R_u^{rel} and R_u^{alg} are the UE u's throughputs solving the *RELAXED-ABS Problem* and the *RELAXED-ABS Problem*, respectively. Then, the optimality gap g is defined as a factor $g < 1$ if $\sum_u \ln(R_u^{alg}) \geq \sum_u \ln(R_u^{rel}(1 - g))$. The smallest value of g that satisfies this can easily be computed.

Table 8.4 demonstrates that the above-described algorithm performance lies within 90% of the optimum in scenarios reflecting a real operational LTE network deployment by a leading operator in a major US metropolitan area.

Figure 8.4 shows the throughput gains that can be achieved with the proposed approach. Compared to a network configuration with no eICIC, the gains are more than 200% for the far-edge UEs (2.5th percentile of the throughputs) and 40–55% for edge UEs (5th–10th percentile of UE throughput). Compared to the case with no small cells present, the gains are even more dramatic and around 300% even for 5th percentile of the throughputs.

Importantly, the throughput gains over a small cell deployment without eICIC configured do not come at an appreciable expense of macrocell UE throughput. Though the macrocell eNodeBs have fewer subframes for transmissions due to ABS

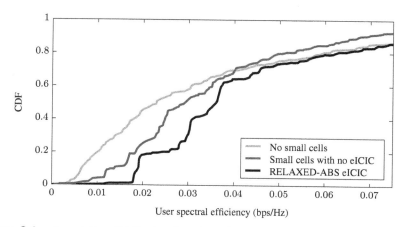

Figure 8.4. Cumulative distribution function of UE throughputs with (i) RELAXED-ABS eICIC, (ii) no eICIC but with small cells, and (iii) no eICIC and no small cells, assuming dense urban macrocell UE density and 4 W small cell transmit power.

offered to small cells, the macrocells have also fewer UEs as many UEs are offloaded by the small cells.

8.5 STOCHASTIC RELAXATION APPROACH

8.5.1 Formulation

The integer relaxation of the *OPT-ABS Problem* (8.1), described in the previous section as the *RELAXED-ABS Problem* (8.2), followed by "rounding" of the obtained solutions to feasible network configurations, is a natural approach to simplifying the NP-hard *OPT-ABS Problem*.

In this section, an alternative stochastic approach is presented. The utility-based formulation of the eICIC-configuration problem remains the same, but to make the system model more practical in comparison to the *RELAXED-ABS Problem*, a few improvements are made. Most importantly, a stochastic interference model is used to reflect the fact that wireless channel gains are by nature volatile due to phenomena such as multi-path propagation. In addition, the notion of cell-selection bias is explicitly incorporated into the system model, while the notion of airtime is modeled by assuming a round-robin scheduler.

8.5.1.1 Statistical Model The notation summarized in Table 8.3 is extended by denoting the set of macrocells that interfere with a small cell p by $\mathcal{I}_p(b) \subseteq \mathcal{M}$ as a function of its REB b. Additionally, $\mathcal{J}_p$ denotes the set of macrocells who could potentially offload UEs to small cell p. In general, some of the elements of $\mathcal{J}_p$ could overlap with $\mathcal{I}_p(b_{\max})$, but no structural relationship between $\mathcal{J}$ and $\mathcal{I}_p(.)$ is generally assumed.

To capture the statistical nature of wireless channels, the vector of all UE locations is assumed to be sampled from a probability space Ω. A specific realization of UE locations is denoted as $\omega \in \Omega$. Accordingly, all the random sets and variables are understood as functions of specific realization of UE location denoted by ω, though dependence on ω is not explicitly shown unless required.

A realization ω of UE locations results in an RSRP vector corresponding to each UE in the realization. Clearly, the evolution of UE locations across time is a stochastic process, whose elements belong to Ω. The only assumption made about the stochastic process is that it is ergodic.

The physical-layer data rate of each UE depends on its serving cell (either a macrocell or a small cell), which in turn depends on the REB of the (nearest) small cell. Let $\mathcal{L}_p(b)$ be the random set of active UEs associated with small cell p when the REB of the small cell is b, and also let $|\mathcal{L}_p(b)| = L_p(b)$.

Other random variables related to UE association are introduced in order to capture the stochastic system model.

Let $\mathcal{L}_{p,m}(b)$ be the random set of small cells that associate with small cell p that also have macrocell m as the best macrocell based on RSRP. Let also $|\mathcal{L}_{p,m}(b)| = L_{p,m}(b)$. Simply put, $\mathcal{L}_{p,m}(b)$ consists of all UEs that associate with small cell p with REB b that would have associated with macrocell m in the absence of the macrocell. Note that $\cup_{m \in \mathcal{I}_p(b)} \mathcal{L}_{p,m}(b) \subseteq \mathcal{L}_p(b)$. Let also

$$\mathcal{O}_{p,m}(b) = \mathcal{L}_{p,m}(b_{\max}) \setminus \mathcal{L}_{p,m}(b) \tag{8.47}$$

be the set of UEs that could have associated with small cell p under the maximum bias but who associate with macrocell m due to REB of p being b. Lastly, let $\mathcal{L}_m$ denote the random set of UEs who associate with macrocell m and who do not have any small cell to associate with (or, it is outside the maximum coverage area of any small cell); also, $|\mathcal{L}_m| = L_m$.

As a function of UE location realization ω, random variables related to UE data rates are defined as follows: $r_u^{(m)}(\omega)$ is the data rate achievable by UE u from its nearest macrocell, $r_u^{(p)}(\omega)$ is the data rate achievable by UE u from its nearest macrocell, $r_u^{(p)}(\omega)$ is the data rate achievable by UE u from its nearest small cell over any ABS, and $\tilde{r}_u^{(p)}(\omega)$ is the data rate achievable by UE u from its nearest small cell over a non-ABS.

To simplify the notational treatment of the optimized logarithmic utility metric, the following useful random variables are defined:

$$X_p(b) = \sum_{u \in \mathcal{L}_p(b)} \ln r_u^{(p)} \tag{8.48}$$

$$\tilde{X}_p(b) = \sum_{u \in \mathcal{L}_p(b)} \ln \tilde{r}_u^{(p)} \tag{8.49}$$

$$X_{p,m}(b) = \sum_{u \in \mathcal{O}_{p,m}(b)} \ln r_u^{(m)} \tag{8.50}$$

$$X_m(b) = \sum_{u \in \mathcal{L}_m} \ln r_u^{(m)} \tag{8.51}$$

As for measurement statistic collection, each UE measurement report to its serving eNodeB is assumed to contain the RSRP values of all cells the UE hears from, that is, all cells with RSRP above a threshold. Denote by $\mathrm{RSRP}_u(c, e)$ the RSRP measurement of cell e reported by a UE u associated to cell c. The random variable $\mathrm{RSRP}(c, e)$ denotes RSRP measurement of cell e from a *random* UE in cell c.

Cell eNodeBs are considered to only report aggregate statistics of the RSRPs. Denote the mean vector and covariance matrix of the RSRP measurement for UEs connected to cell c by $\mathbf{m}_c$ and $\mathbf{K}_c$, respectively. In other words,

$$\mathbf{m}_c[i] = \mathbb{E}[\text{RSRP}(c, i)] \tag{8.52}$$

$$\mathbf{K}_c[i,j] = \mathbb{E}[(\text{RSRP}(c, i) - \mathbf{m}_c[i])(\text{RSRP}(c,j) - \mathbf{m}_c[j])], \tag{8.53}$$

where $\mathbf{m}_c[i]$ is the ith element of $\mathbf{m}_c$ and $\mathbf{K}_c[i,j]$ is the (i,j)th element of the covariance matrix $\mathbf{K}_c$.

It is also assumed that each cell estimates $\mathbf{m}_c$ and $\mathbf{K}_c$ from the sample RSRP values reported by the UEs based on the empirical mean and empirical covariances. With a slight abuse of notation, the notations $\mathbf{m}_c$ and $\mathbf{K}_c$ are used to mean sample estimate values of these parameters. These estimates are reported by each cell c to the centralized self-organizing network (SON) server.

Overall, the measurement data from each cell can be characterized by the following tuple:

$$\langle \mathbf{m}_c, \mathbf{K}_c, \mathbb{E}[L_c] \rangle \tag{8.54}$$

Importantly, all these elements can be measured at the eNodeB using UE radio measurements that contain the RSRP vectors and by tracking the number of active sessions.

8.5.1.2 Measurement-Based Framework   In eICIC, each macrocell offers A_m ABSs out of N_{sf} subframes where N_{sf} is the eICIC periodicity. For a small cell p that has REB set to b, the number of useful ABSs cannot be more than the ABS offered by any interfering macrocell at REB value of b. Specifically, denoting explicitly the dependence of $A_p(b)$ on REB value b,

$$A_p(b) \leq \min_{m \in \mathcal{I}_p(b)} A_m \tag{8.55}$$

Let also $N_p(b) = N_{sf} - A_p(b)$ and $N_m = N_{sf} - A_m$ denote the non-ABSs available to small cell p and macrocell m, respectively. For notational simplicity, the ABS and non-ABS ratios will be denoted by the corresponding lower case symbols. For example, $a_m = A_m/N_{sf}$, $a_p(b) = A_p(b)/N_{sf}$ and similarly $n_m = 1 - a_m$ and $n_p(b) = 1 - a_p(b)$.

For a given UE association, the maximum throughput a UE achieves depends on the underlying MAC, that is, the overall UE load of the cell. A round-robin approach to resource sharing is assumed to make the problem tractable. In particular, denote the vector of small cell REBs $b_p \forall p \in \mathcal{P}$ by $\mathbf{b}_p$. Then, for a given UE association, the throughput of a UE associated with macrocell m is modeled as

$$R_u^{(m)}(a_m, \mathbf{b}_p) = \frac{r_u^{(m)}(1 - a_m)}{\overline{L}_m(\mathbf{b}_p)}, \tag{8.56}$$

where $r_u^{(m)}$ is the data rate of the UE when it associates with macrocell m and $\overline{L}_m$ is the total number of active UE's associated with cell c in the given association. The total number of UEs associated to a macrocell $\overline{L}_m$ is given by

$$\overline{L}_m(\mathbf{b}_p) = L_m + \sum_p O_{p,m}(\mathbf{b}_p), \tag{8.57}$$

a function of the REB b_p of small cell p.

The throughput of a UE associated with small cell p has two parts: ABS throughput ($R_u^{(p)}$) and non-ABS throughput ($\tilde{R}_u^{(p)}$)). It is assumed that the ABS and non-ABS resources are shared separately in a fair manner among competing UEs that have a non-zero rate over the respective subframes. More specifically,

$$R_u^{(p)}(a_p, b_p) = \frac{r_u^{(p)} a_p}{L_p(b_p)}, \; \tilde{R}_u^{(p)}(a_p, b_p) = \frac{\tilde{r}_u^{(p)}(1 - a_p)}{L_p(b_p)}, \tag{8.58}$$

where $L_p(b_p)$ is the total number of active UEs associated with small cell p in the given association with small cell REB set to b_p.

The average utility to be maximized is defined as the average over all possible realizations of UE locations. For a given UE location realizations $\omega \in \Omega$, the utility of macrocell m is defined as the sum of logarithms of throughputs of the associated UEs defined as a function of a_m and $\mathbf{b}_p$

$$U_m(\mathbf{b}_p, a_m) \triangleq \sum_{u \in \mathcal{L}_m} \ln \frac{r_u^{(m)}(1 - a_m)}{\overline{L}_m(\mathbf{b}_p)} + \sum_{p:m \in \mathcal{J}_p} \sum_{u \in \mathcal{O}_{p,m}(b_p)} \ln \frac{r_u^{(m)}(1 - a_m)}{\overline{L}_m(\mathbf{b}_p)} \tag{8.59}$$

$$= \sum_{u \in \mathcal{L}_m \cup \mathcal{O}_{p,m}(b_p)} \ln r_u^{(m)} + \overline{L}_m(\mathbf{b}_p) \ln \frac{(1 - a_m)}{\overline{L}_m(\mathbf{b}_p)} \tag{8.60}$$

Using the definition of the random variable $X_m, X_{p,m}$ for $\overline{L}_m(\mathbf{b}_p)$ being given by (8.57), it follows that

$$U_m(\mathbf{b}_p, a_m) = X_m + \sum_{p:m \in \mathcal{J}_p} X_{p,m}(\mathbf{b}_p) + \overline{L}_m(\mathbf{b}_p) \ln \frac{(1 - a_m)}{\overline{L}_m(\mathbf{b}_p)} \tag{8.61}$$

For a given UE location realization $\omega \in \Omega$, the utility of small cell p is defined as the overall summation of two summation terms: sum of logarithms of ABS-throughputs and sum of logarithms of non-ABS throughputs. In particular, for a small cell p, the small cell utility is defined as a function of its REB b_p and vector of ABS offered by all macrocells $\mathbf{A}_m$

$$U_p(b_p, a_p) \triangleq \sum_{u \in \mathcal{L}_p(b)} \ln \frac{r_u^{(p)} a_p}{L_p(b_p)} + \sum_{u \in \mathcal{L}_p(b)} \ln \frac{\tilde{r}_u^{(p)} (1-a_p)}{\tilde{L}_p(b_p)} \qquad (8.62)$$

$$= \sum_{u \in \mathcal{L}_p(b)} \ln r_u^{(p)} + L_p(b_p) \ln \frac{a_p}{L_p(b_p)} \qquad (8.63)$$

$$+ \sum_{u \in \mathcal{L}_p(b)} \ln \tilde{r}_u^{(p)} + L_p(b_p) \ln \frac{(1-a_p)}{L_p(b_p)} \qquad (8.64)$$

Using the definition of $X_p(b_p)$,

$$U_p(b_p, \mathbf{a}_m) = X_p(b_p) + \tilde{X}_p(b_p) + L_p(b_p) \ln \frac{a_p}{L_p(b_p)} + L_p(b_p) \ln \frac{(1-a_p)}{L_p(b_p)} \qquad (8.65)$$

The overall utility of the system is the expected utility where the expectation is with respect to the probability distribution over the space Ω. It is given by

$$\text{Util}(\mathbf{b}_p, \mathbf{a}_m) = \sum_{p \in \mathcal{P}} \mathbb{E}[U_p(b_p, \mathbf{a}_m)] + \sum_{m \in \mathcal{M}} \mathbb{E}[U_m(\mathbf{b}_p, a_m)] \qquad (8.66)$$

8.5.1.3 Problem Statement

Problem 8.3 *(STOCHASTIC-ABS)*
Given

- *measurement statistics $\langle \mathbf{m}_c, \mathbf{K}_c, \mathbb{E}[L_c] \rangle$ from every cell c*

solve the optimization problem

$$maximize_{\{\mathbf{b}_p, \mathbf{a}_m\}} \sum_{p \in \mathcal{P}} \mathbb{E}[U_p(b_p, \mathbf{a}_m)] + \sum_{m \in \mathcal{M}} \mathbb{E}[U_m(\mathbf{b}_p, a_m)] = Util(\mathbf{b}_p, \mathbf{a}_m) \qquad (8.67)$$

by finding

- *the vector $\mathbf{b}_p$ of small cell REBs,*
- *the vector $\mathbf{a}_m$ of macrocell ABSs.*

To solve the above problem, two questions must be answered:

1. Given measurement statistics $\langle \mathbf{m}_c, \mathbf{K}_c, \mathbb{E}[L_c] \rangle$, how to estimate relevant statistics for $X_p, X_{p,m}, L_p, O_{p,m}, L_m$ that characterize $\mathbb{E}[U_p(b_p, \mathbf{a}_m)]$ and $\sum_{m \in \mathcal{M}} \mathbb{E}[U_m(\mathbf{b}_p, a_m)]$?

2. Given statistics of random variables that characterize $\mathbb{E}[U_p(b_p, \mathbf{a}_m)]$ and $\sum_{m \in \mathcal{M}} \mathbb{E}[U_m(\mathbf{b}_p, a_m)]$, how to find ABSs offering ratio $\mathbf{a}_m$ of macrocells and REB $\mathbf{b}_p$ of small cells that maximize Util$(\mathbf{b}_p, \mathbf{a}_m)$?

To estimate relevant statistics for $X_p, X_{p,m}, L_p, O_{p,m}, L_m$ that are based on measurement statistics $\langle \mathbf{m}_c, \mathbf{K}_c, \mathbb{E}[L_c] \rangle$, Monte Carlo-based sampling can be used as discussed in [20].

It remains to compute optimal values of b_p and a_m so as to maximize $\mathbb{E}[\text{Util}(\mathbf{b}_p, ba_m)]$, given $\mathbb{E}[X_m], \mathbb{E}[L_m], \mathbb{E}[X_p(b_p)], \mathbb{E}[X_{p,m}(b_p)], \mathbb{E}[L_p(b_p)]$, and $\mathbb{E}[O_{p,m}(b_p)]$.

Yet, the first challenge associated with this goal is that it has no convex or concave structure because the moments of X can have, in general, arbitrary dependence on b_p. A second challenge consists of that the problem has to be solved for large networks comprising tens of thousands of cells, thus a computationally light algorithm is preferred.

The following discussion summarizes the approach of [20].

8.5.2 Alternating Maximization

A heuristic based on the principle of alternating maximization [35] is presented here. The main idea of alternating maximization-based algorithms is to iterate the following two steps until a convergence criterion is met:

1. Fix the values of $\mathbf{a}_m$ and maximize the expected utility with respect to REBs $\mathbf{b}_p$.
2. Fix these values of $\mathbf{b}_p$ and maximize the expected utility with respect to the $\mathbf{a}_m$.

In the following, it is shown that this approach lends to an elegant decomposition of the problem that makes it computationally very efficient. Numerical simulations prove a good performance of the resulting algorithm.

8.5.2.1 Maximizing REBs for Fixed ABSs
To define the iterative procedure, let $\mathbf{a}_m^{(t)}$ and $\mathbf{b}_p^{(t)}$ denote the ABS and REB values at the previous iteration. Let also $\{b_p, \mathbf{b}_{\setminus p}^{(t)}\}$ denote the REB vector that has REB values of small cell p set to b_p, and REB values of all other small cells $\mathcal{P} \setminus \{p\}$ set to the values set in iteration t. The REB maximization of small cell p is then set as

$$b_p^{(t+1)} = \arg\max_{b_p} \left\{ \mathbb{E}[U_p(b_p, \mathbf{a}_m^{(t)})] + \sum_{m \in \mathcal{J}_p} \mathbb{E}[U_m(\{b_p, \mathbf{b}_{\setminus p}^{(t)}\}, a_m^{(t)})] \right\} \quad (8.68)$$

Note that the value of $b_p^{(t+1)}$ for each small cell can be carried out independently and simply based on the estimates of $\mathbb{E}[X_p(b_p)]$, $\mathbb{E}[X_{p,m}(b_p)]$, $\mathbb{E}[L_p(b_p)]$ and $\mathbb{E}[O_{p,m}(b_p)]$ for different values of b_p. The values of b_p can be assumed to be discretized, that is, the expectations are available for discrete set of values.

8.5.2.2 Maximizing ABSs for Fixed REB

To maximize the values of ABS for given values of $\mathbf{b}_p^{(t)}$, one can show that the problem of computing

$$\max_{a_m} \left\{ \sum_m \mathbb{E}[U_m(\mathbf{b}_p^{(t)}, a_m)] + \sum_p \mathbb{E}[U_m(b_p^{(t)}, \mathbf{a}_m)] \right\} \tag{8.69}$$

is equivalent to

$$\max_{\mathbf{a}_m \in [0,1]^{|\mathcal{M}|}} \sum_m \overline{L}_m^{(t)} \ln(1 - a_m) + \sum_p L_p^{(t)} \ln a_p (1 - a_p) \tag{8.70}$$

subject to

$$a_p \leq \min_{m \in \mathcal{I}_p(b_p^{(t)})} a_m. \tag{8.71}$$

In the above, $\overline{L}_m^{(t)} = \overline{L}_m(\mathbf{b}_p^{(t)})$ and $L_p^{(t)} = L_p(b_p^{(t)})$ refers to the values at the previous iteration.

In general, the maximization problem given by (8.70) is a convex program and it can be solved using standard non-linear programming techniques followed by discretizing a_m (recall that a_m can take values in steps of $1/N_{sf}$). However, this could be computationally expensive especially for a network with tens of thousands of cells and doing this over and again in every iteration of alternating maximization is prohibitively expensive.

An alternative solution can be proposed that decomposes into a sub-problem for each macrocell. The main observation toward this decomposition is that, from the point of view of each macrocell (small cell), the optimal a_m (a_p) has a closed form if the presence of other macrocells (small cells) in the interference graph is ignored.

It turns out that the solution obtained by taking the average solution from these two views are provably good approximations of optimal solution [20]. These steps are as follows:

1. Each macrocell temporarily reserves a_m^{res} as

$$a_m^{res} = 1 - \frac{\overline{L}_m^{(t)}}{\overline{L}_m^{(t)} + \sum_{p:m \in \mathcal{I}_p(b_p^{(t)})} L_p^{(t)}}. \tag{8.72}$$

Each small cell temporarily reserves a_p^{res} as

$$a_p^{\text{res}} = \frac{L_p^{(t)}}{L_p^{(t)} + \sum_{m \in I_p(b_p^{(t)})} \overline{L}_m^{(t)}} . \tag{8.73}$$

2. Each macrocell sets ABS usable based on a_p^{res} as

$$1 - a_m^{\text{usable}} = \min_{p: m \in I_p(b_p)} (1 - a_p^{\text{res}}), \tag{8.74}$$

and each small cell sets ABS usable based on a_m^{res} as

$$a_p^{\text{usable}} = \min_{m \in I_p(b_p)} a_m^{\text{res}}. \tag{8.75}$$

3. The value of $a_m^{(t)}$ and $a_p^{(t)}$ are set as

$$a_m^{(t)} = \frac{a_m^{\text{res}} + a_m^{\text{usable}}}{2} , \; a_p^{(t)} = \frac{a_p^{\text{res}} + a_p^{\text{usable}}}{2} . \tag{8.76}$$

8.5.3 Performance Evaluation

Using signal propagation maps generated by a commercially available radio network planning tool and using the parameters and drive-test data of real operational LTE network in a major US metropolitan area, the algorithm for solving the *STOCHASTIC-ABS Problem* (8.3) discussed in this section has been evaluated [20].

Similar to the performance evaluation in the previous section, three network configurations are compared: (i) a network with metrocells deployed and eICIC configured based on the discussed algorithm, (ii) a network with metrocells deployed but no eICIC configured, (iii) a network consisting of macrocells only.

Figure 8.5 shows the overall performance for a representative UE snapshot. As observed, the median estimated throughput increases by almost 60% with eICIC configured on the basis of the discussed algorithm. The metrocell density is relatively low (72 metrocells and 132 macrocells). In denser deployments, much higher gains can be expected.

Figures 8.6 and 8.7 shows the gains for two metrocells that receive diverse ABS from the nearby macrocells and that have diverse REB. The first metrocell from Fig. 8.6 receives 18 ABSs out of a 40 consecutive subframes whereby its REB is 7.5 dB. The second metrocell from Fig. 8.7 receives 26 ABSs out of 40 consecutive subframes and has a REB of 0.8 dB. With eICIC, the median estimated throughput of both metrocells is around two to three times higher with respect to no eICIC configured.

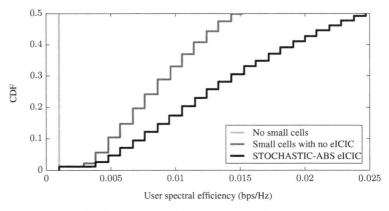

Figure 8.5. CDF of UE throughputs with (i) *STOCHASTIC-ABS* eICIC, (ii) small cells with no eICIC, and (iii) no small cells.

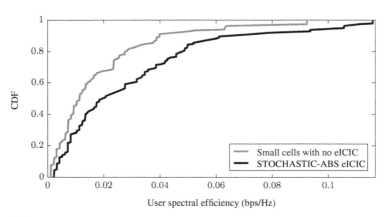

Figure 8.6. CDF of UE throughputs with *STOCHASTIC-ABS* eICIC, and with no eICIC for a first metrocell having 128 UEs.

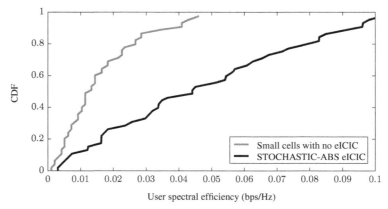

Figure 8.7. CDF of UE throughputs with *STOCHASTIC-ABS* eICIC, and with no eICIC for a second metrocell having 45 UEs.

8.6 SUMMARY AND CONCLUSIONS

The concept of time-domain multiplexing has been introduced to the LTE standards as a way for mitigating inter-cell and inter-tier interference.

In particular, the support for ABS has been introduced in the LTE Release 10. The aggressor cell can be configured to mute their non-critical transmissions in selected subframes that are then reused by victim cell for transmissions to its interference-affected UEs such as UEs at the edge of small cells applying a cell-selection bias to improve its offloading capabilities. The possibility of restricting UE measurements of signal quality to ABSs and non-ABSs helps providing the decision-making eNodeB with reliable feedback on local networking conditions.

From the optimization point of view, the problem of setting ABS muting patterns as well as small cell REBs is challenging due to its NP-hardness. Several heuristics have been discussed, including both empirically inspired techniques and analytically founded approximation algorithms.

Simple techniques such as load balancing require little input but may perform far from the optimum. On the other hand, the presented analytical algorithms require a large amount of inputs that may be difficult or time-consuming to obtain considering that UEs are connected only to one server, but their evaluation by using a real network plan demonstrates large performance gains stemming from joint optimization of ABS and UE association.

A broader implication of these observations is that network optimization is necessary and desirable in the eICIC context. This is especially true in view of the fact that wireless networks undergo significant spatiotemporal traffic load fluctuations observed during the day. Network data on mobility patterns indicate that individual UEs follow regular routes with occasional deviations as intuitively understood from social human behaviors. This regularity can be observed with aggregate cell data over the course of the day and as such cell load together with SINR (or rate) statistics are required to be constantly monitored. Thus, it is imperative to establish a SON-based approach to eICIC parameter configuration of an LTE network.

REFERENCES

1. "FDD home eNode B (HeNB) radio frequency (RF) requirements analysis," 3GPP Technical Specification Group Radio Access Network, Evolved Universal Terrestrial Radio Access (E-UTRA), Tech. Rep. TR 36.921 v12.0.0, Oct. 2014. [Online]. Available: http://www.3gpp.org

2. "FDD home eNode B (HeNB) radio frequency (RF) requirements analysis," 3GPP Technical Specification Group Radio Access Network, Evolved Universal Terrestrial Radio Access (E-UTRA), Tech. Rep. TR 36.922 v12.0.0, Sep. 2014. [Online]. Available: http://www.3gpp.org

3. "Study on small cell enhancements for E-UTRA and E-UTRAN: Higher layer aspects," 3GPP Technical Specification Group Radio Access Network, Evolved Universal

Terrestrial Radio Access (E-UTRA), Tech. Rep. TR 36.842 v12.0.0, Jan. 2014. [Online]. Available: http://www.3gpp.org

4. I. Guvenc, "Capacity and fairness analysis of heterogeneous networks with range expansion and interference coordination," *IEEE Commun. Lett.*, vol. 15, no. 10, pp. 1084–1087, Oct. 2011.

5. H. Claussen, L. T. W. Ho, and L. G. Samuel, "An overview of the femtocell concept," *Bell Labs Tech. J.*, vol. 13, no. 1, pp. 221–245, May 2008.

6. "S1 general aspects and principles," 3GPP Technical Specification Group Radio Access Network, Evolved Universal Terrestrial Radio Access (E-UTRA), Tech. Rep. TR 36.410 v12.1.0, Dec. 2014. [Online]. Available: http://www.3gpp.org

7. "S1 application protocol (S1AP)," 3GPP Technical Specification Group Radio Access Network, Evolved Universal Terrestrial Radio Access (E-UTRA), Tech. Rep. TR 36.413 v12.4.0, Dec. 2014. [Online]. Available: http://www.3gpp.org

8. "Physical layer procedures," 3GPP Technical Specification Group Radio Access Network, Evolved Universal Terrestrial Radio Access (E-UTRA), Tech. Rep. TS 36.213 v11.9.0, Jan. 2015. [Online]. Available: http://www.3gpp.org

9. E. Bjornson, L. Sanguinetti, J. Hoydis, and M. Debbah. (2014). "Designing multi-user MIMO for energy efficiency: When is massive MIMO the answer?" [Online]. Available: http://arxiv.org/abs/1310.3843

10. "Physical layer procedures," 3GPP Technical Specification Group Radio Access Network, Evolved Universal Terrestrial Radio Access (E-UTRA), Tech. Rep. TS 36.213 v12.4.0, Jan. 2015. [Online]. Available: http://www.3gpp.org

11. K. Balachandran, J. Kang, K. Karakayali, and K. Rege, "NICE: A network interference cancellation engine for opportunistic uplink cooperation in wireless networks," *IEEE Trans. Wireless Commun.*, vol. 10, no. 2, pp. 540–549, Feb. 2011.

12. "Further advancements of E-UTRA physical layer aspects," 3GPP Technical Specification Group Radio Access Network, Evolved Universal Terrestrial Radio Access (E-UTRA), Tech. Rep. TR 36.814 v9.0.0, Mar. 2010. [Online]. Available: http://www.3gpp.org

13. "E-UTRA and E-UTRAN: Overall description, stage 2," 3GPP Technical Specification Group Radio Access Network, Evolved Universal Terrestrial Radio Access (E-UTRA), Tech. Rep. TS 36.300 v12.4.0, Jan. 2015. [Online]. Available: http://www.3gpp.org

14. L. Lindbom, R. Love, S. Krishnamurthy, C. Yao, N. Miki, and V. Chandrasekhar. (2011 Jun.). "Enhanced inter-cell interference coordination for heterogeneous networks in LTE advanced: A survey" [Online]. Available: http://arxiv.org/abs/1112.1344

15. "Requirements for support of radio resource management," 3GPP Technical Specification Group Radio Access Network, Evolved Universal Terrestrial Radio Access (E-UTRA), Tech. Rep. TS 36.133 v12.6.0, Jan. 2015. [Online]. Available: http://www.3gpp.org

16. "X2 Application Protocol (X2AP)," 3GPP Technical Specification Group Radio Access Network, Evolved Universal Terrestrial Radio Access (E-UTRA), Tech. Rep. TS 36.423 v12.4.2, Jan. 2015. [Online]. Available: http://www.3gpp.org

17. "Radio Resource Control (RRC) protocol specification," 3GPP Technical Specification Group Radio Access Network, Evolved Universal Terrestrial Radio Access (E-UTRA), Tech. Rep. TS 36.331 v12.4.1, Sep. 2014. [Online]. Available: http://www 3gpp.org

18. R. Madan, J. Borran, A. Sampath, N. Bhushan, A. Khandekar, and T. Ji, "Cell association and interference coordination in heterogeneous LTE-A cellular networks," *IEEE J. Sel. Areas Commun.*, vol. 28, no. 9, pp. 1479–1489, Dec. 2010.

19. S. Deb, P. Monogioudis, J. Miernik, and J. P. Seymour, "Algorithms for enhanced inter-cell interference coordination (eICIC) in LTE Hetnets," *IEEE/ACM Trans. Netw.*, vol. 22, no. 1, pp. 137–150, Feb. 2014.

20. S. Deb, P. Monogioudis, and S. Venkatesan, "Measurement data driven computationally fast eICIC for 4G LTE Hetnets," Tech. Rep., Sep. 2014.

21. D. López-Pérez, H. Claussen, and L. Ho, "Duty cycles and load balancing in HetNets with eICIC Almost Blank Subframes," in *Proc. IEEE PIMRC*, London, United Kingdom, Sep. 8–11, 2013, pp. 1–5.

22. B. Soret and K. I. Pedersen, "Macro transmission power reduction for HetNet co-channel deployments," in *Proc. IEEE Globecom*, Anaheim, CA, Dec. 3–7, 2012, pp. 1–5.

23. P. Mogensen, N. Wei, I. Z. Kovacs, F. Frederiksen, A. Pokhariyal, K. I. Pedersen, T. Kolding, K. Hugl, and M. Kuusela, "LTE capacity compared to the Shannon Bound," in *Proc. IEEE VTC Spring*, Dublin, Ireland, Apr. 22–25, 2007, pp. 1–5.

24. Q. Wu and E. Esteves, *Advances in 3G Enhanced Technologies for Wireless Communications (Chapter 4)*. Norwood, MA: Artech House, Mar. 2002.

25. R. Agrawal, A. Bedekar, R. La, and V. Subramanian, "Class and channel-condition based weighted proportionally fair scheduler," *Teletraffic Sci. Eng.*, vol. 4, pp. 553–567, Sep. 2001.

26. S. Deb, P. Monogioudis, J. Miernik, and J. P. Seymour, "Algorithms for enhanced inter cell interference coordination (eICIC) in LTE Hetnets," Alcatel-Lucent Internal Tech. Rep., 2012. Available upon request.

27. M. R. Garey and D. S. Johnson, *Computers and Intractability: A Guide to the Theory of NP-Completeness*. New York: W. H. Freeman, 1979.

28. D. Lopez-Perez, X. Chu, and I. Guvenc, "On the expanded region of picocells in heterogeneous networks," *IEEE J. Sel. Topics Signal Process.*, vol. 6, no. 3, pp. 281–294, Mar. 2012.

29. D. Lopez-Perez and H. Claussen, "Duty cycles and load balancing in HetNets with eICIC almost blank subframes," in *Proc. IEEE PIMRC*, London, United Kingdom, Sep. 2013, pp. 173–178.

30. M. Vajapeyam, A. Damnjanovic, J. Montajo, T. Ji, Y. Wei, and D. Malladi, "Downlink FTP performance of heterogeneous networks for LTE advanced," in *Proc. IEEE ICC 2011, Workshop Heterogen. Netw.*, Kyoto, Japan, Jun. 2011, pp. 1–5.

31. A. Nedic and A. E. Ozdaglar, "Subgradient methods in network resource allocation: Rate analysis," in *Proc. Inform. Sci. Syst.*, Princeton, USA, Mar. 19–21, 2008, pp. 1189–1194.

32. D. P. Bertsekas, *Nonlinear Programming*. Athena Scientific, Sep. 2004.

33. L. Chen, S. H. Low, M. Chiang, and J. C. Doyle, "Cross-layer congestion control, routing and scheduling design in ad hoc wireless networks," in *Proc. IEEE Infocom*, Barcelona, Spain, Apr. 23–29, 2006, pp. 1–12.

34. R. Srikant, *The Mathematics of Internet Congestion Control (Systems and Control: Foundations and Applications)*. Boston, MA: Birkhauser, 2004.

35. U. Niesen, D. Shah, and G. W. Wornell, "Adaptive alternating minimization algorithms," *IEEE Trans. Inform. Theory*, vol. 55, no. 3, pp. 1423–1429, Mar. 2009.

9

THE SECTOR OFFSET CONFIGURATION

9.1 INTRODUCTION

As introduced in Chapter 1, vendors and operators face the challenge of meeting user equipment (UE) traffic demands in a cost-effective manner. Given the limited amount of available spectrum, this is challenging and requires reusing the available spectrum as much as possible in an efficient manner.

The transition to a heterogeneous network (HetNet) structure where small cells reuse the spectrum locally and provide most of the capacity, while macrocells provide a wide coverage for mobile UEs, can address the mentioned capacity problems, as shown in previous chapters.

Today, indoor small cells are deployed in large numbers, mainly in the form of femtocells, and many vendors have also started to develop outdoor small cells to complement macrocellular coverage (see Chapter 1). Most of today's small cell deployments are configured to transmit on a dedicated frequency carrier, as described in Chapter 4. While this avoids cross-tier interference, it also limits the available spectrum that small cells and macrocells can access and thus is less efficient than co-channel deployments, where small cells and macrocells reuse the whole available spectrum. However, while co-channel deployments provide a better frequency utilization, the additional cross-tier interference can result in coverage, capacity and mobility management issues for some UEs. In Chapters 7 and 8, frequency and time

Small Cell Networks: Deployment, Management, and Optimization, First Edition. Holger Claussen,
David López-Pérez, Lester Ho, Rouzbeh Razavi, and Stepan Kucera.
© 2017 by The Institute of Electrical and Electronic Engineers, Inc. Published 2017 by John Wiley & Sons, Inc.

domain inter-cell interference coordination (ICIC) techniques such as closed access (CA) and almost blank subframe (ABS) were used respectively to mitigate interference in co-channel deployments. However, these techniques still rely on scarifying some frequency/time resources for interference mitigation. Due to the interactions between the small cell and macrocell tiers in co-channel deployments, a joint optimization of the deployment and operation of small cells and macrocells can provide a significantly enhanced network performance. Small cell and macrocell tiers should not be considered as disconnected entities, but planned and deployed as a whole.

In this chapter, a novel sector offset configuration [1, 2] for the macrocell tier is proposed, that can significantly enhance both average and cell-edge macrocell UE throughput without reducing the frequency reuse factor. Moreover, the proposed sector offset configuration provides the basis for an improved ICIC between small cells and macrocells, which in turn results in a significantly enhanced performance for cell-edge small cell UEs. The proposed sector offset configuration also improves mobility performance, since handovers are started at a higher signal quality due to its ICIC properties. Details of two different implementations for the sector offset configuration tailored to multi-carrier and single-carrier HetNets are discussed in detail.

The remainder of this chapter is organized as follows. In Section 9.2, the proposed sector offset configuration for the macrocell tier is presented and its implementation for multi-carrier and single-carrier HetNets is discussed, together with its benefits and drawbacks. In Section 9.3, the proposed sector offset configuration is extended to the vertical domain. In Section 9.4, it is explained how small cells can take advantage of the proposed sector offset configuration and its implications. In Section 9.5, the scenario and system model used for the evaluation of the proposed sector offset configuration are introduced, followed by performance comparisons based on system-level simulations in Sections 9.6 and 9.7 for macrocell-only and HetNet scenarios, respectively. Finally, conclusions are drawn in Section 9.8.

9.2 MACROCELL TIER WITH HORIZONTAL SECTOR OFFSET CONFIGURATION

In order to improve the average and cell-edge UE performance in the macrocell tier, and provide the basis of an improved ICIC between the small cell and the macrocell tiers, a novel horizontal sector offset configuration [1] is proposed, in which:

- The available spectrum is divided into two spectrum fragments.
  In a high-speed packet access (HSPA) network, these spectrum fragments may be two HSPA carriers, while in a long-term evolution (LTE) network, they may also be two different LTE carriers or two subsets of resource blocks (RBs) that belong to the same LTE carrier. If CA is used, these spectrum fragments could be two-component carriers. Technology-dependent implementation details will be addressed in the following sections.

- Each base station (BS) is equipped with two sector configurations so that the antenna patterns in the first configuration are offset with respect to the antenna patterns in the second configuration in a way that the directions of highest gains in the first sector configuration point in the directions of the sector boundaries in the second sector configuration.
- Each sector configuration manages and transmits on a different spectrum fragment.

Figure 9.1 depicts the traditional and the sector offset configurations for a three-sector BS. In the traditional configuration, the coverage of the first and the second spectrum fragments fully overlap with each other in order to reuse the existing hardware, and share the same cell edges. In the sector offset configuration, the two-sector configurations are offset 60 degrees with respect to each other, and thus the direction of the main gain in one spectrum fragment overlaps with the cell edge of the complementary spectrum fragment. In Fig. 9.1, it is also important to note that because the same antenna patterns are used in the traditional and the sector offset configurations, there is a large overlap among the coverage area of neighboring antennas in the sector offset configuration. This is a main difference compared to a six-sector BSs, where neighboring antennas do not overlap much. Such overlap leads to some benefits in terms of handover performance, as will be shown later.

As a result of the sector offset configuration, the macrocell performance is enhanced because:

- There are more antennas pointing in different directions. As a result, more UEs benefit from higher antenna gain.
- ICIC is improved since neighboring antennas transmit in complementary spectrum fragments.

As a consequence, compared to the traditional macrocell sector configuration, the areas suffering from high interference between sectors of a BS do not overlap any more in both spectrum fragments. In other words, in the traditional sector configuration shown in Fig. 9.1a, a UE connected in between two sectors of the traditional BS would suffer from a poor signal-to-interference-plus-noise ratio (SINR). However, in the proposed sector offset configuration shown in Fig. 9.1b, this UE would benefit from an improved SINR because there is an antenna pointing at it, and due to the mentioned ICIC features (i.e., neighboring antennas transmit in complementary spectrum fragments).

It is also important to note that due to the large overlap between neighboring antennas, and the use of complementary spectrum, handovers are started at a higher SINR and thus handover failures (HOFs) are mitigated. UEs have more time to perform the handover procedure.

One way of looking at this sector offset configuration is as a special form of static beamforming [3], where each UE follows the antenna beam via handovers or

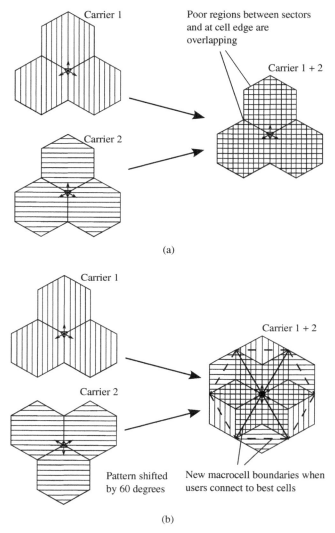

Figure 9.1. Multi-carrier HSPA/LTE HetNet configuration for traditional cell structure with partial frequency reuse (a) and proposed configuration with sector offset (b).

scheduling decisions, instead of having an antenna beamformed in the direction of each UE. The proposed configuration is much simpler and does not require channel amplitude and phase measurements and feedback for multiple transmit antennas.

Alternatively, coordinated multi-point (CoMP) transmissions could also be used to enhance the performance of cell-edge UEs [4]. However, they require intensive inter-BS coordination to ensure that the same signal is transmitted to the targeted cell-edge UE from different BSs. The sector offset configuration is a much lighter approach to improve cell-edge performance, in which the burden of complexity, feedback, and overhead is mitigated.

9.2.1 Network Configuration

At the network level, the sector offset configuration can be realized using two different approaches depending on the number of available carriers.

9.2.1.1 Multi-carrier Network

In a multi-carrier network, a straightforward approach to realize the sector offset configuration is to associate a different frequency carrier to each sector configuration [1]. However, one disadvantage of this sector offset configuration implementation is that connecting each UE to the best sector requires additional handovers compared to the traditional approach. Moreover, since the best neighboring sector is always on a different carrier, these additional handovers are inter-carrier handovers, which require the UEs to perform and report measurements on different carriers. In fairness and in contrast to a traditional cell structure, these handovers occur at higher SINRs when using the sector offset configuration, and are thus less sensitive to delays or other problems that can cause dropped calls.

9.2.1.2 Single-carrier Network

Taking advantage of the flexibility of orthogonal frequency division multiple access (OFDMA), the sector offset configuration can also be realized in LTE networks while using a unique frequency carrier. In this way, the mentioned increased number of handovers in the multi-carrier network implementation can be avoided [2].

In more detail, the proposed configuration for single-carrier LTE networks is as follows (see Fig. 9.2):

- The LTE subframe is divided into two fragments in the frequency domain. For example, if there are R available RBs, the first spectrum fragment may contain the first $R/2$ RBs and the second spectrum fragment may contain the remaining $R/2$ RBs. Other partitioning approaches are also feasible.
- Neighboring offset antennas are grouped in pairs such that each antenna only belongs to one pair and each antenna of a pair transmits on a different spectrum fragment.

In this way, a similar structure as in the multi-carrier network implementation is realized.

As illustrated in Fig. 9.2, it is important to note that the R RBs transmitted by each antenna pair still use the same pilot signal, cell-specific reference symbol (CRS) in

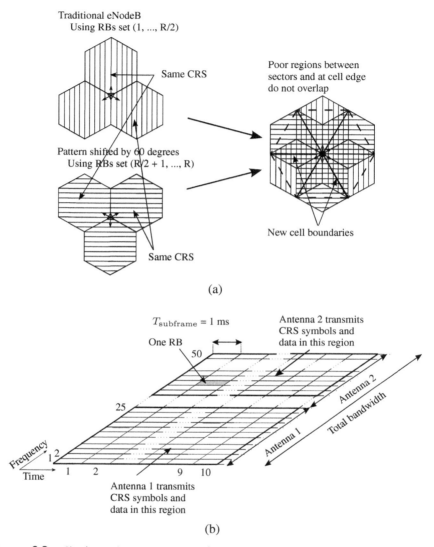

Figure 9.2. Single-carrier LTE HetNet configuration for traditional cell structure with frequency reuse of 1 (a) and proposed configuration with sector offset (b).

LTE jargon, as in the traditional approach, regardless of the orientation of the antennas used to transmit it. As a result, the two offset antennas of a pair appear as a unique antenna port to UEs, and thus handovers between these offset antennas are avoided. The UE only sees one antenna port and is not aware that there are two physical antennas associated with such antenna port.

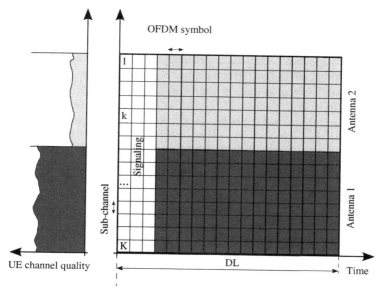

Figure 9.3. UE channel quality and resource block allocation over different antennas.

From a UE perspective, this CRS transmission configuration, in which part of the CRS symbols are transmitted over one antenna (Antenna 1) and the remaining CRS symbols over the offset antenna (Antenna 2), may result in received SINR improvements over part of the RBs and received SINR degradations over the remaining RBs. This very same fact allows appropriate antenna-to-UE assignments using off-the-shelf schedulers, for example, proportional fair (PF). UEs that are under the coverage of Antenna 1 will see higher SINRs in the CRS symbols of the first $K/2$ RBs than in the remaining RBs, and report in the corresponding sub-band channel quality indicators (CQIs) a better signal quality[1] (see Fig. 9.3). Then, due to the better reported SINR, off-the-shelf schedulers will prefer to assign UEs that are under the coverage of Antenna 1 to the first $K/2$ RBs, and UEs that are under the coverage of Antenna 2 to the remaining RBs. Since there is a large overlap among the coverage provided by the Antenna 1 and Antenna 2, some UEs can benefit from a large signal quality in both spectrum fragments (peak throughput can be demonstrated). This proposed sector offset configuration is compliant with current eNodeBs and UEs.

It is important to note that reference signal received power (RSRP) and wideband SINR measurements are taken by the UEs over the CRS symbols across the whole

---

[1] CQI is fed back from UEs to eNodeBs to assist scheduling, and are based on measurements taken over CRS symbols. channel state information-reference signals (CSI-RS) may also be used to increase the reliability of such measurements in the proposed sector offset configuration and aid channel-dependent scheduling [4].

bandwidth. Therefore, the transmission of half of the CRS symbols over one antenna and the remaining CRS symbols over the offset antenna has an impact on such measurements too. In the performance analysis presented in Section 9.6, it will be shown that such impact does not degrade UE performance. Indeed, the split of the transmission of CRS symbols between offset antennas reduces wideband SINRs in the high regime, but improves them in the low regime, thus enhancing performance at the cell edge. This can be seen as a trade-off between cell-edge performance and maximum capacity. The same is true for the physical downlink control channel (PDCCH) messages, which are also scrambled all across the bandwidth. With regard to PDCCH messages, it is worth mentioning that in LTE Release 11 and beyond, PDCCH messages do not need to be scrambled all across the bandwidth, and can be transmitted in specific RBs using enhanced physical downlink control channels (EPDCCHs) [5]. UEs that are under the coverage of Antenna 1 will be configured to receive EPDCCH via RBs in spectrum fragment 1, while UEs that are under the coverage of Antenna 2 will be configured to receive EPDCCH via RBs in spectrum fragment 2. As a result, the mentioned issue vanishes.

9.2.2 Hardware Requirements

An example of how the sector offset configuration can be implemented is shown in Fig. 9.4. Compared to the traditional approach in Fig. 9.4a, the two antenna element columns with a mechanical offset in Fig. 9.4b can be used to realize the sector offset configuration. The two antenna element columns can be fitted in one radome, so that the visual impact at the antenna tower remains similar.

Since the signal on spectrum fragment 1 is transmitted via a different antenna than the signal on spectrum fragment 2, two different radio frequency (RF) chains with lower power amplifiers are also required, instead of the traditional one RF chain with a wideband amplifier (see Fig. 9.5). The energy transmitted on each carrier remains the same.

Moreover, two separated base band units may also be required to process the signals belonging to spectrum fragment 1 and spectrum fragment 2. The digital

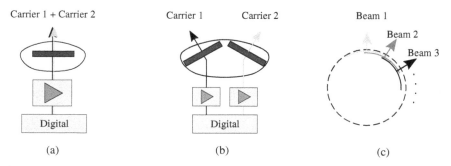

Figure 9.4. Antenna design/implementation. (a) Traditional radome, (b) Sector offset radome and (c) Circular array.

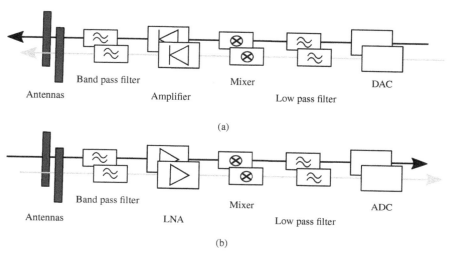

Figure 9.5. RF chain design. (a) Tx RF chain and (b) Rx RF chain.

samples belonging to spectrum fragment 1 can be processed by one base band unit, while the digital samples belonging to spectrum fragment 2 can be processed by the second base band unit. Zero padding can be used to adapt the bandwidth of the signal to the size of the existing fast Fourier transform (FFT)/inverse fast Fourier transform (IFFT) chipset at the base band unit. More advanced approaches in which the digital samples belonging to spectrum fragment 1 and spectrum fragment 2 are fed to the same FFT/IFFT chipset may be possible, but require some updates at the modems.

It is important to note that today's multiple-input multiple-output (MIMO) transceivers are equipped with more than one RF chain and base band unit, thus the only changes required at the BS are at the antennas as well as some software updates. This makes it easy to upgrade existing cellular network deployments. An important point is that this scheme is completely transparent and fully compatible with existing UEs.

Finally, it is worth mentioning that while the sector offset configuration described here is for eNodeBs with three sectors, it can be extended to higher-order sectorizations with smaller angular offsets [1, 2]. For higher-order sectorizations, other antenna structures that are more compact and flexible such as the circular antenna array shown in Fig. 9.4c can be used.

9.2.3 Difference Between the Sector Offset Configuration and Higher-Order Sectorizations

It is important not to confuse the concept of sector offset configuration with that of higher-order sectorization. This is because:

- Antennas with a larger half power beamwidth are used in the sector offset configuration than for higher-order sectorizations. For example, the sector offset configuration in Fig. 9.1 uses 65 degrees half power beamwidth antennas, while a six-sector BS using the traditional configuration would use 33 degrees half power beamwidth antennas. This results in a much larger overlap between neighboring antennas in the sector offset configuration, which facilitates mobility management for mobile UEs (handovers are started at higher SINR). Due to this large overlap among neighboring antennas in the sector offset configuration, UEs in between the two antennas can also be served by both spectrum fragments, and thus can exploit maximum capacity.

- In the traditional configuration, neighboring antennas transmit over the entire spectrum, while in the sector offset configuration, neighboring antennas transmit over complementary spectrum fragments. This provides significant interference coordination at the expense of reducing spectrum reuse in particular areas.

- In the traditional configuration, as many base band units as sectors are required, while in the sector offset configuration, the number of base band units can be halved. Neighboring offset antennas transmitting in complementary spectrum fragments may reuse the same base band unit. This could significantly reduce the cost of the sector offset configuration.

9.3 MACROCELL TIER WITH HORIZONTAL AND VERTICAL SECTOR OFFSET CONFIGURATION

In order to further enhance spatial reuse, the presented horizontal sector offset configuration can be further extended in the vertical dimension. This can be done by adding, for each existing antenna in the horizontal sector offset configuration, an additional static beam pointing toward the same azimuth but with higher downtilt (vertical offset) on the complementing spectrum fragment, as shown in Fig. 9.6a.

In the previously presented horizontal sector offset configuration, in each eNodeB, there were three antenna ports with two physical antennas each, generating six beams. In a given antenna port, Antenna 1 and its horizontal offset Antenna 2 transmitted the same CRS, and both together used the entire bandwidth. In contrast, in the new horizontal-and-vertical sector offset configuration, in each eNodeB, there are six antenna ports with one physical antenna each, generating 12 beams (as explained latter, higher downtilt antenna beams can be generated through beamforming techniques). The lower and higher downtilt antenna beams pointing toward the same azimuth are the ones that transmit the same CRS, thus appearing as a unique antenna port, and both together use the whole bandwidth. As a result, the neighboring antenna beams pointing toward different azimuths now form different sectors and reuse the whole bandwidth in a way that neighboring antenna beams with the same downtilt do not transmit over the same spectrum fragment, but in the complementary one, as shown in Fig. 9.6b.

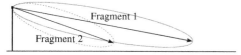

On each spectrum fragment, a second beam on the other spectrum fragment is added with higher downtilt (vertical offset)

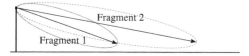

Spectrum fragment 1 uses RBs set (1, ..., R/2)
Spectrum fragment 2 uses RBs set (R/2 + 1, ..., R)

(a)

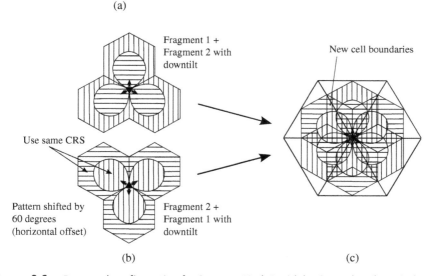

(b)   (c)

Figure 9.6. Proposed configuration for 3 sector eNodeB with horizontal-and-vertical sector offset configuration. (a) Different downtilt for carriers within one sector, (b) View from top on theee-sector cell and (c) Resulting cell structure.

This new horizontal-and-vertical sector offset configuration essentially creates a higher downtilt antenna beam within each existing antenna in the horizontal-only sector offset configuration. This increases spatial reuse at the cost of a larger number of sectors, that is, antenna ports, per eNodeB (and thus handovers). Using a high downtilt in the vertical antenna is important to be able to efficiently reuse the frequency resources of each spectrum fragment locally, while avoiding high interference with sectors of the same eNodeB and neighboring eNodeBs.

In order to achieve good load balancing between the lower and higher down-tilt beams in each antenna port, the higher downtilt beam is prioritized for all UEs, so that UEs are usually offloaded from the lower to the higher downtilt beam. This prioritization is realized at the scheduler, which allocates UEs to the different spectrum fragments so that some load balancing is achieved, for example, the achievable throughputs of the worst UEs (i.e., 5th percentile throughput) in both antenna beams are equalized. The achievable throughputs of UEs are computed based on wideband SINR reports and round robin scheduler assumptions [6]. As a result of this load balancing, some UEs that would be naturally allocated to the lower downtilt antenna beam may be allocated to RBs handled by the higher downtilt antenna beam, which tends to have more available RBs due to its lower initial load (smaller area covered).

This new horizontal-and-vertical sector offset configuration, shown in Fig. 9.6c, retains the properties of the original horizontal-only sector offset configuration:

1. The chances that UEs are served in the high-gain areas of each antenna lobe increase.
2. All neighboring antennas are using a different spectrum fragment. Thus, inter-cell interference is mitigated.

As a result, spatial reuse as well as UEs' SINRs are improved compared to traditional cell configurations, resulting in higher throughput. This is achieved without any increased measurement and reporting overhead due to the fact that the beamforming is static. UEs are connected to the adequate beams through handovers (when moving between beams of different antenna ports) or scheduling decisions (when moving between beams of the same antenna port) instead of having beams formed to follow UEs, as in the traditional dynamic beamforming.

In terms of implementation, similar approaches as for the horizontal-only sector offset configuration can be used. However, the additional vertical offset can be achieved using beamforming with the elements of each antenna column, for example, using a passive feeder network or more flexibly with an active antenna array. This solution can be mounted in the same way as current radomes, and an example of its antenna patterns is shown in Fig. 9.7. This horizontal-and-vertical sector offset configuration described here for eNodeBs with three sectors can also be extended to higher sectorizations with smaller angular offsets. For such higher sectorizations, the circular antenna array shown in Fig. 9.4 [7] can also be used.

9.4 SMALL CELL TIER WITH SECTOR OFFSET CONFIGURATION

Small cells can take advantage of the proposed sector offset configuration for macro-cells, and benefit from an improved ICIC too. This is particularly true with the horizontal-only sector offset configuration where the number of antenna beams is

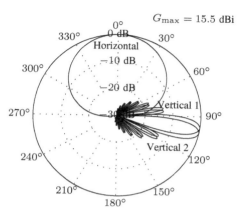

Figure 9.7. Example of antenna patterns for the proposed configuration for three-sector eNodeB with horizontal-and-vertical sector offset configuration.

smaller than in the horizontal-and-vertical sector offset configuration, and thus there can be a better coordination of cross-tier interference. According to the two sector offset configuration realizations presented in Section 9.2.1, multi-carrier and single carrier, and applying the horizontal-only sector offset configuration, small cells may be configured as presented in the following.

9.4.1 Multi-carrier HetNet Configuration

In multi-carrier implementations, a small cell, depending on its location, should be configured such that it only transmits on the spectrum fragment where the macro-cells provide the lowest SINRs [8]. For example, within the cell-edge triangular areas between macrocell sectors shown in Fig. 9.8, small cells should use the spectrum frag-ment complementary to the umbrella macrocell. These small cells may also reuse the same spectrum fragment as the umbrella macrocell with a reduced power to trans-mit to their cell-center UEs. In addition, small cells may be prioritized when their received power or SINR exceeds a predefined value to maximize offloading from the macrocells to small cells. Range expansion can be used for this purpose (see Chapter 6).

The proposed frequency configuration for small cells can be achieved by the following auto-configuration process:

1. Measure the received powers from macrocells at the small cell BS on both carriers and identify the maximum value.
2. Select the carrier with the lowest measured maximum value for small cell use.

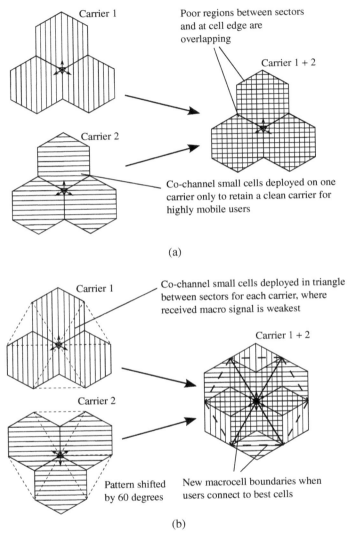

Figure 9.8. Multi-carrier HSPA/LTE HetNet configuration for traditional cell structure with partial frequency reuse (a) and proposed configuration with sector offset (b).

Using this approach, the macrocell performance improves since macrocell UEs are served by the carrier on which the macrocell antenna provides the highest gain, thus avoiding regions with high interference between sectors. A further benefit of this cell structure is that a macrocell carrier always appears clean or with reduced small cell interference at each location, and thus some macrocell UEs can benefit from improved SINRs. Moreover, fast-moving macrocell UEs do not necessarily need to handover to a small cell, but can move through it on the clean macrocell carrier. This prevents dropped calls due to handover delays for fast-moving UEs.

The small cell performance also improves since small cell UEs are served by the carrier on which the macrocell signal is received at lower power. This results in improved coverage, reduced interference and more flexibility at what distance to the macrocell a small cell can be deployed effectively.

9.4.2 Single-carrier HetNet Configuration

As indicated in Section 9.2.1, the proposed sector offset configuration can also be applied to groups of RBs in LTE networks with only one carrier. In this case, eNodeBs transmit half of the RBs over one sector configuration and the remaining RBs over the offset sector configuration. Depending on the small cell location, a subset of RBs may appear clean or with reduced interference from the eNodeB. Therefore, and similarly as before, we propose that small cells use

- the subset of RBs in which the small cell BS measurements report the lowest SINR for cell-edge UEs, and
- the subset of RBs in which the small cell BS measurements report the highest SINR for cell-center UEs [9].

It is important to note that cell-center UEs can afford to suffer from larger interference since they already benefit from good channel conditions due to the short distance between transmitter and receiver. Figure 9.9b illustrates the resulting HetNet structure.

In this case, the scheduler would also automatically assign UEs to the best RBs due to their different experienced signal quality. Using the previously presented example, if a small cell BS is under the coverage of Antenna 1 of its umbrella eNodeB sector, which transmits over the first $K/2$ RBs, its UEs will see higher SINRs in the CRS symbols of the last $K/2$ RBs, where they receive lower cross-tier interference. In this way, the sector offset configuration generates new scheduling opportunities for small cell cell-edge UEs.

9.4.3 Combination of Sector Offset Configuration with Cell Range Expansion

It is also important to note that the proposed small cell operation is compatible with standardized cell range expansion (CRE) and ABSs (see Chapter 8 for more details on ABSs). As illustrated in Fig. 9.9b, when eNodeBs transmit ABSs, small cells can allocate its range-expansion UEs in those subframes that overlap with macrocell ABSs, thus suffering from low interference from the umbrella macrocell. In

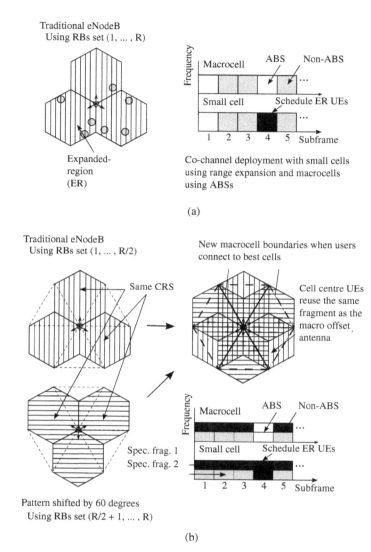

Figure 9.9. Single-carrier LTE HetNet configuration for traditional cell structure with frequency reuse of 1 (a) and proposed configuration with sector offset (b).

contrast, when eNodeBs transmit normal subframes with macrocell UE data, the small cells can just follow the scheduling strategy defined in the previous subsection, that is, cell-edge UEs are allocated to the spectrum fragment with the least interference, while cell-center UEs are allocated to the complementary spectrum fragment.

Moreover, in order to make the most of the scheduling opportunities offered by macrocell ABSs and the sector offset configuration, we propose to equip each small cell with a two-level load balancing scheme.

The first load balancing scheme divides the set of connected UEs to a small cell into two subsets:

- The subset of UEs to be scheduled in subframes that overlap with macrocell ABSs, and
- The subset of UEs to be scheduled in subframes that do not overlap with macrocell ABSs.

The second load balancing scheme works further in the second subset of UEs and divides it into two further subsets:

- Cell-edge UEs, which will be allocated to the RBs benefiting from cross-tier interference mitigation due the sector offset configuration, and
- Cell-center UEs benefiting from good channel conditions due to the short distance between transmitter and receiver, which will be allocated to the remaining RBs.

In both load balancing procedures, the objective is to equalize the throughput of the worst UEs in each subset (e.g., 5th percentile throughput) based on the knowledge of their wideband SINRs and available resources. Equalizing the worst UE performance enhances fairness at the expense of mean UE throughput.

In the following, the proposed algorithms for the first subframe load balancing and the second sector offset RB load balancing levels are presented. For more details on the first and the second load balancing schemes, please refer to [6, 9], respectively.

9.4.3.1 Almost Blank Subframe Load Balancing
In the following, the proposed ABS load balancing algorithm is presented, which is targeted at equalizing the performance of the worst small cell UEs scheduled in subframes non-overlapping and overlapping with macrocell ABSs. Extending the previous notation, we denote $\mathcal{U}_p = \{U_1, \dots, U_n, \dots, U_N\}$ as the set of small cell UEs connected to small cell BS P_p, where N and U_n are the cardinality and one element of the set, respectively. Moreover, we denote $\gamma_n = \{\gamma_n^{\text{noABS}}, \gamma_n^{\text{ABS}}\}$ as the wideband SINRs of UE U_n in subframes non-overlapping and overlapping with macrocell ABSs, respectively. Set $\mathcal{U}_p$ can be divided into two subsets, that is, $\mathcal{U}_p = \mathcal{U}_p^{\text{ABS}} \cup \mathcal{U}_p^{\text{noABS}}$, where $\mathcal{U}_p^{\text{noABS}}$ and $\mathcal{U}_p^{\text{ABS}}$ are the subsets of UEs in the small cell inner and expanded-region coverage areas, respectively, X_p and Y_p are their respective cardinalities, and thus $N = X_p + Y_p$.

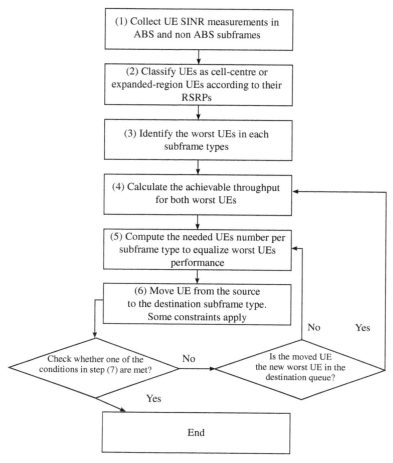

Figure 9.10. ABS load balancing algorithm diagram.

Based on this notation, the proposed ABS load balancing algorithm follows the following steps (see Fig. 9.10):

1. First, the small cell BS estimates for each small cell UE U_n its two wideband SINRs $\gamma_n = \{\gamma_n^{\mathrm{noABS}}, \gamma_n^{\mathrm{ABS}}\}$. The small cell BS can instruct UEs to take measurements on specific subframes.

2. Based on RSRP measurements, the small cell BS classifies small cell UEs as inner coverage area small cell UEs $\mathcal{U}_p^{\mathrm{noABS}}$ and expanded-region small cell UEs $\mathcal{U}_p^{\mathrm{ABS}}$. Thus, small cell UEs are separated into two queues: the non-overlapping ABS queue and the overlapping ABS queue.

3. The small cell BS identifies the worst inner coverage area small cell UE $U_{n_w}^{\text{noABS}}$ and the worst expanded-region small cell UE $U_{n_w}^{\text{ABS}}$ based on their wideband SINR estimates.

4. The small cell BS estimates for both worst small cell UEs, which would be their throughputs $TP_{n_w}^{\text{noABS}}$, $TP_{n_w}^{\text{ABS}}$, if they would be allocated to all RBs handled by the subframes non-overlapping and overlapping with macro-cell ABSs, respectively. $TP_{n_w}^{\text{noABS}} = f(\gamma_{n_w}^{\text{noABS}}) \cdot K$ and $TP_{n_w}^{\text{ABS}} = f(\gamma_{n_w}^{\text{ABS}}) \cdot K$, where f(x) is a mapping function that translates RB SINR into RB throughput and K is the total number of RBs.

5. Once the achievable throughput of both worst small cell UEs is known and assuming a round robin scheduler,[2] the small cell BS computes how many small cell UEs should be there in each subframe-type queue so that the worst small cell UEs in both queues roughly achieve the same throughput performance. Based on a round robin scheduler, this can be computed as follows:

$$X_p' = N - Y_p' \; , \quad Y_p' = \frac{N}{1 + \dfrac{TP_{n_w}^{\text{noABS}}}{TP_{n_w}^{\text{ABS}}}}, \tag{9.1}$$

where X_p' and Y_p' are the total number of small cell UEs that should be in the subframes non-overlapping and overlapping with macrocell ABSs, respectively, to achieve the targeted small cell load balancing.

6. Then, the small cell BS moves small cell UEs one by one from the source subframe type queue to the destination subframe type queue until it achieves the number of desired small cell UEs X_p' and Y_p' in each subframe type.

When moving a small cell UE from the non-overlapping ABS to the over-lapping ABS queue, the small cell BS moves first the worst performing small cell UE in the non-overlapping ABS queue. This is because the performance of this small cell UE will always improve due to ABS interference mitigation, and the target is maximizing the worst small cell UEs performance. However, when moving a small cell UE from the overlapping ABS to the non-overlapping ABS queue, the small cell BS moves first the best performing small cell UE in the overlapping ABS queue. This is because the performance of this small cell UE will always decrease due to macrocell interference, but the best performing UE will be the least impacted. The following constraint should also be considered. All small cell UEs using non-overlapping ABSs should be able to decode their control channels; otherwise, the small cell UE

---

[2] Since when the number of scheduling realizations is sufficiently large, the performance of a PF scheduler tends to be equivalent to that of a round robin scheduler, a round robin scheduler is thus assumed in our load balancing calculations, which gives us a long-term perspective and the possibility of making fast calculations that can be updated regularly.

is not moved. In other words, all small cell UEs using non-overlapping ABS should have a wideband SINR higher than the out-of-synchronization threshold (e.g., $Q_{out} = -8dB$) in that band.

When the small cell BS moves one small cell UE from one subframe-type queue to the other, and as a result, this small cell UE becomes the worst small cell UE in the destination subframe-type queue, the small cell BS must recompute how many small cell UEs should be in each queue according to step (5), and thereafter continue with step (6).

7. The algorithm stops when the small cell BS achieves the number of desired small cell UEs per subframe type, or when the small cell BS needs to start moving small cell UEs from the overlapping ABS to the non-overlapping ABS queue, when in the previous iteration it was moving small cell UEs in the opposite direction, that is, from the non-overlapping ABS to the overlapping ABS queue, or vice versa.

9.4.3.2 Sector Offset Resource Block Load Balancing

Now, we denote $\mathcal{U}_p^{noABS} = \{U_1^{noABS}, \ldots, U_m^{noABS}, \ldots U_M^{noABS}\}$ as the set of small cell UEs connected to small cell BS P_p that are to be scheduled when normal subframes are transmitted at the macrocell, were M and U_m^{noABS} are the cardinality and one element of the set, respectively. Moreover, we denote $\gamma_m = \{\gamma_m^{f1}, \gamma_m^{f2}\}$ as the effective SINRs of UE U_m^{noABS} in the RBs of spectrum fragment 1 and 2, respectively. Now, set $\mathcal{U}_p^{noABS}$ can be divided into two further subsets, that is, $\mathcal{U}_p^{noABS} = \mathcal{U}_p^{f1} \cup \mathcal{U}_p^{f2}$ where $\mathcal{U}_p^{f1}$ and $\mathcal{U}_p^{f2}$ are the subsets of UEs to be allocated in spectrum fragment 1 and spectrum fragment 2, respectively, $\hat{X}_p$ and $\hat{Y}_p$ are their respective cardinalities, and thus $M = \hat{X}_p + \hat{Y}_p$.

Based on this notation, the proposed load balancing algorithm follows the following steps (see Fig. 9.11):

1. First, the small cell BS estimates for each small cell UE U_m^{noABS} its two effective SINRs $\gamma_m = \{\gamma_m^{f1}, \gamma_m^{f2}\}$. Standard UE reports can be used.

2. Based on whether $\gamma_m^{f1} \leq \gamma_m^{f2}$ or vice versa, the small cell BS classifies UEs to be scheduled in subframes that do not overlap with macrocell ABSs as spectrum fragment 1 small cell UEs $\mathcal{U}_p^{f1}$ or spectrum fragment 2 small cell UEs $\mathcal{U}_p^{f2}$. Then, small cell UEs are separated into two queues: the spectrum fragment 1 queue and the spectrum fragment 2 queue.

3. The small cell BS identifies the worst fragment 1 small cell UE $U_{m_w}^{f1}$ and the worst fragment 2 small cell UE $U_{m_w}^{f2}$ based on their effective SINR estimates.

4. The small cell BS estimates for both worst small cell UEs, which would be their throughputs $TP_{m_w}^{f1}$, $TP_{m_w}^{f2}$, if they would be allocated to all RBs of spectrum fragment 1 and 2, respectively. $TP_{m_w}^{f1} = f(\gamma_{m_w}^{f1}) \cdot \frac{K}{2}$ and $TP_{m_w}^{f2} = f(\gamma_{m_w}^{f2}) \cdot \frac{K}{2}$. This computation is straightforward since the small cell BS can

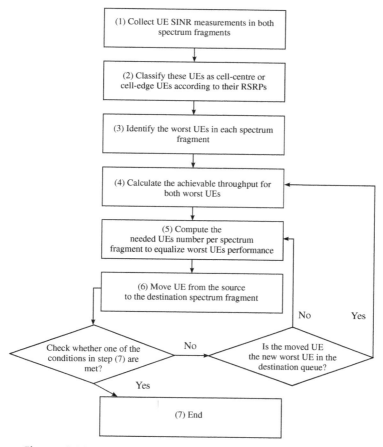

Figure 9.11. Sector offset RB load balancing algorithm diagram.

be aware of the small cell UE wideband SINR in each spectrum fragment if it instructs small cell UEs to take measurements on those.

5. Once the achievable throughput of both worst small cell UEs is known and assuming again a round robin scheduler, the small cell BS computes how many small cell UEs should be there in each spectrum fragment queue so that the worst small cell UEs in both queues roughly achieve the same throughput performance. Based on a round robin scheduler, this can be computed as follows:

$$\hat{X}'_p = N - \hat{Y}'_p \ , \quad \hat{Y}'_p = \frac{N}{1 + \frac{\mathrm{TP}^{f1}_{m_\mathrm{w}}}{\mathrm{TP}^{f2}_{m_\mathrm{w}}}}, \tag{9.2}$$

where $\hat{X}'_p$ and $\hat{Y}'_p$ are the total number of small cell UEs that should be allocated to spectrum fragment 1 and 2, respectively, to achieve the targeted small cell load balancing.

6. Then, the small cell BS moves small cell UEs one by one from the source to the destination spectrum fragment queue until it achieves the number of desired small cell UEs $\hat{X}'_p$ and $\hat{Y}'_p$ in each spectrum fragment. In each iteration, the small cell BS moves the best performing small cell UE in the source spectrum fragment queue, that is, the one with the best effective SINR. Moving the worst performing small cell UE would be non-ideal since its performance would be even worse in the target spectrum fragment, thus preventing worst small cell UE performance equalization. Moreover, when the small cell BS moves one small cell UE from one queue to the other, if this small cell UE becomes the worst small cell UE in the destination queue, then the small cell BS must recompute how many UEs there should be in each queue according to step (5), and thereafter continue with step (6).

7. The algorithm stops when the small cell BS achieves the number of desired small cell UEs per spectrum fragment, or when the small cell BS needs to start moving small cell UEs from the spectrum fragment 1 to the spectrum fragment 2 queue, when in the previous iteration it was moving small cell UEs from the spectrum fragment 2 to the spectrum fragment 1 queue, or vice versa.

9.5 SCENARIO AND SYSTEM MODEL

System-level simulations were performed to evaluate the performance of the proposed sector offset configuration in terms of downlink throughput and HO performance. Both single carrier LTE macrocell-only and HetNet scenarios were considered. Table 9.1 presents the most relevant simulation parameters.

The simulated scenario was downtown Dublin, Ireland, which is described in more detail in Appendix A, and consisted of one LTE carrier centered at 2 GHz with 5 MHz bandwidth and 7 eNodeBs with three sectors each.

The UE distribution consisted of a uniform background distribution with additional hotspots. Hotspots were circular with a 40 m radius and their positions were randomly selected. If a HetNet scenario was considered, an outdoor pico eNodeB was deployed at the center of each hotspot.

UEs were uniformly distributed within the scenario or the hotspots. There were 600 UEs per square kilometer, with 50% of them located in the hotspots. Each hotspot had 20 UEs. Overall, 30 % of all UEs were placed outdoors, and the remaining ones indoors. Such ratio applied to UEs in both the background distribution and the additional hotspots.

CRSs were 3 dB power boosted with respect to other resource elements. The REB of each expanded-region pico eNodeB was individually tailored, according to

TABLE 9.1 Simulation Parameters

Parameters	Value
Scenario	
Macro BS placement	7 eNodeBs, 800 m inter-site distance
Pico BS placement	one at the center of each hotspot
Scenario size	1500 m × 1500 m, around central macrocell
Scenario resolution	2 m
Transmit power	$P_{tx,n} = 21.6$ W (macro), 1 W (picos)
Noise density	-174 dBm/Hz
UEs	
UE density	600 per km², 50% hotspots, 30% outdoors
Hotspots (static)	20 UEs per hotspot within 40 m radius
UE speed	4 km/h (50%), 30 km/h (50%)
Channel	
Carrier frequency	2000 MHz
Bandwidth	5 MHz (1 LTE carrier of 25RBs)
NLOS path loss	$G_{Pn} = -21.5 - 39\log_{10}(d)$ (macro)
	$G_{Pn} = -30.5 - 36.7\log_{10}(d)$ (pico)
LOS path loss	$G_{Pl} = -34.02 - 22\log_{10}(d)$
Shadow fading (SF)	6 dB std dev. [12]
SF correlation	$R = e^{-1/20d}$, 50% inter-site
Environment loss	$G_{E,n} = -20$ dB if indoor, 0 dB if outdoor
Antenna	
Height	25 m (macro), 10 m (pico)
Maximum gain	$G_{max} = 15.5$ dBi (macro), 7.06 dBi (pico)
H. halfpow. beamwidth	$\alpha = 65°$
V. halfpow. beamwidth	$\beta = 11.5°$ (macro)
Front-to-back ratio	$\kappa = 30$ dB (macro)
Downtilt	$\delta_1 = 8.47°$ (macro)
Elements and spacing	4 element dipole, $d_{elem.} = 0.6\lambda$ (pico)
Phase difference	$\delta_{phase} = 95°$ (pico)
Element amplitude	$a_{elem.} = [0.9691, 1.0768, 1.0768, 0.8614]$ (pico)
Handover	
Time-to-trigger (4 km/h / 30 km/h)	480 ms / 160 ms
Hysteresis (4 km/h / 30 km/h)	3 dB / 2 dB
RSRP measurement interval	40 ms
L3 filter reporting interval	200 ms
L3 forgetting factor	1/4
Preparation + Execution time	90 ms
Out of sync. / In sync	$\gamma_{n,k}^{ESSM} < Q_{out} = -8$ dB / $\gamma_{n,k}^{ESSM} > Q_{in} = -6$ dB
Radio link failure timer	$T_{310} = 1$ s
Scheduling	
Time-domain scheduled UEs	$U = 10$ UEs
Average throughput forgetting factor	$1/t_c = 0.5$
UE-targeted throughput	$C_{n,k}^{target} = 128$ kbps

[10], to allow each pico eNodeB to cover its hotspot of 40 m radius. A maximum REB cap of 9 dB was used.

Realistic antenna patterns were used at both the macro and pico eNodeBs, whose most representative features are presented in Table 7.2 and their horizontal and vertical diagrams can be respectively seen in Section A.3. The small cell antenna consisted of a vertical four-element dipole array, providing an omnidirectional coverage in the horizontal plain and a downtilt of 10.83 degrees in the vertical plane. Path losses, shadow and multi-path fading were also modeled according to the details in Table 9.1. More details are provided in Chapter A. UEs were equipped with two receiver antennas. Maximal ratio combining (MRC) was used to combine signals, while exponential effective SINR mapping (EESM) was used to estimate wideband SINRs. Both models are presented in Section A.8.2.

The two-stage time/frequency PF scheduler presented in [11] was adopted, using the Time Domain PF and Carrier over Interference to Average metrics, respectively. 10 ms outdated sub-band CQI reporting with 4 RBs per sub-band was considered at the eNodeB, and EESM was used to derive the effective SINR of each sub-band at the UE.

RSRP and wideband SINR measurements determined handover trigger and radio link failures (RLFs)/HOFs, respectively. A UE suffered from RLFs/HOFs if its wideband SINR was smaller than $Q_{out} = -8$ dB for a time $T_{310} = 1$ s. More details about mobility modeling can be found in Section A.11.

Throughput performance statistics were collected over 10 simulation runs with independent shadow fading channel realizations from UEs connected to the central eNodeB and its overlaid pico eNodeBs, if any. Within each simulation run, 100 uniformly distributed random drops of hotspots and UEs were performed, and within each random drop, 500 subframes were simulated. HO results were calculated based on 500 UE routes through the simulated scenarios, with UEs traveling at 30 km/h. The used outdoor traffic model is described in Appendix A.

9.6 PERFORMANCE COMPARISON FOR MACROCELL-ONLY SCENARIO

In this section, a performance analysis for single-carrier LTE macrocell-only networks is presented. The following reuse approaches over the data channels were used for performance comparison:

- Reuse 1: All eNodeB sectors have access to all RBs [13].
- Reuse f: The bandwidth is divided into f orthogonal fragments, where f is the number of sectors per eNodeB, and each eNodeB sector is assigned one different fragment. This approach avoids interference among adjacent sectors at the expense of reduced resource availability [13].
- Fractional frequency reuse f: Each eNodeB sector divides its UEs into cell-center and cell-edge UEs, depending on whether the difference among the

RSRPs of the best-serving cell and strongest neighboring cell is larger than threshold γ^{edge}. Moreover, as in reuse f, the channel bandwidth is divided into f orthogonal fragments, and each eNodeB sector is assigned one different fragment. Cell-center and cell-edge UEs can access the entire bandwidth, but cell-edge UEs have a higher priority than cell-center UEs on the selected fragments. γ^{edge} was set to 3 dB. The downlink transmit power levels applied to subcarriers reserved for cell-center and cell-edge UEs were tuned according to [14] as

$$P_{n,k,r}^{\text{tx,center}} = \frac{f \cdot P_n^{\text{tx}}}{\alpha + f - 1} \tag{9.3}$$

and

$$P_{n,k,r}^{\text{tx,edge}} = \alpha \cdot P_{n,k,r}^{\text{tx,center}}, \tag{9.4}$$

respectively, where P_n^{tx} is the total sector downlink transmit power and $\alpha = f$ is the cell-edge to cell-center transmit power ratio. In this way, inter-cell interference from cell-center UEs to neighboring cell-edge UEs is reduced.

9.6.1 RSRP and SINR Performance

Figures 9.12 and 9.13 show the cumulative distribution functions (CDFs) of the RSRP and wideband SINR measurements taken by UEs moving at 30 km/h for the

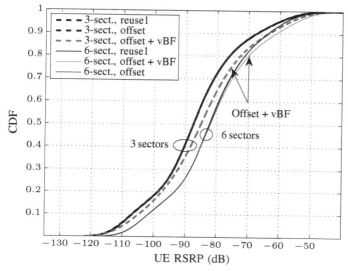

Figure 9.12. RSRP CDF of UEs moving at 30km/h over the CRS.

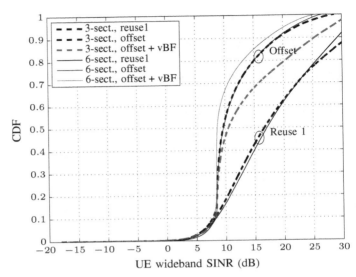

Figure 9.13. Wideband SINR CDF of UEs moving at 30km/h over the CRS.

reuse 1 and the sector offset configurations, horizontal-only as well as horizontal-and-vertical. Both measurements are estimated over CRS symbols, and three-sector and six-sector eNodeBs are considered.

When comparing the performance of reuse 1 and the horizontal-only sector offset configuration, Fig. 9.12 shows that the transmission of half of the CRS symbols over one antenna and the remaining CRS symbols over the offset antenna does not significantly affect the RSRP distribution. Since the RSRP is estimated by the UE as the linear average of the received CRS symbols strengths, and because in the horizontal-only sector offset configuration the reception of half of the CRS symbols is enhanced and the reception of the remaining CRS symbols is degraded, the RSRP distribution barely changes. Both effects compensate each other. In contrast, the horizontal-and-vertical sector offset configuration enhances RSRP estimations with respect to the reuse 1 and the horizontal-only sector offset configurations. This is because in the horizontal-and-vertical sector offset configuration, there are more antennas pointing in different directions than for reuse 1, and because in contrast to the horizontal-only sector offset configuration, the two antennas forming a sector now point in the same direction and have a large degree of overlap.

Figure 9.13 shows that splitting the transmission of CRS symbols between offset antennas degrades wideband SINR estimation in the high regime (wideband SINRs larger than 3 dB)). However, this should not affect mobility performance, since this degradation occurs at wideband SINRs much higher than Q_{out}. In contrast, it is important to note that the sector offset configuration enhances the tail of wideband SINR

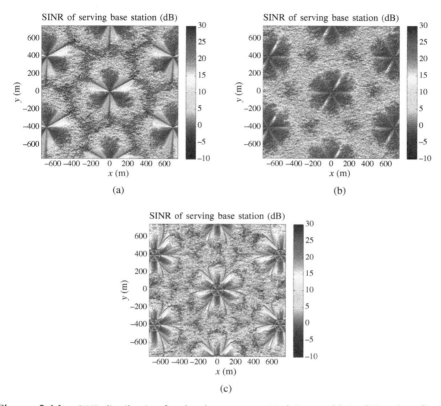

Figure 9.14. SINR distribution for the three-sector eNodeB case. (a) Traditional configuration, (b) Offset configuration and (c) Offset + vBF configuration.

distribution, the SINR performance at the boundaries among sectors of the same eNodeB. This enhances handover performance, as will be shown later.

Figures 9.14a, 9.14b, and 9.14c show, for the three-sector eNodeB case, the resulting spatial SINR distribution of data channels for reuse 1 as well as for the horizontal-only and the horizontal-and-vertical sector offset configurations, respectively—static UEs are assumed. Note that these SINR distributions are different from that of Fig. 9.13. This is because the effects of handover parameters (e.g. hysteresis, time-to-trigger) are not considered, and also because before we focused on CRS performance and now we are focusing on data channel performance. These data channels are not interleaved all across the bandwidth as CRS (benefiting from ICIC). For the traditional configuration in Fig. 9.14a, it is shown that large areas between sectors of the same eNodeB and neighboring eNodeBs suffer from poor SINRs, which translates into low UE throughputs. When the proposed horizontal-only and horizontal-and-vertical sector offset configurations are used, most of these

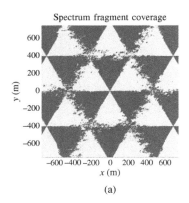 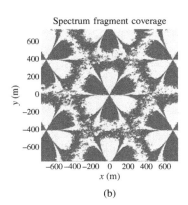

(a)   (b)

Figure 9.15. Spectrum fragment reuse for the three-sector eNodeB case. (a) Offset configuration and (b) Offset + vBF configuration.

poor SINR regions disappear, as shown in Figs. 9.14b and 9.14c, respectively. This results in improved performance at the cell edge due to the chances of UEs being served in the high-gain areas of an antenna lobe increase, and because neighboring antennas are using a different spectrum fragment, with the resulting inter-cell interference mitigation. This latter fact is illustrated in Figs. 9.15a and 9.15b, which show how neighboring antennas do not reuse the same spectrum fragment (the light color denotes the first $K/2$ RBs while the dark color denotes the remaining $K/2$ RBs). Similar SINR improvements can be observed for the six-sector eNodeB configuration.

9.6.2 Capacity Performance

Tables 9.2 and 9.3, which show system-levels results for eNodeBs with three sectors and six sectors, respectively, indicate that previous SINR data channel improvements translate into throughput enhancements.

TABLE 9.2 Performance Comparison for the Three-Sector eNodeB Case

	Reuse 1	Reuse 3	FFR 3	Offset	Offset + vBF
Throughput [kbps]/gain [%]					
Sector throughput	10188/0	5111/ − 50	9191/ − 10	**12,322/ + 21**	**10,242/ + 1**
UE throughput (average)	88.11/0	44.21/ − 50	79.49/ − 10	**107.67/ + 22**	**177.82/ + 102**
Cell-edge UE throughput	26.62/0	24.37/ − 8	28.33/ + 6	**34.99/ + 32**	**68.48/ + 157**
(5th percentile)					
Handovers [event/UE/km²]/gain [%]					
Number	112.82/0	112.82/0	112.82/0	**127.46/ + 13**	**159.23/ + 41**
Failure rate [%]	0.17/0	0.17/0	0.17/0	**0.05/ − 69**	**0.33/ + 94**

TABLE 9.3   Performance Comparison for the Six-Sector eNodeB Case

	Reuse 1	Reuse 6	FFR 6	Offset	Offset + vBF
Throughput [kbps]/gain [%]					
Sector throughput	10270/0	2746/ − 73	9925/ − 3	**12,261/ + 20**	**82,332/ − 20**
UE throughput (average)	181.49/0	48.52/ − 73	175.41/ − 3	**214.01/ + 19**	**287.50/ + 58**
Cell-edge UE throughput	53.13/0	27.37/ − 48	53.29/0	**69.72/ + 31**	**102.51/ + 93**
(5th percentile)					
Handovers [event/UE/km²]/gain [%]					
Number	120.61/0	120.61/0	120.61/0	**137.38/ + 14**	**177.09/ + 47**
Failure rate [%]	0.34/0	0.34/0	0.34/0	**0.13/ − 61**	**0.40/ + 17**

For the three-sector eNodeB case, the horizontal-only sector offset configuration results in a 22% average UE throughput improvement and a 32% 5th percentile UE throughput improvement compared with reuse 1. As indicated earlier, the reasons for such improvements are mainly:

- The chances that UEs are served in the high-gain areas of an antenna lobe increase.

- All neighboring antennas are using a different spectrum fragment. Thus, inter-cell interference is mitigated.

When looking at the performance of the other schemes, Reuse 3 performance is severely affected by the large number of RBs sacrificed for the sake of interference mitigation. Compared to reuse 1, the 5th percentile UE throughput of FFR 3 is enhanced around 6% due to inter-cell interference coordination and the larger power allocated to cell-edge RBs, while average UE throughput is degraded because of the lower transmit power applied to cell-center UEs. This is a common trade-off in FFR schemes, which usually trade overall network performance for cell-edge capacity.

With regard to the horizontal-and-vertical sector offset configuration, it results in a 102% average UE throughput improvement and a 157% 5th percentile UE throughput improvement compared with reuse 1, and a 65% average UE through-put improvement and a 96% 5th percentile UE throughput improvement compared with the horizontal-only sector offset configuration. These gains are due to the more spatial reuse provided by the new sectors and downtilt beam.

For the six-sector eNodeB case, the horizontal-only sector offset configuration results in a 19% average UE throughput improvement and a 31% 5th percentile UE throughput improvement compared with reuse 1. It is important to note that the gains for the six-sector eNodeB case are comparable to but just slightly lower than those for the three-sector eNodeB case. This is due to the increased spatial reuse and thus inter-cell interference. Compared to reuse 1, the cell-edge performance of FFR 6 is significantly degraded because the number of RBs for cell-edge use is now too small.

This shows that FFR schemes require careful optimization, which is not needed in the proposed sector offset configuration.

The horizontal-and-vertical sector offset configuration results in a 58% average UE throughput improvement and a 93% 5th percentile UE throughput improvement compared with reuse 1, and a 34% average UE throughput improvement and a 47% 5th percentile UE throughput improvement compared with the horizontal-only sector offset configuration. With the horizontal-and-vertical sector offset configuration, the gains for the six-sector eNodeB case are also lower than for the three-sector eNodeB case. This is again due to the stronger interference caused by the larger number of cell boundaries and spatial reuse. This stronger interference in the horizontal-and-vertical sector offset configuration, when changing from three to six sectors per eNodeB, can be observed in Fig. 9.13 where the six-sector case cannot follow the wideband SINR trend of the three-sector case, as the other schemes do.

9.6.3 Mobility Performance

For the three-sector eNodeB case, handover results in Table 9.2 show that compared with the traditional configuration (reuse 1, reuse 3, and FFR 3 behave equally), the horizontal-only sector offset configuration slightly increases the total number of handovers by 13%. This is because the transmission of half of the CRS symbols over one antenna and the remaining CRS symbols over the offset antenna slightly increases the patchy areas where ping-pongs occur. However, although the total number of handovers slightly increases, the horizontal-only sector offset configuration decreases the HOF rate by 69%. It is important to note that most of HOFs in the traditional configuration occur when the UEs are nearby the serving eNodeB due to the sharp cell-edges between sectors of the same eNodeB. Due to the larger overlapping and better interference mitigation among sectors of the same eNodeB in the sector offset configuration, handovers are triggered at higher wideband SINRs, which leaves more room to perform the handover before a session drops.

With regard to the horizontal-and-vertical sector offset configuration, Table 9.2 shows that its additional cell boundaries, resulting from adding the new sectors and downtilt beams, lead to an increased number of handovers compared with the traditional and horizontal-only sector offset configurations. The increased frequency reuse, although leading to a much larger network throughput, also results in higher interference. This makes the benefits in terms of HOF rates that were originally achieved with the horizontal-only sector offset configuration disappear. In more detail, compared to the traditional configuration the horizontal-and-vertical sector offset configuration increases the number of handovers by 41% and HOFs by 94%. Note that the absolute values are still very small.

With the traditional sector configuration, increasing the number of sectors per eNodeB from three to six sectors doubles the number of cell boundaries, but not the number of handovers, which just increases by 7%. This is because the total number of handovers is not dominated by the larger number of cell boundaries, but by the ping-pongs in the patchy coverage areas among neighboring eNodeBs generated due to shadowing. These areas are approximately the same in the three- and six-sector eNodeB cases. However, HOFs increase by 104% due to the increase of the number of sharp cell edges between the sectors of the same eNodeB.

For the six-sector eNodeB case, the horizontal-only sector offset configuration also slightly increases the total number of handovers by 14%, and decreases the HOF rate by 61%, compared with the traditional configuration. In contrast, the horizontal-and-vertical sector increases the number of handovers by 47% due to the larger number of cell boundaries, but increases HOFs only by 17%. Gains scale well with higher sectorizations.

9.7 PERFORMANCE COMPARISON FOR HETNET SCENARIO

In this section, a performance analysis for a single-carrier LTE HetNet is presented. The following approaches were used for performance comparison:

- *Case 1*: Small cells without REB and macrocells without ABSs.
- *Cases 2–4*: Small cells with REB and macrocells with an ABS duty cycle of D_{ABS}=[0.00, 0.10, 0.20] (Release 10 eICIC). Small cells are equipped with a balancing algorithm that is targeted at equalizing the worst performance of small cell UEs scheduled in subframes non-overlapping and overlapping with macrocell ABSs [6].
- *Cases 5–7*: Small cells with REB and macrocells with sector offset configuration and an ABS duty cycle of D_{ABS} = [0.00, 0.10, 0.20]. Small cells are equipped with the previously mentioned ABS load balancing as well as the proposed spectrum fragment load balancing described in Section 9.4.3.

Figures 9.16a and 9.16b show the SINR spatial distribution of the best data channels in the presented scenario when macrocells transmit normal subframes and use the traditional and the proposed sector offset configuration, respectively. We can see that the proposed sector offset configuration increases the average SINR of the best data channels due to its better macrocell coverage. Figure 9.16c also illustrates the coverage of spectrum fragment 1 and spectrum fragment 2 when using the proposed sector offset configuration. A good balance between spectrum fragments is achieved. When a small cell is deployed within the spectrum fragment 1 macrocell coverage, it can use spectrum fragment 2, which suffers from low macrocell interference, to serve its cell-edge UEs and the fragment 1 to serve its cell-center UEs, which benefit from good channel conditions. In this way, a large inter-cell interference mitigation is achieved even when the macrocells transmit normal subframes.

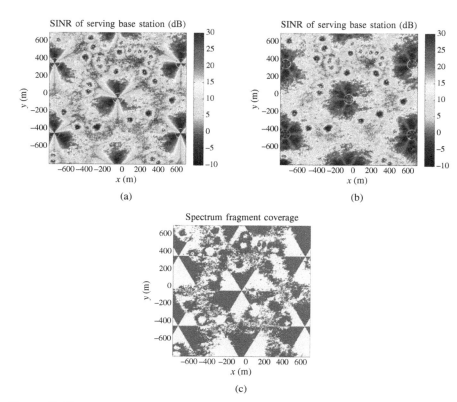

Figure 9.16. SINR distribution and spectrum coverage for the three-sector eNodeB case. (a) Three-sector traditional conf. SINR distribution, (b) Three-sector offset conf. SINR distribution and (c) Three-sector offset conf. spectrum coverage.

9.7.1 Capacity Performance

Table 9.4 shows simulation results for all simulated cases in terms of capacity and mobility, which are analyzed in the following.

Let us first focus on the three cases without ABSs ($D_{ABS} = 0.00$), that is, cases 1, 2, and 5. When we compare the cases with no sector offset configuration with and without range expansion at the small cells (cases 1 and 2), the average macrocell UE throughput increases by 16 % due to UE offloading when activating range expansion but the 5th percentile small cell UE throughput reduces by 233 %. This significant reduction is due to the increased number of small cell UEs, and the high cross-tier interference suffered by the expanded-region small cell UEs. When we compare the cases with cell range expansion with and without sector offset configuration (cases 2 and 5)—no LTE Release 10 eICIC is used—the average macrocell UE throughput

TABLE 9.4 Performance Comparison for a HetNet with Three-Sector eNodeBs and Pico eNodeBs

	(1) No bias	(2-4) Rel. 10 eICIC			(5-7) Rel. 10 eICIC + Sector Offset		
	$D_{ABS} = 0.0$	$D_{ABS} = 0.0$	$D_{ABS} = 0.1$	$D_{ABS} = 0.2$	$D_{ABS} = 0.0$	$D_{ABS} = 0.1$	$D_{ABS} = 0.2$
Throughput [kbps]/ gain-over-case 2 [%]/ gain-over-case 3 or (4) [%]							
Macro sector tp (av.)	8968/ − 3	9274/0	8391/ − 10/0	7489/ − 19/0	10635/ + 14	9630/ + 4/+15	8604/ − 8/+15
Small cell tp (av.)	5137/ + 5	4906/0	5411/ + 10/0	5703/ + 16/0	4529/ − 8	5327/ + 9/−2	5691/ + 16/−1
Macrocell UE tp (av.)	175.3/ − 16	208.0/0	188.2/ − 10/0	168.0/ − 19/0	244.0/ + 17	220.9/ + 6/+17	197.4/ − 5/+18
Macrocell UE tp (5th percentile)	39.7/ − 31	57.2/0	51.1/ − 11/0	45.1/ − 21/0	82.4/ + 44	73.0/ + 27/+43	63.8/ + 12/+41
Small cell UE tp (av.)	463.6/ + 15	402.1/0	443.6/ + 10/0	467.8/ + 16/0	372.7/ − 8	438.6/ − 1/−1	468.6/ + 6/0
Small cell UE tp (5th percentile)	164.0/ + 233	49.2/0	105.0/ + 113/0	133.8/ + 172/0	106.0/ + 115	138.8/ + 182/+32	157.6/ + 220/+18
UE tp (5th percentile)	55.8/ + 2	54.5/0	62.6/ + 15/0	59.5/ + 9/0	91.2/ + 67	93.9/ + 72/+50	85.5/ + 57/+43
Hard handovers [events/UE/km²]/ gain [%]							
Number	242.35/ + 6	227.55/0	254.98/ + 12	254.98/ + 12	240.08/ + 6	263.95/ + 16	263.95/ + 16
Failure rate [%]	3.71/ − 71	12.60/0	2.29/ − 82/0	2.29/ − 82/0	9.06/ − 28	1.24/ − 90/−46	1.24/ − 90/−46

increases by 17 % due to the better cross-tier interference mitigation of the proposed sector offset configuration. This cross-interference mitigation is more obvious at the cell boundaries, where the 5th percentile small cell UE throughput increases by 115 % since cell-edge UEs are scheduled in the spectrum fragment not used by the umbrella sector offset antenna.

Lets us now focus on the cases with cell range expansion and LTE Release 10 eICIC (cases 2–4). We can see that when the ABS duty cycle increases, cross-tier interference is significantly alleviated at the expense of a reduced macrocell performance due to subframe blanking. When D_{ABS} increases from 0 to 0.2 (cases 2–4), the 5th percentile small cell UE throughput increases by 172 %, while the average macrocell UE throughput decreases by 19 %. This is a general trend inherent to ABS operation, which also appears when adopting the proposed sector offset configuration in combination with Release 10 eICIC (cases 5 –7).

In order to assess the gains provided by sector offset while complementing Release 10 eICIC, let us compare the cases with the same ABS duty cycle with and without sector offset. Comparing the cases with $D_{ABS} = 0.1$, case 3 and case 6, the proposed configuration increases the average macrocell UE throughput, 5th percentile macrocell UE throughput and 5th percentile small cell UE throughput by 17 %, 43 %, and 32 %, respectively, while the average small cell UE throughput remains constant (−1 %). Moreover, the overall 5th percentile UE throughput increases by 50 %. Macrocell UE performance is increased due to the better macrocell UE alignment with the antenna beams provided by sector offset and the interference mitigation resulting from sophisticated spectrum fragment reuse. Cell-edge small cell UE performance is also increased since when the macrocells transmit normal subframes, the small cell UEs with worst channel gains are scheduled in the spectrum fragment not used by the umbrella sector offset antenna. Similar gains are observed with $D_{ABS} = 0.2$. Note that the proposed configuration with $D_{ABS} = 0.1$ achieves roughly the same 5th percentile small cell UE performance as Release 10 eICIC with $D_{ABS} = 0.2$— no sector offset configuration. In this case, the proposed sector offset configuration is thus able to decrease the ABS duty cycle up to a factor of 2, and in turn enhance macrocell UE performance by 31 %, while maintaining a targeted 5th percentile small cell UE performance.

9.7.2 Mobility Performance

Because the number of handovers performed in a HetNet is dominated by the number of small cells, one can see from Table 9.4 that all compared approaches incur approximately the same number of handovers per UE per hour. In terms of HOFs, it can be seen that the case with small cell REB and no macrocell ABSs, case 2, provides the worst performance. This is mostly because handovers from small cells to macrocells tend to fail due to high interference conditions. When macrocell ABSs are used, case 3, the HOF rate decreases by 82 % compared to the previous case. This is because small cell to macrocell HOFs are significantly mitigated due to the interference mitigation provided through subframe blanking at macrocells. Note that the ABS duty cycle has no impact on the handover performance, since all UEs performing a handover from a small cell were allocated to subframes overlapping with the macro-cell ABSs with a higher priority. When the proposed sector offset configuration is adopted in combination with macrocell ABSs, case 6, the HOF rate is further decreased by 46 % (compared with the case where macrocell ABSs but no sector offset is used, case 3). Due to the more uniform SINR distribution provided by the sector offset configuration (more macrocell antennas pointing in different directions), macrocell UEs start their handovers at a higher SINR and thus have more time to perform the handover process. The sector offset configuration thus significantly alleviates HOFs in HetNets.

9.8 SUMMARY AND CONCLUSIONS

In this chapter, the concept of sector offset configuration has been presented together with two possible implementations for multi- and single-carrier networks. The implementation aspects of the proposed sector offset configuration at macrocell BSs have also been discussed, considering the network and hardware configuration. Moreover, how small cell BSs can take advantage of the proposed sector offset configuration through self-organization has been explained in detail. Extensive simulations results have been presented to quantify the benefits of the proposed configuration. The proposed HetNet configuration with offset sectorization significantly improves the network performance compared to existing state-of-the-art configurations, both in terms of UE throughput and HOF rate. In more detail, for a single carrier LTE HetNet, the cell-edge UE throughput has been shown to improve by 50 %, while the average and cell-edge macrocell UE throughput have been improved by 17 % and 32 %, respectively. The HOF rate decreased by 46 % when adopting offset sectorization. This comes at the cost of an increased number of antennas at the macrocell site to achieve the sector offset configuration. The concept is fully compatible with current mobile devices and cellular standards, and is, in general, easy to implement.

REFERENCES

1. H. Claussen and L. T. W. Ho, "Multi-carrier cell structures with angular offset," in *Proc. IEEE Pers., Indoor Mobile Radio Commun. (PIMRC)*, Sydney, Australia, Sep. 2012.

2. D. López-Pérez, H. Claussen, and L. Ho, "Improved frequency reuse schemes with horizontal sector offset for LTE," in *Proc. IEEE Pers., Indoor Mobile Radio Commun. (PIMRC)*, London, UK, Sep. 2013.

3. H. Halbauer, S. Saur, J. Koppenborg, and C. Hoek, "3D beamforming: Performance improvement for cellular networks," *Bell Labs Tech. J.*, vol. 18, no. 2, pp. 37–56, 2013.

4. E. Dahlman, S. Parkvall, J. Skold, and P. Beming, *3G Evolution, Second Edition: HSPA and LTE for Mobile Broadband*, 2nd ed. Academic Press, 2008.

5. S. Ye, S. H. Wong, and C. Worrall, "Enhanced physical downlink control channel in LTE advanced release 11," *IEEE Comm. Mag.*, vol. 51, no. 2, pp. 82–89, Feb. 2013.

6. D. López-Pérez and H. Claussen, "3GPP eICIC: Configuring almost blank subframe duty cycles and load balancing in HetNets," in *Proc. IEEE Pers., Indoor Mobile Radio Commun. (PIMRC)*, London, UK, Sep. 2013.

7. H. Claussen and L. T. W. Ho, "Multi-carrier cell structures with vertical and horizontal sector offset using static beamforming," in *Proc. IEEE Int. Conf. Commun. (ICC)*, Budapest, Hungary, Jun. 2013.

8. H. Claussen and L. T. W. Ho, "Multi-carrier cell structures with offset sectorization for heterogeneous networks," in *Proc. IEEE Int. Conf. Commun. (ICC)*, Budapest, Hungary, Jun. 2013.

9. D. López-Pérez and H. Claussen, "Improved frequency reuse schemes through sector offset configuration for LTE heterogeneous networks," in *Proc. IEEE Int. Conf. Commun. (ICC)*, Sydney, Australia, Jun. 2014.

10. D. López-Pérez, X. Chu, and I. Güvenç, "On the expanded region of picocells in heterogeneous networks," *IEEE J. Sel. Topics Signal Process. (J-STSP)*, vol. 6, no. 3, pp. 281–294, 2012.

11. G. Monghal, K. Pedersen, I. Kovacs, and P. Mogensen, "QoS oriented time and frequency domain packet schedulers for the UTRAN long term evolution," in *Proc. IEEE Veh. Technol. Conf.*, May 2008, pp. 2532–2536.

12. 3GPP TR TR 36.839, "Evolved universal terrestrial radio access (E-UTRA); mobility enhancements in heterogeneous networks," 3GPP-TSG R1, Tech. Rep. v 11.0.0.

13. H. Jia, Z. Zhang, G. Yu, P. Cheng, and S. Li, "On the performance of IEEE 802.16 OFDMA system under different frequency reuse and subcarrier permutation patterns," in *Proc. IEEE Int. Conf. Commun. (ICC)*, vol. 24, Jun. 2007, pp. 5720–5725.

14. Z. Xie and B. Walke, "Enhanced fractional frequency reuse to increase capacity of OFDMA systems," in *3rd Int. Conf. New Technol., Mobility Security (NTMS)*, Cairo, Egypt, Dec. 2009.

CONTROL CHANNEL INTER-CELL INTERFERENCE COORDINATION

10.1 INTRODUCTION

From its first release, the standard for long-term evolution (LTE) supports frequency-dependent interference-aware scheduling of the data transmissions. This feature enables satisfactory inter-cell interference coordination (ICIC) in the user equipment (UE) plane on the basis of fractional frequency reuse [1] or multi-point transmission techniques [2]. However, the LTE Release 10 as specified in [3–5] and all previous releases do not define such scheduling flexibility for the control signaling transmissions, which are the data transmission precursors. In particular, the UE-specific control channels, the so-called physical downlink control channels (PDCCHs), are protected from the co-channel interference only by means of static pseudo-random resource interleaving, which is equivalent to interference averaging. The LTE Release 11 [6–8] instead defines a form of flexible scheduling for control channels, the so-called enhanced PDCCHs, but this functionality is not available in legacy UEs, which will still represent a large portion of UEs for years after the rollout of LTE Release 11 networks. Given the importance of backward compatibility in the LTE standards for protecting the network operator's investments, the problem of control channel coexistence is expected to persist over the next few years.

The performance of LTE control signaling under inter-cell interference can be seen as the coexistence bottleneck in co-channel heterogeneous networks (HetNets)

Small Cell Networks: Deployment, Management, and Optimization, First Edition. Holger Claussen,
David López-Pérez, Lester Ho, Rouzbeh Razavi, and Stepan Kucera.
© 2017 by The Institute of Electrical and Electronic Engineers, Inc. Published 2017 by John Wiley & Sons, Inc.

that limits the effective small cell coverage range and UE capacity [9]. The reason is that the scheduling inflexibility of the LTE control channels leads to traffic and location-dependent failures in the control message delivery, and consequently the data-plane performance [10] because the control signaling conveys information on downlink and uplink scheduling grants and power control commands.

Simulations indicate that a UE cannot decode a PDCCH control message from the serving evolved NodeB (eNodeB) if the message signal-to-interference-plus-noise ratio (SINR) drops below approximately −4 dB [11]. This is equivalent to a limit on the small cell range expansion bias (REB) of 4 dB. In practice, the maximum permissible REB can reach up to 5–6 dB before incurring (i) radio link failures (RLFs) of UEs in the small cell expanded region, or (ii) handover failures of macrocell UEs entering the small cell [12, 13].

However, REBs higher than 6 dB are often needed to achieve efficient offloading of macrocell traffic by low-power small cells, especially whenever the small cell eNodeB is offset with respect to the center of the traffic hotspot due to installation and backhaul constraints. Even if an REB of 6 dB is sufficient, the emerging high-capacity LTE applications such as voice over LTE (VoLTE) [14] can be made possible only on the basis of a very efficient management of the control channel resources.

Although easy to implement and robust, the almost blank subframe (ABS) and carrier aggregation (CAG) approaches to control channel coexistence or ICIC in general as described in Chapters 7 and 8, respectively, are inherently sub-optimal in establishing the inter-tier coexistence, because they rely on conservative blanking of entire subframes and the addition of new carrier components, respectively. Moreover, the configuration of ABS cannot resolve the problem of dense small cell deployments, that is, the interference among small cells themselves. The CAG approach is practically limited by the overall scarcity and high costs of the electromagnetic spectrum available to mobile network operators. Likewise, the antenna offset techniques introduced in Chapter 9 can help to balance the SINR distribution, but a better SINR fairness generally cannot guarantee a minimum reception quality, particularly for the control channels of all UEs, required for successful data transfer.

To solve the problem of maximizing the ICIC resource efficiency, while ensuring an elementary reliability of transmissions over shared communication channels, the adaptive nature of the LTE scheduler and its central role in resource management must be exploited. The Third-Generation Partnership Project (3GPP) has standardized the LTE physical layer, communication protocols, and interfaces, but the internal resource management is open for a vendor-specific implementation. It is the efficiency of the algorithms used that determines the overall network performance.

The focus of this chapter is the discussion of the concept of orthogonally filled subframe (OFS), a standards-compliant solution that minimizes the failures of the LTE control signaling in a resource-efficient manner. Taking the known ICIC techniques for flexibly schedulable data transmissions as a prerequisite, its main idea is to optimally activate only the minimum amount of control resources necessary to guarantee a predefined level of the control transmission reliability. Consequently, the

OFS concept can be seen as a generalization of the ABS concept for one shared carrier component. Both proactive and reactive implementations are possible.

As for alternative solutions, the LTE standard offers the possibility for on-demand power-boosting selected PDCCH transmissions, but such an uncoordinated symptomatic approach ultimately causes more inter-cell interference. In [11], the originally data-related concept of fractional frequency reuse [15–17] was extended to the control region to improve the inter-tier coexistence in the control signaling domain. Yet solutions based on the sole usage of power control require extensive and accurate inter-cell coordination, even if the computation period is much larger than the LTE data scheduling interval of one millisecond. For this reason, inter-cell coordination that is not implemented properly can actually lead to an even higher level of co-channel interference.

This chapter is organized as follows. The OFS concept is outlined in Section 10.2. A formal definition of the system model and the related optimization procedures follow in Section 10.3. Implementation issues are discussed in Section 10.4, while the OFS performance benefits are quantified in Section 10.5. Conclusions are given in Section 10.6.

10.2 CONCEPT OF ORTHOGONALLY FILLED SUBFRAMES

To summarize the OFS concept, the terminology and notation of the LTE Release 10 specification [3–5] is adopted including the abbreviations for the primary synchronization channel (PSS), secondary synchronization channel (SSS), physical broadcast channel (PBCH), cell-specific reference symbol (CRS), physical control format indicator channel (PCFICH), physical HARQ indicator channel (PHICH), PDCCH, and physical downlink shared channel (PDSCH).

On a logical level, the LTE control channels can be functionally divided into broadcast channels and unicast channels.

The broadcast channels comprise the *cell-specific* signaling channels that must be received correctly by all cell UEs as a prerequisite of any wireless service. They include the CRS used for channel state information (CSI) measurements, the PCFICH indicating the control-region size in time, the PHICH enabling the hybrid automatic repeat request (HARQ) processes, and the common PDCCHs used for common messaging such as paging. Importantly, the logical broadcast channels are mapped onto the physical orthogonal frequency division multiplexing (OFDM) resources in the available frequency-time space as a pseudo-random but deterministic function of the cell identifier, namely the physical layer cell identity (PCI).

The unicast channels are the *UE-specific* control channels—the dedicated PDCCHs—that carry UE-specific pointers to PDSCH data, uplink scheduling grants, random-access responses and uplink power control commands. The total number of available PDCCHs is determined by the system bandwidth, the PCFICH configuration, and the number of HARQ processes, all of which are system constants broadcast over the broadcast channels.

A UE is associated with a PDCCH for control messaging purposes solely as a function of the subframe number and its identifier, the so-called cell radio network temporary identifier (C-RNTI). Since the PDCCHs can occupy only physical resources that remain after the broadcast channel resources allocation, the physical PDCCH location is also effectively a function of the PCI.

Depending on the UE reception quality as indicated by the channel quality indicator (CQI), a PDCCH can be allocated 1×, 2×, 4×, or 8× more resources than the nominal (minimal) PDCCH size to convey the same control message in a more robust manner. The replication coefficient is referred to as the aggregation level in the LTE terminology. Different PDCCHs in a subframe may use different aggregation levels. To reduce decoding complexity, PDCCHs are constructed in a nested modular manner such that each PDCCH having an aggregation level 2, 4, and 8 is composed of exactly two PDCCHs having a half aggregation level 1, 2, and 4, respectively.

With this in mind, an LTE scheduler can solve the coexistence problem in the control plane by (i) maximizing the number of active UEs whose PDCCH achieves at least $SINR_{min}$, while (ii) maintaining $SINR_{min}$ of the broadcast channels in the entire cell.

In common LTE networks with randomly assigned PCIs and C-RNTIs, such an optimal scheduler would have to explore all subsets of schedulable UEs and compute the associated transmit powers and aggregation levels to determine an optimal UE schedule and powers. Apart from its NP-hard complexity (problem reducible to the k-interval scheduling problem [18]), this approach is practically infeasible in that it implicitly requires extensive real-time inter-cell coordination and a global knowledge of the total channel state information.

The OFS concept, however, keeps the PCIs and C-RNTIs as independent degrees of optimization freedom to take advantage of the fact that PCIs and C-RNTIs are the primary seeds of the standardized pseudo-random algorithms for the management of the physical control channel resources.

In general, it is proposed to activate the UE-specific PDCCHs

- in a predetermined order minimizing the aggregate overlap of interfering active PDCCHs, and
- only to an extent guaranteeing $SINR_{min}$ for both the broadcast and unicast channels in all UE locations under the given small cell REB.

As discussed subsequently, the ordered set of usable PDCCHs and the associated PDCCH powers are obtained for a given scenario as a result of simple off-line combinatorial optimization, whose results are statically stored in the eNodeBs or retrieved on-demand from a server or a cloud service. The optimality metric is generally a combination of capacity and transmission reliability, that is, the size of the active PDCCH subset and the achievable channel SINRs. Depending on the network planning options, PCIs can be treated as constant inputs or as an independent degree of optimization freedom.

Theoretically, collision avoidance in time and frequency (hard orthogonality) can be accompanied by simultaneous power control (soft orthogonality). However, a general multi-level power control implies a variable PDCCH coverage in terms of the $SINR_{min}$. In such a case, the UE location must be monitored and the UE C-RNTI updated if the eNodeB detects an outage possibility, for example, by evaluating the UE reports on the A3 or A5 event concerning the CRS strength of the serving and neighboring cells. Yet, any dependency on mobility monitoring or prediction introduces a certain degree of PDCCH unreliability in view of the practically quantized, outdated, and infrequent CQI measurements. Moreover, updating an C-RNTI is a relatively slow radio resource control (RRC) process (on the order of seconds) and the update cannot be announced over an independent signaling channel – it is standardized to be delivered over the potentially out-of-coverage UE PDCCH itself.

A more practical OFS configuration postulates the broadcast channels to be always active at the maximum power and the PDCCH power control to assume only two states – maximum or zero power. The resulting elimination of one degree of optimization freedom causes the minimized inter-cell interference to be a function of solely the physical resource overlap of the interfering channels, that is, the UE C-RNTIs and the PCIs. Hence, simple C-RNTI control can be advantageously used to implement the optimized restrictions on the PDCCH usage. To this end, only minimal software updates of the scheduler are required in legacy systems.

Given the optimization results, the eNodeB scheduler assigns such C-RNTIs to the active UEs that map the UEs onto the optimized PDCCH pool. As discussed in the following, the UE mapping must be defined as a function of the subframes in which the PDCCHs are to be transmitted. These subframes are then characterized by quality-of-service guarantees thanks to the ensured maximum level of orthogonality of the active PDCCHs, hence the notion of "orthogonally-filled subframes."

The notion of "subframe filling" comes from the perception of the OFS concept as implementing the filling of the standard ABS to the maximum possible capacity with guaranteed $SINR_{min}$ of all the active channels.

Section 10.5 shows that configuring OFSs instead of ABS allows doubling the small cell coverage range compared the non-OFS case, resulting in a better offload of the macrocell traffic or triple the control-channel capacity to enable high-load VoLTE applications. From the UE perspective, a better scheduling fairness of the cell-edge UEs is achieved as well.

10.3 FORMAL DEFINITION

The system model described in the following assumes a co-channel scenario in which the small cell LTE resources direct overlap with the macrocell LTE resources. The case of identical victim-cell and aggressor-cell bandwidth sizes is the most challenging because any non-overlapping sections can serve as "clean" carriers facilitating interference-mitigation optimization. Subsequently, a general optimization frame-work is described to formalize the OFS concept.

10.3.1 System Model

In LTE, both control signaling and data transmissions occur within one subframe of a 1 ms duration. In the time domain, the control resources span across the entire band-width between the first and the (at most) third OFDM symbol as indicated by the PCFICH. The remaining subframe resources are used for data (PDSCH) transmis-sions unless used for the reference and acquisition signals (CRS, PSS, SSS, PBCH).

For the sake of simplicity but with no loss of generality, a two-cell system con-sisting of a single aggressor and a single victim cell is considered. The control chan-nel resources of both cells directly overlap, in the simplest case ensuring frame and subframe synchronization of the two cells. Such a resource configuration does not limit the amount of configurable multicast-broadcast multimedia service (MBMS) subframes but results in direct interference of the cell-search and acquisition signals namely the PSS, SSS, and PBCH.

For ensuring reliable cell discovery, a configuration characterized by offset cell resources may be more desirable as illustrated in Fig. 10.1. Herein, the subframes of both the aggressor and the victim cell are assumed to be synchronized but mutu-ally offset such that their PSS/SSS/PBCH do not collide, although the number of usable synchronous MBMS subframes is reduced accordingly to the intersection of the offset macrocell and small cell MBMS subframes. In addition, both cells blank their PDSCH resources that overlap with the PSS/SSS/PBCH of the other cell. The aggressor- and victim-cell bandwidths overlap at least partially, but there are no other frequency-domain constraints on the bandwidth size and relative direct current (DC) carriers offset.

In general, the aggressor cell may also at least partially blank its PDSCH resources that overlap with the victim-cell PDSCH to enhance the data rates of the

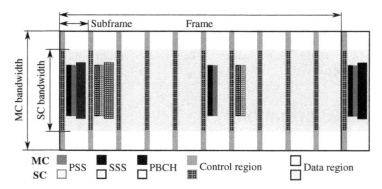

Figure 10.1. Exemplary time-frequency alignment of macrocell and small cell resources (identical CFI, subframe offset of one, identical DC carriers, variable bandwidths).

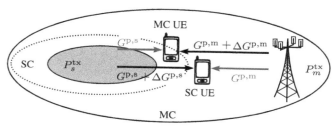

Figure 10.2. System model notation. The letters $G^{p,s}$, $G^{p,m}$, $\Delta G^{p,s}$, $\Delta G^{p,m}$, and P_s^{tx}, P_m^{tx} denote channel gains and transmit powers, respectively.

victim-cell UEs. As observed in the context of Eq. (10.18), the blanking of relatively few CRS resource elements in the data region[1] advantageously permits tolerating maximum interference to all the CRS resource elements in the control region[2] while avoiding RLF, because only a small fraction of *both* the data- and control-region CRS resource elements must achieve a minimum SINR. An RLF is triggered when less-than-minimal wideband CRS SINR is detected.

A spectrally more efficient reactive approach to PDCCH blanking by distributed small cells could be based on detecting the very strong uplink interference from mobile macrocell UEs. Note that whenever the small cell vacates an *arbitrary* PDSCH portion in response to a detection event, the mobile macrocell UE will report significantly improved channel conditions in the particular sub-band and the greedy macrocell scheduler will very likely allocate a PDSCH transmission into the vacated portion without any explicit information exchange with the small cell.

Our notation denotes aggressor-cell and victim-cell parameters by upper and lower-case letters, respectively. Vectors and matrices are represented by bold fonts. In accordance with Fig. 10.2, the channel gains between the aggressor-cell and victim-cell eNodeBs and the aggressor-cell and victim-cell UEs are denoted as $G^{p,m} + \Delta G^{p,m}$, $G^{p,m}$, $G^{p,s} + \Delta G^{p,s}$, and $G^{p,s}$. The effects of fast fading are not considered as they can be compensated by considering inherent performance margins as discussed subsequently. In addition to the UE noise figure n, the transmit powers P_s^{tx}, P_m^{tx} are given as the ratio of the power budget of the cell base station (BS) and the number of the available resource elements, that is, as average "per–resource element" values.

A minimum SINR_{min} is required by the UE to successfully decode a control channel message (typically -8 to -4 dB). To compute the SINR of the worst-off (small cell) UE in the capacity-problem scenario, the received powers $G^{p,s} + \Delta G^{p,s})P_s^{tx}$ and $G^{p,m}P_m^{tx}$ are assumed to be known, while in the mobility-problem scenario, the received powers $(G^{p,m} + \Delta G^{p,m})P_m^{tx}$ and $G^{p,s}P_s^{tx}$ of the worst-off (macrocell) UE are assumed to be known. If $\Delta G^{p,s} = \Delta G^{p,m} = 0$, the explicit knowledge of the UE SINR and signal-to-noise ratio (SNR) is sufficient.

Spanning in time from the xth to 14th OFDM symbols of a subframe where x is announced to UEs by the cell PCFICH broadcast, $1 \leq x \leq 3$.
Spanning in time from the 1st to the xth OFDM symbols of a subframe.

10.3.2 Control Channel Resource Allocation

As part of regular network operation, the operator assigns to each cell a locally unique PCI between 0 and 503. A cell identifies its UEs by a cell-unique C-RNTI, a hexadecimal number ranging from 000A to FFF2.

In the control space, each cell first allocates the physical resource elements of the CRS, PCFICH, and PHICH as a function of its PCI. The parameter N_g, set by the network operator, determines the number of configured HARQ groups, that is, the number of PHICH resource elements. The remaining unoccupied resource elements accommodate the UE PDCCH messages. On a logical level, the PDCCH resource elements are partitioned into control channel element (CCE) consisting of 36 resource elements.

A PDCCH indicates the location of the UE-specific PDSCH data of a scheduled UE bearer (data flow). The UE C-RNTI is used for encoding the PDCCH cyclic redundancy check.

Every subframe, each active UE performs a blind search for its designated PDCCH in the control region. To reduce the search complexity, each UE monitors only a limited UE-specific search space in the control region and attempts to decode a limited number of predefined PDCCH candidates.

The number of PDCCH candidates in a search space depends on the aggregation level L ($= 1, 2, 4, 8$), which is the number of CCEs used for a PDCCH as a function of the CQIs reported by the UE. A higher L implies more robust PDCCH and less PDCCH candidates in the search space.

In a 20 MHz bandwidth, there are totally $N_{\mathrm{CCE},k} = 87, 86, 84, 80$ CCEs for $N_g = 6^{-1}, 2^{-1}, 1, 2$, respectively. Hence up to 10 PDCCH messages of size $L = 8$ CCEs can be transmitted in one subframe while the 7, 6, 4, 0 remaining CCEs can be used for PDCCH messages at a lower L.

The following review of the actual mapping algorithms shows that the overlap of aggressor-cell and victim-cell control channels is a deterministic function of the aggressor-cell and victim-cell PCIs and the UE C-RNTIs.

10.3.2.1 Common Reference Signal   Within the first three OFDM symbols, the CRS of the antenna ports 0 and 1 (abbreviated as ap_0 and ap_1) are mapped on individual resource elements with a time-domain index k and a frequency-domain index l where by Section 6.10.1.2 in [3],

$$l_{\mathrm{ap}_0} = l_{\mathrm{ap}_1} = 0, \tag{10.1}$$

$$k_{\mathrm{ap}_0} = 6n + [(0 + (\mathrm{PCI} \bmod 6)) \bmod 6], \tag{10.2}$$

$$k_{\mathrm{ap}_1} = 6n + [(3 + (\mathrm{PCI} \bmod 6)) \bmod 6], \tag{10.3}$$

$$n \in \left[0, 2N_{\mathrm{RB}}^{\mathrm{DL}} - 1\right]. \tag{10.4}$$

$N_{\mathrm{RB}}^{\mathrm{DL}}$ is the number of resource blocks of 12 subcarriers and 7 OFDM symbols. For a 20 MHz bandwidth, $N_{\mathrm{RB}}^{\mathrm{DL}} = 100$ resource blocks. With no loss of generality, the antenna ports 3 and 4 are omitted in our examples.

10.3.2.2 Physical Control Format Indicator Channel

Defining the resource element group as a unit comprised of four consecutive non-CRS resource elements, the four quadruplets q of PCFICH resource elements are mapped onto four resource element groups, whose representative resource element indices (the "lowest-index" resource element [3]) by Section 6.7.4 in [3] are

$$l_q^R = 0, \tag{10.5}$$
$$k_q^R = (\bar{k} + 6\lfloor q\, N_{RB}^{DL}/2 \rfloor) \bmod 12 N_{RB}^{DL}, \tag{10.6}$$
$$\bar{k} = 6\left(\text{PCI} \bmod 2N_{RB}^{DL}\right), \tag{10.7}$$
$$q \in [0, 3]. \tag{10.8}$$

10.3.2.3 Physical HARQ Indicator Channel

Numbering the available CCEs from 0 to $N_{CCE,k} - 1$ for the subframe index $k \in [0, 9]$, the three resource element quadruplets of each PHICH group p are mapped onto three resource element groups not occupied by PCFICH (totally $9N_{CCE,k}$ resource element groups). Given $N_g \in [1/6, 1/2, 1, 2]$, the representative resource element indices by Section 6.9.3 in [3] are

$$l_q^{R(p)} = 0, \tag{10.9}$$
$$k_q^{R(p)} = (\text{PCI} + p + \lfloor qN_{CCE,k}/3 \rfloor) \bmod N_{CCE,k}, \tag{10.10}$$
$$p \in \left[0, \lceil N_g N_{RB}^{DL}/8 \rceil - 1\right], \tag{10.11}$$
$$q \in [0, 2]. \tag{10.12}$$

10.3.2.4 Physical Downlink Control Channel

Within the set of contiguous $N_{CCE,k}$ CCEs in subframe k, the logical PDCCH search space of a UE identified by an C-RNTI consists of M consecutive logical PDCCH candidates (LPCs) where $M = 6, 6, 2, 2$ for $L = 1, 2, 4, 8$, respectively. By Sections 9.1.1 in [5] and 5.1.4.2.1 in [4], the mth LPC for $m \in [1, M]$ comprises L consecutive CCEs starting at the CCE indexed as

$$\text{CCE}_{m,L}^{\min} = L[(Y_k + m - 1) \bmod \lfloor N_{CCE,k}/L \rfloor], \tag{10.13}$$
$$Y_k = 39827 Y_{k-1} \bmod 65537, \tag{10.14}$$
$$Y_{-1} = \text{C} - \text{RNTI} \in [10, 65522]. \tag{10.15}$$

In a 20 MHz bandwidth, the 10 distinct PDCCH candidates of size $L = 8$ start at the CCEs indexed as $8n$ for $n \in [0, 9]$. The first two PDCCH candidates of size 8 are shared for common paging and access messaging ($Y_k = 0$, $M = 4, 2$ for $L = 4, 8$). In general, each PDCCH candidate of size $L = 8, 4, 2$ divides into two PDCCH candidates of half size.

The logical CCEs are mapped onto the physical resource elements by using common sub-block interleaving followed by a PCI-specific cyclic shift (5.1.4.2.1 in [4]). In particular, all logical resource element groups are written in a top-to-bottom left-to-right manner into a matrix having 32 columns and the minimum necessary number of rows. Unused elements in the left part of the top row are padded by "NIL" elements. The matrix columns indexed from 0 to 31 are then permuted in the order [1, 17, 9, 25, 5, 21, 13, 29, 3, 19, 11, 27, 7, 23, 15, 31, 0, 16, 8, 24, 4, 20, 12, 28, 2, 18, 10, 26, 6, 22, 14, 30]. A column-wise read-out follows (the "NIL" padding is discarded), and a cyclic offset of the permuted CCE space equal to the cell PCI is applied. The final logical-to-physical resource element mapping is carried out in a time-first fashion and ignores already used resource element groups.

10.3.3 Optimization Framework

To formalize the notion of channel overlap, the normalized channel overlap matrix $\Omega_{AGG>VIC}$ is defined as a matrix, whose element in the ith row and the jth column is the ratio of (i) the number of resource elements in which the ith channel of the aggressor cell–aggressor cell (AGG) overlaps with the jth channel of the victim cell (VIC), and (ii) the total resource element number of the jth channel of the victim cell VIC. Both broadcast channels and unicast channels are considered in the definition of Ω.

Although generally $\Omega_{AGG>VIC} \neq \Omega_{VIC>AGG}$, the normalization by the VIC channel size implies that both matrices $\Omega_{AGG>VIC}$ and $\Omega_{VIC>AGG}$ add up row-wise or column-wise to 1, respectively (see the example in Table 10.1). As observed in Section 10.3.2, Ω is a function of the vector **PCI** of the aggressor-cell and victim-cell PCIs.[3]

Let Λ denote the vector of relative weights ranging from 0 to 1, which are used for scaling the maximum aggressor-cell power P_m^{tx} of the aggressor-cell broadcast and unicast channels between 0 and P_m^{tx}. Multiple discrete or continuous-range weight levels can be considered, but this assumption would complicate the C-RNTI management due to possibly limited channel coverage in terms of $SINR_{min}$ as discussed above. Hence, only binary weights are assumed.

Let $\Lambda(N_m)$ denote a binary instance of Λ in which (i) the power weights of the aggressor-cell broadcast channels are equal to 1 (to ensure their correct reception throughout the entire cell), and (ii) exactly N_m weights of the aggressor-cell unicast channels is equal to 1 and the others to 0.

Let $\{\Lambda(N_m)\}$ denote the set of all instances of $\Lambda(N_m)$.

Let λ denote the victim-cell equivalent of the aggressor-cell weights λ.

Let $\Pi(\mathbf{PCI}, N_m, N_s)$ also denote the performance metric that maps the instances $\Lambda(N_m), \lambda(N_s)$ of aggressor- and victim-cell power weights for a given channel

---

[3] The dependency on the PHICH parameter N_g is neglected given its adaptation-limiting relation to the total number of HARQ groups of a cell.

TABLE 10.1   Aggressor-Cell and Victim-Cell Channel Overlap $\Omega_{AGG>VIC}$ in Percent of Victim-Cell Channel Size for Aggressor-Cell and Victim-Cell PCIs 0 and 342

$N_g=1/2$ Victim cell → PCI = 0 Aggr. cell ↓ PCI = 342	CRS AP$_0$	CRS AP$_1$	PCFICH	PHICH	LPC 0	LPC 1	LPC 2	LPC 3	LPC 4	LPC 5	LPC 6	LPC 7	LPC 8	LPC 9	Other 6 CCEs
CRS AP$_0$	·	100	·	·	·	·	·	·	·	·	·	·	·	·	·
CRS AP$_1$	100	·	·	·	·	·	·	·	·	·	·	·	·	·	·
PCFICH	·	·	·	5	·	1	3	·	·	·	·	·	·	·	·
PHICH	·	·	·	·	1	1	4	1	4	3	3	3	1	3	5
LPC 0	·	·	·	10	·	15	54	8	8	10	1	·	·	·	·
LPC 1	·	·	·	10	·	·	13	54	7	8	14	1	·	·	·
LPC 2	·	·	·	14	·	·	·	15	51	8	7	11	3	·	·
LPC 3	·	·	·	10	·	·	·	·	15	54	10	7	10	1	·
LPC 4	·	·	·	5	·	·	·	·	·	15	51	10	10	11	2
LPC 5	·	·	·	14	4	·	·	·	·	·	14	53	10	6	13
LPC 6	·	·	25	·	14	3	·	·	·	·	·	15	53	11	4
LPC 7	·	·	25	10	6	13	3	·	·	·	·	·	14	51	13
LPC 8	·	·	25	19	11	4	14	3	·	·	·	·	·	17	58
LPC 9	·	·	25	5	58	15	4	13	4	·	·	·	·	·	4
Other 6 CCEs	·	·	·	·	6	47	6	6	10	1	·	·	·	·	2

Table entries are rounded to integers ("." represents zero). LPC 0 is considered to be a broadcast channel (in **bold**). Gray cells indicate channels inactivated to reduce the aggregate overlap with active channels (white cells) below 66%.

overlap matrix Ω to scalar values. Practically, the metric Π can be defined as the maximum aggregate overlap between the aggressor-cell and victim-cell channels computed as

$$\Pi(\mathbf{PCI}, N_m, N_s) = \max_{\substack{\forall \Lambda \in \{\Lambda(N_m)\} \\ \forall \lambda \in \{\lambda(N_s)\}}} \mathbb{1}^T \mathrm{diag}(\Lambda)\Omega(\mathbf{PCI})\mathrm{diag}(\lambda), \quad (10.16)$$

where $\mathbb{1}$ is a unit vector, $\mathrm{diag}(\Lambda)$ is a zero matrix whose main diagonal is given by Λ and the operator $[.]^T$ represents a matrix transpose. More generally, $\Pi(\mathbf{PCI}, N_m, N_s)$ can be defined as the lowest control channel SINR at the victim cell determined as

$$\Pi(\mathbf{PCI}, N_m, N_s) = \min_{\substack{\forall \Lambda \in \{\Lambda(N_m)\} \\ \forall \lambda \in \{\lambda(N_s)\}}} \mathrm{SINR}(\Omega(\mathbf{PCI}), \Lambda, \lambda). \quad (10.17)$$

By using the exponential mapping [19] of the "per-resource element" SINRs, we have that

$$
\mathrm{SINR}(\Omega_{\mathrm{AGG>VIC}}, \Lambda, \lambda) = \min_{\forall j | \lambda_j \neq 0} -\beta \ln \left(\sum_{\forall i} \Omega_{\mathrm{AGG>VIC}}[ij] e^{-\frac{\lambda_j (G^{\mathrm{p,s}} + \Delta G^{\mathrm{p,s}}) P_s^{\mathrm{tx}}}{\beta \Lambda_i G^{\mathrm{p,m}} P_m^{\mathrm{tx}} + \beta n}} \right),
$$

(10.18)

where $\beta = 1.57, 1.69, 1.69$ based on the 3GPP recommendations for the quadrature phase-shift keying (QPSK) code rate of $1/2, 2/3, 3/4$, respectively.

To jointly address both capacity and mobility problems, one can consider the worst-case minimum SINR defined as

$$
\mathrm{SINR}(\Omega, \Lambda, \lambda) = \min(\mathrm{SINR}(\Omega_{\mathrm{MC>SC}}, \Lambda, \lambda), \mathrm{SINR}(\Omega_{\mathrm{SC>MC}}), \Lambda, \lambda). \quad (10.19)
$$

Then, to reflect the rather static nature of PCIs[4] in planned cellular networks, the optimum PCIs can be determined first as

$$
\mathbf{PCI}^* = \arg\min_{\mathbf{PCI}} \Theta(\Pi(\mathbf{PCI}, N_m, N_s)) \quad N_m, N_s \in A \qquad (10.20)
$$

to minimize a selection metric Θ over its definition region A.

The selection metric Θ can be defined for $N_m, N_s \in A$ as the sum of distances of $\Pi(\mathbf{PCI}, N_m, N_s)$ from the overall maximum in the **PCI** search space, that is,

$$
\Theta = \sum_{\forall N_m, N_s \in A} \sqrt{\Pi^2 - (\max_{\mathbf{PCI}} \Pi)^2}. \qquad (10.21)
$$

The definition region A is limited by the maximum number of aggressor-cell and victim-cell unicast channel LPCs, that is, $A \subset [0, N_m^{\max}] \times [0, N_s^{\max}]$. If the aggressor-cell and victim-cell UE distribution can be estimated as an (approximate) function $N_m = f(N_s, \mathrm{REB})$ of the small cell REB, A can be reduced to the graph of f, that is, the union of aggressor-cell and victim-cell load pairs $[f(N_s, \mathrm{REB}), N_s]$, for which $\Pi(\mathbf{PCI}, N_m, N_s) \geq \mathrm{SINR}_{\min}$.

Once the network PCI plan with respect to Θ and A is known, the corresponding optimum weights Λ^*, λ^* for the aggressor-cell and victim-cell LPCs are computed subsequently for all N_m, N_s as

$$
[\Lambda^*(N_m), \lambda^*(N_s)] = \arg\max_{\Lambda(N_m), \lambda(N_s)} \Pi(\mathbf{PCI}^*, N_m, N_s). \qquad (10.22)
$$

---

[4] The macrocell PCI can be typically given by the operator as an input constant.

The maximum permissible small cell REB is determined for each $\Lambda^*(N_m)$, $\lambda^*(N_s)$ as the maximum value for which

$$\min \text{SINR}(\Omega(\mathbf{PCI}^*), \Lambda^*(N_m), \lambda^*(N_s)) \geq \text{SINR}_{\min}, \qquad (10.23)$$

where SINR can be defined as in Eq. (10.18), assuming the knowledge of (i) the transmit powers P_s^{tx}, P_m^{tx}, and (ii) the dependencies of $G^{\text{p,s}}$, $G^{\text{p,m}}$, $\Delta G^{\text{p,s}}$, $\Delta G^{\text{p,m}}$ on the small cell REB. Note that in terms of our examples, Eq. (10.23) requires the "worst-case SINR" performance metric Π in the optimized network to always exceed $\text{SINR}_{\min}$ as postulated by the system model.

In view of the standardized C-RNTI-to-LPC mapping (10.13), the obtained power vectors Λ^* and λ^* are used as an input of the UE C-RNTI management to indicate control channels that are "safe" (power weight approaching 1) and "unsafe" (low power weights). Ideally, the UEs are assigned C-RNTIs by the cell eNodeB such that the corresponding LPCs in terms of Eq. (10.13) are characterized by *safe* PDCCH resources to ensure reliable transmission in case of poor networking conditions, as well as *unsafe* PDCCH resources to be used otherwise for overall capacity enhancement and reduction of safe PDCCH resource blocking. The access to either resource type is ideally independent of *time* to reduce scheduling, retransmission complexity, and inter-UE blocking probabilities, as well as *cell location/channel quality* (i.e., the aggregation level) to maximize scheduling capacity.

If the channel SINRs are above the $\text{SINR}_{\min}$ threshold (e.g., the small cell REB is lower than the maximum permissible by Eq. (10.23)), then the zero weights in Λ^* and λ^* that indicate inactive LPCs can be increased up to a level when equality occurs in the minimum SINR condition (10.23). Since LPCs with reduced power may have only limited PDCCH coverage, only zero weights that are non-zero in more populous profiles Λ^* and λ^* having higher N_m and N_s should be increased to simplify the C-RNTI management.

In terms of practical implementation, the optimization of Eq. (10.22) implicitly requires the aggressor cell and the victim-cell to announce the variations of their UE load N_m and N_s because the optimum profiles $\Lambda^*(N_m)$ and $\lambda^*(N_s)$ are functions of N_m and N_s. More importantly, however, an optimum weight profile may not generally be a subset of an optimum weight profile of a larger size, that is, $\Lambda^*(N_m) \not\subset \Lambda^*(N_m + x)$ and/or $\lambda^*(N_s) \not\subset \lambda^*(N_s + x)$ for $x > 0$. In other words, the power weight (coverage) of a given LPC may undesirably vary with the UE cell load. This implies that the UE C-RNTIs must be updated under the optimal Λ^* and λ^* as a function of N_m and N_s to prevent LPC deactivation or PDCCH outages in response to cell load variations.

The aggressor-cell and victim-cell overhead communication and load-dependent C-RNTI updates can be avoided by defining $\Lambda^*(N_m)$ and $\lambda^*(N_s)$ as subsets of $\Lambda^*(N_m^{\max})$ and $\lambda^*(N_s^{\max})$, respectively, for $N_m \leq N_m^{\max}$, $N_s \leq N_s^{\max}$. The ideal LPC activation order, that is, the exact subsets of $\Lambda^*(N_m^{\max})$, $\lambda^*(N_s^{\max})$ as a function of N_m, N_s, can be found by combinatorial search minimizing mutual channel overlap. Such an approximation of Λ^*, λ^* may be theoretically suboptimal but any practically observed performance loss is acceptable if not negligible. Slow periodical corrections of $N_m^{\max}, N_s^{\max}$ reflecting the actual cell loads are desirable.

Table 10.1 indicates the binary $\Lambda^*(N_m)$ and $\lambda^*(N_s)$ for an exemplary load $N_m^{\max} = 6$ and $N_s^{\max} = 7$.

10.4 IMPLEMENTATION

The OFS concept can be implemented by using either a proactive or a reactive implementation philosophy. Before discussing each of these approaches in more detail, general remarks applying to both approaches are described.

10.4.1 General Remarks

Deactivating LPCs in the control region helps prevent RLFs similarly to PDSCH blanking in the data region as discussed in Section 10.3. In general, patterns stretching over multiple (sub)frames can be used to compensate for the fact that UEs can theoretically evaluate the CRS quality in arbitrary subframes. Moreover, the blanking of LPCs for which the worse-off UE does not achieve $SINR_{min}$ under the given small cell REB guarantees that (i) the coverage range of any active LPC comprises the entire cell area, and (ii) the LPC blanking in the OFS subframes can be implemented based on Eq. (10.13) by imposing simple restrictions on the usable C-RNTI pool (in combination with PDCCH candidate selection). This effectively corresponds to the most practical binary power control alternating between maximum and zero LPC power. The example of Table 10.1 indicates how LPCs can be activated at maximum power, while not letting the aggregate channel overlap exceed 66% (see Fig. 10.4 for the associated small cell REBs in various scenarios).

In the context of macrocell CRS interference to the small cell channels, the macrocell two-port multiple-input multiple-output (MIMO) functionality (primarily transmit diversity for cell-edge UEs) can be used without any effect on the OFS operation for small cell REBs typically up to 20 dB. In order to avoid an undesired impact on the CQI measurements, PCI planning is to be used to prevent the exact matching of the CRS resource elements of the macrocell and small cell antenna ports (see Table 10.1). Such matching occurs in one-sixth of the PCI pool. On the other hand, for higher values and only one antenna port at the macrocell or the small cell, it is progressively more desirable to exploit such macrocell and small cell CRS alignment to eliminate the CRS interference contribution. For extreme bias values over 30 dB, such perfect exclusion effectively becomes a necessity.

Dynamic variations of the control region duration in time (as specified by the control format indicator (CFI) in the PCFICH) are possible because the proposed interference-minimizing OFS configurations can be precomputed for each aggressor-cell and victim-cell CFI combination. Yet, the C-RNTI sets of individual OFS solutions may differ. Hence, the CFI variations should not be generally faster than the ability of the RRC layer L3 to reconfigure UE C-RNTIs (units of seconds). The limited RRC responsiveness must be also considered when updating the UE C-RNTIs to ensure sufficient coverage of *power-controlled* UE LPCs in the given UE location, as discussed hereafter.

The discussed optimization framework is based on the worst-case cell-edge UE under full UE load and maximum LPC powers. However, in practice, if the worst-off UE is not at the very edge of the cell, the small cell REB can be set to be smaller than the limit given by the number of active small cell LPCs, and the LPCs of the aggressor cell may be only partially used at lower-than-maximum powers. This general system under-usage creates a performance margin that can be taken advantage of to compensate for the unaccounted fading of communication channels. Depending on the actual UE conditions, it is also always possible to shrink an overly large performance margin by opportunistically reducing the predetermined transmit power in order to temporarily alleviate the overall interference.

10.4.2 LPC Management

The proactive approach is based on preconfiguring the macro and small cells for OFS operation assuming a given maximum numbers of active cell LPCs. Algorithm 10.1 offers a simple example of OFS optimization for the most practical binary "on/off" power control. The knowledge of the expected cell UE load is used to define the maximum LPC numbers N_m^{max}, N_s^{max}, that is, the limits of the area A. Practically, the optimization of LPC blanking is carried out assuming maximum aggregation level $L = 8$, which is also the least computationally demanding case as the search space quadruples for every reduction of L.

In general, only infrequent updates of the macrocell and small cell configuration and variations of the small cell REB, cell PCIs and the CFIs are needed to reflect substantial changes in the networking conditions. OFS reconfigurations are effectuated by means of switching between precomputed optimized solutions. In this way, the inter-cell communication overhead is minimized. The OFS configuration can be determined and disseminated by a background service, for example, running on a cloud-computing platform.

Input : $PCI_m, N_m^{max}, N_s^{max}$ at given agg. level L
Output: PCI_s with associated Λ^*, λ^* and respectively usable C-RNTIs
for *all* PCI_s **do**
$\quad$ find $\Lambda^*, \lambda^* = \underset{\forall \Lambda \in \Lambda(N_m^{max}), \forall \lambda \in \lambda(N_s^{max})}{\arg\min} \Pi(\mathbf{PCI}, N_m, N_s)$
$\quad$ for channel overlap based metric Π from Eq. (10.16);
end
select PCI_s with the highest resource activity factors $\sum \Lambda^*$ and/or $\sum \lambda^*$;
find usable C-RNTIs by using Algorithm 10.2
$\quad$ **Algorithm 10.1:** Algorithm for LPC Orthogonalization.

Accordingly, both the macrocell and the small cell schedulers assign to UEs only C-RNTIs for which at least one associated PDCCH candidate lies entirely within the preselected LPCs, preferably for all aggregation levels. The scheduler ignores for scheduling purposes the PDCCH candidates that fall out of the LPC range.

The alternative reactive approach relies on a timely communication between the cooperating cells to take advantage of resources that are (temporarily) characterized by low interference, or to create such resources on-demand. Events requiring a tailored on-demand approach can arise under higher UE loads (e.g., multiple UEs competing for limited PDCCH resources as function of their C-RNTIs), due to uneven traffic distribution among unpredictably roaming UEs (e.g., different scheduling frequency and aggregation requirements), or stringent requirements on maximum delay of HARQ retransmissions of corrupt packets (e.g., UE schedulability in multiple subframes), as well as whenever an SINR outage is predicted or detected in power-controlled OFS and the time margin is insufficient for an update of the UE C-RNTI.

From the control signaling point of view, the indication of an interference event by the victim cell to the aggressor cell must contain the C-RNTI information of the interference-critical UE. Such a UE is typically the cell-edge UE of the victim cell in a (near)-handover situation to the aggressor cell. For implementation, the victim cell can send the standard LTE x2 (X2) *Handover Requests* message to the aggressor cell over the X2 link as this message contains the RRC context of the source cell, part of which is the UE C-RNTI. Similarly, the aggressor cell can reply to the victim cell by using the standard LTE *Handover Request Acknowledge* message that contains the C-RNTI that the UE would be assigned in case of a successful handover.

As discussed above, the C-RNTI information is in combination with the PCIs sufficient for an optimal unilateral resource planning. The aggressor cell eliminates PDCCH transmissions that contribute to the interference at the victim-cell UE by adapting its scheduling decisions (UE rescheduling or muting, aggregation level, and PDCCH candidate reselection). The interference avoidance period ends when the UE handover is completed, or the UE leaves the critical cell-edge region.

After the handover completion, the source cell can help the target cell to maintain the PDCCH SINR of the handed over UE by adapting its PDCCH scheduling decisions. Depending on how the cells are configured to cooperate in terms of their resource management, various types of master-slave schemes for mutual coordination can be implemented. For example, the target cell can try to maximize the partial reuse of the LPC resources released by the source cell upon the handover completion whereby the source cell can help improve the LPC reuse efficiency by eliminating colliding transmissions.

It is inefficient to use the reactive approach for large-scale interference coordination in the control region since it places a substantial computational and communication burden on the cooperating cells, while in the best case, only the performance of the proactive approach with offline precomputed solutions can be reached.

10.4.3 C-RNTI Management

The hosting macrocell is typically the dominant interferer due to the sector antenna orientation. Hence, the knowledge of the relative aggregate overlap with a dominant aggressor is typically sufficient for reliable ordering of the victim-cell channels in terms of their actual quality, since the binary "on/off" power control (i.e., selective muting) of the aggressor PDCCH resources represents the most practical implementation alternative. If multi-level power control is used or multiple dominant interferers of comparable strength are present, the PDCCH quality can be ranked accurately only on the basis of the SINR experienced by a representative (e.g., worst-case) victim-cell UE.

Assuming binary power control, Fig. 10.3 plots the outage probability of small cell LPCs that are sorted in the ascending order of (i) the outage probability itself (ideal ranking criterion), and (ii) the relative aggregate overlap (approximate ranking criterion). For each LPC, the outage probability was measured in a 3GPP-compliant HetNet from Section 10.5 as the relative fraction of the small cell coverage area in which the SINR of the given candidate is larger than the minimum SINR required for successful decoding. Cell coverage is defined as the set of locations in which the SINRs of the PHICH, PCFICH, and at least one PDCCH at $L = 8$ are at least -4 dB. Minor ranking errors are observed only for the worst-off LPCs having the least-robust aggregation level $L = 1$. Moreover, simulations show that the amplitude of the error-related outage probability fluctuations quickly decreases with the increase of active macrocell LPCs.

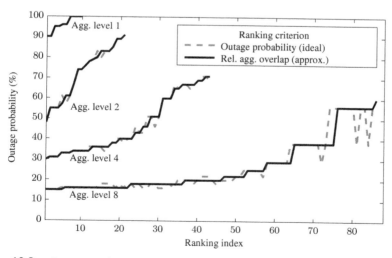

Figure 10.3. Outage probability of logical PDCCH candidates of a small cell, sorted on the order of (i) the relative aggregate overlap with the macrocell channels, and (ii) the outage probability itself. HetNet from Section 10.5.

Input : precision x; percentile n; subframes s; vectors **S, U** of required minimum number of
      safe/unsafe LPCs for each agg. level L (descending order of L)
Output: usable C-RNTIs satisfying **S&U**

for *all aggregation levels L* **do**
    │ **for** *all LPCs excluding common LPCs* **do**
    │   │ quantize aggregate overlap of active LPCs with precision x;
    │   │ mark LPCs in top/bottom nth percentile as safe/unsafe;
    │ **end**
end
$c = 0$
for *all C-RNTIs* **do**
    │ **if** *at least* $S(4\text{-}\log_2(L))$ *safe LPCs and*
    │ *at least* $U(4\text{-}\log_2(L))$ *unsafe LPCs are*
    │ *associated via Eq. (10.13) for all L in all subframes s* **then**
    │   │ select given C-RNTI as usable;
    │ **end**
end

Algorithm 10.2: Algorithm for C-RNTI Search.

Given the relative quality of the control channels, it is possible to select the C-RNTI pool for UE identification, that is, PDCCH scheduling, such that the included C-RNTIs are associated with at least a minimum number of LPCs having a predetermined quality for several aggregation levels and in several subframes.

By predetermined quality is understood either "good" or "bad" quality, that is, low and high overlap with the active aggressor-cell control channels, respectively. In this way, the LTE scheduler is given the flexibility of transmitting a PDCCH message to the given UE independently of the UE networking and mobility conditions.

For example, a C-RNTI can be required to map the UE PDCCH messages via Eq. (10.13) to at least one LPC that belongs to the *least-overlapping* nth percentile of the LPCs in terms of the relative aggregate overlap with the active aggressor-cell channels for multiple higher aggregation levels (e.g., $L = 8, 4, 2$). Simultaneously, the C-RNTIs can be also required to be associated with at least one of the *most-overlapping* nth percentile of the LPCs for the multiple lower aggregation levels (e.g., $L = 2, 1$). Algorithm 10.2 summarizes exemplary search for suitable C-RNTIs.

Once a UE is assigned its C-RNTI, the associated low-interference LPCs are to be used by the scheduler for enabling transmissions to UEs with poor networking conditions (e.g., cell-edge UEs). The minimum "safe" aggregation level should be used to maximize the number of UEs that can be scheduled in one subframe, that is, the control region capacity. In small cells, zero-overlap PDCCHs having $L = 2$ can be typically used to deliver cell-edge messages even for high cell-selection biases. On the other hand, PDCCH messages with higher aggregation levels are more resilient against unexpected signal quality fluctuations due to frequency-selective fading.

Analogically, the simultaneously available high-interference LPCs are to be used for capacity-efficient signaling over complementary PDCCH resources when the networking conditions improve (e.g., cell-center UEs). In this way, inter-UE blocking in access to the safe LPCs is reduced, while improving the control region capacity.

To enable more than one (re)transmission of guaranteed reception quality per subframe, the UE C-RNTIs must be optimized such that the associated LPCs satisfy the required quality constraints in multiple subframes. Preferably, the distance between the optimized subframes should be four which is the HARQ retransmission period.

The number of available C-RNTIs generally decreases with the number of controlled aggregation levels and subframes. From Eq. (10.13), we observe that the LPCs are uniformly spread over the logical PDCCH space. Hence, for a 20 MHz bandwidth having 86 CCEs and $N_g = 1/2$, there is $86/L$ logical search spaces, which corresponds to roughly $(65522 - 10)L/86 = 761L$ C-RNTIs being associated with one LPC of size L. When two or more aggregation levels are controlled, then the total number of available C-RNTIs is at most the minimum of C-RNTIs available for the individually controlled aggregation levels.

Similarly, the lower is the nth percentile of the LPC quality, the lower will be the overall number of available C-RNTI. In this context, recall that each C-RNTI can only be associated with $M = 2, 2, 6, 6$ *consecutive* LPCs for $L = 8, 4, 2, 1$, respectively, typically sharing the same CCEs among different aggregation levels. Generally, both the candidate total number and consecutiveness decrease as function of n.

Clearly, the scarce C-RNTIs guaranteeing good PDCCH quality under most networking conditions should be reserved only to highly mobile UEs experiencing a wide range of networking conditions. Relatively static UEs should be assigned more abundant C-RNTIs with only narrowly tailored PDCCH properties. Yet, it is generally difficult for an LTE scheduler to predict during link setup what will be the mobility/channel conditions of each individual UE. This complicates the targeted selection and reuse of the most suitable C-RNTIs.

If needed, however, a UE C-RNTI can be updated on the basis of the UE channel state information report within units of seconds. Yet apart from the delay and overhead issues in case of frequent wide-scale updates, it is to be noted that the C-RNTI update command is broadcast over the PDCCH channel as part of RRC messaging. In other words, an update of a UE C-RNTI cannot be done in conditions in which the PDCCH channel could fail such as in the case of cell-edge UEs. Hence, additional C-RNTI assignment optimization could be worthwhile to minimize the number of C-RNTI updates and PDCCH failures under a given PDCCH scheduling algorithm.

The knowledge of the LPC quality associated with each C-RNTI may be used for further performance improvements. For example, once a UE schedule for a given subframe is determined, the scheduler can increase the aggregation level of selected LPCs by expanding them into adjacent unused CCEs, and/or minimize the power level necessary to guarantee the required minimum SINR of the PDCCH message at the receiver. In general, the scheduling process can consist of several scheduling iterations employing each such techniques whereby the stopping criterion could be zero or less-than-minimum improvement in the schedule capacity in the last two iterations.

10.5 PERFORMANCE EVALUATION

3GPP-compliant simulations of a typical LTE HetNet deployed in central Dublin, Ireland, have been performed to quantify the range-expansion aspect and the capacity-maximization aspect of the proposed OFS concept. The simulation parameters are summarized in Table A.6 except that an LTE bandwidth of 20 MHz has been assumed. The distance between the macrocell and small cell eNodeBs is 100, 200, 300, 400 m, whereby the power control is binary (LPC power weight set $\{0, 1\}$).

Data are collected for the central macrocell and small cell of the hexagonal grid to prevent edge singularities. The capacity problem is assumed with a macrocell being the aggressor cell and a therein hosted small cell being the victim cell.

10.5.1 Range-Expansion Aspect

To describe the range-expansion aspect of the OFS, Figs. 10.4 and 10.5 show the maximum number of simultaneously active macrocell LPCs at the aggregation level 8 subject to each active small cell LPC achieving at least $SINR_{min} = -4$ dB as a function of the small cell REB, respectively. The macrocell and small cell PCI are 0 and 342, similarly to the case of Table 10.1.

For the assumed $N_g = 1/2$, there are 10 LPCs of aggregation level 8. For simplicity, the figures denote the remaining six CCEs as the 11th LPC. The data are averaged

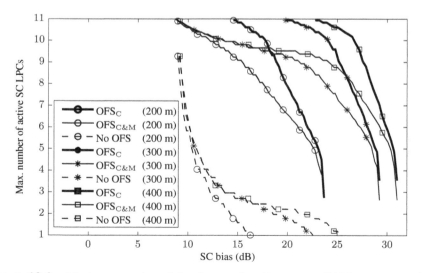

Figure 10.4. Maximum number of simultaneously active macrocell LPCs at aggregation level 8 subject to each active small cell LPC achieving at least $SINR_{min} = -4$ dB as a function of the small cell REB. The macrocell uses 6 LPCs at the aggregation level 8.

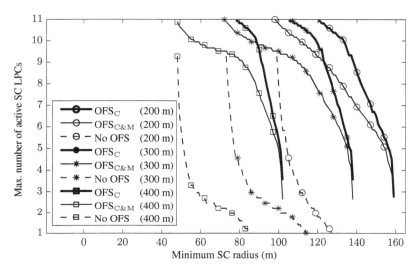

Figure 10.5. Maximum number of simultaneously active macrocell LPCs at aggregation level 8 subject to each active small cell LPC achieving at least SINR$_{min}$ = −4 dB as a function of the minimum small cell radius. The macrocell uses 6 LPCs at the aggregation level 8.

over the top 1 percentile of available PCI (top 50 PCIs from the range 0–503) and shown for three distances between the macrocell and the small cell eNodeBs, namely 200 m, 300 m, and 400 m. The C-RNTI assignment is irrelevant in this case.

Assuming six LPCs of aggregation level 8 to be active at the macrocell, the maximum small cell range is defined by at least one unicast channel or broadcast channel to reach SINR$_{min}$. The only constraint is that the LDPC the LPC 0 or LPC 1 or both must remain always active as they can be used for common PDCCH transmissions in addition to UE-specific messaging.

It is observed from Figs. 10.4 and 10.5 that a small cell REB of at most 5–7 dB ensures the macrocell and small cell coexistence under any macrocell and small cell load (100% LPC overlap is tolerable). Beyond this range, the blanking of macrocell LPCs is necessary to maintain the SINR$_{min}$ of the small cell LPCs but the complete muting of one of the cells occurs for a rather extreme small cell REBs between 27 and 35 dB. By expanding the state-of-the-art small cell REB range beyond the 5–7 dB, the OFS concept allows doubling the small cell radius which significantly improves the macrocell traffic offload by small cells.

10.5.2 Capacity-Maximization Aspect

The capacity-maximization aspect of the OFS is studied for the case of binary LPC power control, the eNodeB distance of 400 m, and the small cell REB of 6.5 dB, for

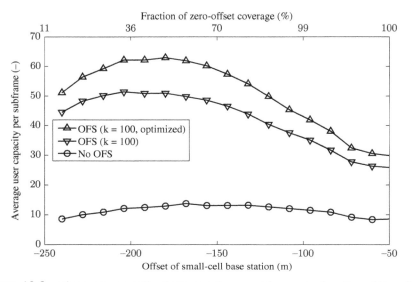

Figure 10.6. The average small cell UE capacity per subframe as a function of the offset of the small cell eNodeB from the center of a hotspot in an OFS-configured system and a non-OFS system.

which all the broadcast channels and unicast channels exceed $SINR_{min} = -4$ dB even if transmitted simultaneously at the maximum power (see Fig. 10.4).

Figure 10.6 shows the average small cell UE capacity per subframe achieved in an OFS-configured system and a non-OFS system as a function of the offset of the small cell eNodeB from the center of a hotspot. The hotspot is defined by the small cell coverage when no eNodeB offset exists (i.e., ideal small cell positioning). The UEs requesting service are distributed uniformly randomly within the hotspot and no UEs are present outside of its boundaries. Data are averaged over 10^5 independent simulation runs.

Two particular scenarios are investigated.

In the non-OFS scenario, a benchmark scheduler using standard random C-RNTI assignment activates at each cell *all* its LPCs for PDCCH transmission purposes, that is, all its 86 available CCEs within the assumed 20 MHz LTE bandwidth.

In the OFS scenario, the scheduler of the macrocell and small cell eNodeB activates only the least interfering 9 and 6 LPCs at the aggregation level 8, respectively. Hence, the small cell scheduler uses only approximately half of its LPC resources. The actual scheduling decisions are made in accordance with the schedule having the highest UE capacity among k randomly generated schedules. An optimization algorithm for reducing inter-UE blocking when generating a schedule can be used to improve the capacity.

As observed in Fig. 10.6, the OFS scheduler can schedule within one subframe a triple amount of UEs than the benchmark scheduler despite muting nearly half of its LPCs. The reason is that the improved interference conditions (i.e., higher SINRs) allow substantially reducing the PDCCH aggregation level, on average four times. The theoretical maximum is an eight-fold reduction.

The rise and fall of small cell capacity as a function of the eNodeB offset from the hotspot center stems from the fact that the cell loses UEs in one segment of its edge due to offsetting, but does not capture new UEs on the opposite cell-edge segment. Consequently, the proportion of cell-center UEs requiring lower PDCCH aggregation level increases implying a small cell capacity increase. Once the cell edge passes the center of the hotspot (approximately the curve peak), then all UEs progressively shift toward the cell edge in response to cell offset and the cell capacity starts decreasing again.

10.5.3 C-RNTI Availability

To evaluate the availability of C-RNTIs associated with LPCs of predefined quality, it is assumed that the macrocell blanks 4 of its 11 LPCs having $L = 8$ such that the maximum aggregate overlap of the remaining PDCCH resources with any of the LPCs of the small cell having $L = 8$ is minimized. A practical constraint to the minimization process consists of that the macrocell activates at least one active common LPC having $L = 8$ to enable common PDCCH signaling.

An LPC of the victim cell is considered as being of high-quality (safe) or low-quality (unsafe) if it belongs to the nth top or bottom percentile of all LPCs in terms of the considered performance metric, respectively. In our simulations, $n = 1$ or $n = 3$.

The safety of each LPC is quantified by means of the outage probability in a random location within the small cell coverage area. A LPC outage is declared if the SINR of the LPC is lower than $-4, -1, 2, 5$ dB for $L = 8, 4, 2, 1$, respectively. The small cell coverage is defined as the set of locations in which the SINRs of the PHICH, PCFICH, and at least one PDCCH at $L = 8$ are at least -4 dB. The measured outage probability is quantized with a step of 5% in order to treat equally LPCs with negligible quality difference.

To indicate the *minimum* number of safe and unsafe LPCs required for each aggregation level (sorted in descending order), two 4×1 vectors denoted as $\mathbf{S}$ (for Safe) or $\mathbf{U}$ (for Unsafe), respectively, are used. For example, the vectors $\mathbf{S} = [1\,1\,0\,0]^T$ and $\mathbf{U} = [0\,0\,0\,2]^T$ indicate that at least one safe LPC is required for two aggregation levels $L = 8, 4$, while at least two unsafe LPCs are required for the aggregation level $L = 1$.

Tables 10.2, 10.3, and 10.4 summarize the total number of C-RNTIs that are associated via Eq. 10.13 for various safety constraint vectors $\mathbf{S}$ and $\mathbf{U}$. For every two instances of $\mathbf{S}$ and $\mathbf{U}$, six C-RNTI selection criteria are simulated by verifying the satisfaction of

- the $\mathbf{S}$-vector constraint in 1, 2, or 3 subframe(s),
- both the $\mathbf{S}$-vector and $\mathbf{U}$-vector constraints in 1, 2, or 3 subframe(s),

TABLE 10.2   Number of Available Golden C-RNTIs: **S/U** Set 1

Selection Criterion	S U	S U	S U	S U	S U	S U	S U	S U
Agg. level $L = 8$	1 0	1 0	1 0	1 0	1 0	1 0	1 0	1 0
Agg. level $L = 4$	1 0	1 0	1 0	1 0	1 0	1 0	1 0	1 1
Agg. level $L = 2$	0 0	0 0	0 1	0 1	0 2	0 2	0 2	* *
Agg. level $L = 1$	0 0	0 2	0 1	0 2	0 0	0 1	0 2	* *
1 subframe + S	11,230	11,230	11,230	11,230	11,230	11,230	11,230	11,230
1 subframe + S&U	11,230	4441	3917	2352	1310	919	786	0
2 subframes + S	1931	1931	1931	1931	1931	1931	1931	1931
2 subframes + S&U	1931	300	226	80	24	12	8	0
3 subframes + S	335	335	335	335	335	335	335	335
3 subframes + S&U	335	20	12	2	0	0	0	0

Criteria **S&U** based on 1st percentile of PDCCH candidates, outage quantization step 5%, The symbol "*" represents any number 0, 1, 2.

subject to

- the exclusion of all CCEs corresponding to the common LPCs of $L = 8$ (index 0,1).

The C-RNTIs that satisfy the last exclusion constraint minimize the blocking with the higher-priority common PDCCH signaling and are referred to as the Golden C-RNTIs.

TABLE 10.3   Number of Available Golden C-RNTIs: **S/U** Set 2

Selection Criterion	S U	S U	S U	S U
Agg. level $L = 8$	1 0	1 0	1 0	1 0
Agg. level $L = 4$	1 0	1 0	1 0	1 0
Agg. level $L = 2$	1 1	1 1	1 2	1 2
Agg. level $L = 1$	0 1	0 2	0 1	0 2
1 subframe + S	5217	5217	5217	5217
1 subframe + S&U	1825	1303	526	393
2 subframes + S	421	421	421	421
2 subframes + S&U	54	28	5	3
3 subframes + S	32	32	32	32
3 subframes + S&U	2	1	0	0

Criteria S&U based on 1st percentile of PDCCH candidates outage probability quantization step 5%, The symbol "*" represents any number 0, 1, 2.

TABLE 10.4 Number of Available Golden C-RNTIs for Different Levels of PDCCH Safety

Selection Criterion	S U	S U	S U
Agg. level $L = 8$	1 0	1 0	1 0
Agg. level $L = 4$	1 0	1 0	1 0
Agg. level $L = 2$	1 0	1 1	1 1
Agg. level $L = 1$	1 1	1 0	1 1
1 subframe + S	128 / 2749	128 / 2749	128 / 2749
1 subframe + S&U	128 / 1835	0 / 922	0 / 789
2 subframes + S	13 / 121	13 / 121	13 / 121
2 subframes + S&U	13 / 54	0 / 14	0 / 10
3 subframes + S	0 / 5	0 / 5	0 / 5
3 subframes + S&U	0 / 1	0 / 0	0 / 0

1st / 3rd percentile of PDCCH candidates, quantization step 5%.

When evaluating the validity of the S and U criteria, the rounded average over all the possible combinations of mutually different subframes is indicated.

Rather strict selection criteria are assumed to explore the feasibility limits of C-RNTI management. In practice, the BS can use C-RNTIs that satisfy various constraints to efficiently address the needs of a diverse UE population.

Assuming the search of Golden C-RNTIs within the 1st percentile of PDCCH safety, we observe from Tables 10.2 and 10.3 that there exists hundreds to thousands of Golden C-RNTIs that satisfy all the represented practical S criteria in up to three subframe(s). In very high-density VoLTE applications, the total UE load is at the order of (lower) hundreds, hence a sufficient number of UE identifiers is available. The simultaneous consideration of the U criterion reduces the C-RNTI pool size to tens to hundreds, but as this criterion is oriented to improve the UE schedulability into unused resources, higher PDCCH safety percentile can be simply assumed to obtain tens/hundreds of Golden C-RNTIs satisfying the $S\&U$ criteria.

In general, increasing the number of controlled subframes by one reduces the number of available Golden C-RNTI around 10 times. The search for four subframes yielded no C-RNTI for any of the listed criteria. To obtain relevant non-zero C-RNTI numbers, the nth percentile of PDCCH safety must be increased substantially (see Table 10.4 for results obtained assuming the 3rd percentile of PDCCH safety).

It is also observed that the number of available Golden C-RNTIs decreases with the number of controlled aggregation levels (i.e., the number of non-zero entries in the S/U vectors) as well as with the minimum number of LPCs required for each aggregation level (the vector element values).

The number of available Golden C-RNTIs is lower when a given number of safe/unsafe LPCs is required for higher aggregation levels compared to the case of lower aggregation levels. This is given by the fact that at lower aggregation levels, there are more LPCs, which increases the chance that there are relatively well co-located LPCs having similar quality characteristic. For example, assuming again the 1st percentile of PDCCH safety, no Golden C-RNTIs were obtained even for single-frame search when at least two safe and/or unsafe LPCs were required for $L = 4$ or $L = 8$, but some Golden C-RNTIs can still be found when at least 4–5 safe and/or unsafe LPCs are required for $L = 1$ and $L = 2$.

10.6 SUMMARY AND CONCLUSIONS

The concept of OFSs was proposed, where the control-region resources for UE-specific PDCCH transmissions are activated (i) in a predetermined order maximizing the macrocell and small cell channel orthogonality, and (ii) only to an extent limiting the macrocell and small cell channel overlap such that the worst-off UEs achieve a minimum channel SINR under the given small cell REB.

This approach generalizes the standardized ABS solutions and outperforms them in terms of capacity and fairness. PCI and C-RNTI optimization are used in combination with power control to maximize the achievable performance limits.

The range expansion aspect of the OFS concept allows expanding the small cell size by trading the number of active PDCCH candidates against the small cell REB. As the REB maximum reaches 27–35 dB, which effectively doubles the small cell radius, the OFSs enable a significantly improved offload of the macrocell traffic by the small cell.

The SINR-maximization aspect allows reducing the PDCCH aggregation level up to 4 times, which effectively triples the system capacity, a feature demanded by the emerging high-load VoLTE applications.

REFERENCES

1. A. L. Stolyar and H. Viswanathan, "Self-organizing dynamic fractional frequency reuse in OFDMA systems," in *Proc. of IEEE INFOCOM*, Phoenix, USA, Apr. 2008.

2. L. Daewon, S. Hanbyul, B. Clerckx, E. Hardouin, D. Mazzarese, S. Nagata, and K. Sayana, "Coordinated multipoint transmission and reception in LTE-Advanced: Deployment scenarios and operational challenges," *IEEE Commun. Mag.*, vol. 50, no. 2, pp. 148–155, Feb. 2012.

3. "Physical channels and modulation," 3GPP Technical Specification Group Radio Access Network, Evolved Universal Terrestrial Radio Access (E-UTRA), Tech. Rep. TS 36.211 V10.7.0, Mar. 2013. [Online]. Available: http://www.3gpp.org

4. "Multiplexing and channel coding," 3GPP Technical Specification Group Radio Access Network, Evolved Universal Terrestrial Radio Access (E-UTRA), Tech. Rep. TS 36.212 V10.8.0, Jun. 2013. [Online]. Available: http://www.3gpp.org

5. "Physical layer procedures," 3GPP Technical Specification Group Radio Access Network, Evolved Universal Terrestrial Radio Access (E-UTRA), Tech. Rep. TS 36.213 V10.10.0, Jun. 2013. [Online]. Available: http://www.3gpp.org

6. "Physical channels and modulation," 3GPP Technical Specification Group Radio Access Network, Evolved Universal Terrestrial Radio Access (E-UTRA), Tech. Rep. TS 36.211 V11.5.0, Mar. 2013. [Online]. Available: http://www.3gpp.org

7. "Multiplexing and channel coding," 3GPP Technical Specification Group Radio Access Network, Evolved Universal Terrestrial Radio Access (E-UTRA), Tech. Rep. TS 36.212 V11.4.0, Jun. 2013. [Online]. Available: http://www.3gpp.org

8. "Physical layer procedures," 3GPP Technical Specification Group Radio Access Network, Evolved Universal Terrestrial Radio Access (E-UTRA), Tech. Rep. TS 36.213 V11.5.0, Jun. 2013. [Online]. Available: http://www.3gpp.org

9. H. H. Puupponen, K. Aho, T. Henttonen, and M. Moisio, "Impact of control channel limitations on the LTE VoIP capacity," in *Proc. Int. Conf. Netw.*, Menuires, France, Apr. 2010.

10. "Mobility enhancements in heterogeneous networks," 3GPP Technical Specification Group Radio Access Network, Evolved Universal Terrestrial Radio Access (E-UTRA), Tech. Rep. TS 36.839 V11.0.0, Sep. 2012. [Online]. Available: http://www.3gpp.org

11. Fujitsu. (2011). "Enhancing LTE cell-edge performance via PDCCH ICIC" [Online]. Available: http://www.fujitsu.com/downloads/TEL/fnc/whitepapers/Enhancing-LTE-Cell-Edge.pdf

12. A. Damnjanovic, J. Montojo, Y. Wei, T. Ji, T. Luo, M. Vajapeyam, T. Yoo, O. Song, and D. Malladi, "A survey on 3GPP heterogeneous networks," *IEEE Trans. Wireless Commun.*, vol. 18, no. 3, pp. 10–21, Jun. 2011.

13. I. G. David López-Pérez, G. de la Roche, M. Kountouris, T. Q. S. Quek, and J. Zhang, "Enhanced inter-cell interference coordination challenges in heterogeneous networks," *IEEE Commun. Mag.*, vol. 18, no. 3, pp. 22–30, Jun. 2011.

14. "Delivering voice in the emerging era of LTE: How to deliver VoLTE with scalability and high performance in the evolved packet core," Alcatel-Lucent Application Note, Tech. Rep., Jun. 2013. [Online]. Available: http://resources.alcatel-lucent.com/asset/167959

15. T. D. Novlan, R. K. Ganti, A. Ghosh, and J. G. Andrews, "Analytical evaluation of fractional frequency reuse for OFDMA cellular networks," *IEEE Trans. Wireless Commun.*, vol. 10, no. 12, pp. 4294–4305, Dec. 2011.

16. T. D. Novlan and J. G. Andrews, "Analytical evaluation of uplink fractional frequency reuse," *IEEE Trans. Commun.*, vol. 61, no. 5, pp. 2098–2108, May 2013.

17. N. Ksairi, P. Bianchi, and P. Ciblat, "Nearly optimal resource allocation for downlink OFDMA in 2-D cellular networks," *IEEE Trans. Commun.*, vol. 10, no. 7, pp. 2101–2115, Jul. 2011.

18. H. Yang, F. Ren, C. Lin, and J. Zhang, "Frequency-domain packet scheduling for 3GPP LTE uplink," in *Proc. of IEEE/ACM Infocom*, San Diego, USA, Mar. 2010.

19. A. K. A. Tamimi, "Exponential effective signal to noise ratio mapping (EESM) computation for WiMax physical layer," Master thesis, Washington University, 2007.

11

UPLINK-ORIENTED OPTIMIZATION IN HETEROGENEOUS NETWORKS

11.1 INTRODUCTION

The operation of wireless cellular networks can be efficiently supported by secondary small cells, deployed at traffic hotspots within the coverage area of the primary macrocells. Yet to benefit from small cell deployments, the co-channel interference in both downlink and uplink transmission modes must be controlled across cells with overlapping coverage.

In previous chapters, it was shown how to configure and optimize small cells on the downlink. In this chapter, the opposite uplink transmission mode is discussed. The focus is on examining the conditions under which the secondary base station (BS)s can share the same communication channel with the primary BSs, given constraints on predefined quality of service such as a minimum signal-to-interference-plus-noise ratio (SINR) for each transmission accommodated in the common interference-limited channel.

To this end, a clear distinction is made between third-generation (3G) code division multiple access (CDMA)-based and fourth-generation (4G) orthogonal frequency division multiple access (OFDMA)-based technologies. In 4G long-term evolution (LTE) systems, the single-carrier FDMA (SCFDMA) technique orthogonalizes on the uplink of different user equipment (UE) transmissions in the same cell by explicit assignments of groups of discrete Fourier transform (DFT)-precoded

Small Cell Networks: Deployment, Management, and Optimization, First Edition. Holger Claussen, David López-Pérez, Lester Ho, Rouzbeh Razavi, and Stepan Kucera.
© 2017 by The Institute of Electrical and Electronic Engineers, Inc. Published 2017 by John Wiley & Sons, Inc.

orthogonal subcarriers. This is fundamentally different from 3G systems based on CDMA where UEs interfere with each other over the carrier bandwidth and advanced receivers, such as successive interference cancellers, are used to suppress same-cell interference.

Another difference between 3G universal mobile telecommunication system (UMTS) and 4G LTE consists of the nature of interference: unlike in 3G wideband code division multiple access (WCDMA) systems, same-cell interference is efficiently mitigated by design in 4G LTE systems. More importantly, however, a wideband approach to interference modeling is sufficient in WCDMA-system studies thanks to the underlying signal spreading, but LTE uplink interference must be analyzed on the level of individual physical resource blocks in view of the used discrete scheduling and frequency hopping.

In the first half of this chapter, the impact of the secondary small cell infrastructure in a 3G WCDMA system is assessed. Of interest is the global achievability of the uplink target SINRs via distributed closed-loop power control—the basic tool for uplink interference control. The effects of cell load and inter-cell coupling are distinguished. Solutions to the problems of BS placement, cell coverage optimization, and target SINR adaptation in WCDMA heterogeneous networks are also offered, as well as a discussion on their compatibility with standard downlink-oriented deployment guidelines.

Subsequently, uplink interference management in 4G LTE networks is considered and a measurement data-driven machine learning paradigm for self-optimizing power control is discussed. The data-driven approach has the inherent advantage that the solution adapts based on network traffic, propagation and network topology, that is, increasingly heterogeneous with multiple cell overlays. Evaluations using radio network plans from a real LTE network operational in a major metro area in the United States shows that, compared to existing approaches, this approach provides substantial gains in UE data rate.

11.2 3G CDMA HETEROGENEOUS NETWORKS

11.2.1 Introduction

The goal of this section is to define explicit uplink-oriented deployment guidelines and self-optimization methods for heterogeneous networks.

Existing studies in this area mainly focus on the coexistence of protected (primary) transmissions with predefined quality of service (QoS) requirements and unprotected (secondary) best-effort transmissions tolerating service outages and using opportunistic spectrum access (see [1] and the references therein). Alternative studies use a game-theoretic approach to ensure simultaneous satisfaction of multiple best-effort transmissions (see [2, 3] and the references therein).

Moreover, rather than defining a medium access scheme guaranteeing QoS only to selected UEs [4], our focus is on finding deployment guidelines and self-configuration methods for co-channel small cells that ensure a target QoS for all UEs within the heterogeneous network. The challenge is in combining the above aspects, as noted also by [5], possibly due to the intractability of the uplink-oriented system models (e.g., non-existent closed-form expressions for matrix eigenvalues [6]).

An extensive survey of the vast body of research in power control can be found in [7]. The pioneering works in [8,9] developed principles and iterative algorithms to achieve a target SINR when multiple UEs simultaneously transmit over a shared carrier. Subsequent studies proposed algorithms for joint optimization of rate and transmit power [10–13]. In particular, [10, 12, 13] consider a utility based framework for joint optimization of rate and transmit power. Uplink power control optimization was made tractable in WCDMA in [14, 15] where the log-convexity of the feasible SINR region was shown. Reliability issues are addressed in [16]. Initial work on deployment guidelines is presented in [17]. Deployment algorithms aiming to maximize coverage are defined in [18, 19].

This section is organized as follows. The considered WCDMA uplink system model is summarized in Section 11.2.2. Adaptive network optimization on the basis of the spectral radius of a network information matrix, mainly by radius minimization, is introduced subsequently in Sections 11.2.3, 11.2.4, and 11.2.5. The analytic framework is developed into practical guidelines and auto-configuration methods for BS placement, cell coverage optimization via pilot signal adjustments, and opportunistic enhancement of target SINRs in Section 11.2.6. Theoretical expectations are numerically validated in Section 11.2.7 by using Third-Generation Partnership Project (3GPP)-compliant simulations of a cellular network deployment. Conclusions follow in Section 11.4.

11.2.2 System Model

11.2.2.1 Definition
A 3G WCDMA cellular network consisting of multiple macrocells is assumed, in which a macrocell can host small cells providing smaller coverage. The location of individual small cells as well as UEs in a macrocell is arbitrary. UEs connect to the BS with the strongest coverage-defining pilot signal (common pilot channel (CPICH)). In accordance with the abilities of typical small cell BS hardware, there is no MC-SC soft handover.

Using the notation from Table 11.1, each of the macrocell and small cell BSs is formally associated with a unique integer index $b \in [1, B]$, where B denotes the total cell number in the network. For simplicity, a cell served by the BS b is also referred to as cell b. The symbol N_b denotes the total number of active UEs being served by the BS b (i.e., the UE load of the cell b). It is assumed that $N_b \geq 1$. The total number of UEs in the network is denoted as $N = \sum_{b=1}^{B} N_b$.

TABLE 11.1   Notation List (in Alphabetical Order)

Symbol	Description
b	Cell index
B	Network cell number
C	Auxiliary constant
D	Auxiliary constant
γ_b	Target SINR of UE served by cell b
$g_{uv}, g_{[b]v}$	Channel gain between UE v and serving cell b of UE u
$G_{c \triangleleft b}$	Coupling of cell b to cell c
I_u	Interference at the serving BS of UE u
k	Time sample index
λ_u	u-th eigenvalue of A
Λ	Spectral radius of A
n_u	Additive noise power of UE u
N	Network UE number
N_b	Number of UEs served by cell b
p_u	Transmission power of UE u
P_b	Power control offset constant
u	UE index
U	Set of network UEs
U_b	Set of UEs served by cell b
A	$N \times N$ matrix of $\gamma_u g_{uv}/g_{uu}$ if $u \neq v$, 0 if $u = v$, $\forall u, v$
B	$N \times 1$ vector of $(P_u + \gamma_u n_u)/g_{uu} \forall u$
E	$N \times N$ identity matrix
p	$N \times 1$ vector of $p_u \forall u$

Similar to the BS indexing, each active UE is assigned a unique integer index u from the integer set $[1, N]$. The set of indices u referring to UEs having the same serving BS b is denoted as U_b where $|U_b| = N_b$. The set of all network UEs is denoted as $U = \bigcup_{b=1}^{B} U_b$, $|U| = N$.

Focusing on uplink transmissions in a single shared channel, a transmission is successful if a minimum SINR is maintained by the UE at the BS throughout its duration. The target SINR γ_u of UE u is assumed to be equal to γ_b, $\gamma_b > 0$, for all UE served by the same BS b, that is, $\forall u \in U_b$. Nevertheless, two different cells can configure different UE target SINRs γ_b.

Importantly, the data bursts of all ongoing transmissions are assumed to be perfectly synchronized (constant mutual interference at maximum level, i.e., the worst-case scenario in terms of achievable channel capacity and UE power usage).

To maintain the target SINRs under time-varying interference, the UEs use iterative power control (PC) having an open and closed loop as defined in the 3G WCDMA standards.

In particular, let us denote $p_u(k)$ as the transmission power of the UE u in discrete (integer) time k, g_{uv} as the channel gain between the UE v and the serving BS b of UE u (hence, the gain between the UE u and its own serving BS b is g_{uu}), or simply $g_{[b]v}$ if the identity of UE u is irrelevant, and $I_u(k) = \sum_{\forall v \in U \setminus u} g_{uv}(k) p_v(k) + n_u$ as the interference experienced at the receiving BS of UE u in time k, where n_u is the additive noise power of UE u at the serving receiver BS. Then, the discrete-time power control of UE u under interference from all other UEs $U \setminus u$ is formalized as

$$p_u(k+1) = \frac{P_u + \gamma_u I_u(k)}{g_{uu}}, \tag{11.1}$$

where $P_u \geq 0 \forall u$ represents an arbitrary power-offset constant.

In practice, the power control step-size can attain only a limited number of values. Yet, this fact is irrelevant because it does not affect the target SINR achievability but only the overall convergence rate.

Since the small cell BSs practically do not support soft handover, the above power control realistically reflects the actual power control operation in WCDMA networks. A limited dynamic range of the power control (11.1) is considered in the practically-oriented numerical simulations, but is avoided in the theoretical analysis for tractability reasons.

The fast power control update rate of 1600 Hz in WCDMA systems (implying a power control update period smaller than the channel coherence time) and the large bandwidth of the communication channel (implying a constant output power for each UE signal at the BS rake receiver) allows multi-path fading of channel gains $g_{uv}(k)$ from the system model to be excluded.

11.2.2.2 Properties
To study formally the impact of small cell deployments on macrocell networks, the $N \times 1$ vector of powers of all UE is denoted as p. Let the uth component of the $N \times 1$ vector B be defined as $(P_u + \gamma_u n_u)/g_{uu}$. Also, let E denote an $N \times N$ identity matrix, and A denote an $N \times N$ matrix, whose element in the uth row and vth column is $\gamma_u g_{uv}/g_{uu}$ if $u \neq v$, and 0 if $u = v$ for all network UEs $u, v \in U = [1, N]$. Suppose naturally that $A - E$ is invertible given the stochastic nature of gains g_{uv}.

Then the conditions for global achievability of target SINRs can be formulated as in [20]. There exists a unique set of N powers

$$\hat{p} = \lim_{k \to \infty} p(k) = -(A - E)^{-1} B \tag{11.2}$$

to which the iterative power control updates (11.1) of all N network UEs converge exponentially fast in time k if and only if

$$\Lambda = \max_{\forall u} |\lambda_u(A)| \leq 1, \tag{11.3}$$

where $\lambda_u(A)$ is the uth eigenvalue of A and any $\lambda_u(A)$ such that $|\lambda_u(A)| = 1$ corresponds to a dimension-one Jordan cell.

Otherwise, the target SINRs are not achievable via power control (11.1) because the updates diverge toward infinite powers situation practically resulting into network-wide transmission outages once the UE maximum power limits are exceeded.

11.2.3 Uplink-Optimum Network Configuration

The spectral radius Λ of the network information matrix A ranges from zero to one in networks with target SINRs achievable by all active UEs via power control (11.1) because (i) $\Lambda > 0$ for $A \geq 0$ and $N > 1$ in general, and (ii) $\Lambda \leq 1$ if target SINRs are achievable.

Hence, the notion of Λ can be interpreted as the fractional load (reuse) of the shared communication channel, that is, the ratio of the channel capacity currently occupied by ongoing transmissions over the theoretically available (although unknown) capacity, defined by the actual network conditions as represented by A, that is, the UE number, channel gain distribution, and target SINRs [21].

Definition 11.1 (Load Factor Λ of Shared Channel). *The load factor of a shared network channel characterized by a given A is defined as $\Lambda = \max_{\forall u} \lambda_u(A)$.*

The notion of Λ is also used to define the uplink channel capacity as the maximum number of simultaneously active UEs satisfying the target SINR constraints. More specifically, the uplink capacity[1] region of a channel shared by B cells each having a target SINR γ_b is given by the set of points $[N_1, \ldots, N_B]$ in the B-dimensional integer space such that $\Lambda \leq 1$ for any possible UE channel gain $g_{uv} \forall u, v \in [1, N]$.

In practice, UE admission control to BS services is carried out locally on a UE-by-UE basis. Optimization considering a longer time perspective or for multiple UEs is difficult due to the unpredictability of UE activity and channel gains.

Ideally, a new UE is admitted into the network only if its presence does not violate the condition $\Lambda \leq 1$ (see [21] for distributed real-time monitoring of Λ). Note that admitting UEs on the basis of power limit satisfaction cannot guarantee the achievability of UE target SINRs.

Hence, managing the network such that Λ is minimized under given operating conditions (e.g., by adaptively adjusting the cell coverage areas) essentially corresponds to the maximization of the probability that a randomly located and activated UE can be allowed to transmit in the shared channel (i.e., the UE blocking probability at serving BSs is minimized).

---

[1] In view of Definition 11.1, and the dependency of Λ on channel gains, capacity is considered to be a property of the shared channel rather than of the network which is understood as a set of BSs and UEs.

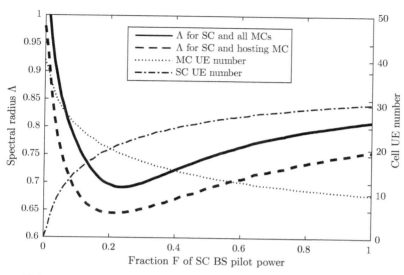

Figure 11.1. Average spectral radius Λ of the matrix A, and average number N of UEs associated to MC/SC BS as a function of the fraction F of the SC BS power budget allocated to the pilot signal.

In view of the UE-by-UE decision-making of the network admission control, the strategy of minimizing Λ then represents a viable approach to maximizing the number of UEs with achievable SINRs in the network, that is, the effectiveness of the usage of the theoretical channel capacity.

Moreover, the magnitude of the equilibrium powers $\hat{p}$ satisfying all achievable target SINRs is inversely proportional to $|1 - \Lambda|$ [21]. Therefore, the minimization of Λ increases the energy efficiency of the network by reducing $\hat{p}$, as well as reduces the probability of transmission outages in case of limited dynamic range of the power control (11.1).

Figure 11.1 shows a typical evolution of Λ as a function of the fraction F of the small cell BS power budget allocated to the pilot signal in a scenario according to *Table 11.2* (data averaged over 10^3 UE topologies). The pseudo-convex shape actually attains only a global minimum well-defined point to be sought by auto-configuration algorithms as visualized in Fig. 11.1.

The process of offloading macrocell UEs to the small cell BS causes initially a steady drop of Λ. As follows from Eq. (11.11), the minimum of Λ is practically achieved when the relative loads of the macrocell and the small cell are equal. If the small cell coverage is further extended, the value of Λ then increases again, although at a rate slower than the rate of the initial drop of Λ as the chance of capturing a new macrocell hotspot UE decreases.

TABLE 11.2 Simulation Parameters

Parameters	Value
Scenario	
BS placement	7 MC sites, 800 m inter-site distance
Scenario size	1500 m × 1500 m, around central MC site
Scenario resolution	2 m
MC BS CPICH power	10% of 21 W
SC BS CPICH power	10% of 1 or 2 W
Noise power	−108 dBm at BS
UEs	
Distribution	Uniformly random, Gaussian in hotspots
UL target SINR	−15 dB for 9.6 kbps voice traffic
Radio propagation	
Carrier frequency	2000 MHz
Bandwidth	3.84 MHz (1 UMTS carrier)
NLOS path gain	$g^{\mathrm{NLOS}} = -21.5 - 29\log_{10}(d)$ [24]
LOS path gain	$g^{\mathrm{LOS}} = -34.02 - 22\log_{10}(d)$ [24]
Shadow fading (SF)	6 dB std dev. [24]
SF correlation	$R = \mathrm{e}^{-1/20d}$ [25], 50% inter-site
Environment gain	$g^{\mathrm{E}} = -20$ dB if indoor, 0 dB if outdoor
MC BS antenna	
Height	25 m
Maximum gain	$g_{\max} = 15.5$ dBi (3 sec.), 19.8 dBi (6 sec.)
H. halfpow. beamwidth	$\alpha = 65°$ (3 sec.), 33° (6 sec.)
V. halfpow. beamwidth	$\beta = 11.5°$(3 sec.), 8.5° (6 sec.)
Front-to-back ratio	$\kappa = 30$ dB
Downtilt	$\delta = 10.57°$(3 sec.), 8.47° (6 sec.)
SC BS antenna	
Height	10 m
Maximum gain	$g_{\max} = 7$ dBi (4-element dipole array)
Element spacing	$d_e = 0.6$ wavelengths
Element phase inc.	$\delta_{\mathrm{phase}} = 1.658$
Element amplitude	$a = [0.97, 1.077, 1.077, 0.86]$

It is also noteworthy that increasing the small cell bias for a given small cell transmission power budget causes a horizontal compression of the Λ-curve in Fig. 11.1, that is, a linear shift of the function minimum to the origin (F is reduced by 50% by a small-cell bias of 3 dB).

11.2.4 Effect of Cell Load on the Global Achievability of Target SINRs

In order to develop the proposed concept of network management on the basis of Λ-minimization to practical deployment guidelines and methods for small-cell

3G CDMA HETEROGENEOUS NETWORKS

auto-configuration, the magnitude of Λ is first investigated with respect to cell load and inter-cell coupling as defined below.

Definition 11.2 *The coupling $G_{c \lhd b}$ of cell b to cell c is*

$$G_{c \lhd b} = \sum_{\forall u \in U_c} g_{[b]u} / g_{[c]u}. \tag{11.4}$$

Clearly, the closer the UEs of the (victim) cell c are to its serving BS (higher $g_{[c]u}$) and farther from the (aggressor) cell b (lower $g_{[b]u}$), the lower is the overall inter-cell coupling $G_{c \lhd b}$. Note that generally $G_{b \lhd c} \neq G_{c \lhd b}$. A cell c is isolated if $G_{b \lhd c} = G_{c \lhd b} = 0$ for all cells b.[2]

The following proposition from [22] allows decoupling the effects of cell load from the subsequently discussed effects of inter-cell coupling on Λ, that is, on the achievability of target SINRs.

Proposition 11.1 (On Λ-Equivalence). *Denoting the cell UE load of a cell b as N_b, the load factor Λ of a shared channel in a network consisting of B cells each serving N_b UEs with a cell-wide common target SINR γ_b is equal to the spectral radius of a $B \times B$ matrix A', whose element A'_{bc} in the bth row and the cth column is defined as*

$$A'_{bc} = \begin{cases} \gamma_b(N_b - 1) & \text{if } b = c, \\ \gamma_b G_{b \lhd c} & \text{if } b \neq c. \end{cases} \tag{11.5}$$

Consequently, the load factor Λ_b of a single isolated cell b with N_b active UEs having a common cell-wide target SINR γ_b can be quantified as

$$\Lambda_b = \gamma_b(N_b - 1) = \frac{N_b - 1}{N_b^{\max} - 1}, \tag{11.6}$$

where the maximum number $N_b^{\max}$ of UEs that can be served by the BS b under the SINR achievability condition (11.3) is

$$N_b^{\max} = \gamma_b^{-1} + 1. \tag{11.7}$$

Clearly, the load factor Λ_b of a single isolated cell b grows linearly with its UE number N_b, independently of the UE channel gains, that is, the UE geographical distribution. Hence, the overall capacity of a set of isolated cells, each having a target

---

[2] Channel gains attaining very low values in absolute terms, or relative to other channel gains can be practically considered as equal to zero.

SINR γ_b, can be quantified as the union of individual cell capacities (11.7), while the overall channel load factor is

$$\Lambda = \max_{\forall b} \gamma_b(N_b - 1) = \max_{\forall b} \frac{N_b - 1}{N_b^{\mathrm{max}} - 1}. \tag{11.8}$$

These observations on isolated cells can also be extended to so-called semi-isolated cells that satisfy for some BS indexing the conditions of (i) no cell b being coupled to a cell c such that $c > b$ (i.e., lower-diagonal A' with $G_{c \lhd b} = 0$ for $c > b$), or (ii) no cell b being coupled to a cell c such that $c < b$ (i.e., upper-diagonal A' with $G_{c \lhd b} = 0$ for $c < b$) [22].

11.2.5 Effect of Inter-cell Coupling on Global Achievability of Target SINRs

As observed above, the optimum strategy for maximizing the capacity of a set of (semi)-isolated cells consists of maximizing the capacity of each cell. Unfortunately, if the network consists of at least five coupled cells, its load factor Λ cannot be described by an exact closed-form formula (*Abel-Ruffini Impossibility Theorem* [6]). Nevertheless, an approximation is possible assuming that the UEs of each cell target the same SINR. If QoS (i.e., SINR) diversification or opportunistic adaptation are required, the results are still applicable but must be combined with numerical real-time monitoring of Λ [21] by network cells.

Proposition 11.2   *The capacity n_b^{max} of a cell b is the maximum N_b satisfying*

$$N_b + \gamma_b G_b \leq N_b^{\mathrm{max}} \tag{11.9}$$

for N_b^{max} from Eq. (11.7) and

$$G_b = \sum_{\forall c \neq b} G_{c \lhd b} = \sum_{\forall v \notin U_b} \frac{g_{[b]v}}{g_{vv}}. \tag{11.10}$$

In the above Proposition [22], the term $\gamma_b G_b$ approximates the capacity loss due to inter-cell coupling. Simulations in Section 11.2.7 indicate that the coupling of a small cell and its hosting macrocell is much stronger than the coupling between two macrocells, making the latter macrocell-to-macrocell coupling effect negligible.

Reducing the system model to the case of a macrocell hosting a single small cell, the macro/small cell coupling can be studied in a more concrete manner. More specifically, let us assume that the two cells are served by BS 1 and BS 2, having the total UE number N_1 and N_2 UEs as well as target SINRs γ_1 and γ_2, respectively.

Then, by Proposition 11.1, the load factor Λ of the two-cell cluster is then given by the magnitude-dominant solution

$$\Lambda = \frac{1}{2}\left(\Lambda_1 + \Lambda_2 + \sqrt{(\Lambda_1 - \Lambda_2)^2 + 4C}\right) \tag{11.11}$$

of the characteristic polynomial

$$\det(A' - \lambda E) = \det\left(\begin{vmatrix} -\lambda + \Lambda_1 & A'_{12} \\ A'_{21} & -\lambda + \Lambda_2 \end{vmatrix}\right) = \tag{11.12}$$

$$= \lambda^2 - (\Lambda_1 + \Lambda_2)\lambda + \Lambda_1\Lambda_2 - C$$

of the matrix A' from Eq. (11.5) where $\Lambda_b = \gamma_b(N_b - 1)$ is the channel load factor of the isolated cell b, and

$$C = A'_{12}A'_{21} = \gamma_1 G_1(n_1)\gamma_2 G_2(n_2) = \tag{11.13}$$

$$= \sum_{v=\min U_2}^{\max U_2} \gamma_1 g_{[1]v}/g_{vv} \sum_{u=\min U_1}^{\max U_1} \gamma_2 g_{[2]u}/g_{uu}. \tag{11.14}$$

The coupling term C equals zero if the two cells are not coupled. Then, $\Lambda = \Lambda_1$ if $\Lambda_1 \geq \Lambda_2$, and $\Lambda = \Lambda_2$ if $\Lambda_1 \leq \Lambda_2$, that is, the SINR achievability is limited only by cell load.

For analytical purposes, the random variable C can be approximated by a deterministic product $C(N_1, N_2) = \gamma_1\gamma_2 N_1 N_2 E[G_1(N_1)G_2(N_2)] = \gamma_1\gamma_2 N_1 N_2 E[G_1(N_1)]$ $E[G_2(N_2)]$, where $E[.]$ denotes the expected value of a random variable [23]. The multiplicative relationship of the expected values of G_1, G_2, and $G_1 G_2$ stems from the fact that G_1 and G_2 are uncorrelated because UE location distributions in cells 1 and 2 are mutually independent.

Subsequently, the next proposition defines the strategy for achieving the capacity limit of a two-cell cluster.

Proposition 11.3 *Assume the inter-cell coupling $C(N_1, N_2)$ as defined in Eq. (11.11) is*

$$C = \gamma_1\gamma_2 N_1 N_2 D \tag{11.15}$$

for a constant $D > 0, D \neq 1$.[3] Let us also assume that normalized reuse limit L is given where $0 < L \leq 1$. Then, the total capacity $N_1 + N_2$ of two cells 1 and 2 having

---

The singular case $D = 1$ is practically irrelevant due to the volatility of g_{uv}.

total UE numbers $N_1, N_2 > 1$ and target SINRs γ_1, γ_2, respectively, is maximized for $\Lambda \leq L$ if

$$N_1 = N_2 - \frac{L}{1-D} \frac{\gamma_1 - \gamma_2}{\gamma_1 \gamma_2}. \tag{11.16}$$

11.2.6 Guidelines for Small Cell Deployment and Self-Configuration Methods

The insight into network management based on the dynamic minimization of Λ can be developed into methodological solutions to the problems of (i) BS placement, (ii) cell coverage definition, and (iii) opportunistic adaptation of uplink SINRs. The antenna-related issues such as beamforming and tilt-angle optimization are applicable as well.

In the following text, centralized as well as decentralized methods for the minimization of Λ are proposed. The centralized methods (referred to as Methods 2a and 3a) are characterized by optimum performance but suffer from large data exchange and computational overhead. More practical are distributed methods (referred to as Methods 1a, 2b, and 3c) which, however, reach only suboptimal performance. Methods referred to as Methods 1b, 2c, and 3b are both optimum and practical but applicable only to the case of two interfering cells.

11.2.6.1 Base Station Placement
Given that a cell b is usually characterized by $\sum_{\forall c \neq b} G_{c \triangleleft b} \ll 1$, that is, $N_b - 1 \gg \sum_{\forall c \neq b} G_{c \triangleleft b}$ even for a low UE number N_b, it follows from Ineq. (11.9) that, in general, network cells should serve active UEs such that the cell-specific load factors $\Lambda_b = \gamma_b (N_b - 1)$ are equal.

Consequently, in the MC-SC scenario, a small cell must be able to offload a part of the macrocell traffic such that the difference between Λ_{SC} and Λ_{MC} is minimized. In view of the macrocell planning by the network operator, all macrocells can be seen as having a similar UE load and targeting the same uplink SINRs. As already mentioned, numerical real-time monitoring of Λ is necessary if SINR diversification is required [21].

Constraining the small cell BS placement to the region in which sufficient macrocell UEs can be captured ($\Lambda_{SC} \sim \Lambda_{MC}$), the ideal capacity-maximizing placement of a cell b minimizes the average coupling factor $\sum_{\forall c \neq b} G_{c \triangleleft b}$ in Ineq. (11.9). Clearly, the contribution $G_{c \triangleleft b}$ is relevant only in the case of physically close (neighboring) cells. More accurately, in case of an MC-SC sub-network, the small cell BS placement must minimize the average joint coupling $D = G_{SC \triangleleft MC} G_{MC \triangleleft SC}$ in Eq. (11.16).

Denoting the term $G_{c \triangleleft b}$ obtained for N_c UEs of cell c as $G_{c \triangleleft b}(N_c)$, the operator can practically choose a BS location such that the expected values of the inter-cell coupling terms are minimized for a given n.

Method 1a (*Suboptimum for B Cells*). Place the BS b to minimize $E[\sum_{\forall c} G_{c \triangleleft b}(n)]$ subject to $|\Lambda_b - \Lambda_c| \leq \epsilon \forall c \neq b$ for $\epsilon \geq 0$ and $\gamma_b, \gamma_c > 0$.

Method 1b (*Optimum for Two Cells*). Place the small cell BS to minimize $E[G_{SC \triangleleft MC}(n)G_{MC \triangleleft SC}(n)] = E[G_{SC \triangleleft MC}(n)]E[G_{MC \triangleleft SC}(n)]$ subject to $|\Lambda_{MC} - \Lambda_{SC}| \leq \epsilon$ for $\epsilon \geq 0$ and $\gamma_{MC}, \gamma_{SC} > 0$.

Setting $n = 1$ is sufficient because the UE locations (i.e., channel gains) are independent and identically distributed random variables, that is, the distribution of $G_{c \triangleleft b}(n)$ converges for $n \to \infty$ by the *central limit theorem* to a normal distribution, whose mean value is an n-multiple of $E[G_{c \triangleleft b}(1)]$. The convergence process is further accelerated if the standard log-normal model of the shadowing of g_{uv} is taken into account. In [23], the conditions for the approximability of $G_{c \triangleleft b}$ by a log-normal distribution are discussed in more detail.

Network monitoring and data mining techniques can be used in conjunction with numerical simulations to determine the distribution of $G_{c \triangleleft b}(n)$, that is the UE distribution in the network area (small cell hotspot) for a given radio propagation model (see Fig. 11.3 for realistic examples of such distributions).

On a practical level, numerical simulations in Section 11.2.7 indicate that deploying the small cell BS in the center of a hotspot results in a lower coupling G, and up to 20% higher capacity compared to the case of the small cell BS being located in an offset position with respect to the hotspot center (see Fig. 11.6). Similarly, serving more indoor UEs than outdoor UEs results in a higher MC-SC capacity (see Fig. 11.4).

It is noteworthy that adaptive variations of small cell coverage directly affect the distribution of $G_{SC \triangleleft MC}(n)$ and $G_{MC \triangleleft SC}(n)$, and generally speaking, it should be accounted for in the inter-cell coupling minimization. Nevertheless, given the high (low) power budget and antenna gain of macrocell (small cell) BSs, the actual variations of $G_{SC \triangleleft MC}(n)$ and $G_{MC \triangleleft SC}(n)$ with respect to small cell coverage variations are rather negligible (see Fig. 11.4).

11.2.6.2 Cell Coverage Self-Configuration and Optimization

Although general in nature, the concept of minimizing Λ is primarily applicable to self-configuration and optimization of small cell coverage. In this context, the variations of the pilot signal by a (small cell) BS b must efficiently change the number N_b of attached UEs. Otherwise, the inability to capture new UEs for higher pilot signal outputs would result in a degradation of the downlink transmission due to the reduction of the corresponding transmission power budget.

By Proposition 11.2, the distributively implementable balancing of the products $\Lambda_b = \gamma_b(N_b - 1)$ essentially minimizes $\Lambda(N_1, \ldots, N_B)$. In the two-cell scenario, Λ attains a unique global minimum for $\Lambda_1 = \Lambda_2$ if $\gamma_1 = \gamma_2$.

Method 2a (*Optimum for B Cells*). Adjust cell coverage to minimize $\Lambda(A)$ subject to $\Lambda \leq L$ for given $0 < L \leq 1$.

Numerically, it is preferable to compute Λ based on the matrix A' instead of A, given the matrix rank relationship $N \gg B$. The minimization of Λ based on A generated for time-averaged cell load and inter-cell coupling reduces the dynamic range of cell coverage variations in response to UE admission events and handovers—a feature desirable in terms of lower handover (failure) rate, but ma: generally cause an overall increase of Λ, that is, a suboptimal capacity usage.

In addition or alternatively, real-time measurements of Λ using the technique of [21] can be used for heuristic or blind corrections of the pilot signal output, fo example, using a heuristic pilot adjustment as a function of the first derivative o $\Lambda(k)$.

More applicable is the following distributed implementation based on Proposi tion 11.2. The shared radio network controller can be used in WCDMA networks t enable the data exchange.

Method 2b (*Suboptimum for B Cells*). Adjust cell coverage to maximize $n_b \forall$ subject to $n_b \leq \frac{L_b}{\gamma_b} + 1 - x_b$ for given $0 < L_b \leq 1$ and $x_b \geq \left[\sum_{\forall c \neq b} G_{c \triangleleft b} \right]$.

Theoretically, x_b should equal the maximum (average) value of $\sum_{\forall c \neq b} G_{c \triangleleft b}$. Yet given the approximative nature of *Method 2b*, the term x_b can be practically reduce even below this threshold (often by a factor of up to two if the UE locations ar distributed in a regular, e.g., grid-like, fashion) without jeopardizing the stability o the power control (11.1). Otherwise, typical or average values of x_b can be measure or modeled.

By Proposition 11.1, the small cell is required to offload the macrocell UEs o the basis of strategy (11.16).

Method 2c (*Optimum for Two Cells*). Adjust cell coverage to keep $N_1 = N_2 - \frac{L}{1-G} \frac{\gamma_1 - \gamma_2}{\gamma_1 \gamma_2}$ for given G and $0 < L \leq 1$.

If the target SINRs are identical for all cells, all the above uplink-oriente schemes essentially fairly distribute the UEs among the cells. Such load balancin is typical for downlink-oriented coverage optimization schemes, implying that th uplink and downlink coverage preferences can be seen as compatible. However, in . network with heterogeneous demands on cell SINRs, the downlink and uplink opti mization strategies are in conflict proportional to the target SINR difference.

The uplink (UL) and downlink (DL) antagonism can be resolved by weightin; the outputs of the respective coverage optimizers, that is, the preferred pilot signa powers p_{DL}^{CPICH} and p_{UL}^{CPICH}, depending on the (average) share of the downlink an uplink traffic volumes T_{DL} and T_{UL} of the cell. For example,

$$
\begin{aligned}
p^{CPICH} &= \frac{1}{1 + \alpha} p_{DL}^{CPICH} + \frac{\alpha}{1 + \alpha} p_{UL}^{CPICH} = \\
&= \frac{T_{DL}}{T_{DL} + T_{UL}} p_{DL}^{CPICH} + \frac{T_{UL}}{T_{DL} + T_{UL}} p_{UL}^{CPICH}
\end{aligned}
\tag{11.17}
$$

for $\alpha = T_{UL}/T_{DL}$. The issue of coverage stability and integrity of this conflic resolution scheme requires further study.

In general, increasing the power of the small cell BS pilot signal degrades th effective downlink throughput of the small cell UEs while improving the down link throughput of the remaining macrocell UEs. Similarly, the maximum achievabl uplink SINR of the small cell (macrocell) UEs decreases (increases) with a growing small cell coverage. Hence, the small cell should maximize its coverage only if the operator deploys the small cell with the primary goal to offload part of the macrocell traffic and improve the macrocell services. The coverage maximization process is limited only by a degradation of the small cell downlink performance to a level that is comparable to the macrocell BS performance level.

11.2.6.3 Opportunistic Adaptation of Target SINRs

By Proposition 11.2, if a cell b varies its target SINR γ_b such that the term $\gamma_b(N_b - 1) + \gamma_b \sum_{\forall c \neq b} G_{c \lhd b}$ is smaller or equal to the network maximum of such cell-specific terms, then the post-adaptation magnitude of the channel reuse Λ is smaller or equal to the original preadaptation Λ level. In other words, the post-adaptation equilibrium powers $\hat{p}$ remain essentially the same or smaller compared to the preadaptation power output (recall that $\hat{p} \sim |1 - \Lambda|^{-1}$).

As normally $\gamma_b(N_b - 1) \gg \gamma_b \sum_{\forall c \neq b} G_{c \lhd b}$, one can state that, for an $x_b \geq 0$, the cell b can opportunistically increase its target SINR to equalize its product $\gamma_b(N_b - 1) + x_b$ with the network maximum $\max_{\forall b} \gamma_b(N_b - 1) + x_b$, while incurring none or only a minimal increase of network powers $\hat{p}$ owing to the invariance of Λ (see Fig. 11.7).

The adaptation of target SINRs is particularly suitable for small cells that fail to offload the planned or required portion of the macrocell traffic (e.g., due to inferior antenna gain and power output, or unfavorable UE distribution) as it allows the small cells to fully exploit the available capacity of the shared channel represented by Λ without affecting the power usage of the current UEs and the SINR-constraint admissibility of the prospective UEs.

Method 3a (*Optimum for B Cells*). Define $\Delta(\Lambda)$ as the difference of Λs after and before a target SINR update. Then, given the channel gains $g_{uv} \forall u, v$, the maximum target SINR of UE u is defined for given target SINRs of other UEs by (i) $\Lambda = L$ for given $0 < L \leq 1$, or (ii) $\Delta(\Lambda) = \Delta L$ for given $\Delta L \geq 0$.

Method 3b (*Optimum for Two Cells*). Given the channel gains $g_{uv} \forall u, v$ and the target SINR γ_1 of cell 1, the maximum target SINR γ_2 of cell 2 is defined by $\gamma_2 = \gamma_1 L(\gamma_1(1 - G)(n_2 - n_1) + L)^{-1}$.

The *Method 3a* (as well as its distributed version *Method 3b* based on Proposition 11.2) can be readily extended to the case of multiple UEs, with or without requiring the satisfaction of both conditions (i) and (ii). Setting $\Delta L = 0$ always ensures no increments of the UE transmission powers.

Method 3c (*Suboptimum for B Cells*). Define $\Delta(\gamma_b)$ as the difference of γ_bs after and before an update of γ_b. Then, for given channel gains $g_{uv} \forall u, v$, the maximum target SINR γ_b of a cell b is defined for $x_b \geq 0$ by (i) $\gamma_b = L/(N_b - 1 + x_b)$ for $0 < L \leq 1$, or (ii) $\Delta(\gamma_b)(N_b - 1) = \Delta L$ for $\Delta L \geq 0$.

Practically, one can set x_b to zero or to even a slightly negative value and monitor the satisfaction of the condition $\Lambda \leq 1$ by means of an independent direct evaluation of Λ using the schemes of [21]. To avoid outages, power limit violations by the power control (11.1) should be also monitored (a standard feature of radio network controllers) because $\hat{p} \to \infty$ for $\Lambda \to 1^-$.

In general, the more balanced the cell products $\gamma_b(N_b - 1)$ are, the higher is the raise of Λ due to inter-cell coupling (a higher magnitude of the cell terms $\sum_{\forall c \neq b} G_{c \triangleleft b}$ compared to the maximum of $\gamma_b(N_b - 1)$). This effectively prohibits the setting of x_b to constants significantly lower than the actual maximum or average magnitude of the inter-cell coupling.

Opportunistic SINR adaptation is therefore to be considered when the small cell is deployed to cover a specific hotspot area (e.g., the outdoor premises of a café) with the focus on the small cell performance. A minimum necessary (fixed) small cell BS pilot signal is to be used to ensure a proper coverage of the desired area (and the achievability of uplink target SINRs if the number of macrocell UEs is generally high) so that the small cell uplink target SINR can be maximized.

11.2.7 Performance Evaluation

11.2.7.1 Network Setup The deployment of a macrocell network in Dublin is simulated. The network setup consists of seven sites with three-sector macrocells per site as in [25]. The central site is located at the WGS84 coordinates N 53.340494 and E 6.264374 which are coordinates [0,0] in Fig. 11.2. The macrocell downlink transmission power of 21 W is set to achieve an average indoor SNR at the cell edge of 5 dB. Other simulation parameters are summarized in *Table 11.2*. Channel gains including building penetration loss are computed based on 3GPP recommendations as described in more detail in Appendix A.

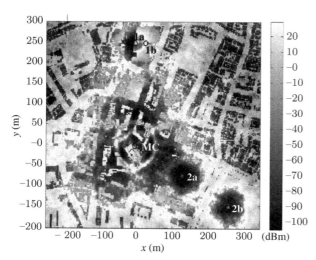

Figure 11.2. Example of received pilot CPICH powers in the central part of the simulated network in Dublin, Ireland. SC downlink transmission power is 1 W. Simulated SC BS locations denoted as *1a*, *1b*, *2a*, and *2b*.

Two different types of co-channel small cell deployment in the proximity of a macrocell traffic hotspot are studied. For each scenario, 10^5 random independent UE placements are generated.

In the first set of experiments, a traffic hotspot is created 248 m north of the central site having the center at coordinates $[0, 248]$ m in Fig. 11.2. This location has a good balance of indoor and outdoor spaces. A small cell BS, whose downlink power is set to 1 W, is deployed either at the hotspot center (*"centric SC BS"* scenario denoted as *1a* in Fig. 11.2), or with an eastward offset of 30 m (*"offset SC BS"* scenario denoted as *SC 1b* in Fig. 11.2). The UE CPICH sensitivity is either -115 dBm (CPICH reception in both indoor and outdoor spaces—*"outdoor/indoor"* scenario), or -95 dBm (CPICH reception essentially in outdoor spaces only—*"outdoor"* scenario).

In the second set of experiments, the hotspot is created in the southeast-oriented macrocell sector of the central site with a varying distance to the macrocell BS. The hotspot center coordinates are $[130, -74]$ m (*"MC/SC close"* scenario denoted as *2a* in Fig. 11.2), or $[260, -150]$ m (*"MC/SC far"* scenario denoted as *2b* in Fig. 11.2). The small cell BS is always placed at the hotspot center and its downlink transmission power can be either 1W or 2W (subscenarios *"SC BS 1 W"* and *"SC BS 2 W,"* respectively). Both locations lack major indoor spaces whose presence is undesired in this scenario set.

There are two UE types: non-hotspot UEs distributed uniformly randomly throughout the network, and hotspot UEs distributed around a hotspot center with a Gaussian distribution having a variance of 40 m. All UEs are equally likely to request service. The distribution parameters with respect to cell dimensions are such that the MC-SC traffic ratio is 1:1.

In accordance with the system model, the data bursts of all ongoing transmissions are assumed to be perfectly synchronized (the worst-case co-channel interference scenario).

Figure 11.3 shows the cumulative distribution function (CDF) of the single-UE coupling terms $G_{SC \triangleleft MC}(1)$, $G_{MC \triangleleft SC}(1)$ and $G_{SC \triangleleft MC}(1)G_{MC \triangleleft SC}(1)$ for the small cell BS being is placed either in the hotspot center (small cell BS *1a* in Fig. 11.2), or 30 m west of the hotspot center (small cell BS *SC 1b* in Fig. 11.2).

11.2.7.2 Results The actual degradation of the MC-SC capacity region in response to inter-cell coupling is illustrated in Figs. 11.4 and 11.6. For validation purposes, a black bold line indicates ideal load balancing (11.16) for zero inter-cell coupling ($G = 0$, independent cells). This bold line crosses the depicted capacity regions very close to their "knees," characterized by the capacity maximum $N_{MC} + N_{SC}$. The non-zero coupling term G causes a shift of the maximum away from this line, yet this shift is rather negligible and hardly exceeds one UE per cell.

Assuming the maximum threshold $L = 1$ and no inter-cell coupling $G = 0$, the equality in (11.9) indicates that the macrocell can serve at most 32 UEs for $\gamma_{MC} = -15$ dB, while the small cell can serve at most 16 UEs for $\gamma_{SC} = -12$ dB.

Figures 11.4 and 11.6 indicate a $10 - 20\%$ decrease of the MC/SC pole capacity as the macrocells and small cells can simultaneously serve up to $28 - 31$ and $11 - 14$

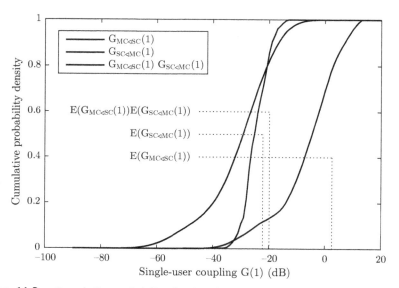

Figure 11.3. Cumulative probability density of single-UE coupling $G_{SC⊲MC}(1)$, $G_{MC⊲SC}(1)$, and $G_{SC⊲MC}(1)G_{MC⊲SC}(1)$ in an MC-SC sub-network having the SC BS placed (i) in the hotspot center (*SC BS 1a* in Fig. 11.2), and (ii) 30 m off the hotspot center (cell *SC 1b* in Fig. 11.2).

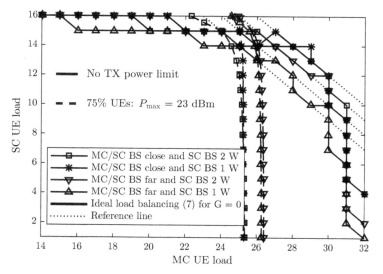

Figure 11.4. Uplink capacity regions of a co-channel MC-SC sub-network with perfectly synchronized UE transmissions (worst-case scenario) for variable *MC-SC BS distance* and *SC BS output* ($\gamma_{MC} = -15$ dB, $\gamma_{SC} = -12$ dB). To indicate the transmit power raise as a function of the cell load, two cases are distinguished: (i) unlimited transmit powers, and (ii) transmit power limit of 23 dBm for 75% of the UE total, and no limit for the remaining 25%.

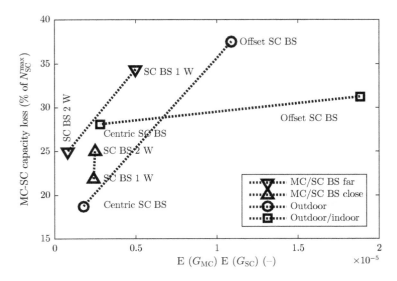

Figure 11.5. Total capacity loss $N_{MC}^{max} + N_{SC}^{max} - n_{MC}^{max} - n_{SC}^{max}$ of the MC-SC cluster due to inter-cell coupling, expressed as the percentage of the load-limited capacity N_{SC}^{max} of the small cell (i.e., the double of the "per-cell" loss) and shown as a function of $D = E[G_{MC}(N_{MC})]E[G_{SC}(N_{SC})]$.

UEs, respectively, depending on the simulated scenarios. Nevertheless, in total, the small-cell deployment still significantly increases the total MC-SC capacity.

Figure 11.5 summarizes the total capacity loss of the MC-SC cluster in all scenarios as a function of the inter-cell coupling.

Clearly, a small cell offset and a larger coverage area due to higher small cell power lead both to a higher inter-cell coupling. A similar effect can be observed in case of relatively large distances between macrocell and small cell BSs. The role of the small cell BS output and the CPICH sensitivity, both defining the small cell coverage, is of a rather secondary nature.

The maximum achievable small cell target SINR is plotted in Fig. 11.7 for the hotspot-centric cell *SC BS 1a* covering both the indoor and outdoor spaces. Assuming a maximum power limit of 23 dBm, the assumed standard SINR of −15 dB can be opportunistically increased by up to 8 dB, an increase of up to 23 dB. Similar results are obtained in the other studied scenarios.

11.3 4G LTE HETEROGENEOUS NETWORKS

11.3.1 Introduction

The uplink transmission mode of 4G LTE is fundamentally different from 3G WCDMA systems for two reasons—the absence of intra-cell interference among UEs

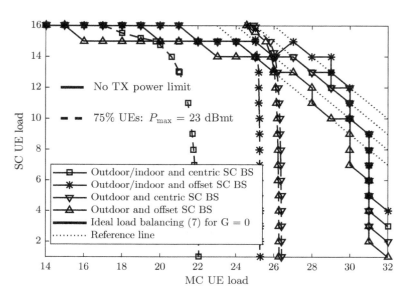

Figure 11.6.   Uplink capacity regions of a co-channel MC-SC sub-network with perfectly synchronized UE transmissions (worst-case scenario) for variable *SC BS offset* and *indoor coverage* ($\gamma_{MC} = -15$ dB, $\gamma_{SC} = -12$ dB). To indicate the transmit power raise as a function of the cell load, two cases are distinguished: (i) unlimited transmit powers, and (ii) transmit power limit of 23 dBm for 75% of the UE total, and no limit for the remaining 25%.

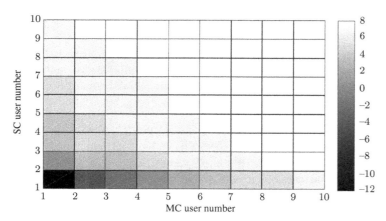

Figure 11.7.   Maximum UL target SINR for a co-channel SC for $\gamma_{MC} = -15$ dB assuming perfectly synchronized UE transmissions (worst-case scenario) and maximum power limit of 23 dBm. The SC BS is placed in hotspot center and covers both indoor and outdoor spaces (CPICH sensitivity -115 dBm).

connected to the same serving cell, and the independent scheduling with resource-block granularity by neighboring cells. Consequently, the LTE uplink interference is far more variable in both time and frequency, requiring a fresh approach to its management, taking into account its unpredictable and unstable nature. Moreover, the LTE standard mandates a power control mechanism that is based on the principle of fractional power control (FPC) [26–29].

From a practical point of view, the LTE standard also specifies a set of unique cell-specific parameters that enable the operation of FPC. The configuration of such parameters is characterized by trade-offs between the efficiency of path loss and interference compensation by power control means and fairness of resource allocation among cell UEs [30].

Given the complexity of FPC and its direct connection to random variables such as path loss and cell load, self-organizing network (SON)-based solutions to FPC configuration have gained much research attention. The importance of uplink-oriented SON is exacerbated by the emergence of heterogeneous networks and small cell deployments. Unlike in a macrocell-only environment, UEs located at the macrocell edge are a major source of interference to UEs served by a nearby small cell, hence the need for load-aware timely optimization of the uplink conditions for optimum network-wide performance.

Small cells are deployed on demand in traffic hotspots or areas with poor macrocell coverage. Consequently, their number can exceed the number of hosting macrocells even by orders of magnitude. Hence, solutions to uplink interference management should satisfy the following requirements: (i) adaptiveness to network traffic, propagation geometry, and topology, (ii) scalability with the size of the network (typically up to tens of thousands of cells) in terms of both data processing power and network control performance, and (iii) architectural compliance and adherence to LTE standards.

One approach to SON solutions for the LTE uplink consists of using closed-loop power control algorithms for dynamically adjusting SINR targets so as to achieve a fixed or given interference target at every cell [31]. Yet, the constraint of hard SINR targets and assumption of closed-loop power control bring up the issues of SINR feasibility (i.e., power control stability) similar to the above discussed 3G WCDMA case.

Preferred approaches strive not only to provide a basic fractional power control configuration, but also pursue a network-wide optimization objective, typically expressed in terms of coverage, capacity, or fairness. Practically available inputs for automated optimization of FPC parameters would be the in-network aggregated statistics of UE path loss and traffic load.

In this context, it is to be noted that the study of LTE SON algorithms has been focusing mainly on downlink-related problems and uplink-oriented studies are scarce. In [32,33] the problem of downlink inter-cell interference coordination (ICIC) for LTE is studied, while in [34] downlink transmit power profiles are optimized in different frequency carriers. In [35–38], various forms (downlink, uplink, mobility-based etc.) of traffic-load balancing with LTE SON are studied. For an extensive collection of material and presentations on the latest industry developments in LTE SON, we refer the reader to [39,40].

The following text introduces efficient learning-based SON algorithms for adaptive uplink power control in LTE that are driven by in-network data measurements. The section is organized as follows. The analytic system model and proposed SON architecture are discussed in Subsections 11.3.2 and 11.3.3. It is shown that the processing of raw UE measurement reports can be used to derive a measurement statistics succinct yet expressive enough to optimize the LTE uplink performance. This allows the definition of a measurement data-driven framework for setting power control parameters for optimal uplink interference management in an LTE network as formulated in Section 11.3.4.

Then, a learning-based algorithm for optimal setting of cell-specific power control parameters in LTE networks is presented in Section 11.3.5. The algorithm provably converges to optimal solutions. The evaluation using a radio network plan from a real LTE network deployed in a major metropolis demonstrates substantial performance gains for the considered network in Section 11.3.6. Conclusions follow in Section 11.4.

11.3.2 System Model

A typical LTE network of heterogeneous cells is assumed, all of which share the same carrier frequency implying frequency reuse 1. The distinction between macrocells and small cells is not relevant for the subsequent discussion but clearly plays a key role in performance evaluations.

The uplink model conforms to the LTE standard. Accordingly, UEs are assumed to use M-point DFT precoding of otherwise OFDMA uplink transmissions, a so-called SCFDMA multiple access scheme, to reduce the UE peak-to-average power ratio. The mapping on resource blocks to subcarriers is done through frequency hopping mechanisms [41].

Importantly, the equalization of the received signal is performed separately for each resource block leading to the SINR being averaged only across subcarriers of a resource block. Thus, the SINR in a resource block is a direct measure of UE performance together with the selected modulation and coding scheme. As for the interference component of the SINR, there is at most one interfering UE per neighboring cell per resource block as resource blocks cannot be shared by UEs of the same LTE cell.

Furthermore, the LTE fractional power control is assumed as a tool for mitigating uplink interference from other cells, while allowing the UEs to flexibly make use of favourable channel conditions.

More specifically, each cell sets at a slow time-scale (order of minutes) two cell-specific parameters used by its UEs to set their average transmit power and a target SINR as a predefined function of local path-loss measurements.

The two parameters are the nominal UE transmit power $P^{(0)}$ and fractional path-loss compensation factor $\alpha < 1$. A UE u with an average path-loss PL to its serving cell transmits at a power spectral density (PSD)

$$P^{TX}_{(u)} = P^{(0)}(PL)^{\alpha} \tag{11.18}$$

expressed in Watt per resource block. The parameter α can be interpreted as a fairness parameter that controls the SINR magnitude as function of the UE proximity to the serving evolved NodeB (eNodeB) [30].

The total UE transmit power is capped to the hardware-limited maximum P^{max}_{total}. In other words, the total UE transmission power is $\min(MP^{TX}_{(u)}, P^{max}_{total})$ where M is the number of uplink resource blocks assigned to a UE by an uplink scheduling grant.

At the same time, each UE is closed-loop power-controlled around the mean transmit power PSD of Eq. (11.18) to ensure that a suitable average SINR target is achieved for given interference and path loss. This closed-loop power control is carried out at a relatively fast time scale based on explicit power control adjustments via uplink grants transmitted in the physical downlink control channel (PDCCH). The power adjustments can be either absolute or relative.

To define interfering cells and their interference, let us index network cells by $c \in C$ and denote by J_c the set of cells that interfere with cell c. Table 11.3 lists other important parameters and variables that will be introduced subsequently.

Ideally, a cell $e \in J_c$ if the uplink transmission of some UE associated with cell e is received at eNodeB of cell c with received power above the noise floor. Yet, in practice, interfering cells must be defined based on the measurable quantities. To this

TABLE 11.3 Notation List

Symbol	Description
$c, C \ (c \in C)$	Cell index and set of cells, respectively
$P^{(0)}_c$	Nominal transmit power (LTE FPC parameter of cell c)
α_c	Path-loss compensation factor (LTE FPC parameter of cell c)
π_c	Nominal transmit power in logarithmic scale, that is, $\ln(P^{(0)}_c)$
U_c	Set of UEs associated with cell c
J_c	Set of cells that interfere with cell c
$l^{(u)}_c$	Mean path loss of UE u to its serving eNode c
L_c	Path loss of a random UE $u \in U_e$ to its serving eNodeB c
$l^{(u)}_{e \to c}$	Mean path loss of UE $u \in U_e$ to eNodeB of cell c
λ_c	$\lambda_c = \ln L_c$
$\lambda_{e \to c}$	$\lambda_{e \to c} = \ln L_{e \to c}$
$l_c(b), p_c(b)$	Bin value and bin probability, respectively, for path-loss histogram bin b of cell c
$\gamma_c(b)$	Expected SINR in log-scale of UE of cell c with path loss $l_c(b)$

end, we therefore say that $e \in J_c$ if some UE u associated to cell e reports downlink reference signal received power (RSRP) from cell c (i.e., RSRP is above the minimum threshold of 140 dBm). This is an indication that the uplink signal of UE u could interfere with the eNodeB of cell c. This assumption is based on the symmetry of uplink and downlink path losses.

11.3.3 SON Architecture and Measurement Statistics Inputs

The LTE standard leaves unspecified how each cell sets the value of $P_{(0)}$, α, and the setting of UE SINR targets. Clearly, aggressive (conservative) parameter setting in a cell will improve (degrade) the performance in that cell, but will cause high (low) interference in neighboring cells. Since the FPC scheme and its parameters are cell-specific, its configuration lies within the scope of SON framework, that is, the solution should adapt the parameters based on suitable periodic network measurements.

An applicable SON architecture is shown in Fig. 11.8. Herein, network monitoring happens in a distributed manner across the radio access network, and the heavy duty algorithmic computations happen centrally at the Network Management Servers. As opposed to a fully distributed (across eNodeBs) computation approach, this kind of hybrid SON architecture is the preferred by many operators for complex SON use-cases [39, 40] for two key reasons: capability to work across BSs from different vendors as is typically the case, and not having to deal with convergence issues of distributed schemes (due to asynchrony and message latency).

The first main building block of the SON architecture are the monitoring components at the cells that collect network measurements, appropriately process them

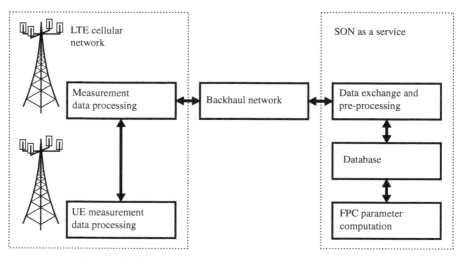

Figure 11.8. SON architecture for LTE uplink optimization.

to create key performance indicators (KPIs), and communicate the measurement and KPIs to the central server. Another main building block is the algorithmic computation engine (cloud servers) at a central server that makes use of the KPIs and compute the SON parameters that are then fed back to the network.

The KPIs from the network are typically communicated periodically. The period should be such that it is short enough to capture the changing network dynamics that call for network reconfiguration, and is long enough for accurate estimation the relevant statistics. In general, most networks have periodic measurement reports with a frequency of 5–15 minutes [42], but much faster minimization of drive tests (MDT) data and UE trace information can be collected. In addition, measurement reports can be triggered by events like traffic load above a certain threshold beyond normal.

Given this architecture, we proceed with the definition of a minimal set of network measurements that are required to configure FPC parameters, to periodically compute the FPC parameters to maximize the network performance in the uplink. In the next section it will be discussed how to adapt α, $P_{(0)}$, and the average interference threshold for every cell based on the measurement statistics.

To define the minimum SON-driving measurement statistics, consider a network snapshot with a collection of UEs U, in which a generic UE is indexed by the index $u \in U$. In accordance with Table 11.3, other notation is introduced as follows.

Let U_c denote the set of UEs u associated with cell c, and $l_c^{(u)}$ denote the path loss from $u \in U_c$ to its serving cell c. The term L_c without the UE superscript denotes the path loss of a random UE associated with cell c. The mean path loss from a UE $u \in U_e$ to a cell $c \in J_e$ is denoted by $l_{e \to c}^{(u)}$, while the term $L_{e \to c}$ without the UE superscript denotes the path loss from cell e to cell c of a random UE belonging to cell e.

All the path-loss variables are considered as time averages with averaged out fast fading and frequency selectivity thanks to receiver SCFDMA equalization. This assumption is also in line with the fact that fast fading occurs at a time scale much faster (milliseconds) than the FPC parameter optimization (minutes).

Assume also that a UE u transmits to its serving cell c over a set S of subcarriers, assigned by the eNodeB scheduler. The subcarriers within S can be used in a neighboring cell by at most one UE, while the resource block utilization in time depends on the cell load.

The following binary random variable is used to derive the SINR over S:

$$O_{e \to c} = \begin{cases} 1 & \text{if cell } e \text{ schedules a UE over a resource block also used by cell } c, \\ 0 & \text{otherwise.} \end{cases} \quad (11.19)$$

It is assumed that $O_{e \to c}$ does not depend on the specific choice of S and has identical distribution for all S.

Denoting by v the UE that occupies S at cell e and ignoring fast fading, the interfering signal at the serving cell c of UE u for transmissions over S in cell e can be expressed as

$$I_{e \to c} = O_{e \to c} P_{(v)}^{\mathrm{TX}} l_{e \to c}^{(v)} = O_{e \to c} P_e^{(0)} \left(l_e^{(v)} \right)^{\alpha_e} \left(l_{e \to c}^{(v)} \right)^{-1} \tag{11.20}$$

Note that the interference $I_{e \to c}$ is random even without fluctuations due to fast fading because the interfering UE v transmitting over S in cell e could be any random UE in cell c. In other words, the quantities $l_e^{(v)}, l_{e \to c}^{(v)}$ can be viewed as a random sample from the joint distribution of $(l_e^{(v)}, l_{e \to c}^{(v)})$.

Thus, the total interference at cell c over frequency block S is a random variable I_c given by

$$I_c = \sum_{e \in J_c} O_{e \to c} P_e^{(0)} (L_e)^{\alpha_e} (L_{e \to c})^{-1} \tag{11.21}$$

The superscript of the loss variables dropped to indicate that the loss is from a random UE in an interfering cell using the resource block S. As mentioned previously, this randomness is induced by medium access control (MAC) scheduling of the interfering cell.

Hence, the SINR of UE u in cell c can be expressed as function of path loss $l_c^{(u)}$ of UE u as

$$\mathrm{SINR}(l_c^{(u)}) = \frac{P_{(u)}^{\mathrm{RX}}}{I_c + N_0} = \frac{P_c^0 \left(l_c^{(u)} \right)^{-(1-\alpha_c)}}{\sum_{e \in J_c} O_{e \to c} P_e^{(0)} (L_e)^{\alpha_e} (L_{e \to c})^{-1} + N_0} \tag{11.22}$$

It follows from Eq. (11.22) that $\mathrm{SINR}(l_c^{(u)})$ is a random variable that is fully characterized for any $l_c^{(u)}$ by joint distributions of $(L_e, L_{e \to c})$ (i.e., the joint path-loss distribution of uplink UEs for every interfering cell e and c), and $O_{e \to c}$ (i.e., the probability that cell e schedules a UE interfering with cell c on a given resource block) for all $e \in J_c$. To compute the average network performance over the SON computation period, the mean uplink load ρ_c and path-loss distribution L_c in cell c will be required as well (see Eq. (11.27)).

The above joint distributions can be computed from histograms of UE RSRP measurements. The resource block occupancy probabilities $Pr(O_{e \to c} = 1)$ can be estimated as the fraction of UEs in cell e that interfere with cell c, that is, as the fraction of UEs in cell e that report measurements from cell c, assuming that all UEs in a cell use the radio resources uniformly as a result of proportional-fair MAC scheduling.

11.3.4 Problem Formulation

In order to define the LTE uplink optimization problem, let us now formulate a suitable performance metric that accounts for the SINR randomness as well as strikes a balance between aggregate cell throughputs and fairness.

Importantly, its average would be taken over UE path losses. Toward this end, we first define a performance metric V for a typical UE u in cell c, whose path loss to its serving cell is given by $l_c^{(u)}$. The metric V is a concave increasing function of

$$\gamma_c \left(l_c^{(u)} \right) = E \left[\ln \text{SINR} \left(l_c^{(u)} \right) \right], \tag{11.23}$$

which is the expected logarithmic-scale SINR of a typical UE u in cell c.

The dependence of V on $\ln(\text{SINR}_u)$ indicates the relationship to UE data rates. The log-scale conversion makes the problem tractable since the feasible power region is log-convex [15], and thus, elements of convex optimization theory can be used for defining our SON optimization solutions.

In order to define a network-wide performance metric obtained by averaging over all UE path losses, additional notations are needed for the histogram of L_c variables. Suppose the path-loss histogram of L_c is divided into k disjoint intervals with midpoint of the intervals given by $l_c(1), l_c(2), \dots, l_c(k_c)$. Suppose also that the empirical probability of items in histogram bin b (i.e., the number of its data items binned into bin b divided by the total number of histogram items) in cell c are given by $p_c(b)$.

Then, the overall system utility is defined as the expected utility of all UEs in all the cells where the expectation is over the empirical path-loss distribution given by the measurement data. Being a function of vectors $\boldsymbol{P}^{(0)}$ and $\boldsymbol{\alpha}$ consisting of elements $P_c^{(0)}$ and α_c, respectively, the overall network utility UTIL is

$$\text{UTIL}(\boldsymbol{P}^{(0)}, \boldsymbol{\alpha}) = \sum_{c \in C} E[\text{total utility of UEs in cell } c] \tag{11.24}$$

$$= \sum_{c \in C} \sum_{b=1}^{k_c} E[\text{total utility of UE with path loss } l_c(b) \text{ in cell } c] \tag{11.25}$$

$$= \sum_{c \in C} \sum_{b=1}^{k_c} E[\text{number of UEs with path loss } l_c(b)] \times V(\gamma_c(l_c(b))) \tag{11.26}$$

$$= \sum_{c \in C} \sum_{b=1}^{k_c} p_c p_c(b) V(\gamma_c(l_c(b))), \tag{11.27}$$

where $\gamma_c(l_c(b))$ is the expected SINR in logarithmic scale for any UE with path loss $l_c(b)$ to its serving eNodeB in cell c. The last step follows because the expected number of UEs with path loss $l_c(b)$ is a product of the expected number of UEs in cell c and the probability of a UE with path loss $l_c(b)$.

An example of V can be the Shannon data rate in logarithmic scale, $V(\gamma) = \ln(\ln(1 + \exp(\gamma)))$, a concave function that is well known to strike the right balance between fairness and overall system performance.

In summary, the uplink optimization problem can be then formulated as follows.

Problem 11.1   *Given the path-loss histograms of L_c, $(L_e, L_{e \to c})$, the empirical distribution $O_{e \to c}$, and the average traffic load ρ_c at every cell c, find the optimal $P_c^{(0)}$ and α_c for every cell c such that* $\mathrm{UTIL}(\boldsymbol{P}^{(0)}, \boldsymbol{\alpha})$ *from Eq. (11.27) is maximized.*

To make the problem tractable in the sense of [15], all powers and path losses are converted to logarithmic scale and new variables in accordance with Table 11.3 are introduced as follows.

$$\theta_c = \mathrm{E}\left[\ln\left(\sum_{e \in J_c} O_{e \to c} P_e^{(0)} (L_e)^{\alpha_e} (L_{e \to c})^{-1} + N_0\right)\right] \qquad (11.28)$$

$$\pi_c = \ln P_c^{(0)}, \lambda_c(b) = \ln l_c(b), \lambda_{e \to c}(b) = \ln l_{e \to c}(b) \qquad (11.29)$$

$$\lambda_c = \ln L_c, \lambda_{e \to c} = \ln L_{e \to c} \qquad (11.30)$$

Then, $\gamma_c(l_c(b))$ can be rewritten as

$$\gamma_c(l_c(b)) = \pi_c - (1 - \alpha_c)\lambda_c(b) - \theta_c, \qquad (11.31)$$

where

$$\theta_c = \mathrm{E}\left[\ln\left(\sum_{e \in J_c} O_{e \to c} e^{\pi_e + \alpha_e \lambda_e - \lambda_{e \to c}} + N_0\right)\right] \qquad (11.32)$$

Accordingly, the problem of maximizing $\mathrm{UTIL}(\boldsymbol{P}^{(0)}, \boldsymbol{\alpha})$ can be stated as the problem of maximizing

$$\mathrm{UTIL}(\boldsymbol{P}^{(0)}, \boldsymbol{\alpha}) = \sum_c \sum_b \rho_c p_c(b) V(\gamma_c(l_c(b))) \qquad (11.33)$$

subject to equality constraints of Eqs. (11.31) and (11.32).

By rewriting the equality constraints as inequalities [43], while imposing upper bounds for maximum transmit power per resource block and maximum average interference, the LTE uplink optimization problem can be summarized as a convex nonlinear program with inequality constraints [43].

Problem 11.2   *Given the measurement statistics of L_c, $(L_e, L_{e \to c})$, $O_{e \to c}$, and the average traffic load ρ_c at every cell c, the problem of maximizing* $\mathrm{UTIL}(\boldsymbol{P}^{(0)}, \boldsymbol{\alpha})$ *from Eq. (11.27), subject to maximum transmit power constrain on resource block, is equivalent to the following convex program:*

$$\max_{\{\pi_c\},\{\alpha_c\},\{\gamma_c(b)\}} \sum_{c,b} \rho_c p_c(b) V(\gamma_c(b)) \tag{11.34}$$

subject to

$$\forall\, c \in C, b \in [1, k_c] : \gamma_c(b) \leq \pi_c - (1 - \alpha_c)\lambda_c(b) - \theta_c, \tag{11.35}$$

$$\forall\, c \in C : \theta_c \geq \mathrm{E}\left[\ln\left(\sum_{e \in J_c} O_{e \to c} e^{\pi_e + \alpha_e \lambda_e - \lambda_{e \to c}} + N_0\right)\right], \tag{11.36}$$

$$\forall\, c \in C, b \in [1, k_c] : \pi_c + \alpha_c \lambda_c(b) \leq \ln P_{\max}, \tag{11.37}$$

$$\alpha_c \in [0, 1], \gamma_c(b) \in [\gamma_{\min}, \infty), \theta_c \in [\ln N_0, \ln I_{\max}]. \tag{11.38}$$

Herein, the constraint of Eq. (11.35) states that the expected SINR in log-scale can be no more than the receiver UE power minus the expected uplink interference at the eNodeB. The constraint Eq. (11.36) states that the expected interference at each cell has to be at least the expected sum of interferences from the neighboring cells. The constraint Eq. (11.37) limits the maximum power allowed per resource block, while the constraint Eq. (11.38) states the valid domain of the variables α_c, θ_c, and π_u, where $\gamma_{\min}$ is the minimum decodable SINR. Note that the valid domain of θ_c is $\ln N_0, \ln I_{\max}]$ because the maximum average interference in the cells could need a cap due to hardware design constraints.

The SINR target of an arbitrary UE in cell c with average path loss $l_c^{(u)}$ is

$$\mathrm{SINR}_{target}(l_c^{(u)}) = \frac{\min\left[P_{\max}\left(l_c^{(u)}\right)^{-1}, P_c^{(0)*}\left(l_c^{(u)}\right)^{-(1-\alpha_c^*)}\right]}{I_c^*}, \tag{11.39}$$

where π_c, α_c^*, θ_c^* represent the optimal (linear scale) values for cell c in terms of the uplink optimization Problem 11.2 and $P_c^{(0)*} = \exp(\pi_c^*)$ and $I_c^* = \exp(\theta_c^*)$, respectively.

11.3.5   Optimization Algorithm Based on Stochastic Learning

The uplink optimization Problem 11.2 is challenging due to the need of computing the expected value of the interference constraint (11.36)—a practically nearly impossible task for iterative gradient-based solvers in view of the high number of interferer combinations per cell and input histogram bins.

In this context, in [43], a computationally manageable approximation of the prob
lem is proposed.

For the purpose of explanation, the original optimization variables $\gamma_c(b)$, π_c, α_e
are denoted by a vector z and will be referred to as the primal variables. Similarly
the set of constraints of the Problem 11.2 is denoted by the random function

$$
h(z, \boldsymbol{O}, \lambda) = \begin{cases}
[\gamma_c(b) - \pi_c - (1 - \alpha_c)\lambda_c(b)]_{c \in C, b \in [1, k_c]}, \\
\left[\ln\left(\sum_{e \in J_c} O_{e \to c} e^{(\pi_e + \alpha_e \lambda_e - \lambda_{e \to c})} + N_0\right) - \theta_c\right]_{c \in C}, \\
[\pi_c + \alpha_c \lambda_c(b) - \ln P_{\max}]_{c \in C, b \in [1, k_c]}.
\end{cases}
\tag{11.40}
$$

Then, Problem 11.2 can be reformulated as

$$
\max_z \sum_{c,b} V(\gamma_c(b)) \quad \text{subject to} \quad \mathrm{E}[h(z, \boldsymbol{O}, \lambda)] \leq 0,
\tag{11.41}
$$

where the expectation is with respect to the random variables $O_{e \to c}$ and $(\lambda_e, \lambda_{e \to c})$.
Let us define the random Lagrangian function as

$$
\mathrm{L}(z, h(z, \boldsymbol{O}, \lambda), p) = \sum_u V(\gamma_u) - p^t h(z, \boldsymbol{O}, \lambda),
\tag{11.42}
$$

where $p \geq 0$ denotes the so-called Lagrange multiplier vector, whose dimension i$
equal to the number of constraints. Since the objective function of Problem 11.2 i$
concave and the constraint set is convex, one can readily show from convex optimiza
tion theory [44] that

$$
OPT = \max_z \min_{p \geq 0} \mathrm{E}[\mathrm{L}(z, h(z, \boldsymbol{O}, \lambda), p)]
\tag{11.43}
$$

$$
= \min_{p \geq 0} \max_z \mathrm{E}[\mathrm{L}(z, h(z, \boldsymbol{O}, \lambda), p)],
\tag{11.44}
$$

where OPT denotes the optimal value of Problem 11.2. The goal is then to solve
Problem 11.2 by tackling the saddle-point problem (11.45). The bounds on feasible
primal variables are

$$
\pi_{\min} \leq \pi_c \leq \pi_{\max},
\tag{11.45}
$$

$$
\gamma_{\min} \leq \gamma_c(b) \leq \gamma_{\max},
\tag{11.46}
$$

where

$$
\pi_{\max} = \ln P_{\max}, \pi_{\min} = \gamma_{\min} + N_0, \gamma_{\max} = \ln P_{\max} - N_0
\tag{11.47}
$$

The equivalent saddle-point problem (11.44) can be solved by using standard primal-dual techniques [45]. In such an approach, the primal variables and dual variables are updated alternatively in an iterative fashion, and each update uses a gradient ascent of $E[L(.)]$) for the primal variables and gradient descent of $E[L(.)]$) for the dual variables.

To reduce the computational complexity outlined above, one can perform primal-dual iterations by substituting the random variables with random samples of the random variables in every iteration. Thus the expectation computation is replaced by a procedure to draw a random sample from the histograms, which is a computationally light procedure. The idea of replacing a random variable by a sample in an iterative scheme is well known [46].

In particular, the stochastic approximation algorithm consists of the following steps (see Algorithm 11.1 for a summary):

Initialization:
Primal variables z_n, dual variables p_n, average primal variables $\hat{z}_n$, and average dual variables $\hat{p}_n$ in iteration n are initiated to positive values within bounds (11.46).

Iterative steps:
Denote $a_n = 1/n^\zeta$, $0.5 < \zeta \leq 1$ as the step size in iteration n. Repeat the following steps for $n = 0, 1, 2, \ldots$:

- *Random sampling of interferers:*
  This step is performed for each interfering cell pair (c, e). For each cell $e \in J_c$, the following steps are performed:
  - Toss a coin with probability of head $Pr(O_{e \to c} = 1)$. Denote the outcome of this coin toss by the $0 - 1$ variable χ_{ec} where χ_{ec} takes value one if there is a head.
  - If $\chi_{ec} = 1$, then draw a random sample from the joint distribution of $(L_e, L_{e \to c})$ based on the histogram of this joint distribution. Denote the random sample, which is a 2-tuple by $(sa[1], sa[2])$. Let $(\xi_{ec}[1], \xi_{ec}[2])$ denote the logarithm of this random sample, that is, $\xi_{ec}[i] = \ln(sa[i])$ for $i = 1, 2$.
  - In the function $h(.)$ given by Eq. (11.40) representing the problem constraints, replace the random variables $(\lambda_e, \lambda_{e \to c})$ by the random sample $(\xi_{ec}[1], \xi_{ec}[2])$.

  Denote χ, ξ as the vectors of χ_{ec}, ξ_{ec} samples. Denote $h(z_n, \chi, \xi), p_n$ as the functions of these random samples along with the primal and dual variables in iteration n.

- *Primal update:*
  The primal variables are updated as

$$z_{n+1} = z_n + a_n \nabla_z L(z_n, h(z_n, \chi, \xi), p_n), \qquad (11.48)$$

where $\nabla_z L(.)$ denotes the partial derivative of $L(.)$ with respect to z. Next project updated values of $\alpha_e, \pi_e, \theta_e, \gamma_u$'s within intervals $[0, 1]$, $[\pi_{min}, \pi_{max}]$, $[\ln N_0, \ln I_{max}]$, $[\gamma_{min}, \gamma_{max}]$, respectively. For example, if the value of α_e in iteration $n + 1$, $(\alpha_e)_{n+1} > 1$, then set $(\alpha_e)_{n+1} \leftarrow 1$; and, if any $(\alpha_e)_{n+1} < 0$ set $(\alpha_e)_{n+1} \leftarrow 0$.

- *Dual update:*
  The dual variables are updated as

$$p_{n+1} = [p_n + a_n h(z_n, \chi, \xi)]^+. \tag{11.49}$$

- *Updating average iterates:*
  The current value of the solution is given by average iterates

$$\hat{z}_{n+1} \leftarrow \frac{1}{n} z_n + \left(1 - \frac{1}{n}\right) \hat{z}_n, \tag{11.50}$$

$$\hat{p}_{n+1} \leftarrow \frac{1}{n} p_n + \left(1 - \frac{1}{n}\right) \hat{p}_n. \tag{11.51}$$

It follows from [43] that the iterates generated by the above algorithm converge to the optimal solution for "almost" every sample path of the iterates ($z_n \to z^*$ with probability 1). The iterative steps are repeated until $\hat{z}_n$ and $\hat{p}_n$ converge within a desirable precision. There are many techniques for testing the convergence of stochastic iterations. The step-sizes a_n can be adapted to speed up the convergence [47].

A low-complexity heuristic approximation of the Algorithm 11.1 can be found in [43].

The basic idea of the heuristic consists of modifying the problem into a standard non-linear program that can be solved using off-the-shelf solvers. The first modification is based on the notion of certainty equivalence —the random log-scale interference is replaced by the logarithm of the expected interference in (11.36) such that any solution with this modified constraint is a feasible solution of the uplink optimization. The second modification is regression-based, where the fitting

---

1 Initialize the primal variables to z_0 and the dual variables to p_0. ;
   for *iterations* $n = 1$ to $n = N_{it}$ **do**
2 | For each interfering cell pair (e, c), draw a random path-loss sample from the joint distribution histogram of $(L_e, L_{e \to c})$ and then replace the path-loss random variables by these samples in the expression of $h(.)$ in Eq. (11.41). ;
3 | Update the primal variables according to gradient ascent based update rule in Eq. (11.49). ;
4 | Update the dual variables according to gradient descent based update rule in Eq. (11.50). ;
5 | Maintain average of the primal and dual variables over all the iterations ;
   end
6 The average of the primal variables over all the iterations is the output. ;

Algorithm 11.1: Stochastic Primal-Dual Algorithm for Problem 11.2.

of distribution parameters using measurement data is carried out using a known parametric distribution as model for the path-loss statistics.

11.3.6 Performance Evaluation

11.3.6.1 Simulation Setup For evaluation purposes, a real operational LTE network in a major metropolitan area is simulated. The selected area consists of around 9 km^2 in the central business district of the city with 115 macrocells. This part of the city has a very high density of macrocells due to high volume of mobile data traffic. Since small cells were not yet deployed in reality in the area of interest, 10 small cell locations were manually embedded into the network planning tool using its built-in capabilities.

The terrain category, cell site locations, and drive-test propagation data from the network were fed into a commercially available radio network planning (RNP) tool that is used by operators for cellular planing [48]. The carrier bandwidth is 10 MHz in the 700 MHz LTE band. The macrocells have transmit power 40 W and small cells have transmit power 4 W.

To generate UE measurement data required for the evaluation, synthetic mobile locations were generated using the capabilities of the RNP tool. Then, using the drive-test calibrated data, terrain information and statistical channel models, the RNP tool was used to generate a signal propagation matrix in every pixel in the area of interest.

The RNP tool was subsequently used to drop thousands of UEs in several locations where the density of dropped UEs was 450 active mobile per square km (dense-urban). In addition, mobile hotspots were defined around some of the small cells where the mobile density was doubled. Based on the signal propagation matrix in every pixel and mobile drop locations, the RNP tool generated all path-loss data from UEs to it serving cell and neighboring cells, and also the mobile to cell association matrix.

Finally, the mobile path-loss data were used to generate the histograms and the other measurement data which were then fed into the algorithms.

Network configurations generated by the compared algorithms were evaluated for random UE snapshots generated by the RNP tool where the data rate of each UE was based on its SINR target given by Eq. (11.39).

The proposed Algorithm 11.1 was compared with the performance of standard LTE fractional power control characterized by the same value of α for all cells and $P^{(0)}$ set such that the SINR of every mobile UE is above the LTE decoding threshold. As for the proposed algorithm, the histograms were kept in decibels (logarithmic scale) with a bin size of 1 dB. Also, $P_{max} = 100$ mW per resource block.

11.3.6.2 Results Figures 11.9, 11.10, and 11.11 show that the proposed Algorithm 11.1 provides a data rate improvement over the best FPC schemes, of 4.9×, 3.25×, 2.12× for the 20th, 50th, 80th percentile, respectively. In other words,

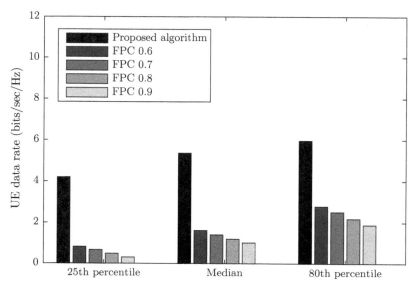

Figure 11.9.   Network-wide data rate comparison.

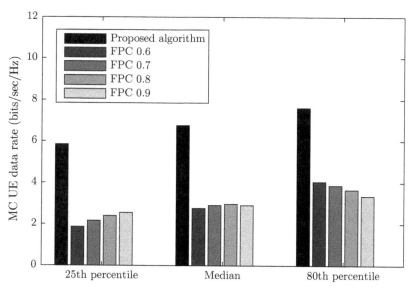

Figure 11.10.   Macrocell UE data rate comparison.

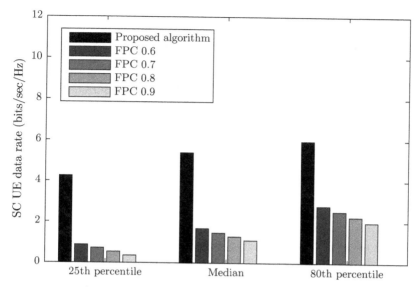

Figure 11.11. Small cell UE data rate comparison.

with the proposed algorithm, at least half the UEs see at least three-fold data rate improvement and 80% of the UEs see at least a two-fold improvement.

The larger gain at lower percentiles indicate that the proposed algorithm is really beneficial to UEs toward the cell edge, which are most affected by inter-cell interference.

As seen in Figs. 11.10 and 11.11, the proposed algorithm provides higher improvements to macrocell UEs as compared to small cell UEs. The median improvement of small cell UE data rate is 2.3×, whereas it is 3.3× for macrocell UEs. This can be explained by the fact that macrocells use higher power and have larger coverage leading to more UEs affected by inter-cell interference and thus more UEs can benefit from the proposed algorithm.

11.4 SUMMARY AND CONCLUSIONS

Uplink power control is a primary determinant of wireless system uplink performance. BSs typically use parametrized closed/open-loop control for setting the uplink powers of mobile UEs they serve to levels that guarantee a desired target SINRs or compensates for experienced path losses.

Traditional fast SINR balancing using closed loop mechanism are typical for 3G WCDMA systems but can be implemented in LTE systems as well. The LTE power control philosophy puts emphasis on fractional path-loss compensation by using broadcast-based parameter distribution, UE-specific performance monitoring, and opportunistic power adjustments when granting medium access.

Although simple in terms of implementation, uplink power control is, however, difficult to optimize due to its inherent trade-offs between overall network capacity and fairness with respect to cell-edge UEs as well as target SINR achievability. Other conflicting objectives include power consumption minimization and resource reuse. The randomization effects of bursty data traffic pose technical challenges of their own kind. In heterogeneous networks, the asymmetry between macrocell and small cell sizes generally further distorts the interference relationships between connected cells.

It is possible to define relatively accurate models of uplink power control that lend themselves to an analytic study. Stability analysis, primal-dual optimization and stochastic approximation have been used to derive an efficient algorithm for uplink optimization as well as guidelines for uplink-oriented small cell deployment and (self)-configuration. Using realistic network simulations, substantial benefits in heterogeneous environments have been quantified.

Given the challenges associated with the planning and optimization of heterogeneous networks in practice, SON-based implementations of the proposed schemes are of particular interest to commercial operators.

REFERENCES

1. T. Yucek and H. Arslan, "A survey of spectrum sensing algorithms for cognitive radio applications," *IEEE Commun. Survey Tuts.*, vol. 11, no. 1, pp. 116–130, Jan. 2009.

2. M. R. Musku, A. T. Chronopoulos, D. C. Popescu, and A. Stefanescu, "A game-theoretic approach to joint rate and power control for uplink CDMA communications," *IEEE Trans. Commun.*, vol. 58, no. 3, pp. 923–932, Mar. 2010.

3. T. Alpcan, T. Basar, R. Srikant, and E. Altman, "CDMA uplink power control as a noncooperative game," in *IEEE Trans. Wireless Netw.*, vol. 8, no. 6, pp. 659–670, Dec. 2002.

4. S. Kucera, "Enabling co-channel small-cell deployments in SINR-constraint networks by distributed monitoring of normalized network capacity," *IEEE/ACM Trans. Netw.*, vol. 22, no. 5, pp. 1577–1590, Oct. 2014.

5. Small Cell Forum, "Enterprise femtocell deployment guidelines," A Small Cell Forum White Paper, Tech. Rep., Feb. 2012. [Online]. Available: http://www.smallcellforum.org/

6. E. W. Cheney and D. R. Kincaid, *Linear Algebra: Theory and Applications.* Jones & Bartlett Learning, 2009.

7. M. Chiang, P. Hande, T. Lan, and C. W. Tan, *Power Control in Wireless Cellular Networks Power Control in Wireless Cellular Networks: Foundation and Trends in Networking.* Delft, The Netherlands: Now Publishers, 2008.

8. R. D. Yates, "A framework for uplink power control in cellular radio systems," *IEEE J. Sel. Areas Communi.*, vol. 13, no. 7, pp. 1341–1347, Sep. 1996.

9. G. J. Foschini and Z. Miljanic, "A simple distributed autonomous power control algorithm and its convergence," *IEEE Trans. Veh. Technol.*, vol. 42, no. 4, pp. 641–646, Nov. 1993.

10. C. Saraydar, N. B. Mandayam, and D. J. Goodman, "Pricing and power control in a multicell wireless data network," *IEEE J. Sel. Areas Commun.*, vol. 19, no. 10, pp. 1883–1892, Oct. 2001.

11. M. Chiang and J. Bell, "Balancing supply and demand of bandwidth in wireless cellular networks: Utility maximization over powers and rates," in *Proc. INFOCOM*, Hong Kong, China, Mar. 2004, pp. 2800–2811.

12. P. Hande, S. Rangan, M. Ciang, and X. Wu, "Distributed uplink power control for optimal SIR assignment in cellular data networks," *IEEE/ACM Trans. Netw.*, vol. 16, no. 6, pp. 1420–1433, Dec. 2008.

13. M. Xiao, N. B. Shroff, and E. K. P. Chong, "A utility-based power-control scheme in wireless cellular systems," *IEEE/ACM Trans. Netw.*, vol. 11, no. 2, pp. 210–221, Apr. 2003.

14. H. Boche and S. Stanczak, "Strict convexity of the feasible log-SIR region," *IEEE Trans. Commun.*, vol. 56, no. 9, pp. 1511–1518, Sep. 2008.

15. C. W. Sung, "Log-convexity property of the feasible sir region in power-controlled cellular systems," *IEEE Commun. Lett.*, vol. 6, no. 6, pp. 248–249, Jun. 2002.

16. O. Simeone, E. Erkip, and S. S. Schitz, "Robust transmission and interference management for femtocells with unreliable network access," *IEEE J. Sel. Areas Commun.*, vol. 28, no. 9, pp. 1469–1478, Nov. 2010.

17. S. Kucera and H. Claussen, "Uplink-oriented deployment guidelines and auto-optimization algorithms for co-channel small cells in W-CDMA heterogeneous networks," in *Proc. IEEE Global Commun. Conf.*, Atlanta, USA, Dec. 9–13, 2013.

18. M. F. Khan and B. Wang, "Effective placement of femtocell base stations in commercial buildings," in *Proc. Int. Conf. Ubiquitous Future Netw.*, Shanghai, China, Jul. 8–11, 2014, pp. 176–180.

19. J. Liu, Q. Chen, and H. D. Sherali, "Algorithm design for femtocell base station placement in commercial building environments," in *Proc. IEEE INFOCOM*, Orlando, USA, Mar. 25–30, 2012, pp. 2951–2955.

20. A. Halanay and V. Rasvan, *Stability and Stable Oscillations in Discrete Time Systems*. Amsterdam, The Netherlands: Gordon and Breach Science Publishers, 2000.

21. S. Kucera, L. Kucera, and B. Zhang, "Efficient distributed algorithms for dynamic access to shared multi-user channels in SINR-constrained wireless networks," *IEEE Trans. Mobile Comput.*, vol. 11, no. 12, pp. 2087–2097, Dec. 2012.

22. S. Kucera and H. Claussen, "Uplink-oriented deployment guidelines and auto-configuration algorithms for co-channel W-CDMA heterogeneous networks," *IEEE Trans. Wireless Commun.*, vol. 14, no. 7, pp. 3752–3763, Jan. 2014.

23. S. Kishore, "Capacity and coverage in two-tier cellular CDMA networks," Doctoral thesis, Princeton University, 2003.

24. 3GPP TR 36.814, "Evolved universal terrestrial radio access (E-UTRA); further advancements for E-UTRA physical layer aspects," Tech. Rep. v 9.0.0.

25. H. Claussen and L. T. W. Ho, "Multi-carrier cell structures with angular offset," in *Proc. IEEE Int. Symp. Pers., Indoor Mobile Radio Commun.*, Sydney, Australia, Sep. 2012, pp. 1–6.

26. M. Amirijoo, L. Jorguseski, R. Litjens, and R. Nascimento, "Effectiveness of cell outage compensation in LTE networks," in *Proc. IEEE Cons. Commun. Netw. Conf.*, Las Vegas, USA, Jan. 2011, pp. 642–647.

27. M. Boussif, N. Quintero, F. D. Calabrese, C. Rosa, and J. Wigard, "Interference based power control performance in LTE uplink," in *Proc. IEEE Wireless Commun. Syst.*, Reykjavik, Iceland, Oct. 2008, pp. 698–702.

28. M. Coupechoux and J. M. Kelif, "How to set the fractional power control compensation factor in LTE?" in *Proc. IEEE Sarnoff Symp.*, Princeton, USA, May 2011, pp. 1–5.

29. C. U. Castellanos, D. L. Villa, C. Rosa, and K. I. Pedersen, "Performance of uplink fractional power control in UTRAN LTE," in *Proc. IEEE Veh. Technol. Conf. Spring*, Singapore, May 2008, pp. 2517–2521.

30. A. Simonsson and A. Furuskar, "Uplink power control in LTE - overview and performance, subtitle: Principles and benefits of utilizing rather than compensating for SINR variations," in *Proc. IEEE Veh. Tech. Conf. Fall*, Calgary, Canada, Sep. 2008, pp. 1–5.

31. H. Zhang, N. Prasad, S. Rangarajan, S. Mekhail, S. Said, and R. Arnott, "Standards-compliant LTE and LTE-A uplink power control," in *Proc. IEEE Int. Commun. Conf.*, Ottawa, Canada, Jun. 2012, pp. 1–5.

32. M. Rahman, H. Yanikomeroglu, and W. Wong, "Enhancing cell-edge performance: a downlink dynamic interference avoidance scheme with inter-cell coordination," *IEEE Trans. Wireless Commun.*, vol. 9, no. 4, pp. 1414–1425, Apr. 2010.

33. R. Madan, J. Borran, A. Sampath, N. Bhushan, A. Khandekar, and T. Ji, "Cell association and interference coordination in heterogeneous LTE-A cellular networks," *IEEE J. Sel. Areas Commun.*, vol. 28, no. 9, pp. 1479–1489, Nov. 2010.

34. A. Stolyar and H. Viswanathan, "Self-organizing dynamic fractional frequency reuse in OFDMA systems," in *Proc. IEEE Infocom*, Phoenix, USA, Apr. 2008, pp. 691–699.

35. A. Obinger, S. Stefanski, T. Jansen, and I. Balan, "Load balancing in downlink LTE self-optimizing networks," in *Proc. IEEE Conf. Veh. Technol. (Spring)*, Taipei, Taiwan, May 2010, pp. 1–5.

36. J. Turkka, T. Nihtila, and I. Viering, "Performance of LTE SON uplink load balancing in non-regular network," in *Proc. IEEE Pers. Indoor Mobile Radio Commun.*, Toronto, Canada, Sep. 2011, pp. 162–166.

37. H. Hu, J. Zhang, X. Zheng, Y. Yang, and P. Wu, "Self-configuration and self-optimization for LTE networks," *IEEE Commun. Mag.*, vol. 48, no. 2, pp. 94–100, Feb. 2010.

38. H. Son, S. Lee, S. Kim, and Y. Shin, "Soft load balancing over heterogeneous wireless networks," *IEEE Trans. Veh. Technol.*, vol. 57, no. 4, pp. 2632–2638, Jul. 2008.

39. SON LTE. (2015) [Online]. Available: http://www.sonlte.com/technology/

40. LTE World SON. (2015) [Online]. Available: http://lteworld.org/category/tags/son

41. E. Dahlman, S. Parkvall, J. Skold, and P. Beming, *3G Evolution: HSPA and LTE for Mobile Boradband*. Academic Press, 2008.

42. J. T. J. Penttinen, *The LTE/SAE Deployment Handbook*. John Wiley & Sons, Jan. 2012.

43. S. Deb and P. Monogioudis, "Learning-based uplink interference management in 4G LTE cellular systems," *IEEE/ACM Trans. Netw.*, vol. 23, no. 2, pp. 398–411, Apr. 2015.

44. D. Bertsekas, A. Medic, and A. Ozdaglar, *Convex Analysis and Optimization*. United States: Athena Scientific, Apr. 2003.

45. A. Nedic and A. Ozdaglar, "Subgradient methods for saddle-point problems," *J. Optim. Theory Appl.*, vol. 142, no. 1, pp. 205–228, Jul. 2009.

46. V. Borkar, *Stochastic Approximation: A Dynamical Systems Viewpoint*. Cambridge University Press, Sep. 2008.

47. A. P. George and W. B. Powell, "Adaptive stepsizes for recursive estimation with applications in approximate dynamic programming," *Mach. Learn.*, vol. 65, no. 1, pp. 167–198, Oct. 2006.

48. "Atoll 3.2, radio network planning tool" (2015) [Online]. Available: http://www.forsk.com/atoll

PART IV

MOBILITY MANAGEMENT AND ENERGY EFFICIENCY

12

MOBILITY MANAGEMENT

12.1 INTRODUCTION

In order to realize the potential coverage and capacity benefits of heterogeneous networks (HetNets), operators are facing new technical challenges as already discussed in previous chapters. Among these technical challenges, mobility management is of special importance. The deployment of a large number of small cells will increase the complexity of mobility management. Mobile user equipments (UEs) may trigger frequent handovers (HOs) when they move across the small cell coverage areas, increasing core network signaling. The risk of handover failure (HOF) may also increase due to the nature of the small cell coverage.

In cellular networks, the HO process allows a mobile UE to transfer its active connections from its serving cell to a target cell in connected mode, while maintaining quality of service [1]. Compared to other wireless networks, such as wireless fidelity (Wi-Fi), where mobility management is not provided, the HO process is a key advantage and differentiator of cellular networks. It ensures continuous connectivity to mobile UEs.

In conventional homogeneous cellular networks, UEs typically use the same set of HO parameters throughout the network. However, in HetNets, where macrocells and small cells have different coverage area sizes, using the same set of HO parameters for all cells and UEs may degrade mobility performance. For example, the use

Small Cell Networks: Deployment, Management, and Optimization, First Edition. Holger Claussen,
David López-Pérez, Lester Ho, Rouzbeh Razavi, and Stepan Kucera.
© 2017 by The Institute of Electrical and Electronic Engineers, Inc. Published 2017 by John Wiley & Sons, Inc.

of range expansion with different range expansion bias (REB) in different small cells will affect when and where the HO process is initiated in each cell. Therefore, in HetNets, there is a need for a cell-specific HO parameter optimization. Moreover, high-mobility macrocell user equipments (MUEs) may run deep inside the small cell coverage area before the HO is completed, thus incurring HOF due to degraded wideband signal-to-interference-plus-noise ratio (SINR). HOs performed by high-mobility MUEs may also be unnecessary, when they quickly pass through the small cell coverage area. These facts also impose the need for a UE-specific HOs parameter optimization.

Due to its importance, mobility management in HetNets have attracted a lot of interest from the wireless industry, research community, and standardization bodies [2–5]. Results in studies performed by the 3GPP RAN2 working group [2] indicate that mobility performance in HetNet deployments is not as good as in macrocell-only deployments, and that performance enhancements are needed for highly mobile UEs.

Before starting, it is important to note that the challenge in indoor small cell deployments is different than in outdoors ones. In the former, a main objective of mobility management is confining the coverage of indoor small cells to the shape of their hosting households, and thus prevent the occurrence of frequent macro-to-small cell HOs and vice versa. In the latter, the target is not only to avoid unnecessary HOs from occurring, but also to facilitate the necessary ones. With regard to mobility management in indoor small cell deployments, different approaches for initial configuration and self-optimization of coverage for both residential femtocells and enterprise femtocells deployments were presented in Chapter 5. The mobility management in outdoor small cell deployments is the focus of this chapter.

12.2 THE HANDOVER PROCESS

In cellular networks, the HO is the process through which a mobile UE communicating with one base station (BS) is switched to another BS during a call or data service [6–8]. The HO is an essential process in UE mobility management. Mobile UEs can be handed over to different radio access technologies (RATs), to different frequency bands within the same RAT, or to different cells within the same frequency band. In this chapter, we only consider the intra-RAT and intra-frequency HOs, where the source BS and the target BS use the same RAT and frequency. The same principles apply to the other HO types.

Based on the connectivity between the UE and the source BS, two types of HOs can be defined:

- Hard HO, where the UE cuts connection with the source BS before establishing connection with the target BS.
- Soft HO, where the UE keeps connection with both the source and target BSs during the HO process.

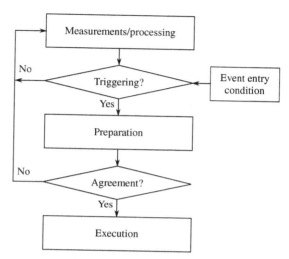

Figure 12.1. The HO stages.

Since long-term evolution (LTE) networks are based on a flat architecture without radio network controllers (RNCs), they do not support macro-diversity combining methods. As a result, soft HO is not available. Therefore, we only consider hard HOs in this chapter.

The decision to hand over from one cell to another is based on various criteria that take channel degradation, Erlang capacity, and blocking considerations into account. In LTE networks, event-triggered measurement reports (MRs) based on reference signal received power (RSRP) and reference signal received quality (RSRQ) measurements have been specified to aid the HO process triggering. Filtering of measured RSRP and RSRQ samples, together with hysteresis and time-to-trigger (TTT) mechanisms are also specified to support efficient HO decisions and avoid frequent and unnecessary HOs. For an intra-mobility management entity (MME) HO, the entire HO process is mostly confined between the source and target eNodeBs, using X2 interfaces. There may be interactions with the MME to switch the user plane.

As shown in Fig. 12.1, the HO process in LTE can typically be divided into four stages: measurements/processing, triggering, preparation, and execution. In the following, these four stages are presented in more detail, together with descriptions of their major roles and involved signaling at the radio access network (RAN) level, as depicted in Fig. 12.2.

12.2.1 Handover Measurements and Processing

In order to take HO decisions, UEs continuously perform RSRP and RSRQ measurements. These measurements are standardized and defined as follows:

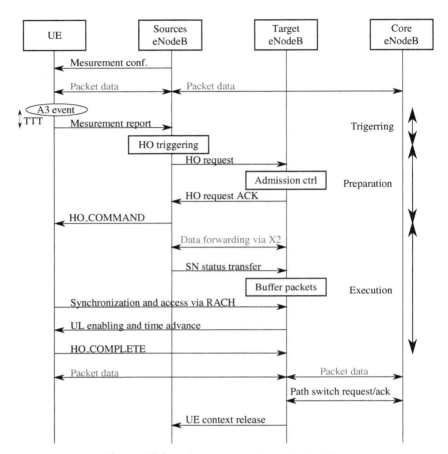

Figure 12.2.   The HO procedure and signaling.

- RSRP [9] is defined as "the linear average over the power contributions (in W of the resource elements that carry cell-specific reference signals within the considered measurement frequency bandwidth. For RSRP determination, the cell-specific reference signal R0 shall be used. If the UE can reliably detect that cell-specific reference signal, R1 is available, it may use R1 in addition to R0 to determine RSRP. The reference point for the RSRP shall be the antenna connector of the UE."

- RSRQ [9] is defined as "the ratio $N \times \frac{RSRP}{E\text{-}UTRAcarrierRSSI}$, where N is the number of resource blocks (RBs) of the E-UTRAcarrierRSSI measurement bandwidth and evolved UTRA (E-UTRA) carrier received signal strength indicator (RSSI is the linear average of the total received power (in W) observed by the UE only in the orthogonal frequency division multiplexing (OFDM) symbols containing reference symbols for antenna port 0, from all sources, including co-channel serving and non-serving cells, adjacent channel interference, thermal noise, etc. The measurements in the numerator and denominator shall be made over the same set of RBs. The reference point for the RSRQ shall be the antenna connector of the UE. If receive diversity is in use by the UE, the reported value shall not be lower than the corresponding RSRQ of any of the individual diversity branches."

In order to mitigate the effects of multi-path fading and Layer 1 measurement/estimation imperfections, UEs perform a filtering of RSRP and RSRQ measurements. The processed measurements are then reported back to the source eNodeB in a periodic or event-triggered manner through MRs. Based on such fed back MRs and when a criteria is met, the HO is triggered. In the following, the measurement and filtering procedures are described in more detail, paying special attention to RSRP measurements. RSRQ ones follow a similar procedure.

RSRP measurements and processing are performed in Layer 1 and Layer 3. Figure 12.3 shows an example [10].

First of all, the UE computes an RSRP sample by estimating the linear average over the power contributions (in Watts) of the resource elements that carry cell-specific reference signals within the considered measurement bandwidth. The specified bandwidth can range from a minimum bandwidth of 6 RBs to the maximum available bandwidth.

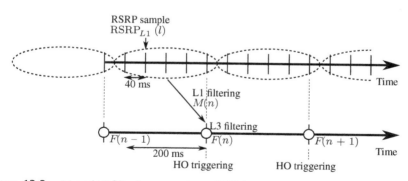

Figure 12.3. L1 and L3 filtering procedures. RSRP is measured over one subframe every, for example, 40 ms and recorded as $RSRP_{L1}(l)$. L1 filtering performs linear averaging over every 200 ms to provide $M(n) = 1/5 \sum_{k=0}^{4} RSRP_{L1}(l - k)$, with $RSRP_{L1}$ and $M(n)$ in linear units. Finally, L3 filtering performs averaging over every 200 ms to obtain $F(n) = (1 - a)F(n - 1) + a10 \log_{10}\{M(n)\}$, where a is the L3 filter coefficient, and $M(n)$ is in linear units and $F(n)$ is in dB. Note that $a = 1/2^{k/4}$ and k is specified in Table 12.1.

In order to improve the reliability of the measurements, the UE linearly averages several RSRP samples to obtain an L1 RSRP sample. Such linear averaging is performed in Layer 1, and hence is known as L1 filtering. For a typical setup, as shown in Fig. 12.3, RSRP samples may be taken every 40 ms, and then 5 RSRP samples may be averaged together to produce an L1 RSRP sample.

In order to further improve reliability, the UE also averages L1 RSRP samples through a first-order infinite impulse response (IIR) filter. Such exponential averaging is performed in Layer 3, and hence is known as L3 filtering. A typical L3 filtering period could be 200 ms, as shown in Fig. 12.3. These values are considered in [11], but they depend on the UE implementation, and thus other values may apply. For example, Since successive log-normal shadowing samples are spatially correlated, the L3 filtering period would be preferred to be adaptive to the degree of shadowing correlation in the cell-specific reference signals. For example, higher-mobility UEs will measure less correlated L3 RSRP samples than low-mobility UEs, and thus they would benefit from a shorter L3 filtering period.

12.2.2 Handover Triggering

The UE will send an MR to the serving eNodeB in a periodic or event-triggered manner to alert it of a potential HO. Event triggered MRs are sent to the serving eNodeB when the L3 RSRP sample, the output of the first-order IIR filter, meets an HO event entry condition. The serving eNodeB will decide whether to initiate an HO-based on the reception of this MR.

Eight types of HO event entry conditions have been defined in LTE (see [12], Section 5.5.4):

- Event A1: Server becomes better than threshold
- Event A2: Server becomes worse than threshold
- Event A3: Neighbor becomes offset better than server
- Event A4: Neighbor becomes better than threshold
- Event A5: Server becomes worse than threshold1 and neighbor becomes better than threshold 2
- Event A6: Neighbor becomes offset better than secondary server (this condition applies to carrier aggregation configurations)
- Event B1: Inter-RAT neighbor becomes better than threshold
- Event B2: Server becomes worse than threshold1 and inter-RAT neighbor becomes better than threshold 2

Events A1 to A6 are used for intra-LTE system report triggering, and events B1 to B2 are used for inter-system report triggering. It is important to note that HOs between eNodeBs operating in the same frequency band are usually triggered by event A3

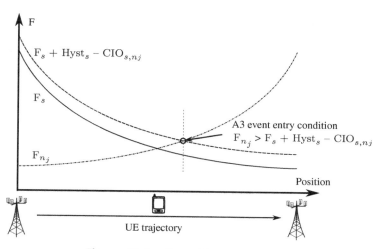

Figure 12.4. Event A3 entry condition.

entry condition, and thus this condition will be the focus of our studies in the rest of the chapter.

For illustrations purposes, Fig. 12.4 shows the event A3 entry condition, where the inequality of event A3 is given by:

$$F_{n_j} > F_s + \text{Hyst}_s - \text{CIO}_{s,n_j}, \quad (12.1)$$

where F_s is the L3 RSRP sample of the serving cell s, not taking into account any offset, F_{n_j} is the L3 RSRP sample of neighboring cell n_j, not taking into account any offset, Hyst_s is the hysteresis parameter, and CIO_{s,n_j} is the cell individual offset (CIO) of the neighbor cell n_j with respect to the serving cell s [12]. The CIO is an offset that will be applied by the UE to the measurement results of neighboring cells before it determines whether or not an event A3 has occurred.

Once the event A3 entry condition (12.1) is met, the UE starts the TTT timer. Otherwise, no other steps are taken. In order to mitigate fading effects, only when the event A3 entry condition is satisfied through the whole TTT window, the UE reports the event A3 entry condition via a MR to the serving cell, which may or may not trigger an HO. This is to be decided by the source and target cells.

Optimization of the hysteresis Hyst_s or TTT parameters could be used to mitigate HOFs and unnecessary HOs caused by too frequent HOs, and thus reduce the signaling overhead. For example, small values of TTT may lead to too early HOs, increasing unnecessary HOs, while large values of TTT may result in too late HOs, increasing HOFs. Different from hysteresis Hyst_s and TTT, which are used to adjust parameters in the serving cell, CIO is used to regulate the target cell chosen by the UE.

12.2.3 Handover Preparation

If, according to the filtered measurements, a certain HO event entry condition is met during the TTT, the UE alerts the serving cell through an MR where the HO measurements are fed back. Then, the preparation phase starts. The source cell issues a HO request message to the target cell, which carries out admission control procedures according to the quality of service requirements of the UE [13]. After admission, the target cell prepares the HO process, and sends a HO request acknowledgment to the source cell. When the HO request acknowledgment is received at the source cell, data forwarding from the source cell to the target cell starts, and the source cell sends a HO_COMMAND to the UE. The source or the target cell may abort the HO process at any of these steps, if some conditions are not met.

12.2.4 Handover Execution

In the HO execution phase, the UE synchronizes with the target cell and accesses it [13]. The UE sends a HO_COMPLETE message to the target cell when the HO procedure is finished. The target cell, which can then start transmitting data to the UE, sends a path switch message to inform the network that the UE has changed its serving cell. Thereafter, the network sends a UE update request message to the serving gateway, which switches the downlink (DL) data path from the source cell to the target cell. The network also sends end marker packets through the old path to the source cell, asking it to release any resources previously allocated to this UE.

12.3 RADIO LINK FAILURE, HANDOVER FAILURE AND UNNECESSARY HANDOVERS

The main goal of mobility management in HetNets is to reduce the radio link failure (RLF) caused during a HO, that is, HOFs, and avoid the waste of network resources caused by unnecessary HOs. In the following, the concepts of RLF, HOF, and unnecessary HOs are defined.

12.3.1 Radio Link Failure

A UE declares RLF [11, 14] when it loses the synchronization with the network. As a result, the UE releases its network resources and any ongoing service will be dropped. There may be many reasons for RLF, for example, coverage holes, adverse channel conditions. A primary target of network optimization is to avoid RLFs.

A key measurement to understand RLF is the wideband SINR, which is defined as the ratio of the average power received from the serving cell to the average interference power received from other cells plus noise. In system-level simulations, wideband SINR can be computed as described in Section A.8.

A model for evaluating RLF is presented in [15], where Q_{out} is the wideband SINR threshold for an out-of-sync event and Q_{in} is the wideband SINR threshold for an in-sync event. T310 is a timer that is triggered when the number of out-of sync events have reached a certain value. N310 is the threshold for the number of out-of-sync events to start the T310 timer. N311 is the threshold for the number of in-sync events to abort the T310 timer. In order to mitigate fading effects, when a UE tracks RLFs, Q_{out} is monitored with a 200 ms window and Q_{in} is monitored with a 100 ms window. Both windows are updated once per frame, that is, once every 10 ms, with the measured wideband SINR value. These values are used in [11], but they depend on the UE implementation, and thus other values may apply. If a UE detects that its average wideband SINR value is lower than Q_{out}, it will report an out-of-sync event. If it detects that its average wideband SINR value is higher than Q_{in}, it will report an in-sync event. When the out-of-sync event has been reported for N310 times, the UE starts the T310 timer, which is the time limit to decide whether an RLF occurs. If the in-sync event is detected for less than N311 times when the T310 timer expires, an RLF occurs. Otherwise, the T310 timer is aborted. After detecting a RLF, T311 is started, which is the RLF recovery timer and is defined as the maximum time allowed for the recovery of the radio link actively performed by the UE. This capability is not considered in this chapter.

12.3.2 Handover Failure

For the purpose of modeling and classification of HOFs, the HO process is divided into three states according to [2]:

- State 1: Before the event A3 entering condition is satisfied.
- State 2: After the event A3 entering condition is satisfied, but before the HO_COMMAND is successfully received by the UE.
- State 3: After the HO_COMMAND is successfully received by the UE, but before the HO_COMPLETE is successfully sent by the UE.

An HOF occurs if an RLF occurs in State 2 or State 3 [2]:

- In State 2: When the UE is attached to the source cell, an HOF is counted if one of the following criteria is met:
  - Timer T310 has been triggered or is running when the HO_COMMAND is received by the UE (see Fig. 12.5).
  - RLF is declared in State 2, for example, when TTT is running (see Fig. 12.6).

  For monitoring RLF in State 2, the wideband SINR is measured once every 10 ms, and wideband SINRs are filtered by a linear filter with a sliding window of 200 ms (i.e., 20 samples), as indicated earlier.

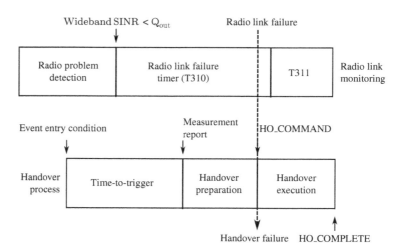

Figure 12.5. Timers in radio link monitoring and HO processes. Timer T310 is running while HO_COMMAND is received.

- In State 3: After the UE is attached to the target cell, an HOF is counted if the following criteria is met:
  - The target cell filtered wideband SINR is less than Q_{out} at the end of the HO execution time.

For monitoring RLF in State 3, the wideband SINR should be measured at least twice every 40 ms, that is, the HO execution time, and averaged over the samples.

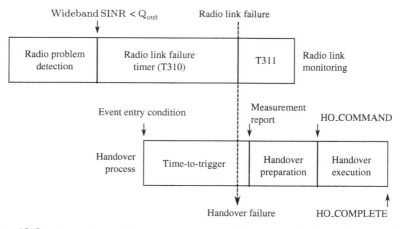

Figure 12.6. Timers in radio link monitoring and HO processes. RLF is declared while TTT timer is running.

The HOF rate is defined as the ratio of the number of HOFs to the total number of HO attempts.

12.3.3 Unnecessary Handovers

Whether or not an unnecessary HO occurs is determined by the time duration that a UE stays connected to a cell directly after an HO, namely time-of-stay. The time-of-stay starts when the UE sends a HO_COMPLETE message to a cell, and ends when the UE sends another HO_COMPLETE message to another cell. If a UE has a time-of-stay less than a threshold T_p, then the HO that terminates this time-of-stay is considered as an unnecessary HO. The recommended T_p is 1 s [2].

An unnecessary HO is considered as a ping-pong, if the prior-to-source and target cells are the same cell, where the prior-to-source cell is the cell to which the UE was connected before handing over to the source cell. Figure 12.7 shows an example of one unnecessary HO classified as ping-pong and two other cases, which cannot be classified as ping-pongs.

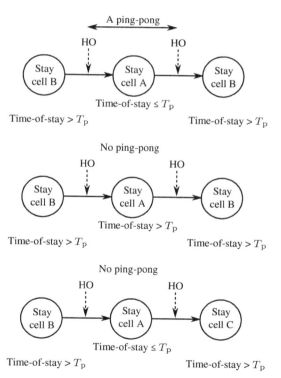

Figure 12.7.   Timers in radio link monitoring and HO processes. Example of unnecessary HO classified as ping-pong and two other cases, which cannot be classified as ping-pongs.

The ping-pong rate is defined as the ratio of the number of ping-pongs to the total number of successful HOs (excluding HOFs).

12.4 MOBILITY CHALLENGES IN HETEROGENEOUS NETWORKS

Small cells pose a number of challenges for the mobility management. As will be explained in the following section, small cells may contribute to an increase in HOFs and unnecessary HOs, due to the small cell coverage nature. This fact threatens one of the main advantages of a cellular network, the ability of a UEs to move around while being connected to the network and carrying ongoing services. This will affect customers' perception of cellular networks, and may increase churn. As a result, large efforts have been made in evaluating HetNet mobility performance taking range expansion and enhanced inter-cell interference coordination (eICIC) features, for example, almost blank subframe (ABS), into account. Based on such knowledge, a number of robust mobility management techniques are being proposed, aiming at enhancing mobility performance such that the mobility performance of a HetNet is at least as good as that of legacy macrocell-only networks. In the rest of this chapter, the mobility performance in HetNets is analyzed and different techniques to enhance it are presented.

12.4.1 The Small Cell Coverage Area Issue

Recent reports show that HO performance in HetNet deployments is not as good as in legacy macro-only deployments. For example, the results in [2] indicate that the HOF rate in HetNets can be twice higher than in legacy macro-only deployments. The reasons for such HOF increase are explained below.

From the discussion in Section 12.2, it is obvious that the HO process takes some time due to measurements and processing, triggering, preparation, and execution. Since the UE triggers TTT before sending the MR, the time elapsed may be well above 500 ms, depending on the HO parameter configuration. While this time may not be an issue when performing a macrocell to macrocell HO due to the large distance between eNodeBs, it is a threat when performing a macrocell to small cell HO and vice versa. As depicted in Fig. 12.8, due to the small size of the small cell coverage area, a mobile UE handing over from a macrocell to a small cell may quickly move deep inside the small cell coverage area, while connected to the macrocell. As a result, the macrocell-connected UE will observe a large interference from the small cell. Then, it may happen that its wideband SINR is significantly degraded, thus losing synchronization with the network and incurring RLF, or that the UE cannot decode the HO_COMMAND sent from the macrocell to signal the switch of serving cell. Both cases result in a HOF. Due to the delay of the HO process, it may also occur, in the same manner, that a mobile UE connected to a small cell and getting out of it may lose synchronization or not decode its HO_COMMAND, since by the time the HO

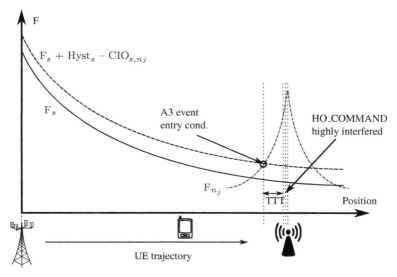

Figure 12.8.   The HOF issue in HetNets.

preparation phase is completed, the UE may have already left the small cell coverage area and receive high interference from the macrocell.

12.4.2 Range Expansion and Almost Blank Subframes

In order to address problems caused by the DL transmit power difference between eNodeBs and pico eNodeBs in HetNets, cell selection methods allowing UEs to associate with cells that do not provide the strongest RSRP are necessary. A widely considered approach is range expansion [16], as presented in Chapter 6, in which the UE adds a positive REB to L3 HO measurements to artificially increase picocells DL coverage footprints. Although range expansion is able to mitigate uplink (UL) intercell interference and provide load balancing in HetNets, it degrades the DL signal quality of picocell user equipments (PUEs) in the expanded region, since they are not connected to cells that provide the strongest RSRPs.

In order to alleviate this issue, ABSs [16] (see Chapter 8 for detailed description), in which no control or data signals but just reference signals are transmitted, can be used to mitigate DL inter-cell interference to range expanded PUEs. Specifically, macrocells can schedule ABSs, and picocells can schedule range-expanded PUEs in the subframes that overlap with the macrocell ABSs, so that their performance can be enhanced.

Both picocell range expansion and macrocell ABSs have an impact on the mobility performance [17]. This is illustrated in Fig. 12.9 and Fig. 12.10, where an example

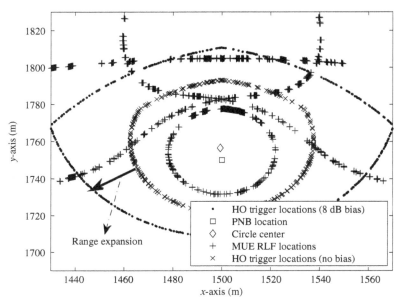

Figure 12.9. Coverage areas of macrocell/picocell with and without range expansion and RLF locations from macrocell perspective ($Q_{out} = -8$ dB).

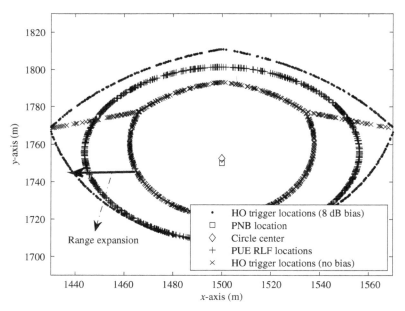

Figure 12.10. Coverage areas of macrocell/picocell with and without range expansion and RLF locations from picocell perspective ($Q_{out} = -8$ dB).

for HO trigger locations and HOF locations are generated by a simulator that implements Third-Generation Partnership Project (3GPP) simulation assumptions according to [2]. The eNodeB is located at (1500, 1500) meters, while the pico eNodeB is located at (1500, 1750) meters. The eNodeB is surrounded by two tiers of interfering eNodeBs, that is, 18 eNBs and 54 sectors, with an inter-site distance of 500 meters. An HOF is observed if the UE wideband SINR is lower than a threshold (i.e., $Q_{out} = -8$ dB). Under the assumption of neglecting HO preparation (50 ms) and execution (40 ms) times, this captures the combination of RLF in HO State 2. In the figure, it can be observed that the HO trigger locations are scattered around the perimeter of a circle. This is because UEs are moving at 60 km/h in this simulation, while MRs are produced at intervals of 200 ms, which causes delays in triggering HOs. In contrast, the HOF locations resemble a deformed circle, where deforming occurs due to sector antennas deployed at the eNodeBs. No shadowing or multi-path fading is considered in this example to facilitate the illustration.

From Figs. 12.9 and 12.10, it can be inferred that if a MUE or PUE crosses the RLF boundary before the TTT expires when moving in or out of the picocell coverage area, then an RLF occurs and the UE experiences an HOF [17]. These figures also show that, with range expansion at the picocell, the event A3 positions are pushed away from the pico eNodeB location, thus increasing the picocell coverage area and potentially allowing for a better spatial reuse. In Fig. 12.9, since the gap between the event A3 boundary and the RLF boundary is larger with picocell range expansion, it is more likely that the TTT will expire before the UE SINR falls to Q_{out}, and the HO is completed successfully [4]. Range expansion thus facilitates the macro-to-pico HO. In contrast, in Fig. 12.10, range expansion challenges the pico-to-macro HO, since the space between the event A3 boundary and the RLF boundary gets smaller, and thus it is more likely that the UE SINR falls to Q_{out} before the TTT expires. In such a case, a HOF occurs [4]. This issue can be avoided if ABSs are scheduled at the macrocells, and the picocell schedules the HO_COMMAND in a subframe overlapping with an umbrella macrocell ABS. It is interesting to observe that in Fig. 12.10, RLF may occur even earlier than the event A3, stressing the need for macrocell ABSs to support expanded region picocells.

12.5 MOBILITY PERFORMANCE IN HETEROGENEOUS NETWORKS

In order to characterize the mobility management performance in a HetNet in terms of HOFs and ping-pongs, system-level simulations have been performed [17] using the large area simulation recommendations in [2]. Further details on these large area simulations can be found in Section A.11, and the most important simulation parameters are presented in Table 12.2.

In these simulations, a hexagonal eNodeB layout with 19 eNodeBs, 57 sectors, and an inter-eNodeB distance of 500 meters was considered. Four pico eNodeBs

TABLE 12.1   HO Parameter Sets in [2]

Profile	Set 1	Set 2	Set 3	Set 4	Set 5
TTT (ms)	480	160	160	80	40
A3 offset (dB)	3	3	2	1	−1
RSRP L3 Filter K	4	4	1	1	0

were randomly distributed within each eNodeB sector coverage area. Five different HO parameter configurations were taken into account, as summarized in Table 12.1 [2], where the longest and shortest TTT durations are 480 ms in simulation Set 1 and 40 ms in simulation Set 5, respectively. For each simulation set, an L1 and L3 filtering period of 200 ms was used, along with full cell-loading. UEs were randomly distributed over the entire simulation scenario, and each UE moved along a straight line toward a randomly selected direction. A UE did not change direction until it hit the border of the scenario. When a UE hit the border of the scenario, it is bounced back and moved toward another randomly selected direction.

Figure 12.11 presents the simulated HOF and ping-pong rates, respectively, for two different HetNet cases:

1. Picocells without range expansion and macrocells without ABSs.
2. Picocells using an 8 dB bias for range expansion and macrocells using ABSs.

Simulation results show that for both cases considered, reducing TTT reduces HOFs, but increases ping-pongs; and conversely, increasing TTT increases HOFs, but reduces ping-pongs. Among the five HO profiles, Set 3 with an intermediate TTT of 160 ms yields the best trade-off. Moreover, the results show that using picocell range expansion and implementing macrocell ABSs decreases HOFs, but increases ping-pongs. Again, the same trade-off between HOFs and ping-pongs can be seen; that is, optimizing HO parameters to reduce HOFs would increase ping-pongs, and vice versa [17]. This makes HO optimization as shown in previous sections an intricate problem, which could be significantly exacerbated by a large number of deployed small cells in a HetNet.

In the following, HOFs and ping-pongs are analyzed in more detail. In terms of HOF rates (see Fig. 12.11(a)), results are in line with [2], indicating that:

- A higher UE velocity reduces the time window for the UE and the network to complete the HO procedure, and leads to higher HOF rate.
- A larger hysteresis or longer TTT increases the HO process time, which makes the UE to move deeper inside the target cell coverage area while connected to the source cell, and leads to higher HOF rate.

The case of picocells with no range expansion and macrocells with no ABSs results in the worst HOF performance. This is because the picocell coverage areas

TABLE 12.2   Simulation Parameters

Parameter	Value
Carrier frequency	2.0 GHz
System bandwidth	10 MHz
Number of eNB/sectors	19/57, with 500 m ISD
NB antenna patterns (TR 36.814)	3D pattern
eNB antenna patterns (TR 36.814)	Omni-directional pattern
NB antenna tilt	15 degree
NB antenna gain	15 dB
eNB antenna gain	5 dB
UE antenna gain	0 dB
Macrocell path-loss model	$128.1 + 37.6\log_{10}(R)$ dB
Picocell path-loss model	$140.7 + 36.7\log_{10}(R)$ dB
Shadowing standard deviation	8 dB (macrocell), 10 dB (picocell)
Correlation distance of shadowing	25 m
Macrocell shadowing correlation	0.5 (1) between cells (sectors)
Picocell shadowing correlation	0.5 between cells
Transmit power	46 dBm (eNB), 30 dBm (PeNB)
Penetration loss	20 dB
Antenna configuration	1×2
Picocell range expansion bias	8 dB (whenever applicable)
Cell loading	100 %
UE velocities	3, 30, 60, 120 km/h
UE noise figure	9 dB
Thermal noise density	-174d Bm/Hz
Channel model	Typical urban (6 rays)
HO metric	1 Rx for RSRP measurement
SINR metric	2 Rx, MRC and EESM
RSRP measurement bandwidth	25 RBs
L3 filter coefficient (a)	0.5
HO preparation (execution) delay	50 ms (40 ms)
Q_{out} (Q_{in})	-8 dB (-6 dB)
T310	1 s
Min. eNB-UE (PeNB-UE) distance	35 m (10 m)
Min. eNB-PeNB (PeNB-PeNB) distance	75 m (40 m)

without range expansion are small, and thus UEs may quickly run deep inside the target cell coverage area before the TTT expires, significantly degrading its wideband SINRs before the HO process is completed. Pico-to-macro HOF rate is the highest for all the configuration sets and speeds.

The case of picocells with range expansion and macrocells with ABSs results in a better HOF performance. This is because range expansion pushes away the event A3 event entry condition boundary from the pico eNodeB location, and the UE has

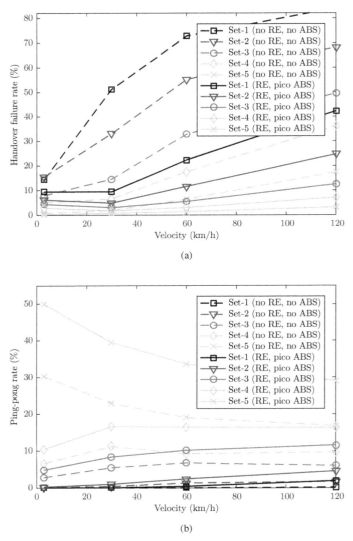

Figure 12.11. Simulated HOF and ping-pong rates with four randomly deployed pico-cells per macrocell sector, with and without an 8 dB range expansion bias, and ABS. (a) HOF rate and (b) Ping-pong rate.

more time to perform the HO process. Pico-to-macro HOF is not an issue due to the interference mitigation provided by the macrocell ABS.

In terms of ping-pong rates (see Fig. 12.11(b)), results are also in line with [2], indicating that ping-pong rates are relatively high for low-velocity UEs with parameter Set 5, while ping-pong rates are relatively low for low-velocity UEs with parameter Set 1. The case of picocells with no range expansion and macrocells with no ABSs performs better than that of picocells with range expansion and macrocells with ABSs. When the picocell coverage area is increased through range expansion, cell selection oscillation caused by fading occurs in a larger area (just because the small cell is bigger with cell range expansion). As a consequence, the number of ping-pongs increases. In both cases, ping-pongs are alleviated with larger TTTs, for example, Set 1 and Set 2. This is because a better L1 and L3 filtering can be performed to mitigate fading and Layer 1 estimation imperfection.

From the above results, it is important to note that for low-mobility UEs (i.e., velocity ≤ 30 kmh), no significant problems have been observed in terms of HOF and loss of connectivity (some issues with ping-pongs have been identified). However, it becomes hard to simultaneously achieve good HOF and ping-pong performance for higher UE velocities, which fortunately are less likely for PUE.

12.5.1 Impact of Range Expansion and Almost Blank Subframes

The impact of picocell range expansion and macrocell ABSs was studied in detail through system-level simulations by different companies in [2], Section 5.5.6., and references therein, where it is important to note that two different ABS configurations as described below were considered:

- Ideal ABS coordination (denoted as "perfect eICIC"): The ABS patterns of all the macrocells are synchronized in time, that is, the ABS from all macrocells occur at the same subframes, and all the macrocells are subframe-aligned. No cell-specific reference signal collision between macrocells and picocells is assumed either. As a result, a UE served by a picocell in the range expanded area does not observe interference from any macrocell.
- Non-ideal ABS coordination (denoted as "imperfect eICIC"): The ABS patterns of the macrocells are not synchronized in time. A PUE is protected by the ABSs from the overlay macrocell only. Therefore, the radio link monitoring of the PUE is affected by interference from all other neighbor macrocells. This represents a worst-case scenario.

From the simulations results in [2] Section 5.5.6, it can be seen that in comparison to the baseline HetNet without picocell range expansion and macrocell ABSs, the ideal ABS coordination can improve mobility performance across different REBs. This is in line with the results presented in Fig. 12.11. The macro-pico and pico-macro HOF rates reduce significantly with ideal ABS coordination [18]. With this ideal ABS coordination, the reduced interference from macrocells improves the macro-to-pico HO performance due to the more reliable random access channel (RACH) process to the target picocell. Moreover, the reduced interference also improves the pico-to-macro HO performance due to the HO_COMMAND being more reliably delivered from the source picocell. The larger the REB, the higher the performance gain.

The non-ideal ABS coordination can also reduce the HOF rate when the REB is small. However, when REB becomes large (i.e., > 6 dB), the mobility performance becomes worse than that of the baseline HetNet without picocell range expansion and macrocell ABSs. This performance degradation is due to increased interference that UEs in larger range expanded regions experience from the neighboring macrocells. However, it is important to note that the non-ideal ABS coordination assumed in this simulations is a worst case scenario, where a PUE is protected by the ABSs from its umbrella macrocell only, and thus there is room for improvement.

In general, it is important to bear in mind that, with ideal ABS coordination the HO performance between the macrocells and picocells is improved significantly and macro-pico and pico-macro HO become less of an issue in HetNets. However this is not the case for non-ideal ABS coordination when REB are larger than 6 dB. This shows the importance of a proper network configuration of ABS patterns, and calls for further optimization of mobility management in HetNets. For further results please review Section 5.5.6 in [2].

12.6 MOBILITY OPTIMIZATION IN HETEROGENEOUS NETWORKS

The HO performance of high-velocity UEs deteriorates significantly due to the small size of the small cell coverage area, and thus high-velocity UEs have a higher risk of service interruption in HetNets. Previous simulation results in Section 12.5 showed that HOF rates are much higher for high-velocity UEs, for example, 120 km/h than that for low-velocity UEs, for example, 3km/h. From these results, it can be derived that UE velocity has a significant impact on the HO performance in HetNets. Moreover, if successful, the frequent HOs of high-velocity UE toward small cells may incur large overheads between eNodeBs as well as between eNodeBs and UEs due to the short time-of-stay. As a result, it can be thus concluded that small cells cannot cost-efficiently handle the traffic load of high-velocity UE, and thus new UE velocity dependent HOs that treat high- and low-velocity UEs differently are highly regarded [2].

12.6.1 Handover Parameter Optimization

High-velocity UEs are likely to experience a RLF before the HO is completed because they may not be able to connect soon enough to the target cell due to a lengthy HO process. Therefore, optimizing HO parameters, for example, hysteresis and TTT such that the duration of the HO process is a function of the UE velocity is a widely studied approach to enhance mobility performance in HetNets.

When using HO parameter optimization, the main idea is to scale hysteresis and TTT using a UE-specific factor, which is selected as a function of the UE velocity, such that high-mobility UEs are handed over to the target cell earlier than

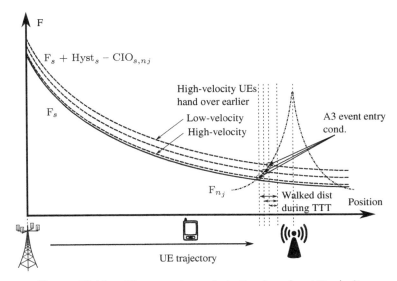

Figure 12.12. HO parameter optimization based on UE velocity.

low-mobility UEs (see Fig. 12.12). In this way, HOFs are mitigated for high-mobility UEs due to the shorter HO duration (e.g., shorter TTT), while ping-pongs are reduced for low-mobility UEs due to a better measurement filtering (e.g., longer TTT).

In order to verify the performance of the proposed UE velocity-dependent HOs, based on HO parameter optimization, large area system-level simulations were performed. Three UE velocities were considered: 30 km/h, 60 km/h, and 120 km/h. For the UE velocity 60 km/h, the HO parameters were set as hysteresis $= 2$ dB, TTT $= 160$ ms, and $K = 1$ (Set 3 in Table 12.1). For the UE velocity 30 km/h, these mobility parameters were scaled up by 1.5, while for the UE velocity 120 km/h, they were scaled down by 0.5 [19].

The simulations results in Fig 12.13 show that the number of HOFs per UE per second significantly reduces with HO parameter optimization in all cases, macro-to-pico, pico-to-macro, and macro-to-macro HOs. When using such optimization, macro-to-pico HOF/UE/s were reduced by 60.58 %, while pico-to-macro HOF/UE/s were reduced by 44.86 %. Macro-to-macro HOF/UE/s were also reduced by 57.91 %, indicating that HO parameter optimization is always beneficial.

As a drawback, this method does not prevent high-mobility UEs to hand over to small cells, which may be unnecessary due to short time-of-stay. Moreover, if the network is in charge of selecting the UE-specific factor, cell-to-UE signaling would be required. In contrast, if the UE selects it, more processing would be required at the handset and no degree of freedom would be left to the operator. Mobility state estimation by the UE without relying on the global positioning system (GPS) is also

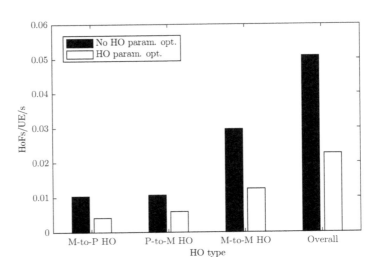

Figure 12.13. HOF performance with UE velocity-dependent HO parameter optimization.

a challenge. All these issues still have to be solved before making this HO parameter optimization approach successful.

12.6.2 Inter-cell Interference Coordination

In order to minimize the number of unnecessary HOs, it would be desirable to keep the high-velocity UE with the macrocell, while making sure that they do not incur HOFs. In this line, the authors in [17] proposed a mobility-based inter-cell interference coordination scheme that combines HO parameter optimization with eICIC. In this scheme, high-velocity UEs are kept with the macrocell and low-velocity UEs are handed over to the small cells, while HOFs and ping-pongs for high- and low-velocity UEs are respectively mitigated. On the one hand, in order to protect high-mobility UEs from HOFs, small cells release certain resources, so that macrocells can schedule their high-velocity UEs in these resources without inter-cell interference from small cells. Since high-velocity UEs are not handed over the small cells, ping-pongs are also mitigated. Resources can be coordinated in time (e.g., subframes), frequency (e.g., component carriers) or space (e.g., beam directions) domains. On the other hand, in order to protect low-velocity UEs from ping-pongs, and using the same concept as in the previously presented HO parameter optimization, a large TTT is selected for low-velocity UEs. It is important to note that when a small cell is placed at a coverage hole of a macrocell, it is better to allow high-velocity UEs to handover to the small cells. The umbrella macrocell should know which small cells are located at the macrocell coverage hole, and should be able to make HO decisions accordingly.

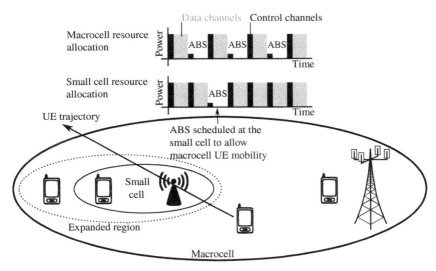

Figure 12.14. UE velocity dependent HO.

In order to support the proposed scheme, first, UE velocity information should be available at the eNodeB, and thereafter, based on that knowledge, the interference from the small cell when the UE is attached to the macrocell and moving across the small cell coverage area should be mitigated. For this purpose, a possible implementation is the use of ABSs scheduled at the small cell, as illustrated in Fig. 12.14. In this case, the macrocell schedules the high-velocity MUE traversing the small cell coverage in the macrocell subframes overlapping with the picocell ABSs. In addition, UL interference mitigation should also be used to not only protect the DL of the passing by MUE but also its UL.

As one can easily expect, the mobility performance improvement through the proposed inter-cell interference mitigation comes at the expense of releasing resources by the small cells. In order to minimize the throughput loss of the small cells, only high-velocity MUEs should be scheduled in subframes overlapping with the picocell ABSs, while the rest will be handled through HO parameter optimization using a larger TTTs to suppress ping-pongs. Moreover, it may also be possible to semi-dynamically adjust the duty cycle of ABSs at a picocell based on the percentage of high-mobility UEs within a given time window.

In order to verify the performance of the proposed UE velocity dependent HOs, based on mobility-based inter-cell interference coordination, large area system-level simulations were performed. Three UE velocities were considered: 30 km/h, 60 km/h, and 120 km/h. If the UE velocity was either 60 km/h or 120 km/h, those UEs were not handed over the small cells. Set 3 in Table 12.1 was used as HO configuration for UEs moving at 30 km/h [19]. In this case, the HO parameters of UEs moving at 30 km/h are not scaled up by 1.5 in order to access the gains due to interference mitigation.

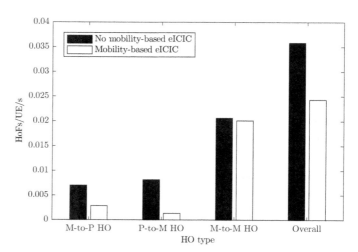

Figure 12.15. HOF performance with mobility-based inter-cell interference coordination.

The simulations results in Fig 12.15 show that the number of HOFs per UE per second significantly reduces in terms of macro-to-pico and pico-to-macro HOFs. When using the proposed scheme, macro-to-pico HOF/UE/s were reduced by 59.32 %, while pico-to-macro HOF/UE/s were reduced by 83.98 %. Macro-to-macro HO performance was not impacted, since the proposed interference coordination only applies to macrocell and small cell interactions.

From the presented results, it can be observed that although mitigated, HOFs issues still higher than desirable for a truly mobile network. This highlights the need for new network architectures that consider small cells and the mobility management issue as a key design driver. New architectures based on dual connectivity and carrier aggregation functionalities are envisioned to provide a better mobility support.

12.7 MOBILITY STATE ESTIMATION

As discussed in the previous section, UE mobility state estimation may play a key role in the HO parameter optimization, in which HO parameters are fine-tuned according to UE velocity. For example, high-velocity UEs may use large TTTs to prevent HOFs, while low-velocity UEs may use short TTTs to mitigate ping-pongs.

A procedure for detecting UE mobility states in LTE systems is described in [20], where three UE mobility states are defined: normal, medium, and high. The serving cell may identify the mobility state of a UE by comparing the number of cell reselections or HOs with a cell reselection threshold (CRT), N_{CR_M} for normal velocity and N_{CR_H} for high velocity, within a given time window T_{crMax} [21].

As reported in [2], the UE mobility state estimation performs well in macro-only networks, providing a good correlation between the velocity of UEs and the mobility state estimation. This is because, in most macro-only networks, it is easy to choose appropriate mobility state estimation thresholds, $N_{CRMedium}$ and N_{CRHigh}, for the mobility state estimation algorithm to distinguish the UE mobility states. However, using such mobility state estimation algorithm in HetNets produces unreliable results, which are biased by the small cells.

One way to improve the mobility state estimation performance in HetNets is to scale each mobility state estimation count with a weight that is directly proportional to the size of the cell involved in the HO [22]. Another approach is to use higher mobility state estimation thresholds for higher small cell densities [5]. For a more accurate estimation of UE mobility state, Doppler frequency measurements may also be used in combination with cell reselection count [23]. Typically, a UE already measures Doppler frequency for the purpose of channel estimation, and thus measurement of the Doppler frequency would not be a new function and may also be used for mobility state estimation purposes.

In a separate study on improving the UE operation considering small cell search and discovery, the UE velocity could also be considered. UEs should decide whether to search for neighboring small cells based on their velocity. If a UE velocity is high, it should not handover to small cells, and thus should not search for small cells. This would also prevent high-velocity UEs to perform small cell measurements and reporting them, thus saving UE power and radio resources. This is especially important for inter-frequency load balancing, where the UE time-of-stay in the small cell may be relatively short to benefit from the load balancing [19].

12.8 SUMMARY AND CONCLUSIONS

In this chapter, the mobility management challenges in HetNets have been presented. The HO process in LTE systems has been reviewed, paying special attention to its main four stages, measurements/processing, triggering, preparation, and execution, together with important concepts such as RLF, HOF, and ping-pongs. The mobility performance in HetNets has also been analyzed, taking into account different UE velocities, network configurations, and HO parameters. Important insights and trade-offs have been highlighted. Based on the significant impact of UE velocity on the HO performance in HetNets, a UE velocity-dependent HO scheme has been presented and its advantages quantified. The need for a better mobility state estimation and new network architectures that consider small cells and the mobility management issue as a key design driver has also been discussed to conclude the chapter.

REFERENCES

1. TS 23.009, "Handover procedures," 3GPP Tech. Rep., v.10.0.0, Apr. 2011.
2. TR 36.839, "Mobility enhancements in heterogeneous networks," 3GPP Tech. Rep., v.0.2.0, Sep. 2011.

3. Samsung, "Mobility support to pico cells in the co-channel HetNet deployment," 3GPP Standard Contribution (R2-104017), Stockholm, Sweden, Mar. 2010.

4. D. López-Pérez, I. Guvenc, and X. Chu, "Theoretical analysis of handover failure and ping-pong rates for heterogeneous networks," in *Proc. IEEE Int. Workshop Small Cell Wireless Netw. (co-located with IEEE ICC)*, Ottawa, Canada, Jun. 2012.

5. D. López-Pérez, I. Guvenc, and X. Chu, "Mobility enhancements for heterogeneous wireless networks through interference coordination," in *Proc. IEEE Int. Workshop Broadband Femtocell Technol. (co-located with IEEE WCNC)*, Paris, France, Apr. 2012.

6. L. K. L. Bajzik, P. Horvath, and C. Vulkan, "Impact of intra-LTE handover with forwarding on the user connections," in *16th Mobile Wireless Commun. Summit*, Jul. 2007, pp. 1–5.

7. Y. Yuan and Z. Chen, "A study of algorithm for LTE intra-frequency handover," in *Proc. Int. Conf. Comput. Sci. Service Syst. (CSSS)*, Jun. 2011, pp. 1986–1989.

8. D. Singhal, M. Kunapareddy, V. Chetlapalli, V. B. James, and N. Akhtar, "LTE-advanced: Handover interruption time analysis for IMT-A evaluation," in *Proc. Int. Conf. Signal Process., Commun., Comput. Netw. Technol. (ICSCCN)*, Jul. 2011, pp. 81–85.

9. TS 36.211, "Long term evolution physical layer; general description," 3GPP Tech. Rep., v.1.0.0, Mar. 2009.

10. M. Anas, F. D. Calabrese, P. E. Ostling, K. I. Pedersen, and P. E. Mogensen, "Performance analysis of handover measurements and layer 3 filtering for UTRAN LTE," in *Proc. IEEE Int. Symp. Pers., Indoor, Mobile Radio Commun. (PIMRC)*, Athens, Greece, Sep. 2007, pp. 1–5.

11. TS 36.133, "Evolved universal terrestrial radio access (E-UTRA); requirements for support of radio resource management," 3GPP Tech. Rep., v.8.8.0, Dec. 2009.

12. TS 36.331, "Radio resource control; protocol specification," 3GPP Tech. Rep., v.10.4.0, Dec. 2011.

13. D. Pacifico, M. Pacifico, C. Fischione, H. Hjalrmasson, and K. Johansson, "Improving TCP performance during the intra LTE handover," in *Proc. IEEE Global Telecommun. Conf. (GLOBECOM)*, Dec. 2009, pp. 1 –8.

14. TS 36.508, "Common test environments for user equipment (UE) conformance testing," 3GPP Tech. Rep., v.9.3.0, Dec. 2010.

15. N. DOCOMO, "Evaluation model for Rel-8 mobility performance," 3GPP Standard Contribution (R1-091578), Seoul, Korea, Mar. 2009.

16. D. López-Pérez, I. Güvenç, G. de la Roche, M. Kountouris, T. Q. Quek, and J. Zhang, "Enhanced inter-cell interference coordination challenges in heterogeneous networks," *IEEE Wireless Commun. Mag.*, vol. 18, no. 3, pp. 22–31, Jun. 2011.

17. D. López-Pérez, I. Guvenc, and X. Chu, "Mobility management challenges in 3GPP heterogeneous networks," in *IEEE Comm. Mag.*, Dec. 2012, pp. 70–78.

18. Alcatel-Lucent, "Evaluation of mobility performance in HetNet with ABS at macro cells," 3GPP Standard Contribution (R2-122829), Prague, Czech Republic, May 2012.

19. Alcatel-Lucent, "UE speed-based methods and mobility state estimation for improving the mobility performance in HetNets," 3GPP Standard Contribution (R2-122813), Prague, Czech Republic, May 2012.

20. "User equipment (UE) procedures in idle mode," 3GPP Tech. Rep., TS 36.304 v.12.3.0, Sep. 2014.

21. J. Puttonen, N. Kolehmainen, T. Henttonen, and J. Kaikkonen, "On idle mode mobility state detection in evolved UTRAN," in *Proc. IEEE Int. Conf. Inform. Technol.: New Generations*, Las Vegas, NV, Apr. 2009, pp. 1195–1200.

22. R. M. Europe, "Summary of email discussion [77bis#32] LTE/HetNet mobility: MSE," 3GPP Standard Contribution (R2-122474), Prague, Czech Republic, May 2012.

23. NTT DOCOMO, "Enhanced mobility state estimation by Doppler frequency measurements," 3GPP Standard Contribution (R2-124007), Qingdao, China, Aug. 2012.

13

DORMANT CELLS AND IDLE MODES

13.1 INTRODUCTION

While the telecommunication industry is considered as the backbone of the global economy, it also represents an environmental challenge. A report from Ofcom suggests that information and communications technologies account for 2% of global CO_2 emission with 0.7% contribution from mobile and fixed communication devices [1]. As an example, British Telecom consumes 0.7% of all electricity usage in the United Kingdom, which represents 1.7% of all industrial consumption [2].

With more legislation being introduced to protect the environment such as the Carbon Reduction Commitment Energy Efficiency Scheme in the United Kingdom, service providers are seeking for solutions to improve their green credentials. Moreover, considering that the electricity bill accounts for 20–30% of the network operational expense [3], there is a strong motivation for operators to become more energy-efficient.

In addition, while cost and performance remain the main purchasing decision metrics, the growing public awareness of environmental issues is expected to change the behavior of consumers toward companies with improved sustainability credentials. According to a survey [4], more than 30% of their business consumers stated that the sustainability credentials of their telecommunication suppliers were a significant influence on spending.

Small Cell Networks: Deployment, Management, and Optimization, First Edition. Holger Claussen, David López-Pérez, Lester Ho, Rouzbeh Razavi, and Stepan Kucera.
© 2017 by The Institute of Electrical and Electronic Engineers, Inc. Published 2017 by John Wiley & Sons, Inc.

Considering the above motives, the adoption of solutions for reducing energy consumption becomes even more likely if that solution can simultaneously address other performance-related concerns of the telecom operators. Owing to the ever-increasing data traffic demand in today's mobile networks, small cells are considered by operators as an effective solution to improve the network's capacity through spatial reuse of wireless resources. Small cells also have a high potential for improving the energy efficiency of the network. This chapter presents a framework that is used to quantify the energy saving gains of deploying small cells using a parametric model. Moreover, two solutions are presented to reduce the energy consumption of small cell base stations (BSs) by effectively enabling idle mode procedures for such BSs.

13.2 POTENTIAL ENERGY SAVING GAINS OF DEPLOYING SMALL CELLS

In this section, the potential energy saving gains of deploying small cells are quantified by using a parametric power consumption model.

The energy saving gain refers to the reduction of the total network-wide energy consumption when deploying a heterogeneous network consisting of both macrocell and small cell BSs compared to a similar network in terms of capacity and coverage which only consist of macrocell BSs.

The analysis presented here is more related to urban and dense urban deployments where small cell BSs are most likely to be used. Moreover, the effect of future growth of the traffic demand is considered.

13.2.1 Traffic Demand and Base Station Capacity Models

In order to quantify the energy consumption of the network, a realistic distribution map of the traffic demand is required.

Figure 13.1 illustrates the normalized traffic demand map of the urban area of Wellington, New Zealand, which is used as a reference for the rest of the analysis presented in this section. Moreover, the year-to-year growth of the traffic demand is assumed according to the model presented in [5]. To capture the effect of the traffic variations during the day time, the model developed by the EARTH project is used [6]. Figure 13.2 illustrates the average intra-day variations of the traffic demand according to this model. Considering such temporal variations is essential for quantifying the total energy consumption of the network.

In addition, macrocell and small cell BSs are associated with a maximum throughput quantity denoted by $T_{\max}^{m}$ and $T_{\max}^{sc}$, respectively. Evidently, a BS will not serve traffic demand that exceeds its maximum capacity.

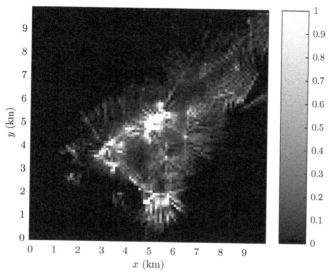

Figure 13.1. Normalized traffic demand map of Wellington, New Zealand.

The ratio of the maximum throughput of small cell BSs to their corresponding macrocell counterparts is parametrized as ζ. Theoretically, a small cell BS can provide the same throughput as a macrocell, implying $\zeta = 1$. However, to provide further insights, variations of ζ have been studied in the presented analysis.

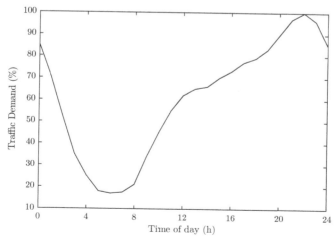

Figure 13.2. Intra-day variations of the traffic demand.

13.2.2 Placement of Base Stations

For the deployment of macrocell BSs, an algorithm was developed with the objective of finding the minimum number of BSs that can satisfy both coverage and capacity constrains. This implies that at least 98% of the total traffic demand is served by the BSs and the distance between the user equipments (UEs) and their serving BSs does not exceed 1150 m [7].

For the placement of macrocell BSs, a heuristic algorithm was used which starts from the lowest possible number of BSs and continues to insert more BS iteratively until the coverage and capacity requirements are satisfied [8].

Since the analysis concentrates on power efficiency performance, a greenfield deployment is considered where BSs were nonexistent before. This green field considers the set of all location bins in the map, U, and a set of potential macrocell BSs, B, with the maximum pilot transmit power of PP_{max}. Moreover, the following notation is used to describe the steps of the placement algorithm for macrocell BSs:

- u_x and u_y denote the coordinates of location bin $u \in U$.
- b_x and b_y denote the x-coordinates and y-coordinates of macrocell BS $b \in B$.
- $PP_b < PP_{max}$ is the pilot power of BS b.
- $S_b \subseteq U$ is the set of location bins at which the received pilot signal power from BS b is the strongest.
- C_b is a subset of S_b where members of C_b also satisfy the maximum distance criteria. In other words, $\forall u \in C_b : u \in S_b \wedge \text{distance}(u, b) < 1150$ m, where distance (u, b) represents the distance between location bin u and BS b.
- T_u is the traffic demand corresponding to bin location u, and
- T^b is the sum of traffic demand served by BS b, and can be expressed as

$$T^b = \max\left(\sum_{\forall u \in C_b} T_u, T^m_{max} \right).$$

The macrocell BS placement algorithm consists of the following steps:

- **Step 1:** Calculate the theoretical minimum number of macrocell BSs to accommodate 98% of the total traffic demand.

$$N_{min} = \left\lceil \frac{0.98 \sum_{\forall u \in U} T_u}{T^m_{max}} \right\rceil$$

- **Step 2:** Set the iteration counter, I_c, to zero and place each BS randomly with a pilot power also chosen uniformly at random.

- **Step 3:** Considering $k > 1$ as a constant design parameter, update the location of all BSs as follows.

$$\forall b \in \mathcal{B},\ b_x \leftarrow \frac{\sum_{\forall u \in C_b} T_u u_x + k \sum_{\forall u \in S_b \setminus C_b} T_u u_x}{\sum_{\forall u \in C_b} T_u + k \sum_{\forall u \in S_b \setminus C_b} T_u}$$

and

$$\forall b \in \mathcal{B},\ b_y \leftarrow \frac{\sum_{\forall u \in C_b} T_u u_y + k \sum_{\forall u \in S_b \setminus C_b} T_u u_y}{\sum_{\forall u \in C_b} T_u + k \sum_{\forall u \in S_b \setminus C_b} T_u}.$$

- **Step 4:** Calculate the average traffic demand of all BSs, T_{ave}, and update the pilot power of BSs respectively. T_{ave} is simply calculated as $T_{ave} = \sum_{\forall u \in U} T_u / |\mathcal{B}|$.
  Then considering the constant design parameter, $\sigma > 1$, update the pilot power of each BS:

$$\forall b \in \mathcal{B}: \begin{cases} PP_b \leftarrow PP_b\, \sigma & \text{if } \sum_{\forall u \in S_b} T_u < T_{ave} \\[2ex] PP_b \leftarrow \dfrac{PP_b}{\sigma} & \text{Otherwise.} \end{cases}$$

- **Step 5:** Check for the coverage and capacity criteria. In other words, if $\sum_{\forall b \in \mathcal{B}} T^b / \sum_{\forall u \in U} T_u < 0.98$, go to Step 8, otherwise go to Step 6.
- **Step 6:** Update the iteration count, $I_c \leftarrow I_c + 1$, and check if the maximum iteration count is reached. That is to say, if $I_c < I_c^{\max}$, go to Step 3, otherwise go to Step 7.
- **Step 7:** Add a new BS and go to Step 3.
- **Step 8:** Report the number of Macrocell BSs, $N_m = |\mathcal{B}|$, and the average normalized load of all macrocell BSs, L_m, where $L_m = \sum_{\forall b \in \mathcal{B}} T^b / N_m T_{\max}^b$.

There are two design parameters in the algorithm, namely k and σ. Here, $k > 1$ defines the weighting between the coverage and capacity considerations in locating the macrocell BSs. Higher k values grant more weight toward coverage (weighting toward those bin locations that are unreachable by their serving BSs). On the other hand, $\sigma > 1$ is used to balance the speed of convergence and the step size of the BSs pilot powers, where higher σ results in faster convergence but simultaneously coarser power adjustments. Placement of small cell BSs follows a simpler procedure, in which small cell BSs are placed in hotspot locations where the traffic demand is highest. It is assumed that each small cell BS can cover a bin in the demand map implying a coverage radius of approximately 60 m.

After placing a small cell BS, the demand map is updated by taking into account the offloading effects of the previous BS, and the next BS will be then placed into the scenario. Since a greenfield deployment consisting of both macrocell and small cell BSs is considered, small cell BSs are deployed first at the hotspot locations. Afterward, the traffic demand map is updated, and then the algorithm for the placement of macrocell BSs is used to determine the minimum number and the associated average load of macrocell BSs.

13.2.3 Power Consumption Model

The power model considered consists of two components, namely a fixed element which refers to the no load power consumption, P_n, and a load-dependent component [6]. The maximum power consumption refers to the case when the BS is fully loaded. P_f denotes the power consumption in the fully loaded condition. A linear relationship between the power consumption and the BS load is assumed.

Figure 13.3 illustrates the power consumption model as described. Subsequently, P_n^m and P_f^m refer to the no load and full load state power consumption of macrocell BSs, and P_n^{sc} and P_f^{sc} are the corresponding quantities for small cell BSs.

Moreover, since the load dependability of the power consumption is considered as an important aspect, γ is defined as a parameter to represent the ratio of P_n to P_f. Similarly, γ_m and γ_{sc} represent this ratio for macrocell and small cell BSs, and they are set to default values of 0.6 and 0.8 respectively [6]. Generally, lower values of γ

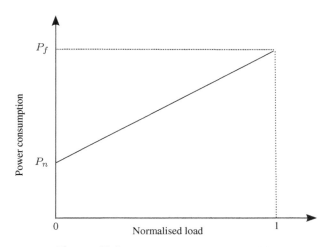

Figure 13.3.   Power consumption model.

imply a more energy-efficient BS, where the fixed power consumption component is less significant.

Finally, the ratio of P_f^m to P_f^{sc} is parametrized as α. Considering LTE as the wireless access technology, the authors of [6] provide a detailed breakdown of power consumption components both for macrocells and small cells, where the full load power consumption is 1350 W for a typical macrocell BS with three sectors and 14.7 W for a small cell BS. Subsequently, unless otherwise stated explicitly, these values are used in the presented analysis. Considering a network consisting of only macrocells, the evaluation of the total power consumption for a greenfield deployment includes the following steps. Given the maximum capacity of BSs and the traffic demand map, the minimum number and the average load of BSs that are sufficient to satisfy the coverage and capacity requirements are derived using the macrocell placement algorithm presented in the previous section. Then, the total network-wide power consumption of the macrocell tier, P_{total}^m can be then written as:

$$P_{total}^m = N_m \left(P_n^m + L_m \left(P_f^m - P_n^m \right) \right), \tag{13.1}$$

where N_m is the number of macrocell BSs and L_m is the average load of macrocell BSs. Similarly, the total power consumption of the small cell tier, P_{total}^{sc}, can be expressed as

$$P_{total}^{sc} = N_{sc} \left(P_n^{sc} + L_{sc} \left(P_f^{sc} - P_n^{sc} \right) \right), \tag{13.2}$$

where N_{sc} is the number of small cell BSs and L_{sc} is the average load of small cell BSs.

In a heterogeneous network consisting of both macrocell and small cell BSs, the total power consumption of the network, P_{total}^{all}, is the sum of the power consumption of the two tiers. Finally, the power improvement factor, PI (denoted as PI_m and PI_{sc} for macrocell and small cell BSs, respectively), is defined as the total ratio by which the overall power consumption of BSs is expected to improve over the years. This parameter will be helpful when introducing the "what...if" analysis on the power consumption of the network over the years.

13.2.4 Evaluation of Network-Wide Power Consumption

Figure 13.4 illustrates the ratio of the traffic demand being offloaded by small cell BSs as a function of the number of small cell BSs being deployed in the Wellington scenario. The general observation is that significant traffic offloading can be achieved by deploying a relatively small number of small cell BSs. This is mainly because small cells are deployed in hotspots where the traffic demand is high, and thus, despite their limited coverage area, their offload gain is significant. However, the curve starts to saturate as more small cell BSs are deployed. This is because the residual traffic demand becomes more homogeneous as the hotspots are being offloaded, which

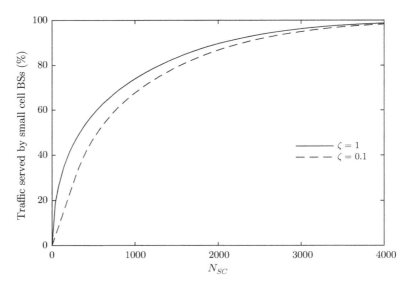

Figure 13.4.   Percentage of the traffic demand that is served by small cell BSs.

means that small cells with limited coverage may not have significant offload impac
anymore. The results are additionally shown for two values of ζ, namely 1 and 0.1
These values can be viewed as extreme cases, one implying the same level of capacity
provided by small cell BSs and a macrocell sector and the other one representing the
case where the small cell capacity is only 10% of that of a macrocell sector. Interest
ingly, there is not a significant difference between the two curves. This is because
due to their rather limited coverage, small cell BSs rarely require the same capacity o
a macrocell sector to serve the traffic demand of their coverage area. To better under
stand this point, Fig. 13.5 shows the average load of all small cell BSs as more BS
are deployed. The figure confirms the low average load of small cell BSs, especially
when more cells are deployed in the areas where the traffic demand may not be nec
essarily high. Such low average daily load of small cells also implies a short active
cycle of BSs, which is also intuitively expected due to their reduced coverage range
As a result of this, improving the idle mode power consumption of such BSs may
have notable impact on the overall energy consumption of the network when smal
cells are widely deployed.

 Accordingly, Fig. 13.6 shows the total network power consumption when
increasing the number of deployed small cells. The results are shown for two val
ues of γ_{sc}. As shown, when small cell BSs have a low γ_{sc}, an aggressive deploymen
results in improved energy efficiency. This is because with more small cells being
deployed, the number of macrocell BSs required to provide coverage and capacity
for the residual demand is reduced, and thus the overall network power consumption

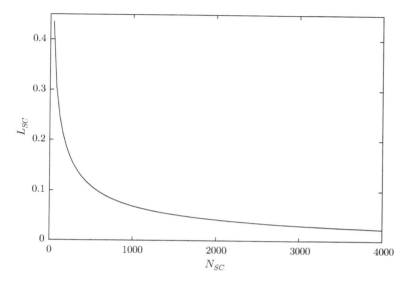

Figure 13.5. Average load of small cell base stations.

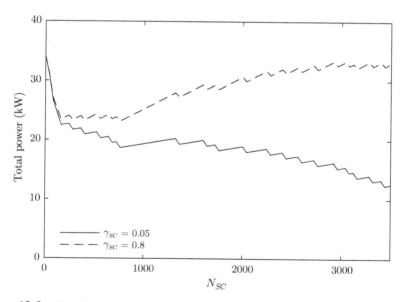

Figure 13.6. Total power consumption as a function number of small cell base stations deployed.

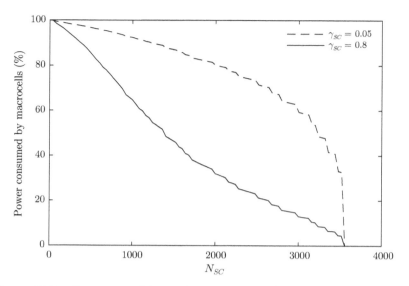

Figure 13.7. Percentage of the total power consumed by macrocell base stations.

is smaller. Note that this is the case when considering a greenfield deployment where there is no existing macrocell in the scenario. Even with less efficient small cell BSs where γ_{sc} is set to 0.8, there is a notable reduction in the overall power consumption after deploying few small cell BSs, as these small cells BSs will offload hotspots and thus some of the macrocells can be dismissed. However, the overall power consumption starts to increase after a certain point. This is because beyond this point, the offload gain of small cells starts to flatten (see Fig. 13.4), and adding more BSs only results in increased overall power consumption.

To gain an understanding of the power consumption contribution from macrocells and small cells, Fig. 13.7 shows the percentage of the power consumed by macrocell BSs. Again, the results are presented for two values of γ_{sc}. As it is shown, there is up to 50% difference between the contribution of the power consumption when γ_{sc} is varied from 0.05 to 0.8 and the same number of small cells is deployed. It is also interesting to note the sharp drop of the curve from 35% to 0% when $\gamma_{sc} = 0.05$, meaning that no macrocell BS is required in the network. The reason for this observation is that even with very few remaining macrocell BSs and when the number of small cells is nearly 3500, the contribution of the power consumption of those few macrocells compared to the more power efficient small cells are still considerable (nearly 35%).

Another important parameter which impacts the optimal combination of macrocells and small cells in a mixed deployment is the relative overall power consumption of macrocell BSs to that of the small cell BSs, which is represented by α. Figure 13.8

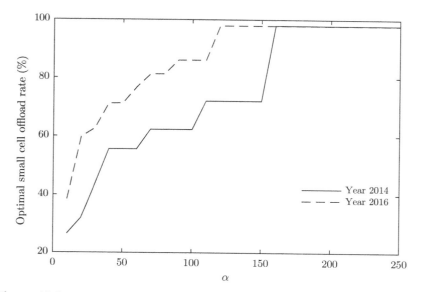

Figure 13.8. Optimal offload rate of small cells as the function of the α parameter.

shows the optimal small cell offload rate for different values of α. Based on the values reported in [6] for exiting long-term evolution (LTE) BSs, the current value of α is approximately 90 when considering outdoor deployed BSs. Obviously, higher α results in an increased number of small cells required to achieve the optimal mixed configuration. Moreover, there is a stronger tendency toward small cells when the traffic demand is higher. For example, in 2016 and with α set to 75, more than 80% of traffic demand is to be served by small cells. However, this quantity is just above 60% in 2014.

Finally, to evaluate the combined effect of all parameters, Fig. 13.9 shows the total power consumption for a set of mixed deployments and compares that to the macrocells-only scenario. Figure 13.9 shows very significant power reduction gains when considering mixed deployments compared to the marocell-only scenario. For example, when PI is set to 50% for both macrocell and small cell BSs and γ_m and γ_{sc} are set to 0.2 and 0.05, respectively, a power reduction gain of 46 fold is observed for 2016 when compared to the baseline macrocell-only scenario.

The results presented in this section clearly highlight the fact that migrating to a heterogeneous network architecture consisting of both macrocell and small cell BSs can significantly enhance the energy efficiency of the network.

While the analysis presented here has been mainly focused on the downlink and the power consumption of the network, the effect of data offloading from macrocells combined with the reduced transmission distance in small cells provides further energy savings in the uplink direction resulting in longer UE battery life.

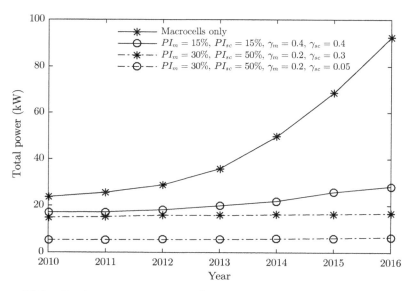

Figure 13.9. Total power consumption of the network over the years and for different deployment configurations.

Moreover, the results confirm that the active duty cycle of small BSs is rather short, hence, improving the energy consumption of small cells in the idle mode is expected to have high energy saving rewards.

13.3 DISTRIBUTED IDLE MODE PROCEDURE FOR SMALL CELL BASE STATIONS

With the expected wide-spread deployments of small cells, the power efficiency of these cells in idle mode becomes an important aspect. Not only that, but also from the performance and interference perspective, muting the small cell BSs when they are not serving a UE is beneficial. More specifically, the continuous pilot signal transmission of small cell BSs causes interference to neighboring cells, even when the BS is not serving any UE. Although the pilot signal transmit power is adjusted to only around 10 percent of the maximum BS transmit power, it can still be detrimental, particularly in dense urban deployments.

In this section, an idle mode procedure for small cell BSs is presented, that detects active UEs within range of a BS based on a rise in the measured uplink noise level. The procedure allows disabling the pilot transmissions and most processing operations when they are not needed to support an active call.

13.3.1 Algorithm Description

As described earlier, the small cell BS that is located within the coverage of a macro-cell disables its pilot transmissions and the associated radio processing when no active calls are being made by UEs.

This is achieved by a low-power sniffer capability in the BS that allows the detection of an active call from a UE to the underlying macrocell. When a UE located inside the coverage range of the BS makes/receives a call to/from the macrocell, the sniffer detects a rise in the received power on the uplink frequency band.

If this noise rise exceeds a predetermined threshold, the detected UE is deemed close enough to be potentially covered by the small cell BS. The rise in the noise floor is easily detectable since the UE transmits at high power to the macrocell while it is in close proximity to the small cell BS [9].

This technique allows the small cell BS to switch off all pilot transmissions as well as the processing associated with wireless reception, when no UE is involved in an active call. The hardware components that are required to keep the BS connection active on the core network remain in place, along with the noise rise sniffing elements. Once it is in active mode, the UE reports the received pilot signal strength from the small cell BS to the macrocell to which it is connected, and the UE is then handed over from the macrocell to the small cell BS.

The noise rise detection threshold is an important metric for the performance of the algorithm. An incorrectly set detection threshold can hamper small cell usability by either not identifying calls from UEs or by being overly sensitive to sniffing for UEs that lie outside the coverage of the small cell BS. Moreover, the detection threshold is dependent on the location of the small cell BS within the macrocell as well as on its own coverage profile. Therefore, it is important to pair automatic configuration of the detection threshold with the ability to fine tune it during operation.

One possible way of adjusting the noise rise detection threshold is through obtaining macrocell path-loss information from the UE. The transmit power required for a UE to set up a call with a macrocell depends on the path loss to the macrocell. During operation, a small cell BS can obtain typical path losses to the macrocell (provided that the macrocell transmit power is known) via UE measurements at different locations within its coverage. Then, based on the own path losses of small cells to the same location points, a suitable noise rise detection threshold can be set to sense a macrocell call from a UE from any location within the small cell coverage area.

For initialization purposes, a baseline measurement performed by the small cell can be used to set the detection threshold, which is a useful starting point due to the limited range of the small cell coverage. During operation, as UE measurements are obtained, this initial threshold value is fine-tuned accordingly. Figure 13.10 illustrates the operational flowchart of the small cell BS incorporating the idle mode procedure.

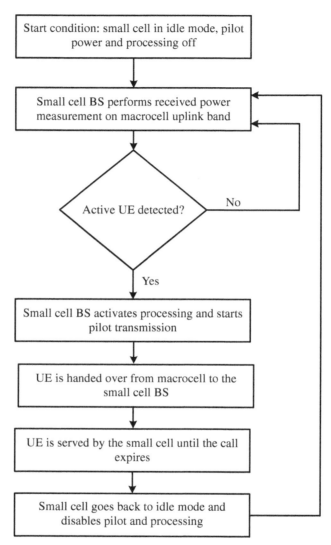

Figure 13.10.   Flowchart of distributed idle mode procedure for small cell base stations.

Note that this procedure requires macrocellular coverage since it relies on detecting transmissions from a UE to a macrocell. Therefore, the small cell BS needs to identify if sufficient macrocell coverage is available. This can be detected by measuring the macrocell pilot channels at the small cell BS and by UE measurement reports. Idle mode cannot be activated in the small cell BS without sufficient macrocell overlay coverage.

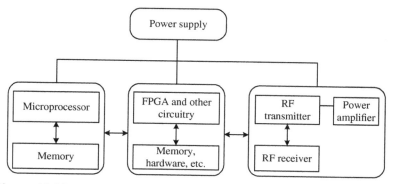

<u>Figure 13.11.</u> Schematic hardware design for a typical femtocell base station.

13.3.2 Numerical Evaluation

In this section, the potential energy saving gain of deploying the described distributed idle mode procedure is evaluated.

The presented case study considers residential femtocells as a type of small cell BSs.

Figure 13.11 shows a high-level schematic representation of a typical femtocell hardware design. It includes a microprocessor that is responsible for implementing and managing the standardized radio protocol stack and the associated base-band processing. One or more random access memory components that are required for various data handling functions are connected to the microprocessor. The design also contains a field-programmable gate array (FPGA) and some other integrated circuitry to implement a host of features, such as data encryption, hardware authentication, and network time protocol. The radio component within the FPGA acts as an interface between the microprocessor and radio frequency (RF) transceiver. There are separate RF components for packet transmission and reception, each consuming a certain amount of power. An RF power amplifier is also present to pass a high-power signal to the transmitting antenna.

Table 13.1 shows the energy consumption profile of the femtocell hardware components highlighted in Fig. 13.11. When fully active, the hardware circuits consume

TABLE 13.1 Energy Consumption Profile of Femtocell Hardware

Hardware Component	Energy Consumption (Watts)
Microprocessor and associated memory	2.2
FPGA-associated memory	2.5
Other circuitry	2
RF transmitter	1
RF receiver	0.5
RF power amplifier	2

a total of $P_{\text{Act}} = 10.2$ W, and with the power supply at 85% efficiency, the power drawn from the power socket amounts to 12 W. Now, let P_{Saved} be the power saved by switching off the hardware components in the femtocell BS idle mode. The power consumption in idle mode thus becomes $P_{\text{Idle}} = P_{\text{Act}} - P_{\text{Saved}}$. By denoting the average duty cycle of the femtocell η, the average reduction in the femtocell's power consumption can be characterized as a percentage of the total power:

$$\Omega = \left(\frac{P_{\text{Saved}}}{P_{\text{Act}}} (1 - \eta) \right) \times 100. \tag{13.3}$$

By switching to idle mode, the femtocell BS switches off the power amplifier, RF transmitter, RF receiver, and miscellaneous hardware components related to non-essential functions in the idle mode, such as data encryption and hardware authentication. However, a low-power radio sniffer ($P_{\text{Sniff}} = 0.3$ W) is switched on to perform received power measurements on the macrocell uplink band. Based on the hardware ratings provided in Table 13.1, this results in a power saving of 4.2 W, which corresponds to the upper bound on η equaling 41.2%.

In order to calculate the average duty cycle, $\hat{\eta}$, simulations were performed to model both voice and data traffic generation in femtocell-deployed households. In this model, the voice traffic is characterized by variable grades of usage during different time periods of the day.

The voice call generation by each UE is modeled as a Poisson process resulting in an exponentially distributed inter-arrival times with mean λ_v. Table 13.2 shows the categorization of residential femtocell voice traffic at different periods of the day and the corresponding mean inter-arrival times.

The data traffic model is based on the web browsing traffic model in [10] where each data session comprises a sequence of packets. The number of packets in a data session is modeled using the Poisson distribution with a mean of 30, and the packet inter-arrival time is modeled using the exponential distribution with a mean value of 20 seconds. In each session, a mean of 720 kB of data are downloaded. Using these parameters, each data session length varies between 5 minutes and 30 minutes, with a mean of 16 minutes.

Figure 13.12 shows the simulation results corresponding to the percentage of reduction in the power consumption, Ω when using the idle mode. As expected, Ω starts to decline with the increase of the number of UEs. One shortcoming

TABLE 13.2   Femtocell Voice Traffic Model

Time Period	Grade of Usage	$\lambda_v(1/\text{min})$	Traffic/user (erlang)
22:00–8:00	Low	1/240	0.0125
8:00–18:00	Medium	1/60	0.05
18:00–22:00	High	1/30	0.1

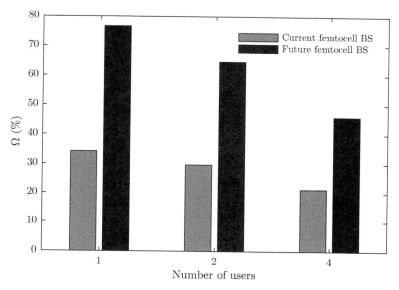

Figure 13.12. The percentage of reduction in the power consumption, Ω, when using the idle mode.

of the current femtocell hardware design described in Fig. 13.11 is the dependence of different hardware components on each other, which prevents switching off elements individually. For instance, the FPGA acts as a communication interface between the RF transceiver and the microprocessor and cannot be switched off independently.

A modular hardware design approach, which provides the flexibility of handling components individually, can offer a significant improvement over the current model. Moreover, with the ever-increasing processing capabilities, additional functions can be incorporated in the multi-core microprocessor, thereby removing the redundant portions of the hardware and driving significant savings in the operational energy expenses.

By focusing on the idle mode algorithm and leveraging the modular hardware design principle mentioned above, the low-power sniffer in conjunction with a control element can offer substantial energy saving benefits. The control element can be perceived as a low-power switching circuitry that can trigger the activation of the rest of the hardware components, upon the detection of a noise rise above the detection threshold by the sniffer.

In this scenario, it is reasonable to assume a femtocell power consumption of 0.8 W (including 0.3 W of the sniffer) in idle mode, which can increase the upper bound on energy reductions to 92%. Using this model, Fig. 13.12 additionally shows the power saving gains that can be achieved in future femtocell BSs.

13.4 CENTRALIZED IDLE MODE PROCEDURE FOR SMALL CELL BASE STATIONS

The decentralized idle mode procedure introduced in the previous section is simple, but suffers from a number of drawbacks. For example, since the approach is decentralized, it cannot take information on other cells into account. As a result, upon detection of an active UE, the small cell BS may wake up to serve the UE even if the underlying macrocell network is not fully loaded and can handle the call request. This becomes more problematic when considering the future k-tier overlapping networks. For example, if a call is within coverage of a femtocell, which is located within a picocell, which again is located within a metrocell, the UE communication to the macrocell may wake up all of them. Moreover, it would be difficult to adjust the threshold for waking up the small cells optimally. This is especially true since in many radio access technologies including wideband code division multiple access (WCDMA), the uplink transmit power of a UE would not only be a function of the absolute path loss to the serving macrocell BS, but also the macrocell load (i.e., noise floor) will impact the required transmit power on the uplink. Furthermore, depending on the timer settings, the BS may wake up for a passing by UE. In this case, in addition to a false wake up which impacts the overall power efficiency, there will be the overhead of handover signaling and increased risk of call drop as a result of premature handovers. Finally, deeper sleep modes can be realized if the small cell BS does not need to use the sniffer for monitoring potential UEs.

A centralized idle mode procedure can address many of the shortcomings of the decentralized approach. In this scheme, as the name implies, small cell BSs can enter deep sleep modes and a centralized entity is responsible for waking them up, when it is necessary. In doing so, the controller can account for the existing load level of different tiers of the network, thus can make a more informed decision. However, the main challenge here is to estimate which UE will fall into the coverage range of which small cell BS.

13.4.1 Algorithm Description

In this section, a centralized idle mode procedure scheme where the macrocell BS acts as the central entity is introduced.

The procedure consists of the following steps.

Considering all small cell BSs being in sleep mode, the serving macrocell BS evaluates if there is a need for offloading some of the traffic to any of the small cell BSs.

If the answer is positive, the macrocell BS tries to identify a set of UEs that are in the coverage area of small cell BSs. Upon successfully identifying those small cell BSs, the macrocell BS sends a wake up signal to those cells (using the X2 interface for example).

From this point, the procedure is similar to the distributed approach in a way that upon activation of the target small cell BS, the UE is handed over to the small cell until the call expires or the UE moves out of the small cell coverage.

The critical part of the algorithm is the task for the macrocell BS to identify if a UE is in the coverage area of any of the small cells, and if that is the case, which small cell BS. This task is closely related to geo-locating UEs.

While there exist different approaches to the problem, a practical, radio access technology agnostic and standard compliant solution based on RF finger-printing is introduced here. Formally, RF fingerprinting is defined as the process which is used to identify the locations of UEs from the characteristics of their radio signal(s) received from a number of BSs including the serving cell. Received pilot signal strength is considered as the most common feature for RF fingerprints and is subsequently used in the presented scheme.

As a routine part of their operation, UEs periodically send measurement reports indicating the received signal strength of their serving and neighboring cells to their serving BSs. In addition, a BS can explicitly request UEs to perform and report measurements where it can additionally specify the measurements' characteristics. These measurement reports are critical for making handover decisions (see Section 12). This implies that collecting RF-fingerprints from UEs does not require additional or nonstandard operations and overheads.

RF fingerprints have been frequently used to provide location information. Generally, this procedure can be decomposed into two steps where first the RF fingerprints with appropriate features are collected and associated to known geographical coordinates, and then a machine learning-based classifier estimates the coordinates of other measurement points.

Similarly, in the context of using RF fingerprints for waking up small cell BSs, the macrocell BS collects a set of RF fingerprints where the association of the RF fingerprints to the serving cells is known. In other words, the macrocell BS knows which RF fingerprint corresponds to which serving cell, be it the macrocell BS itself or any of the overlaid small cells.

This set can be constructed by exchange of information between the small cells and the macrocell. This data set will then be used as the training set to estimate the possible serving cells of other RF fingerprint observations. Once the size of the training set is sufficiently large, small cell BSs can enter sleep mode.

From this point, the exploitation phase starts where the macrocell BS uses this training set to decide if a UE with a given RF fingerprint falls into the coverage area of any of the small cell BSs or not. This is known as the supervised classification task in the machine learning terminology. Moreover, the process of refining the training set continues during the operation. For example, in case of a false wake-up signal where the small cell BS wakes up and finds no active UE in its coverage area, the macrocell BS can update the training set by including the given RF fingerprint in the training set for future references. Figure 13.13 shows the steps involved in the centralized idle mode procedure.

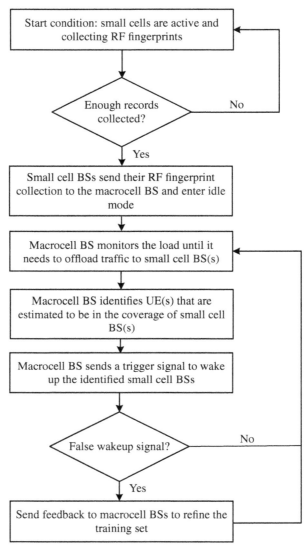

Figure 13.13. Flowchart of centralized idle mode wake-up procedure for the small cell base stations.

Now considering the above steps in more detail, the macrocell BS first needs to decide at what point the training set is constructed sufficiently in order to be used for classification of new observations. This can be known using cross-validation tests. The cross-validation is a method for evaluating the robustness and accuracy of a statistical model. A round of cross-validation consists of dividing the training set into two complementary subsets where one is used for training and the other subset is used for testing and validation.

Moreover, to reduce variability, multiple rounds of cross-validation are performed using different partitions, and the validation results are subsequently averaged. Considering misclassification rate as the measure of fitness, the macrocell BS can perform a cross-validation test and once the misclassification falls below a certain threshold and the results are statistically significant, the training set can be used for classifying unseen samples.

Another important aspect is the choice of the classification technique. Broadly speaking, supervised learning classification techniques can be categorized into two groups of deterministic and probabilistic methods.

Deterministic methods estimate the class of an observation by directly considering observation values, whereas probabilistic methods consider observations as part of a random process. Considering the incompleteness of the fingerprint vectors in practice, fitting probabilistic distributions to the observation data is likely to be inaccurate.

Therefore, a class of deterministic classifiers known as k-nearest neighbors (k-NN) can be used for the classification of the RF fingerprints to identify their relevant serving BSs, when small cell BSs are in the idle mode. k-NN methods are computationally simple, which makes them suitable for dynamic updates in classification decisions. The method simply finds the k-most similar observations to the test data based on a distance metric, d. A typical nearest neighbor method uses the pair-wise Euclidean distance to define the similarities. Moreover, the following practical issues should be taken into account when using the algorithm:

- RF fingerprint measurements by the UEs could be noisy and some measurement points may be dropped at random by the device.
- The initial training set might have been constructed from biased samples and may not be representative of the entire scenario.
- The number of visible neighboring BSs might be limited especially during early stages of the deployment.

In addition to the classification decision, the classifier can be modified to additionally return a level of confidence behind the classification predictions. This will be very helpful to cope with scenarios when the training set consists of biased samples and when the classifier has difficulty matching a given RF fingerprint observation against the training set. In these cases, the certainty level associated with the classification decision is expected to be low. Therefore, the macrocell BS can treat such classification decisions as inaccurate and discard them. For example, if a UE is identified to be in the coverage of a given small cell BSs with a low certainty level, the macrocell BS may decide not to trigger that small cell BS to wake up. Even if the small cell BS is triggered and understands that no active UE is located in its coverage area, this information can be fed back to the macrocell BS to be incorporated into the training set. Finally, the issue of the variable number of visible cells can be addressed by applying a two-stage classification where in the first stage a Jaccard similarity metric [11] is used to filter RF fingerprints with similar visible cells and then the k-NN classification is applied.

13.4.2 Numerical Evaluation

To examine the practical feasibility of the described centralized idle mode procedure, RF fingerprint measurements were collected from the city center of Dublin, Ireland. Figure 13.14 shows the path where fingerprints were collected.

For the purpose of evaluation of the algorithm in the absence of actual deployed small cells in the scenario, the RF fingerprints consisting of the received signal power from a set of all visible neighboring macrocells were collected along with the global positioning system (GPS) coordinates corresponding to each RF fingerprint.

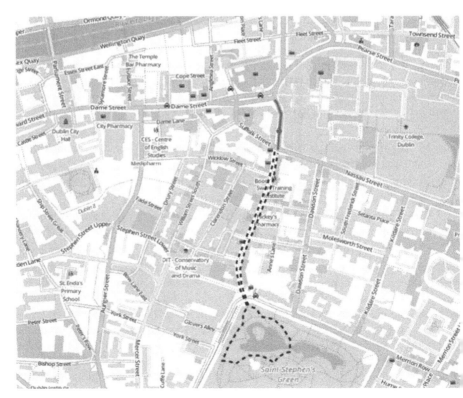

Figure 13.14. Fingerprint measurement path in the city center of Dublin, Ireland.

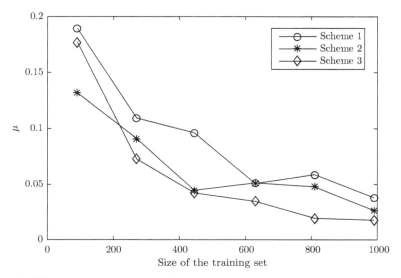

Figure 13.15. The average misclassification error rate, μ, against the number of RF fingerprints available in the training set.

Subsequently, in the post-processing stage, one can consider the deployment of small cell BSs in given locations and subsequently can estimate the coverage range of those BSs. With this information, the RF fingerprints can now be associated to the relevant serving cells, which is indeed equivalent to labeling the data. For example, given a particular location of a small cell BS and the estimated coverage range of that BS, one can identify whether a particular RF fingerprint record with the given GPS coordinates would fall into the coverage of the said small cell BS or not. Now, having all RF fingerprints labeled, similar to the cross-validation test, one can divide the set into the training and test subsets in order to evaluate the accuracy of the classification model.

Using the 10-fold cross-validation test, Fig. 13.15 shows the average misclassification error rate, μ, against the number of RF fingerprints available in the training set.

- **Scheme 1:** The search is conducted over the entire training set.
- **Scheme 2:** The search is conducted over samples of the training set that have at least one common neighboring BS with a given RF fingerprint record.
- **Scheme 3:** The search is conducted using the proposed two stage search scheme where relevant samples in the training set are first identified using the Jaccard similarity metric.

From Fig. 13.15, it is apparent that the misclassification rate starts to decrease with the increase in the number of observations in the training set. More importantly,

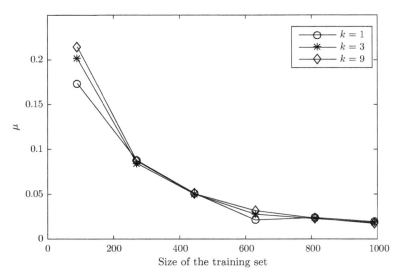

Figure 13.16. The misclassification performance of k-NN method with different value of k.

the results confirm the feasibility of the centralized idle mode procedure since th. misclassification rates are at acceptably low levels. Indeed, few false wake up trigger may not significantly impact the network, and furthermore, the continuous learnin; will result in enhancing the performance of the proposed algorithm over time. More over, in Fig. 13.15, the performance of the three classification schemes is shown From the results, it is apparent that the proposed two-stage classification method i more efficient. Considering this proposed classification method, Fig. 13.16 shows th. misclassification performance for different values of k in the k-NN method. Again while the results are more or less similar, the figure suggests that choosing highe k values may results in increased misclassification as the classifier may additionall; consider more training samples, which may not necessarily be relevant.

13.5 SUMMARY AND CONCLUSIONS

Energy consumption and environmental sustainability of small cells will become pressing issue in the future. In addition to the environmental factors, operators con tinuously seek to find solutions to improve their cost base by minimizing their opera tional expenditure (OPEX). This is particularly important as operators need to expan. their network infrastructure to accommodate the ever-increasing growth of the traf fic demand in the coming years. Given that the electricity consumption accounts fc nearly 30% of the network OPEX, and that small cell BSs are expected to play central role in offloading the increased traffic demand, the energy efficiency of such small cell BSs becomes an important aspect.

In addition to the energy consumption aspects, the continuous pilot signal transmission of small cell BSs causes interference to their neighboring cells, even when no active UE is being served. The issue becomes more pronounced when considering future dense deployments of small cells.

The analysis presented in this chapter suggests that compared to a network architecture consisting of only macrocell BSs, migrating to a heterogeneous network model consisting of both small cell and macrocell BSs results in significant energy saving gains. This is indeed due to the fact that small cell BSs in general have similar capabilities as their macrocell counterparts in terms of capacity, but have rather a smaller coverage range and energy consumption.

Given that the spatial distribution of the traffic demand is extremely uneven (especially in urban and dense-urban areas), where traffic hotspots account for a major bulk of the overall demand, small cell BSs can be very beneficial if deployed optimally. Since the power consumption of these small cell BSs is at a much lower level to that of macrocell BSs, a heterogeneous network architecture can provide significant power efficiency enhancement. Since the traffic demand varies significantly over hours of a day, appropriate idle mode procedures can further improve the energy efficiency of the network significantly.

Two approaches for enabling idle mode procedures were presented in this chapter. The first scheme was a distributed method which is implemented in the small cell BSs. The main principle here is that, when serving no active UE, the small cell BSs enter the idle mode where it can switch off most of the radio parts and the processing circuitry. The BS will then attempt to detect the presence of active UEs in its coverage vicinity. Upon detection of such an active UE, the BS wakes up by reactivating the radio parts and processing circuitry. The UE will then be handed over to the small cell BS where it is served until the call expires or the UE moves out of its coverage area. Detection of active UEs is based on monitoring the uplink band, and once the detected energy on this band exceeds a predefined threshold, the presence of an active UE is presumed.

In spite of its simplicity and practicality, the distributed idle mode procedure suffers from a number of shortcomings. First of all, configuring the UE detection threshold optimally may not be a straightforward task since the UE uplink transmission power depends on the number of factors including the load of the cell. This uncoordinated scheme may also result in several small cells waking up to serve the same UE. Additionally, there are scenarios where the underlying macrocell BS may well be capable of handling the traffic demand without the need for assistance from the small cell BSs.

In order to address this issues, a centralized idle mode procedure was introduced, which takes into account a number of criteria that can be defined by the operator for deciding whether or not and which small cell BSs to activate. In this scheme, small cell BSs enter the idle mode and are only activated upon receiving a triggering signal from the macrocell BS. It is then up to the macrocell BS to determine which active UE is located in the coverage area of which small cell BS (if any), and if any small cell BSs should be activated. It was shown that this is feasible by constructing a training database consisting of RF fingerprints measured by UEs when small cells are all active. Subsequently, using a machine learning classification scheme, the macrocell BS can identify if a UE with a given RF fingerprint record can be served by any of the small cell BSs that are in idle mode. If the macrocell BS decides to benefit from small cells offloading assistance and can identify UEs that are estimated to be under the coverage of small cells, a wake-up message is sent to those small cell BSs, and the UEs are handed over to the small cell BSs.

REFERENCES

1. "Understanding the environmental impact of communication systems," Ofcom White Paper, Tech. Rep., Apr. 2009.
2. "The line goes green," Delloite, White Paper, Tech. Rep., May 2010.
3. I. Ashraf, F. Boccardi, and L. Ho, "Sleep mode techniques for small cell deployments," *IEEE Commun. Mag.*, vol. 49, no. 8, pp. 72–79, Aug. 2011.
4. "Green quadrant: Sustainable telecoms 2011," Verdantix Report, Tech. Rep., Feb. 2011.
5. D. Kilper, G. Atkinson, S. Korotky, S. Goyal, P. Vetter, D. Suvakovic, and O. Blume, "Power trends in communication networks," *IEEE J. Sel. Topics Quantum Electron.*, vol. 17, no. 2, pp. 275–284, Mar. 2011.
6. G. Auer, O. Blume, V. Giannini, I. Godor, M. Imran, Y. Jading, E. Katranaras, M. Olsson, D. Sabella, P. Skillermark and W. Wajda, "D2. 3: Energy efficiency analysis of the reference systems, areas of improvements and target breakdown," INFSOICT-247733 EARTH (Energy Aware Radio and NeTwork TecHnologies), Tech. Rep, 2010.
7. R. Razavi and H. Claussen, "Urban small cell deployments: Impact on the network energy consumption," in *2012 IEEE Wireless Commun. Netw. Conf. Workshops (WCNCW)*, Apr. 2012, pp. 47–52.
8. H. Claussen, "Autonomous self-deployment of wireless access networks," *Bell Labs Tech. J.*, vol. 14, no. 1, pp. 55–71, Spring 2009.
9. H. Claussen, I. Ashraf, and L. T. Ho, "Dynamic idle mode procedures for femtocells," *Bell Labs Tech. J.*, vol. 15, no. 2, pp. 95–116, Sep. 2010.
10. ETSI, "101 112 selection procedures for the choice of radio transmission technologies of the UMTS (UMTS 30.03 version 3.2. 0), v 3.2. 0," 1998.
11. H. Seifoddini and M. Djassemi, "The production data-based similarity coefficient versus Jaccard's similarity coefficient," *Comput. Ind. Eng.*, vol. 21, no. 1, pp. 263–266, 1991.

PART V

SMALL CELL DEPLOYMENT

14

BACKHAUL FOR SMALL CELLS

14.1 INTRODUCTION

The deployment of small cells can provide very large gains in capacity due to the higher spatial reuse and the better propagation conditions owing to the base stations (BSs) being located closer to the user equipment (UE). While this helps to solve the capacity problems on the access side, it introduces many challenges with respect to the provision of backhaul for these small cells.

Described simply, the purpose of backhaul is to provide connectivity between the small cells and the core network nodes with a desired quality of service (QoS) level. Figure 14.1 shows a high-level view of a small cell backhaul network, where the last mile link involves the connection of the small cell to a point of presence (POP).

A POP is defined here as a central access point, where the traffic from different small cells is aggregated and then sent to the core network through a transport network. In a large number of cases, small cells are deployed as overlays on top of macrocells in a HetNet configuration, either to provide high capacity in UE traffic hotspots or coverage in black spots. In this case, backhaul infrastructure is already in place to provide connectivity for macrocell BSs back to the core network.

The operator can take advantage of this by connecting small cell BSs directly to the macrocell sites, that is, the POP is co-located at the macrocell site. Alternatively, the macrocell sites can be used to provide backhaul connectivity to dedicated remote

Small Cell Networks: Deployment, Management, and Optimization, First Edition. Holger Claussen,
David López-Pérez, Lester Ho, Rouzbeh Razavi, and Stepan Kucera.
© 2017 by The Institute of Electrical and Electronic Engineers, Inc. Published 2017 by John Wiley & Sons, Inc.

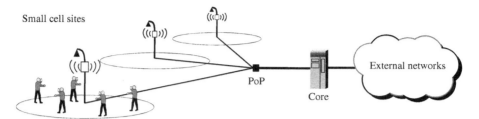

Figure 14.1. High-level backhaul architecture.

POP, that is, the POP is not located at the macrocell site. In this case, the aggregated traffic from the POP is sent back to the macrocell site, or if this option is not available or practical, through an alternative transport network.

Currently, well-known solutions used for macrocell backhaul such as line-of-sight (LOS) microwave wireless links, and fiber or copper wired connections are being adapted to meet the needs of small cell backhaul. However, small cell deployments have different requirements to macrocell deployments.

In order to maximize their efficiency, small cells have to be deployed as close as possible to targeted traffic hotspots. This means that small cell BSs require a lot of flexibility in terms of where they can be placed, such as on street furniture and building facades. In these locations, the availability of wired backhaul such as fiber or copper connections are often not present, and providing new connection points to these locations can be relatively costly and requires a long lead time to implement as it involves the laying of cables.

Additionally, small cell BSs are typically mounted at street level 3 to 10 m above ground, which is below building rooftops. This significantly reduces the probability of small cell sites having an unobstructed LOS to existing macrocell sites or a remote backhaul POP, making the use of high-capacity LOS microwave more difficult. Due to their smaller coverage areas and higher numbers, small cells are also much more cost-sensitive, and need to be easier to deploy and more scalable compared with traditional macrocell backhaul solutions.

In order to enable widespread small cell deployments, new technologies as well as planning and optimization techniques are required. This chapter focuses on the requirements and challenges for outdoor small cells, as this scenario introduces new challenges for backhaul provisioning.

14.2 WIRELESS BACKHAUL

There is a wide array of solutions available with different advantages and disadvantages with respect to coverage, capacity, and costs, which will have a significant impact on their suitability in different scenarios.

14.2.1 Description of Wireless Backhaul Options

Wireless backhaul solutions can be divided in NLOS and LOS solutions:

Non-line-of-sight Wireless Backhaul   Non-line-of-sight (NLOS) backhaul systems operate in carrier frequencies below 6 GHz in order to take advantage of favourable propagation characteristics. They are usually orthogonal frequency division multiplexing (OFDM)-based to cope with multi path fading, and utilize 10 or 20 MHz channel bandwidths.

NLOS backhaul systems typically have a point-to-multi-point (PMP) configuration as illustrated in Fig. 14.2, although point-to-point (PTP) configurations are also an option that can be used. This configuration involves the use of a hub module that is connected to the core network, and serves as the POP for small cell BSs. This means that the hub module has to be placed in a location where a backhaul connection to the core network is available, for example, through fiber or high-capacity LOS wireless solutions. This can either be at an existing macrocell site, or any other location where backhaul to the core is available.

The antenna of the hub module is typically deployed at rooftop level in order to increase the chances of having better propagation conditions to the small cell BSs and cover a plurality of them. The hub module can be placed at locations lower than the rooftop level, but this would obviously result in higher path losses and worse propagation conditions, resulting in a smaller possible range between the hub and small cell BSs.

The hub module communicates with a number of small cell BSs that are within its range. This is usually done using a sector widebeam antenna similar to those used at macrocells. In a similar approach to sectorized macrocell BS configurations, multiple

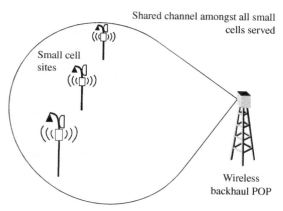

Figure 14.2.   Point-to-multipoint configuration using NLOS wireless backhaul.

hub modules can be deployed at a single site using a three or more sectors in order to serve more small cell BSs.

Alternatively, other antennas such as more advanced beamforming array antennas can be used. Using these array antennas, narrowbeams can be pointed in order to obtain higher antenna gains and lower interference, resulting in higher performance.

Point-to-Point Wireless Backhaul   PTP wireless backhaul is a well-established solution that is currently widely used to provide backhaul for macrocell BSs. It utilizes carrier frequencies that are higher than 6 GHz.

At such frequencies, a clear LOS is typically required between the small cell BS and the POP, since blockage by obstacles may significantly reduce the signal strength. At lower frequencies, near-line-of-sight propagation is possible.

LOS PTP solutions rely on the use of high-gain narrowbeam antennas. Higher frequency bands have the added advantage of having physically smaller antennas, since the required antenna aperture size becomes smaller with higher frequencies (assuming no variations in antenna efficiency, the size is inversely proportional to the square of the frequency [1]).

Examples of PTP wireless backhaul that have been put forward for small cell backhaul include PTP microwave systems that operate typically between 18 to 42 GHz, and 60 GHz mm-wave solutions that operate in unlicensed bands. Because of their LOS propagation requirement, these wireless backhaul solutions that use high-frequency carriers will be collectively called LOS backhaul solutions. Figure 14.3 shows an example of a PTP LOS link, where a small cell BS is connected directly to the POP.

Because of their LOS requirement, PTP LOS links can be linked together in order to establish a connection between the small cell BS and the POP. When

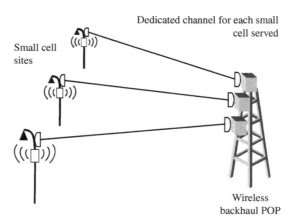

Figure 14.3.   Point-to-point line-of-sight wireless backhaul link.

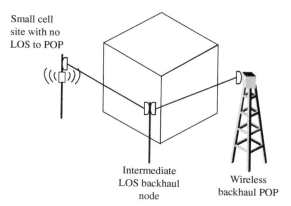

Figure 14.4. Small cell BS connected to POP through an intermediate node.

a LOS connection to the POP is not available, two available options can be considered:

- Hop scenario: This scenario, as illustrated in Fig. 14.4, involves linking a small cell BS with the POP through intermediate locations using PTP LOS connections. The links involved can have the same capacity, since it is used to carry the traffic of one small cell.
- Daisy chain scenario: This scenario, as illustrated in Fig. 14.5, involves linking a small cell BS with the POP through secondary locations where other small cell BSs are also located. In this case, the links closer to the POP need to be of a higher capacity than those further down the chain, since they need to support the traffic from multiple small cells.

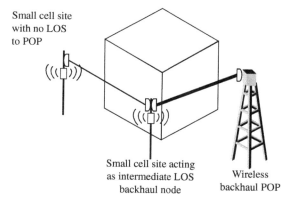

Figure 14.5. Small cell BS connected to POP through an intermediate node that includes another small cell BS.

Point-to-Multipoint Wireless Backhaul   PMP wireless backhaul is another tech-nology that has been proposed as a solution for small cell wireless backhaul. The fundamental difference between PTP and PMP LOS systems relates to the way fre-quency channels are used when aggregating links at the POP. In a PTP LOS system, each link uses a pair of transceiver modules with narrowbeam antennas at either end and has a statically allocated frequency channel. This means that each small cell BS has a dedicated frequency channel that is only used for the link between itself and the POP. In a PMP LOS system, by contrast, the links between the small cell BSs and POP use a shared frequency channel.

The POP typically uses a widebeam antenna, and some form of medium access control is required to coordinate the transmissions. This technology has the advantage over PTP LOS microwave systems in terms of spectrum licensing costs and having a lower number of modules at the POP. However, since the radio resources are shared, the overall capacity per link is lower when traffic from multiple small cell BSs needs to be backhauled at the same time.

14.2.2 Issues to Consider for Wireless Backhaul Solutions

It is clear from the overview of different wireless backhaul solutions presented above that each solution has different advantages and disadvantages. When designing back-haul for small cells, the network designer needs to choose solutions that best suit the deployment scenario. Here, the different aspects that needs to be considered are discussed in more detail.

14.2.2.1 Coverage   Wireless backhaul coverage of a POP is defined as the area in which the a POP can connect with a small cell BS, and this is dependent on several different aspects. When designing LOS wireless network links, engineers have to consider several different aspects that influence the coverage area of the link. One of the aspects is the attenuation by gasses in the atmosphere. These losses are higher at certain frequencies, coinciding with the mechanical resonant frequencies of the gas molecules.

Figure 14.6 shows the atmospheric gaseous losses by dry air and water vapor at sea level under normal conditions [2] for frequencies up to 300 GHz. The resonances for frequencies below 100 GHz occur at 24 GHz for water vapor and 60 GHz for oxygen, resulting in absorption peaks centered around these two frequencies. While these losses can be significant for links over 1 km long, in small cell scenarios, the distances between nodes are typically much shorter. Therefore, they are usually not a major factor that limit the coverage in LOS wireless backhaul.

The most obvious aspect limiting LOS wireless backhaul coverage for small cells is, as its name implies, the requirement for a clear LOS between the POP and the small cell BS in order to achieve connectivity in the first direct link. In urban environments, the probability of LOS existing between a POP located at rooftop level and a small cell location at street level can be low. For example, in [3], the probability of LOS,

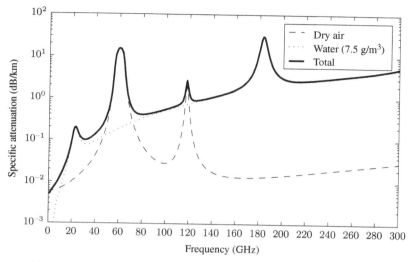

Specific attenuation by atmospheric gases at sea level.

P_{LOS}, between two antennas with heights of 5 m and 32 m respectively that are d meters apart in an urban environment is modeled by the equation:

$$P_{LOS} = \min\left(\frac{18}{d}, 1\right)(1 - e^{\frac{-d}{72}}) + e^{\frac{-d}{72}} \tag{14.1}$$

This model is based on relatively limited data sets and specific assumptions and approximations, and should therefore not be considered exact since LOS probabilities in urban environments can vary drastically in different locations. However, it is used here to provide an averaged approximation of LOS probabilities. Figure 14.7 shows the plot of P_{LOS} against d, showing that LOS probabilities are less than 50% when d is higher than 75 m. This limits the use of LOS wireless backhaul in urban areas.

In co-channel deployments, due to the high levels of interference, small cells would not likely be placed in close proximity to macrocells in the first place, particularly if there is LOS between the small cell and macrocell sites. These contradicting requirements means that the LOS wireless backhaul is not well-suited for providing links between small cells and POP at macrocell sites in co-channel deployments.

At the lower microwave frequencies, however, some near LOS propagation is possible. In such situations, the signals from a POP can reach a small cell by diffracting or reflecting around obstacles using pre-existing buildings and structures. Diffraction occurs when an electromagnetic wave hits the edge of a building, where the energy of the wave is scattered in the plane perpendicular to the edge of the building.

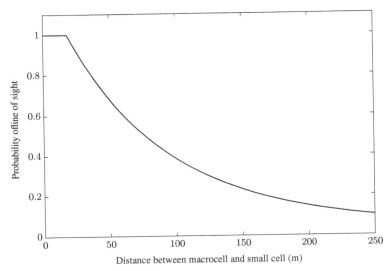

Figure 14.7. Probability of line of sight between two antennas with heights of 5 m and 32 m, respectively.

A microwave POP located at the rooftop of a building can achieve near LOS connectivity to a small cell using diffraction by aiming the antennas at the edge of the rooftop of a building close to the small cell BS.

The diffraction loss, L_d, can be approximated using a knife edge diffraction calculation [4] as

$$L_d = 6.9 + 20\log_{10}\left(\sqrt{(v - 0.1)^2 + 1} + v - 0.1\right), \tag{14.2}$$

where v is the Fresnel diffraction parameter.

$$v = h\sqrt{\frac{2}{\lambda}\left(\frac{1}{d_1} + \frac{1}{d_2}\right)}, \tag{14.3}$$

where λ is the wavelength, and h, d_1, and d_2 are the distances between the points shown in Figure 14.8 in meters.

Using Eqs. 14.2 and 14.3, the diffraction loss for the scenario illustrated in Fig. 14.8 for a 28 GHz microwave transmission can be calculated as approximately 45 dB. The reflection loss, on the other hand, is dependent on the material of the surface being reflected on and the angle of incidence.

In [5], the reflection loss for a 28 GHz transmission against a brick wall at an incidence angle of 15 degrees was measured as 24 dB. It is clear that the use of diffractions

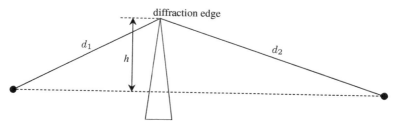

Figure 14.8. Geometrical elements for diffraction loss calculation.

and reflections causes significant losses in the energy of the signal, but this can be overcome by the advantages obtained from the high antenna gains and large available bandwidths. These gains can make microwave solutions achieve higher throughput performance in near LOS scenarios compared with solutions that utilize frequency carriers below 6 GHz that have non-line-of-sight (NLOS) propagation characteristics over short distances [6].

Figure 14.11a and Figure 14.11b show the relationship between the maximum antenna gain G_M^a and half-power beamwidth φ_{3dB} of antennas with circular apertures, calculated using the equations [1]:

$$G_M^a = \eta \left[\frac{\pi D}{\lambda} \right]^2 \tag{14.4}$$

$$\varphi_{3dB} = 69.3 \frac{\lambda}{D}, \tag{14.5}$$

where η is the antenna efficiency, set to 0.7, and D is the diameter of the antenna aperture. The difference in achievable antenna gain between sub-6 GHz frequencies and higher frequencies, particularly at millimetre wave frequencies, is significant. For example, referring to Fig. 14.11a, the maximum gains for an antenna with a 0.3 m aperture antenna for 5 GHz is 22.3 dBi, while an antenna with a similar size is 44 dBi for 60 GHz.

The exploitation of diffractions and reflections allows engineers to connect POP to small cells that are within one or two reflections, or one diffraction away, which helps increasing the coverage area of the POP, albeit to a limited extent.

As an example, Fig. 14.9 shows a simulated scenario of an urban environment, with a POP antenna placed 3 m above the rooftop of a 24 m-high building at a main road junction. Areas at 5 m height that have a clear LOS to the POP are illustrated by the shaded areas in Fig. 14.9a, showing limited coverage area along the streets of the junction. However, this does not take into account the effect of reflections and refractions, and to have an insight into this, a ray-tracing propagation model is used.

Figure 14.9b shows the areas at 5 m above ground level with received powers above −100 dBm for a POP operating at 28 GHz with a transmit power of 250 mW,

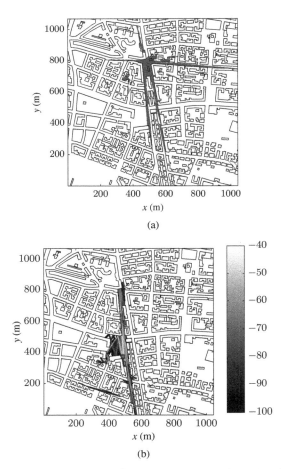

Figure 14.9. Coverage areas in an urban environment with LOS and wall reflections.
(a) Areas with direct line-of-sight to a rooftop point of presence and (b) Extension of coverage using directional antenna and with wall reflections.

using the circular antenna with a diameter of 0.3 m that has a maximum gain and half-power beamwidth presented in Fig. 14.11.

The antenna is aligned to deliberately cause as much reflections off a wall about 400 m to the south of the POP, and the resulting reflections are able to extend the coverage of the POP beyond LOS areas. The challenge in doing this, however, requires some effort in aligning the antennas so that they are pointed in the right direction. It would also only be possible when the deployment scenario includes POPs that are located at the highest vantage points and a high availability of surfaces in which the signal can be diffracted and reflected on.

Based on this simple analysis of LOS probabilities, LOS wireless backhaul coverage does appear to be small and applicable only in a limited number of deployments. In order to overcome the limitations of the LOS requirement, the use of passive reflectors and active repeaters can be used to circumvent obstacles. Alternatively, daisy chaining small cells together can be used to reach small cells that are in difficult locations.

These techniques can extend LOS wireless backhaul coverage significantly, but there are a few drawbacks to consider. Using passive reflectors or active repeaters introduces additional costs due to the increased number of sites and equipment required. Daisy chaining, on the other hand, would only be possible if the small cell densities were high enough to be located within reach of each other, and may require some compromises to be made in the optimality of the placement of small cells in order to enable daisy chaining.

NLOS solutions, on the other hand, are able to achieve a relatively larger coverage in areas without clear LOS due to more favourable propagation characteristics that allows better reflections, diffraction, and penetration. However, the trade-off involved with this is that antennas for NLOS wireless backhaul do not have the same high gains and directivity compared with those available at higher frequencies. In order to obtain the same high gains and directivity, the size of the antennas becomes impractically large for small cell deployments.

Figure 14.10 shows the received powers at 5 m above ground level from a POP deployed at the same location, using the same transmit power and antenna size as the POP in the scenario illustrated in Fig. 14.9, but utilizing a 5.8 GHz frequency, a lower antenna gain, and wider beamwidth, as shown in Fig. 14.11.

It is apparent from this example scenario that the achievable coverage area in NLOS wireless backhaul is higher than that of LOS wireless backhaul in urban

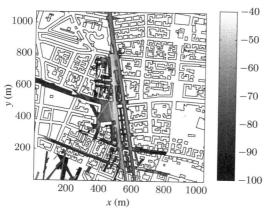

Figure 14.10. Received powers for NLOS point of presence.

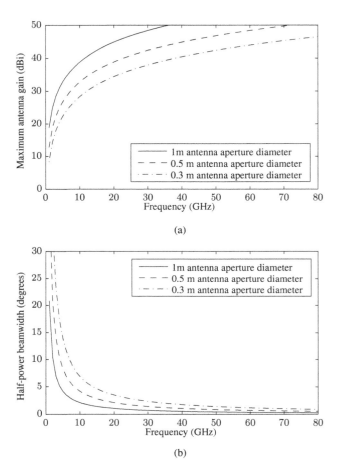

Figure 14.11. Maximum antenna gain and half-power beamwidth for circular antenna apertures of different diameters. (a) Maximum antenna gain and (b) Total half-power beamwidth.

environments, even when the differences in antenna gains are taken into consideration. The increased beamwidths of NLOS wireless backhaul also has the advantage of removing the need for precise antenna alignments, which eases the deployment procedure.

From a coverage point of view, the lower frequency bands used for NLOS backhaul place less restrictions in terms of the possible locations in which small cells can be placed, which is a highly desirable property in terms of deployment optimization. However, they have significant drawbacks in terms of capacity, which will be elaborated in the next section.

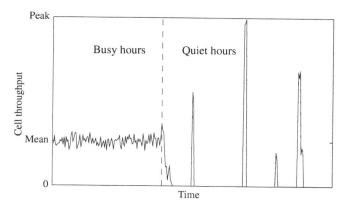

Figure 14.12. Illustration of busy time mean and quiet time peak cell throughputs.

14.2.3 Capacity

When dimensioning backhaul to small cells, the aim of network architects is to ensure that the capacity of the backhaul link is sufficient to support cell throughputs that offer acceptable levels of service to UEs. To gain an insight into this, cell throughput can be characterized under two loading conditions: busy time mean throughput during busy hours and peak cell throughput during quiet hours, as illustrated in Fig. 14.12.

At quiet hours, the number of UEs that are using the cell's radio resources is low, and most of the time only one active UE needs to be served. The BS can therefore allocate the cell's entire radio resources to that single UE. If the single UE is located in an area with very good signal-to-interference-plus-noise ratio (SINR) conditions, then the peak rates of the access technology can be achieved.

During busy hours, there are several active UEs that are sharing the cell's radio resources. As the location of the UEs would likely be distributed across the whole cell's coverage area from the center to the edge, there would be a corresponding variation in the SINR experienced by the UEs. Since there would be UEs located in both high SINR and low SINR areas, the total cell throughput in these conditions results in a mean traffic that is less than the peak occurring during quiet hours.

The achievable peak rates are dependent on the configuration of the small cell BS and UE. For example, Table 14.1 shows results from [7] of the peak and mean cell throughputs of different LTE configurations. The peak figures represent the maximum device capability, and the mean figures were obtained through simulations.

The results show that the mean cell throughputs during busy hours is around five to seven times lower than the peak rates achieved during quiet hours. As small cells have a smaller coverage area, they also typically serve a smaller number of

TABLE 14.1   Mean and Peak UE Plane Traffic for Different LTE Configurations

Direction	Configuration	Cell Throughputs [Mbps]	
		Mean	Peak
Uplink	1 × 2, 10 MHz, category 3	8.0	25.0
	1 × 2, 20 MHz, category 3	15.0	50.0
	1 × 2, 20 MHz, category 5	16.0	75.0
	1 × 2, 20 MHz, category 3, MU-MIMO	14.0	50.0
	1 × 4, 20 MHz, category 3	26.0	50.0
Downlink	2 × 2, 10 MHz, category 2	10.5	50.0
	2 × 2, 10 MHz, category 3	11.0	75.0
	2 × 2, 20 MHz, category 3	20.5	100.0
	2 × 2, 20 MHz, category 4	21.0	150.0
	4 × 2, 20 MHz, category 4	25.0	150.0

UEs compared to macrocells, and given that small cell BSs are usually deployed at hotspots, the UEs are clustered close to the cell sites. Small cells therefore have a higher likelihood of achieving peak throughputs. In order to support these peak throughput levels, the backhaul should be provisioned with a capacity in the region of around 200 Mbps (when both U-plane and C-plane traffic are considered). If the small cell site supports multiple access technologies such as wireless fidelity (Wi-Fi) and 3G, the required capacity is increased even further.

The capacities of wireless backhaul links are generally dependent on several well known factors common in any wireless communication system. This includes the amount of bandwidth available, the modulation scheme which is dependent on the SINR, and the sharing of radio resources.

In general, LOS backhaul solutions are able to provide high capacity due to the large amount of spectrum bandwidth available, with bandwidths of 50 MHz or more possible for a link. They also have less issues with interference due to the use of narrowbeam antennas and higher signal attenuation, which enables stable high throughput levels to be maintained. LOS links can be designed to achieve capacities in excess of 1 Gpbs [8], which easily meets the requirements for peak cell throughputs.

Achieving such high capacity using NLOS solutions however, is challenging due to the limited available spectrum. Because of propagation characteristics that provide good coverage, frequencies below 6 GHz are widely used for many different applications, such as cellular radio access network (RAN) and in Wi-Fi networks. As a consequence of this high level of demand, most of the spectrum in these bands is already utilized.

Due to the use of wider beam antennas and a lower attenuation of signals, interference is also an issue, which can lead to unpredictable levels of throughput and lower link reliability [5], although the use of interference management techniques can alleviate this to a certain extent. Currently, NLOS solutions for small cell backhaul have claimed peak capacities of up to 250 Mbps [9], although in practice this is usually lower when the impact of interference and losses due to the environment are taken into account.

Typical throughput figures for NLOS backhaul are in the region of 100 Mbps or less [8], which is lower than the peak cell throughputs possible in LTE. In this case, the backhaul would potentially be the bottleneck limiting the throughput of UEs in the small cell.

The throughput figures above would be applicable to PTP configurations, where a whole channel is dedicated to a single link. In a PMP configuration, the capacity of the channel is shared with multiple transceivers using some form of multiplexing methodology, and therefore the available capacity to each transceiver is dependent on the traffic from the small cells, with lower capacity available when multiple small cells are backhauling traffic at the same time.

While LOS wireless backhaul is capable of providing significantly higher capacity per link compared to NLOS backhaul, and at a higher reliability, there are trade-offs involved with regard to the use of coverage as discussed in the previous section, and cost, which will be discussed in the following sections.

14.3 WIRED BACKHAUL

Wired backhaul for small cells can be in the form of fiber or copper based solutions. The digital subscriber line (DSL) family of technologies uses legacy copper twisted pair telephone infrastructure, and is widely used to provide last mile Internet connections with bitrates of up to hundreds of Mbps.

Fiber technologies such as a passive optical network (PON) provides a very high performance connection with bitrates of up to multiple Gbps. Although wired backhaul solutions can provide reliable, high-capacity connections for small cells, the cost and time needed to install the required cables can be high relative to wireless backhaul solutions.

One of the factors that can impact the practicalities of using wired backhaul for small cells is the terminating option used, that is, where the cable between the small cell and the POP has to be connected. The POP for copper connections of DSL type would for instance be to a digital subscriber line access multiplexer (DSLAM), and an optical distribution unit (ODU) or optical line termination (OLT) for fiber PON connections.

If the distance between the small cell site and the POP is high, the most obvious issue this causes is the increase of cost for installing the cables, which also includes obtaining legal permits from relevant authorities to perform the required civil works involved.

For DSL type connections, the distance from the DSLAM also has a significant impact on the data rates of a loop, which is why loop lengths are commonly kept below 500 m [10]. With more recent developments in DSL technology, such as the use of vectoring in the recent G.fast DSL standard [11], data rates of up 1 Gbps can be achieved with wire pairs shorter than 100 m, with the rates reducing to 500 Mbps at 100 m, 200 Mbps at 200 m and 150 Mbps at 250 m [12].

Techniques such as bonding can also be used to aggregate multiple DSL loops to achieve the bitrates required by the small cell site. PON connections, on the other hand, can have a physical reach of up to 60 km [13], although currently, fiber terminals are not as widely deployed compared to DSL terminals.

The availability of existing copper or fiber infrastructure is therefore an important factor that influences the use of wired backhaul for small cells. In developed, urban environments, DSL infrastructure is often already widely deployed, with high densities of DSLAMs.

As the demand of high-speed Internet connections rises, the coverage of fiber infrastructure is also starting to become widely available as well as operators begin rolling out fiber to the x (FTTx) topologies that brings fiber increasingly closer to the customer premises. The use of wired backhaul can be a good option for small cells that are deployed to provide additional capacity, as they tend to be deployed in locations with high UE demand, for example, urban areas with high availability of fiber and DSL-type connections.

14.4 COST

Apart from capacity and coverage, the third aspect to consider when choosing which backhaul solution to use is the cost.

Operators have expressed the desire for the costs of a small cell site to be 10% of the cost of a macrocell [14], and as backhaul can take up a significant portion of the overall cost of a small cell site, the cost constraint on small cell backhaul is very tight. In order to gain an insight into the costs of different backhaul solutions, a high-level calculation of the cost of providing backhaul to a small cell is presented here, and the effect of different deployment parameters are examined.

It is important to note that the cost calculation approach given below is provided purely as an example, as it makes several simplifications, and does not include many additional cost items. The costs given are also estimated values, and the reader should be aware that cost figures can vary significantly in reality.

Five different options for providing backhaul are considered: sub-6 GHz NLOS and 60 GHz LOS wireless backhaul, Gigabit passive optical network (GPON) and very-high-bit-rate digital subscriber line (VDSL) wired backhaul, and leased carrier Ethernet services. Tables 14.2 and 14.3 show the assumed parameters and associated costs for the different backhaul options.

The assumptions used in this calculation are that the wireless backhaul solutions use an existing macrocell site as a POP, and that LOS to the macrocell site is available. For the NLOS solution, a PMP setup is used if the capacity per small cell is less than 75 Mbps, otherwise a PTP setup is used up to 150 Mbps.

TABLE 14.2 Key Assumptions for Wireless Small Cell Backhaul Cost Calculation

Type	Item	Cost [$]
General Wireless Costs	Power connection cost per site	1790
	Cell site router	2500
	Annual backhaul power use per end point	72
	Annual maintenance cost	8% of CapEx
60 GHz LOS Wireless	Capital expenditure per end site	2600
	Installation per end site	600
Sub-6 GHz NLOS Wireless	Hub module	2000
	Remote module	1500
	Point to point module per end site	2000
	Installation per site	400
	RF engineering per link	205

Beyond 150 Mbps, the NLOS solution is assumed to be infeasible. A PTP setup is assumed for the LOS solution, with a maximum link capacity of 1 Gbps. For the GPON and VDSL backhaul solutions, it is assumed that the mobile operator owns an installed GPON or DSL infrastructure, and the capacity for a single VDSL loop is assumed to be 100 Mbps and the capacity of the GPON link is 1 Gbps.

TABLE 14.3 Key Assumptions for Wireline Small Cell Backhaul Cost Calculation

Type	Item	Cost [$]
General Wireline Costs	Buried cable cost per meter	60
	Right of way permit	4200
	Power connection cost per site	1200
	Site survey and installation cost	1200
	Annual backhaul power use per site	72
	Splicing cost per site	20
GPON	Capital expenditure per site	158
	Fiber cost per meter	10
	Annual maintenance cost	0.8% of CapEx
VDSL	DSL outdoor modem	600
	DSLAM port per loop	80
	Material cost per site	150
	Annual maintenance cost	1.6% of CapEx
Leased Carrier Ethernet	One time connection cost per site	3000
	Monthly bandwidth charge for first 50 Mbps	300
	Monthly bandwidth charge for subsequent 50 Mbps	100

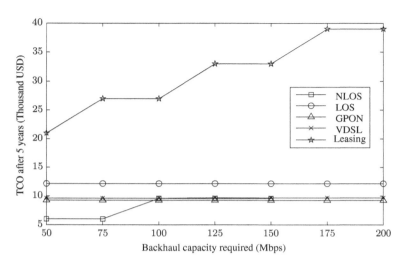

Figure 14.13. Total cost of ownership after 5 years of a small cell backhaul with different capacities.

Figure 14.13 shows the total cost of ownership (TCO) over 5 years for providing last-mile backhaul to a small cell for the different backhaul solutions with different throughput capacity requirements, with a 30 m distance between the small cell to the fiber and VDSL connection points. NLOS wireless backhaul has an appreciably lower cost relative to other solutions below 75 Mbps where a PMP solution is used and the multiplexing of multiple small cells to one hub results in cost savings. Figure 14.14

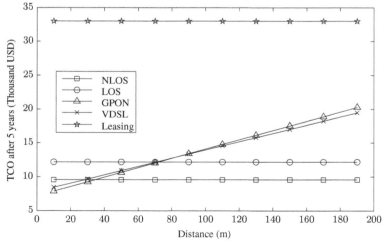

Figure 14.14. Total cost of ownership after 5 years of a small cell backhaul with different distances.

shows the TCO over 5 years for the different backhaul solutions to provide 40 Mbps with the distance between the small cell site and the POP, varying between 10 m and 190 m. The cost of the wireless backhaul and leasing solutions do not vary with distance, but the cost of the wired backhaul increases due to the cost of laying the required cables.

While the cost calculations shown here are simplified examples, the main message that is being conveyed here is that there are cost and capacity trade-offs involved when selecting backhaul solutions. Wired backhaul can be cost-effective when the distance to the POP is low, but becomes costly if large distances are involved. NLOS wireless backhaul may have the lowest cost in many cases but has limitations in capacity, while LOS wireless backhaul provides a high capacity at relatively low cost, but as described above, has limited coverage in urban environments.

14.5 SUMMARY AND CONCLUSIONS

In order to enable the capacity gains offered by the densification of small cells, the provision of backhaul to small cells in an economically viable manner is required. In this chapter, different backhaul options with their respective advantages and disadvantages were discussed.

Wireless solutions have the advantage of providing backhaul without the costly and time consuming process of installing cables. However, LOS wireless backhaul does have the limitation that the small cell site has to be within LOS to the POP, which can limit its range, or necessitate the use of multi-hop configurations in urban environments. But with its use of high frequency bands, it can achieve high data rates due to the larger bandwidths available, and is less prone to interference.

NLOS wireless backhaul, as its name implies, does not suffer from the LOS requirement, and can be used to provide backhaul to multiple small cell sites using a PMP configuration in a very cost-effective manner. On the other hand, NLOS backhaul uses sub-6 GHz frequency bands, which are expensive to license, and limited in the achievable data rates because of the lower bandwidth available.

There are, however, very promising developments in the use of large-scale antenna systems (LSASs) or massive multiple-input multiple-output (MIMO) that can partially overcome both the coverage and capacity issues associated with current LOS and NLOS backhaul solutions.

Massive MIMO is a promising concept in cellular networks proposed in [15] and [16]. The basic idea behind massive MIMO is to exploit the benefits of conventional MIMO, but on a much greater scale by increasing the number of antennas at the BS to a number that is significantly higher than the number of UEs or small cell BSs served. Through the utilization of measured channel responses, as the number of antennas increases, the performance of the system increases due to favourable propagation (where communications channels of different UEs become more orthogonal) and channel hardening effects (where the effects of uncorrelated noise and fast fading are averaged out). The potential benefits of massive MIMO include greater selectivity in transmitting and receiving the data streams, which leads to greater throughput, a reduction in the required radiated power, effective power control that provides uniformly good service throughout the cell, and greater simplicity in signal processing [17].

While massive MIMO was initially conceived to enhance the performance in the cellular access (between BSs and UEs), it is also highly suited for the application of providing backhaul to small cells [18]. However, massive MIMO is currently a relatively new concept, and its implementation is subject to many challenges such as the reduction of hardware and software complexity, low power requirement in transceiver design, and pilot contamination [19]. These, and many other challenges, are currently active research areas in industry and academia, but it remains a promising way of solving the issues of wireless backhaul for small cells in the future.

Wired backhaul has the advantage of providing high capacity, reliable backhaul links to small cells, and is currently widely used for small cells. However, the time and cost needed to install new cables to small cell sites can be relatively high, particularly in areas where existing wireline infrastructure is not widely deployed. In urban environments, DSL infrastructure is often already widely available, and as continued investments are made to increase the available capacities, the densities of POP such as DSLAM and FTTx cabinets is likely to become higher. This can make the use of wired backhaul a cost-effective option for large-scale small cell deployments.

In general, there are many possible options available to provide backhaul to small cells, and the choice is dependent on many factors such as the cost, availability, and QoS requirements of the small cell. Small cell deployments will therefore feature a mixture of different backhaul solutions that best suits the location and requirements of the small cell. Future wireless communications will begin to pose major challenges to the provision of small cell backhaul. This includes the increasing densification of small cells, the advent of new radio access technologies producing ever higher throughputs, and new architectures requiring close coordination between cells and the centralization of base band processing.

Many of these challenges are highly dependent on the availability of high-capacity and low-cost backhaul, so a major effort in developing new and innovate techniques for small cell backhaul is required for the demands of future 5G networks.

REFERENCES

1. G. Barué, *Microwave Engineering: Land & Space Radiocommunications*, ser. Wiley Survival Guides in Engineering and Science. John Wiley & Sons, 2008.
2. "Attenuation by atmospheric gases," International Telecommunication Union, Geneva, Recommendation ITU-R P.676 3, Sep. 2013.

3. "Further advancements for E-UTRA physical layer aspects," 3GPP Technical Specification, Evolved Universal Terrestrial Radio Access (E-UTRA), Tech. Rep. TR 36.814, Mar. 2010.

4. "Propagation by diffraction," International Telecommunication Union, Geneva, Recommendation ITU-R P.526 13, Nov. 2013.

5. J. Hansryd, J. Edstam, B.-E. Olsson, and C. Larsson, "Non-line-of-sight microwave backhaul for small cells," *Ericsson Rev.*, vol. 2013, no. 3, Feb. 2013.

6. M. Coldrey, J.-E. Berg, L. Manholm, C. Larsson, and J. Hansryd, "Non-line-of-sight small cell backhauling using microwave technology," *IEEE Commun. Mag.*, vol. 51, no. 9, pp. 78–84, Sep. 2013.

7. NGMN, "Small cell backhaul requirements," A Deliverable by the NGMN Alliance, Tech. Rep., Jun. 2012.

8. M. El-Sayed and P. Gagen, "Mobile data explosion and planning of heterogeneous networks," in *2012 XVth Int. Telecommun. Netw. Strategy Planning Symp. (NETWORKS)*, Oct 2012, pp. 1–6.

9. "Radwin 5000 jet scb brochure" (2016) [Online]. Available: http://www.radwin.com/contentManagment/uploadedFiles/case-studies/small%20cell%20brochure_2016a_web.pdf

10. C. Leung, S. Huberman, K. Ho-Van, and T. Le-Ngoc, "Vectored dsl: Potential, implementation issues and challenges," *IEEE Commun. Surveys Tuts*, vol. 15, no. 4, pp. 1907–1923, Fourth 2013.

11. ITU Std. G.9701: Fast access to subscriber terminals (G.fast)—Physical layer specification. ITU-T (Dec. 2014). [Online]. Available: https://www.itu.int/rec/T-REC-G.9701-201412-I/en

12. M. Timmers, M. Guenach, C. Nuzman, and J. Maes, "G.fast: evolving the copper access network," *IEEE Commun. Mag.*, vol. 51, no. 8, pp. 74–79, Aug. 2013.

13. F. Effenberger, D. Clearly, O. Haran, G. Kramer, R. D. Li, M. Oron, and T. Pfeiffer, "An introduction to pon technologies," *IEEE Commun. Mag.*, vol. 45, no. 3, pp. S17–S25, Mar. 2007.

14. D. Mendyk and M. Donegan, "Wireless backhaul for small cells: Who's doing what," *Heavy Reading 4G/LTE Insider*, vol. 4, no. 4, Aug. 2013.

15. T. Marzetta, "Noncooperative cellular wireless with unlimited numbers of base station antennas," in *UCSD Inf. Theory Applicat. Workshop*, Feb. 2010.

16. T. Marzetta, "Noncooperative cellular wireless with unlimited numbers of base station antennas," *IEEE Trans. Wireless Commun.*, vol. 9, no. 11, pp. 3590–3600, Nov. 2010.

17. T. Marzetta, "Massive mimo: An introduction," *Bell Labs Tech. J.*, vol. 20, pp. 11–22, 2015.

18. E. Larsson, O. Edfors, F. Tufvesson, and T. Marzetta, "Massive mimo for next generation wireless systems," *IEEE Commun. Mag.*, vol. 52, no. 2, pp. 186–195, Feb. 2014.

19. Z. Zhang, X. Wang, K. Long, A. Vasilakos, and L. Hanzo, "Large-scale mimo-based wireless backhaul in 5g networks," *IEEE Wireless Commun.*, vol. 22, no. 5, pp. 58–66, Oct. 2015.

15

OPTIMIZATION OF SMALL CELL DEPLOYMENT

15.1 INTRODUCTION

In the previous chapters, many different techniques were presented that aim to increase the performance of small cells. However, one of the factors that can have a very large influence on the overall benefit of small cells is the location in which they are deployed. There are many factors which contribute to a good location for small cells, but the main ones to consider are interference, offload, backhaul, and costs.

In order to maximize the benefits of small cell deployments, they should ideally be placed in locations where the received signal strength from the macrocell underlay is low in order to minimize co-channel interference and maximize the small cell coverage area. On the other hand, they should also be placed in areas with user demand hotspots so that they can offload as many user equipments (UEs) as possible. Moreover, the availability of backhaul that is able to support the offloaded traffic is of course required. Finally, the cost of deploying a small cell at a specific location needs to be considered.

These factors needs to be considered together when selecting suitable locations to deploy small cells that provide a good investment for network operators. With their low transmit power and coverage area, the margin of error involved is small and the

Small Cell Networks: Deployment, Management, and Optimization, First Edition. Holger Claussen, David López-Pérez, Lester Ho, Rouzbeh Razavi, and Stepan Kucera.
© 2017 by The Institute of Electrical and Electronic Engineers, Inc. Published 2017 by John Wiley & Sons, Inc.

difference in offload and performance between two small cell candidate locations just tens of meters apart can be significant. In addition to that, the selection of small cell locations can be very labor-intensive if performed through manual cell planning, due the high number of potential sites relative to macrocells.

Therefore, for future wide-scale deployments with high densities of small cells, the way in which small cell locations are identified requires more accuracy and automation compared to typical approaches used for macrocell networks. In this chapter, an overview of the different steps involved in optimizing the position of small cells is presented, including the creation of UE demand maps, the calculation of small cell offload potential, and the calculation of costs. Finally, an example of how these are put together in order to select small cell locations is given.

15.2 OVERVIEW OF THE SMALL CELL SITE SELECTION PROCESS

Figure 15.1 illustrates the processes involved in selecting candidate sites for small cells. The first step is the collection of network data such as information on UE call traces and measurements, and information of existing cells within the deployment area. The location of the small cell candidate sites is also required.

These are then used to prepare the inputs needed to perform the site selection process. This includes estimation of the UE traffic demand and propagation maps of the area where the small cells are to be deployed. The different options to provide backhaul to the candidate sites also needs to be evaluated, and this is used to calculate the costs of deploying small cells at each site.

Once those maps are available, the offload potential and cost of each candidate small cell site can be calculated, taking into account the backhaul options available

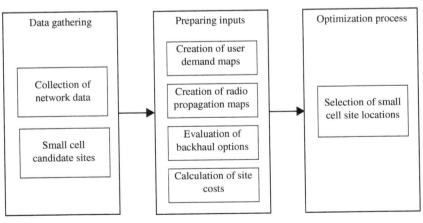

Figure 15.1. Overview of the small cell placement process.

for each site. This is used to perform the optimization and decide on which sites are selected, based on the criteria of cost and performance.

15.3 UE TRAFFIC DEMAND MAP GENERATION

In order to provide an accurate estimation of the offload potential of a small cell candidate site, a UE traffic demand map has to be calculated to provide information on the geographical location and volume of the traffic demand. The UE demand tends to be distributed unevenly, particularly in dense urban environments, where a significant portion of traffic is generated at traffic hotspots. It is therefore important to identify these traffic hotspots accurately so that small cells can be placed in targeted locations where it can achieve the highest offload and increase the overall network performance.

There exists many different ways in which information on UE demand can be obtained, and a few examples are given below that have different advantages and disadvantages. Demand information derived from measurements and call traces from the mobile operator's network is able to provide the actual volume of UE traffic, but currently does not provide accurate geo-location information. On the other hand, other techniques that provides more accurate geo-location information do not provide full information of the volume of UE traffic. Therefore, information from multiple sources has to be combined to create a demand map that provides sufficient accuracy of both the location and volume of demand.

15.3.1 Call Traces

The first technique is using call trace data from the existing macrocell underlay network. This is a well-established approach that uses network measurements to determine UE location and the corresponding data usage or session attempts. The information contained within the call traces collected varies depending on the software used, but they typically contain time-stamped information on the UE session's data usage and measurements that can be used for geo-location, including received signal strengths from surrounding cells such as reference signal received power (RSRP) and timing information such as the round trip time (RTT) measurement between the base station (BS) and the UE.

While the information of the volume of traffic load created by UEs can be easily derived from call traces, the accuracy of UE geo-location using the received signal strength and timing information is limited. The geo-location of UEs using these measurements fundamentally relies on triangulation. Put simply, triangulation involves the calculation of the distances between the UE and three or more BS antennas, and the estimated location of the UE is the point at which the distance circles intersect. Triangulation using timing information relies on calculating the RTT between the UE and the base station.

In long-term evolution (LTE), the RTT can either be calculated using timing advance measurements, either by taking the sum of the receive-transmit time difference at the BS and the receive-transmit time difference at the UE (Type 1), or by using the receive-transmit time difference estimated from receiving a physical random access channel (PRACH) preamble during the random access procedure (Type 2). The distance between the UE and BS is calculated simply as $d = c \times \mathrm{RTT}/2$. The accuracy that can be achieved with such techniques is dependent on the radio environment. Even if the BS locations are precisely known and the cable lengths between the BS and the transmit antennas are properly calibrated, the effects of reflections introduces errors when the UE is not within line-of-sight to the base stations, which causes a degradation of accuracy. In urban environments, the accuracy is typically 150 m or worse, particularly if the UE is at the edge of the cell. It also requires the UEs to be able to make measurements of three or more base stations. This can result in poor accuracy in areas of weak macrocell coverage, which are good candidates for small cell deployment sites.

When using received signal strength measurements such as RSRP for triangulation, the distance between the UE and the BS is calculated using path-loss models. This approach, however, is highly inaccurate, particularly in urban environments, as they do not take into account the actual wall losses between the UE and the base station. Therefore, RSRP is rarely used for triangulation. However, it can be used for radio frequency (RF) fingerprinting techniques, where the location of the UE can be estimated by comparing its RSRP measurements with an RSRP map of the area. RF fingerprinting techniques can offer accuracies in the range of $1 - 5$ m [1]. However, an up-to-date RSRP map with sufficient accuracy is difficult to obtain.

15.3.2 Minimization of Drive Testing Server

Another technique for locating hotspots is by using the minimization of drive tests (MDT) feature, which was introduced in Third-Generation Partnership Project (3GPP) LTE Release 10 [2]. MDT is a feature that provides the network operation, administration, and maintenance (OAM) system the ability to collect field measurements from the UEs, along with location information if available. As its name implies, the aim of MDT is to reduce the amount of conventional manual drive testing performed by engineers in order to perform network optimization. Among the measurements that can be collected are downlink and uplink data volume measurements, which was an enhancement introduced in LTE Release 11. This measurement can be used to construct the UE traffic demand map.

As with the call trace technique described above, the measurements sent back by the UEs can also be used to derive the location of the UE using techniques such as triangulation or RF fingerprinting. However, the MDT feature also enables the UE to tag the measurements using global navigation satellite system (GNSS) positioning such as global positioning system (GPS) functionality that are now commonly available at the UE hardware. This GNSS location information has a relatively high accuracy, and does not require any post-processing of data or RF fingerprint databases. However, it is unreliable in built-up urban areas with tall buildings and unavailable indoors. In addition, the MDT feature only enables a subset of UEs to be monitored at any one time, so the resulting traffic demand information obtained will be incomplete.

15.3.3 Sniffer-Based Measurements

Apart from collecting measurements directly from the cellular network, another possible approach is to employ sniffers. Sniffers are devices that are placed at strategic locations and perform measurements of transmissions originating from mobile devices in order to calculate estimates of their location. Devices with wireless local area network (WLAN) or Bluetooth capabilities enabled transmit periodic messages that allows a sniffer to detect their presence within its vicinity, and using the received powers of the messages, a coarse localization can be performed. Sniffing WLAN and Bluetooth messages can be done passively, so it can be performed in a manner that is transparent to the UE. However, it only provides an indication of the number of UEs present within the vicinity of a sniffer, as it does not capture the data traffic being consumed by the UE, or even if the UE is subscribed to the mobile operator's network. Another drawback of utilizing sniffers is that in order to provide demand information over a large area, many sniffers must be deployed.

15.3.4 Social Network Data

Another possible source of geo-located UE demand information is social network data. In many social networking services, the use of geo-location is sometimes used to tag a post to a specific location. These geo-located posts can be collected to create heat maps of the usage of social networking services. If we assume that volume of geo-located social networking posts is correlated with UE density, these data can provide an indication of UE traffic distribution. However, this assumption may not be correct, and also does not give information on the total traffic of a specific mobile operator's network.

15.4 SMALL CELL DEPLOYMENT OPTIMIZATION

In this section, an example of the optimization of the small cell locations using the information on UE demand, radio environment, backhaul availability, and costs is presented.

15.4.1 Simulation Scenario

An outdoor scenario based on a 1 km² area of Manhattan is used. Figure 15.2 shows the street map of the scenario, with the locations of existing macrocell base base

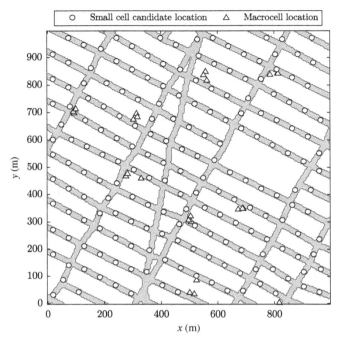

Figure 15.2.   Outdoor small cell deployment scenario.

stations and small cell BS candidate locations. The area is served by a total of 25 macrocells, with rooftop mounted antennas and a maximum transmit power of 46 dBm. The locations and configurations of the macrocell base stations and antennas are based on a real network deployment. A total of 194 candidate sites for small cells are specified. They are located outdoors, and are placed along the side of the roads at regular intervals with a minimum distance of 50 m between each site, with an antenna height of 6 m, which is below the rooftops. This was done to emulate the assumption that the small cell base stations are to be deployed on street light posts. The small cell base stations are assumed to have a maximum transmit power of 30 dBm, with an omni-directional antenna.

Using this scenario, the path-loss maps for each small cell candidate location and macrocell BS are generated. This was done using a propagation model that combines three approaches:

- A waveguide model that treats the urban canyon as a waveguide.
- An over-the-top model that treats diffractions over the top of buildings as waves propagating over a slab.
- A diffusion model that is used to model the outdoor to indoor propagation.

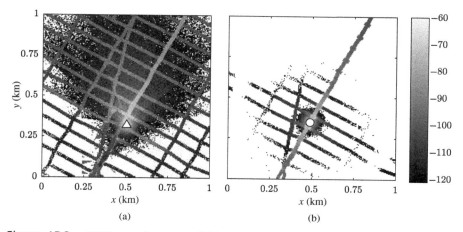

Figure 15.3. RSRP map of a macrocell (a) and small cell (b) generated using the path-loss models.

The model is an adaptation of [3] to the case of outdoor and outdoor to indoor propagation, and has a standard deviation of 9.5 dB compared with measurements. Other propagation modeling approaches can be used to generate the path-loss maps such as ray tracing, which can offer higher accuracy. The advantage of the approach used here is that it does not require a database of detailed building information, and requires only a relatively short computational time, while giving reasonably accurate results. Figure 15.3 shows the RSRP levels of a macrocell in the scenario, and a small cell placed at a candidate location.

To obtain the UE demand map, 600 active UEs are assumed to be present in the scenario, that is, the overall UE density is 600 UEs per km^2. There are 20 UE demand hotspots located randomly within the area, with 70 % of the UEs placed within 100 m of the hotspot centers. Figure 15.4 shows the resulting UE demand map used in the example, which is assumed to be known.

In terms of backhaul availability, we assume that each small cell site requires a 100 Mbps backhaul link, and that there are wireless and wired backhaul options available. With wired backhaul, we consider digital subscriber line (DSL) and fiber backhaul connections, with DSL and fiber end points that are placed randomly throughout the scenario, but with a minimum distance of 300 m between end point sites. In order to provide wired backhaul to a small cell candidate site, it is assumed that cables have to be laid between the candidate site and the nearest wired backhaul end point.

With wireless backhaul, the locations of the wireless points of presence are assumed to be at existing macrocell sites, with non-line-of-sight (NLOS) and line-of-sight (LOS) wireless backhaul options available. The feasibility of LOS wireless backhaul at a small cell candidate site is determined by whether the small cell BS has LOS to a macrocell site.

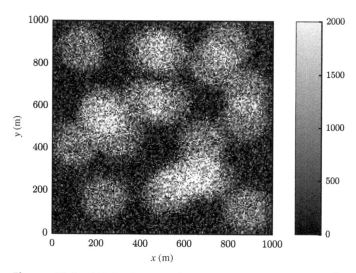

Figure 15.4. UE density map, shown in number of UEs per km².

Figure 15.5 shows the areas in which LOS backhaul is available as gray areas, which covers 38 out of the 194 candidate small cell sites. Only point-to-point setup is assumed for LOS wireless backhaul. For the NLOS backhaul, the feasibility of usage at a candidate site is driven by the achievable bitrates, which is calculated based on the signal-to-interference-plus-noise ratio (SINR), and how many small cell sites are supported by each NLOS hub.

For each small cell candidate site, each backhaul option is provisioned to achieve at least 100 Mbps if possible, and the 5-year total cost of ownership (TCO) of each feasible backhaul option is calculated based on the figures in Chapter 14. The option that provides the lowest cost is selected as the backhaul technology for a candidate site.

In addition to the backhaul costs, the site leasing costs are included. Here, we assume that the small cell base stations would be deployed on street furniture such as streetlights, utility poles, and bus stops, and on the facade of buildings. To reflect this variety of sites, the candidate sites are randomly assigned one out of three different site leasing costs: $100, $200, and $300 per month. All other site costs, such as BS equipment, installation and maintenance are assumed to be the same for all candidate sites. Therefore, only the backhaul and site leasing costs are used to determine the cost difference between candidate sites. Figure 15.6 shows the cumulative distribution function (CDF) of the 5-year TCO for the backhaul and site leasing costs calculated for all the candidate sites in the scenario. The site costs vary from around $14,000 to $60,000, with an average site cost of $34,000. It should be noted that the cost calculations used here are exemplary, as costs are influenced by many factors and can vary widely.

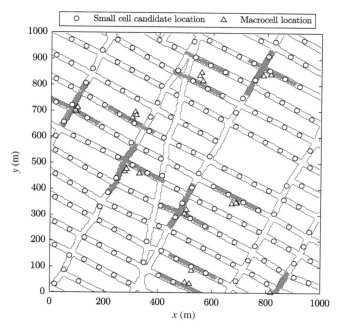

Figure 15.5. Areas with wireless LOS backhaul availability.

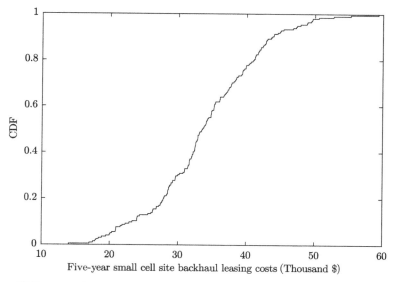

Figure 15.6. Cumulative distribution function of backhaul and site leasing costs of candidate small cell sites.

15.4.2 Overview of the Deployment Optimization Problem

Using the information of the radio environment, UE demand, backhaul availability, and costs described above, the process of optimizing the deployment location of small cells can be performed. The small cell deployment for the scenario is represented by the vector $S = [S_1, S_2, \ldots, S_n]$ where n is the total number of candidate sites. S_i is a binary variable where $S_i = 1$ if a small cell is deployed at candidate site i, and $S_i = 0$ otherwise. The number of small cells being deployed, n_d, is therefore equal to $\sum_{i=1}^{n} S_i$.

The objective of the optimization here is to select the small cell deployment S that maximizes the median UE throughput gain relative to a macrocell-only deployment, divided by the total backhaul and site leasing cost of the small cells.

In order to calculate the UE throughput for a given deployment S, the path-loss maps and UE demand maps are used. First, the path-loss maps are used to calculate the SINR map of the scenario. The location of 600 UEs are then randomly generated based on the UE distribution of the UE demand maps. The association of UEs to cells is performed, with the assumption that the UE is served by the cell with the highest RSRP. The worst-case SINR experienced by the UEs are mapped to the LTE peak throughput. Assuming round robin scheduling, the UE peak throughputs are then divided by the total number of UEs served by their associated cell in order to obtain the average UE throughputs. This process is repeated with 100 different UE placements to obtain sufficient statistics of UE throughputs. Details of the throughput calculation process is given in Appendix A.

The cost of the deployment used for the optimization is simply obtained as a summation of the backhaul and site leasing costs of the sites where the small cells are deployed.

15.4.3 Site Selection Using Greedy Algorithm

An example approach that can be used to select the small cell deployment is a greedy algorithm. This is done by iteratively selecting candidate sites that produces the highest throughput gain per cost metric until the desired number of small cells have been deployed. In detail, the steps involved are as follows:

1. The simulation is performed without any small cells deployed at the candidate site, that is, just with the macrocells. The median UE throughput for this baseline scenario, $T_{baseline}$ is obtained. At this point, all small cell candidate sites are empty, that is, $\sum_{i=1}^{n} S_i = 0$.

2. A small cell is then placed at the first empty candidate site and the scenario is simulated to obtain the median UE throughput. The difference between the throughput $T_{baseline}$ is calculated to obtain the gain in median UE throughput. The small cell is then removed from the first empty candidate site and placed at the next one, the throughput gain is calculated again. This is repeated until all empty candidate sites have been evaluated.

3. The gain in median UE throughput of each empty site is divided by its back-haul and site leasing cost to obtain the throughput gain per cost metric. The empty candidate site which has the highest metric is selected, a small cell is placed there and the deployment S is updated. The median UE throughput of the updated deployment is set as the new baseline throughput $T_{baseline}$.

4. Steps (2) and (3) are repeated until the desired number of small cells have been deployed.

As an example, Fig. 15.7 shows the metric of the 194 candidate sites in the scenario calculated in the first iteration of the greedy algorithm. Due to its combination of high throughput gain and low site cost, site 13 has the highest metric of 0.83 bps/\$, and is therefore selected for a small cell deployment. It is interesting to note that some sites have a metric that is negative. This is because deploying small cells in some sites, particularly those close to macrocell sites, causes a degradation in UE throughput due to interference and low UE offload.

The algorithm is run to select 20 sites. For comparison, a result for a greedy selection based only on the median UE throughput is also shown, where the cost of deployment is not considered in the optimization. For the sake of brevity, the selection based on throughput only is denoted by S_{Gtput} and the selection on the throughput per cost metric is denoted by S_{Gcost}. Figure 15.8 shows the CDF of the UE throughputs for the scenario with no small cells deployed, and when small cells are deployed according to S_{Gtput} and S_{Gcost}, while Fig. 15.9 shows the CDF of the small cell site

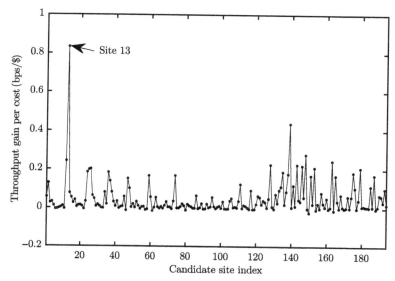

Figure 15.7. Median UE throughput gain per cost of candidate sites calculated at the first greedy algorithm iteration.

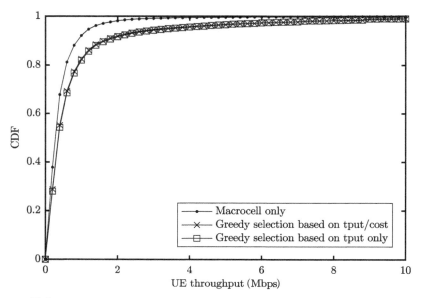

Figure 15.8. Cumulative distribution function of UE throughputs for macrocell-only deployment, and with 20 small cells deployed using the greedy algorithm.

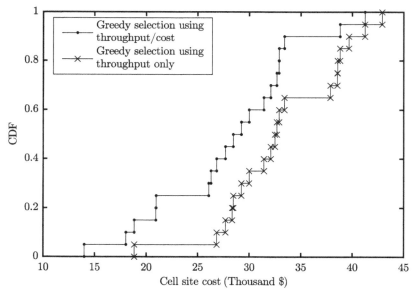

Figure 15.9. Cumulative distribution function of cost, and with 20 small cells deployed using the greedy algorithm.

TABLE 15.1 Throughput and Cost of Deployments Selected Using Greedy Algorithm

	UE Throughput Statistics				
	Median	5th Percentile	95th Percentile	Total	Cost
Macrocell Only (no small cells)	263 kbps	66.2 kbps	1229 kbps	263 Mbps	–
S_{Gtput}	357 kbps	77.7 kbps	3396 kbps	537 Mbps	\$662 k
S_{Gcost}	349 kbps	76.1 kbps	3341 kbps	534 Mbps	\$563 k

costs. We can observe that the difference in throughput performance between the two small cell deployments is small, while the difference in costs of the sites chosen is appreciably significant.

To delve into the results in more detail, the throughput statistics and costs of the deployments S_{tput} and S_{cost} are given in Table 15.1.

As expected, deploying 20 small cells introduces significant increases to UE throughput performance, where the median UE throughput is increased by 32%, and the total UE throughput is more than doubled relative to the macrocell only baseline. The difference between S_{Gtput} and S_{Gcost} is relatively small, with S_{Gcost} achieving a median UE throughput that is just 8 kbps, or 2.2% lower than S_{Gtput}. There is also less than 2% difference between the two selections for the total, 5th and 95th percentile UE throughputs.

However, the total backhaul and site leasing cost for S_{Gcost} is \$563k, while S_{Gtput} has a total cost of around \$662k. S_{Gcost} was therefore able to provide a similar performance in UE throughput, using a deployment cost that is 15% lower relative to S_{Gtput}.

15.4.4 Site Selection Using Genetic Algorithms

One of the drawbacks of using the greedy algorithm is that the selection of sites is done iteratively, which may lead to sub-optimal deployments due to its tendency to converge to a local minimum. Figure 15.10 shows a highly simplified example illustrating a UE hotspot area that requires multiple cells to cover it. In this example, the best approach would be to deploy small cells at the two locations shown. Individually, the two sites chosen would not offer the best solution, but it is the combined deployment of the two that is important in this case. As the greedy algorithm only evaluates the effect of deploying at sites individually, the greedy algorithm would choose to deploy a small cell at the center of the hotspot first, before selecting neighboring locations to cover the remaining hotspot areas, leading to a sub-optimal deployment.

Techniques that evaluate the deployment of a combination of multiple small cells at the same time, rather than one at a time, are therefore more likely to lead to more

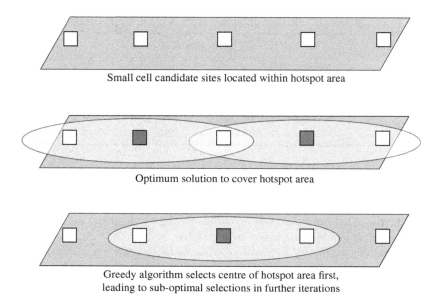

Figure 15.10.   Illustration of the disadvantages of greedy algorithm.

optimal deployments. The total number of possible permutations when trying to select n_d sites out of a total of n sites is given by $n!/((n - n_d)!n_d!)$. This means that the use of a global search of all possible combinations to find the global optimum is only feasible in cases where the number of candidate sites are small. In the Manhattan scenario, for example, the total number of possible permutations when selecting 20 out of 194 candidate sites is over 8×10^{26}. There are several techniques available that can be used to search for good solutions in this case. One such technique that can be applied is genetic algorithm (GA) [4].

GA is an optimization technique that mimics the process of natural evolution, where a population of candidate solutions to an optimization problem are evolved toward better solutions through the use of the selection of the fittest solutions for mutation and crossover combinations to create next generations.

A GA requires a genetic representation of a solution, and a fitness function to evaluate the solution. A genetic representation of a solution is basically a way of expressing a possible solution such that GA operations of mutation and combinations can be applied. In the context of the deployment optimization problem, a solution is a deployment combination, and we use the vector **S**, as described earlier, as the genetic representation. This representation of a vector of ones and zeros is a common one used in GA, as it is a convenient form in which to perform genetic operations. The fitness function used to drive the evolution is the same metric used in the greedy algorithm, that is, the median UE throughput divided by cost of the deployment.

The overall process of searching for solutions using GA is as follows:

1. A population of N different solutions is randomly generated.
2. The fitness of each of the solutions in the population is evaluated.
3. The selection of "parent" solutions are chosen based on their fitness. In order to reduce the selection bias of high-performing solutions and introduce more randomness to the search, the selection is performed using the stochastic universal sampling technique [5]. Crossover and mutation operations are performed on the parents to generate offspring and populate the next generation.
4. Steps (2) to (3) are repeated for a set number of generations, and the solution with the highest fitness found throughout the evolution is selected.

Figure 15.11 shows an example of the crossover operation being performed on two parent solutions. In this example, the number of candidate sites is $n = 11$, and the number of sites to select is $n_d = 6$. Therefore, the solutions are represented by 1×11 vectors, and the sum of the elements of the vectors $\sum_{i=1}^{n} S_i = 6$. The crossover operation involves randomly selecting a crossover point, and swapping the elements of the two parent solutions at the crossover point to create two offspring, as illustrated in Fig. 15.11. However, the sum of the elements of the resulting offspring vectors may not be equal to n_d. In the example, the sum of the vectors Offspring 1 and Offspring 2 are 7 and 5, respectively.

In order to correct this, an additional step is taken after the crossover operation, and is illustrated in Fig. 15.12. As the sum of Offspring 1 is 7, one of the elements of the vector with the value 0 is selected randomly and changed to 1. For Offspring 2, the sum of the vector is 5, so one of the elements of the vector with value 1 is selected randomly and changed to 0.

In addition to crossovers, a mutation operation is also used to create new offspring. The mutation operation involves randomly choosing two elements of a parent

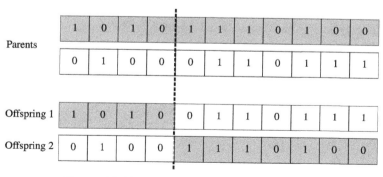

Figure 15.11. Example of a crossover operation.

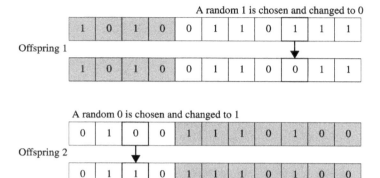

Figure 15.12.   Correction performed after crossover operation.

solution with the value 0 and 1, and changing their values to 1 and 0 respectively. The purpose of performing the mutation in this way is to preserve the requirement that the sum of the offspring vectors has to be equal to n_d.

15.4.5   Site Selection Using Steepest Ascent

In addition to GA, another possible approach that searches through different combinations of deployments is one that uses a steepest ascent (STA) technique:

1. To initialize the search, a deployment **S** is generated randomly, but with $\sum_{i=1}^{n} S_i = n_d$. The performance metric (median UE throughput/cost) of the deployment is calculated.

2. The performance metrics of the deployment when the status of a single site is changed is calculated. This step is illustrated in Fig. 15.13. In the example, the deployment **S** has 6 out of 11 candidate sites with a small cell deployed. Then, S_1 is changed from 0 to 1 (i.e., a cell is added at site 1), and then evaluated. This evaluation is repeated for all sites.

3. The occupied site that has the highest metric when the small cell is removed, and the empty site with the highest metric when a small cell is added is identified. In the example, this is sites 3 and site 6, respectively. The corresponding elements of **S** for these two sites are changed, and the metric of this new deployment is evaluated.

4. Steps (2) to (3) are repeated until the performance metric of **S** stops increasing. At this point, the algorithm has converged upon a solution.

The steepest ascent approach is similar to the greedy algorithm in the sense that, the trajectory of the search is dictated by the direction with the highest gain, and would therefore converge toward the nearest local optima. In order to overcome this, once

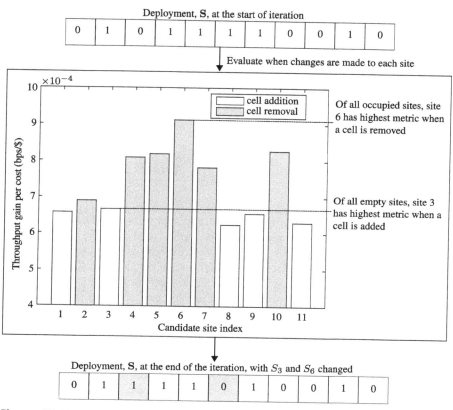

Figure 15.13. Illustrative example of the evaluation performed in one iteration of the steepest ascent approach.

the STA algorithm has converged to an optima, it is restarted with another random starting point. After a set number of restarts have been performed, the converged solution that has the highest metric is chosen as the final solution.

15.4.5.1 Performance of Genetic Algorithm and Steepest Ascent Small Cell Site Selection
The GA and STA approaches described were applied to the Manhattan scenario, to select 20 sites out of the 194 possible candidate sites to deploy small cells. For the GA approach, the population size, N, is set to 100. Half of the population is generated through mutation and the other half through crossovers, and the evolution is performed over 300 generations. For the STA approach, the algorithm converges to a solution after an average of 20 iterations. The search was performed with 20 restarts. The deployment obtained from the GA and STA approaches are denoted by S_{GAcost} and S_{SAcost}, respectively.

TABLE 15.2   Throughput and Cost of Deployments Selected Using Genetic Algorithm and Steepest Ascent

	UE Throughput Statistics				Cost
	Median	5th Percentile	95th Percentile	Total	
Macrocell-Only (no small cells)	263 kbps	66.2 kbps	1229 kbps	263 Mbps	–
S_{Gtput}	357 kbps	77.7 kbps	3396 kbps	537 Mbps	$662 k
S_{Gcost}	349 kbps	76.1 kbps	3341 kbps	534 Mbps	$563 k
S_{GAcost}	351 kbps	77.1 kbps	3293 kbps	536 Mbps	$568 k
S_{SAcost}	351 kbps	77.1 kbps	3293 kbps	536 Mbps	$568 k

In this scenario, the deployments that the two approaches produced were actually identical. Although the two algorithms were run for a relatively large numbers of iterations to ensure that the search space has been explored sufficiently, they actually found their solutions quickly. The GA approach evolved the solution after 45 generations, and the STA approach found it at its first attempt, after 19 iterations.

Table 15.2 shows the UE throughput and site costs resulting from the STA and GA approaches. Since the two approaches produced the same deployments, the results are the same for both. While S_{GAcost} and S_{SAcost} have a higher throughput gain/cost metric of 0.1549 bps/$ compared to 0.1528 bps/$ with S_{Gcost}, this difference is small, and the throughput and costs achieved is broadly similar.

However, when the number of small cell sites to select is increased, the advantage of using the GA and STA techniques starts to increase. Table 15.3 shows the

TABLE 15.3   Throughput and Cost of Deployments of Selected Deployments with Different Number of Small Cells Deployed

		Number of Small Cells Deployed			
		20	60	100	140
S_{Gtput}	Median throughput	357 kbps	523 kbps	664 kbps	773 kbps
	Cost	$662 k	$2055 k	$3453 k	$4829 k
S_{Gcost}	Median throughput	349 kbps	516 kbps	661 kbps	767 kbps
	Cost	$563 k	$1910 k	$3324 k	$4674 k
S_{GAcost}	Median throughput	351 kbps	514 kbps	660 kbps	765 kbps
	Cost	$568 k	$1893 k	$3245 k	$4613 k
S_{SAcost}	Median throughput	351 kbps	516 kbps	660 kbps	765 kbps
	Cost	$568 k	$1901 k	$3249 k	$4616 k

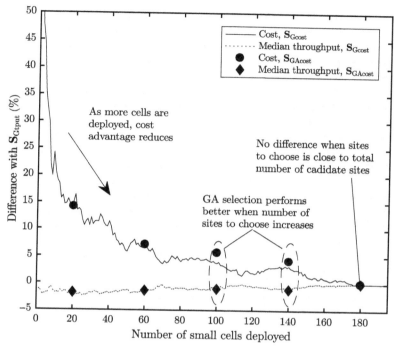

Figure 15.14. Difference in median throughput and cost of S_{Gcost} and S_{GAcost} relative to S_{Gtput}.

throughput and cost of the different deployment optimization approaches, and Figure 15.14 shows the difference in the throughput and cost of S_{Gcost} and S_{GAcost} (which are optimized for throughput and cost) relative to S_{Gtput} (which is optimized just for throughput) when the number of small cells to deploy, n_d, increases. We can see that when n_d is low, the cost of S_{Gcost} is significantly lower than S_{Gtput}. However, this cost advantage quickly decreases as the number of small cells deployed increases. This is because sites that have a large throughput to cost ratio are usually chosen first, and as more cells are deployed, sites that have a lower ratio are used. Eventually, the cost advantage disappears if the number of cells deployed is close to the total number of candidate sites, n. On the throughput side, as expected, S_{Gcost} achieves a lower median throughput than S_{Gtput}. However, the difference between the two remains low regardless of the number of cells deployed, with the maximum difference of 2.4% and an average difference of less than 1%.

When the number of small cells deployed is low compared to the total number of candidate sites (at 20 and 60 small cells deployed), we can also see that the difference between S_{Gcost} and S_{GAcost} is small. However, when the number of small cells increases, S_{GAcost} has an appreciably better performance in terms of cost, with minimal differences in throughput. With 100 and 140 small cells, S_{GAcost} has a cost that is 6% ($208k) and 4.5% ($216k) lower than S_{Gtput}, respectively, while S_{Gcost} has a difference of 3.7% ($129k) and 3.2% ($155k), respectively.

This is because when larger numbers of small cells are deployed, the choices that have to be made become more difficult, as the obvious sites with low cost and high throughputs are taken. In cases like these, the GA and STA approaches have the advantage of being able to explore the solution space more comprehensively to provide better deployments.

15.4.6 Analysis of Small Cell Placements

In this section, an analysis of the deployment sites chosen by the algorithms is done to gain some insights into the factors determining good small cell sites.

In order to give a better picture of the impact of deployments on the UE throughput, a macrocell throughput map is generated. The map is generated by calculating the UE throughput that can be achieved at each location on the map, if there are no small cells deployed and the UE distribution follows the UE demand map shown in Fig. 15.4. This throughput calculation takes into consideration the macrocell SINR and the macrocell load. The resulting throughput map calculated is shown in Fig. 15.15. The areas with high macrocell throughputs tend to be located close to the macrocell sites due to the higher levels of SINR, although areas covered by highly loaded macrocells have relatively lower throughput levels.

Combining the macrocell throughput map with the UE demand map shown in Fig. 15.4, we obtain a combined map shown in Fig. 15.16. The map shows areas where macrocell throughput is less than 300 kbps, and UE density is more than 600 UEs per km^2. The areas highlighted in the map are broadly similar to the UE demand

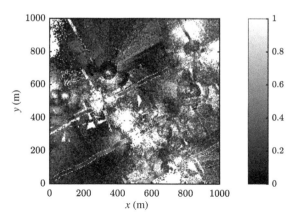

Figure 15.15. Map of macrocell throughput, units are in Mbps.

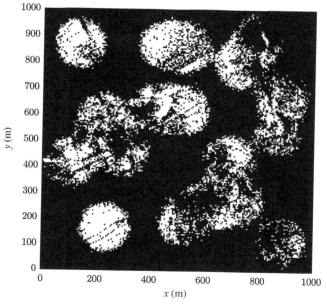

Figure 15.16. Map of areas where macrocell throughput is less than 300 kbps and UE density is more than 600 UEs per km² (white).

hotspot areas, but with areas close to the macrocell removed, as these areas have high macrocell SINR levels. These are areas where the deployment of small cells is desirable, as it highlights areas with many UEs that are experiencing low throughput levels. For the sake of brevity, we define these areas as high offload areas.

Figure 15.17 shows the S_{GAcost} deployment for 100 cells overlaid on top of the high offload area map in Fig. 15.16. The white circles indicate the sites where a small cell is deployed, the white triangles show the macrocell antenna locations, and the white crosses show the empty candidate sites. The white lines show the borders of the small cell coverage areas. We can see that the small cells have been deployed largely to cover the high offload areas. While the average coverage area of a small cell is around 2900 m², there are small cells deployed that have really small coverage areas, with the smallest covering an area of 950 m². These small cells, however, are deployed in sites that have a combination of low cost, and are within high offload areas. So, while it may be a good rule of thumb to avoid placing small cells at sites close to macrocells to achieve higher coverage areas, these sites can become feasible when taking cost and offload into consideration. Conversely, there are small cells that are deployed in locations outside the high offload areas. These tend to be sites where costs are relatively low and the small cell coverage achieved is high.

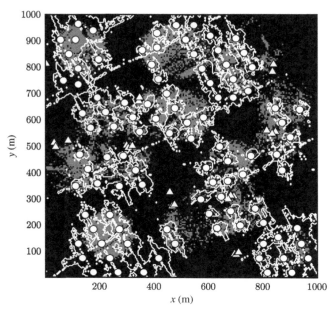

Figure 15.17.   High offload areas with the S_{GAcost} deployment with 100 small cells.

15.5 SUMMARY AND CONCLUSIONS

The correct placement of small cells is important to maximize the gains of small cell deployments. Placing small cells in the wrong location can not only result in low gains in capacity, but in the worst case can even cause a degradation in capacity due to increased interference. However, capacity gains is just one of the factors that needs to be considered when planning where to place small cells. The site costs also have to be considered in order to maximize the overall return of investment of a deployment.

In this chapter, an example of the optimization of a small cell deployment that can be performed is given, using several different optimization algorithms. In the scenario used, it was shown that the deployment optimization that considers cost and throughput can achieve very similar throughput gains but with significant cost savings compared to deployment optimizations that only consider throughput. The deployments chosen by the optimization algorithms are fairly consistent with the perceived wisdom of placing small cells at UE demand hotspots, or away from macrocell sites to achieve larger coverage areas. One can place a small cell close to a macrocell site if the small cell is located within an area with many UEs, and conversely, a small cell can be deployed in an area with a low number of UEs if it has a large coverage area due to being far away or shielded from the macrocell sites.

However, these deployment optimization decisions can only be performed if the UE demand and radio propagation information has sufficient accuracy. Obtaining these inputs for the optimization can be considered one of the main challenges behind small cell deployment optimization. Due to the small coverage area of small cells, the UE demand maps needs to be derived with accurate UE geo-location, of at least tens of meters. Current cellular-based geo-location approaches have an accuracy on the order of 100 m, so more accurate but practical techniques are needed to geo-locate where UE calls and sessions are made. Accurate radio propagation maps are also needed. While techniques such as ray tracing can achieve very good accuracy, the high computation resources required to cover large areas with many candidate sites can make this difficult. Alternative techniques that can provide sufficient accuracy with lower computation time are needed to perform optimization of large-scale deployments.

As the deployment of small cells becomes more widespread, the deployment optimization process, and in particular the generation of the inputs for the optimization, needs to be automated as much as possible. This remains the subject of ongoing research, but is an essential part of enabling the rapid deployments of small cells by reducing or eliminating the time and cost of manual cell site surveys.

REFERENCES

1. H. Liu, H. Darabi, P. Banerjee, and J. Liu, "Survey of wireless indoor positioning techniques and systems," *IEEE Trans. Syst. Man Cybern., Part C: Appl. Rev.*, vol. 37, no. 6, pp. 1067–1080, Nov. 2007.

2. "Radio measurement collection for minimization of drive tests (MDT): Overall description, stage 2," 3GPP Technical Specification, TS 37.320, v10.4.0, Jan. 2012.

3. D. Chizhik, J. Ling, and R. Valenzuela, "Self-alignment of interference arising from hallway guidance of diffuse fields," *IEEE Trans. Wireless Commun.*, vol. 13, no. 7, pp. 3853–3862, Jul. 2014.

4. D. Goldberg, *Genetic Algorithms in Search, Optimization and Machine Learning*. Addison-Wesley, 1989.

5. J. E. Baker, "Reducing bias and inefficiency in the selection algorithm," in *Proc. Second Int. Conf. Genet. Algorithms Genet. Algorithms Appl*, Hillsdale, NJ: L. Erlbaum Associates Inc., 1987, pp. 14–21. [Online]. Available: http://dl.acm.org/citation.cfm?id=42512.42515

PART VI

FUTURE TRENDS AND APPLICATIONS

ULTRA-DENSE NETWORKS

16.1 INTRODUCTION

Driven by a new generation of wireless user equipments (UEs) and the proliferation of bandwidth-intensive applications, mobile data traffic is predicted to grow exponentially over the coming years (see Section 1.1). In order to meet these increasing traffic demands, cellular operators are adopting various approaches to further enhance network capacity using, for example,

- more spectrum (new frequency bands, whitespace, and cognitive radio)
- more antennas (multiple-input multiple-output (MIMO) techniques)
- higher-order modulation schemes (256 quadrature amplitude modulation (QAM) being studied in Third-Generation Partnership Project (3GPP) [1])
- inter-base station (BS) coordination (coordinated multi-point (CoMP) schemes)

However, while these, along with other various techniques, can help to improve the network capacity, the most significant gains in cellular networks have been achieved from the higher spatial reuse of spectrum through the deployment of smaller and

Small Cell Networks: Deployment, Management, and Optimization, First Edition. Holger Claussen,
David López-Pérez, Lester Ho, Rouzbeh Razavi, and Stepan Kucera.
© 2017 by The Institute of Electrical and Electronic Engineers, Inc. Published 2017 by John Wiley & Sons, Inc.

smaller cells (see Chapter 2). Increasing capacity in line with the predicted growth of traffic demand may not require the deployment of additional expensive macrocells if small cells are deployed in hotspots where the offloading gains are maximized. Moreover, small cells are also able to provide significant power savings. Due to its small coverage area, a small cell BS can be switched on and off depending on the current traffic distribution, resulting in a reduced network power consumption. The short distances between transmitters and receivers also translate to lower uplink transmit powers, resulting in enhanced UE battery life.

In this chapter, the progression of further exploiting the capacity gains offered by small cells through the use of very small cells in indoor scenarios is explored, and the concept of an attocell is proposed, which is characterized by a very low-power BS (relative to femtocells) that aims at providing high capacity and tailored coverage to individual indoor UEs.

The remainder of the chapter is structured as follows. In Section 16.2, different scenarios where attocells can be employed are discussed, and compared with other existing solutions. In Section 16.3, the various challenges involved in deploying attocells and the potential achievable gains are outlined. In Section 16.2, results of long-term evolution (LTE) system-level simulations of an attocell deployment are presented, showing the potential gains in terms of capacity enhancements, while in Section 16.5, the impact of attocell deployments on UE mobility is examined in more detail. In Section 16.6, the potential performance gains of attocell deployments with advanced enhancements in LTE such as higher-order modulations and higher bandwidths are discussed. Finally, the conclusions are given in Section 16.7.

16.2 THE ATTOCELL

Within the scope of this chapter, attocell BSs are defined as very low-power, low-cost BSs that are deployed in high densities indoors to provide a very high wireless capacity to targeted areas with a coverage of typically less than 10 m in diameter. Attocells aim at enhancing the achievable throughput of UEs and addressing operator's capacity deficits thought effective indoor data offloading.

Figure 16.1 illustrates the attocell concept along with the different tiers of a cellular network. At the top tier, in terms of intended coverage size, macrocells provide a wide coverage, while metrocell overlays provide capacity at targeted indoor/outdoor UE hotspot areas (e.g., shopping malls and bus stops). Femtocells overlay macrocells and metrocells to provide indoor coverage in residential and enterprise scenarios. Attocells lie at the bottom of these tiers, and their main objective is to provide the highest level of capacity to covered UEs. This is achieved by exploiting spatial reuse gains since the number of UEs connected to each attocell is very small due to its highly localized coverage area. Table 16.1 presents a comparison of attocell BSs features against other existing types of BSs in heterogeneous cellular networks (HCNs).

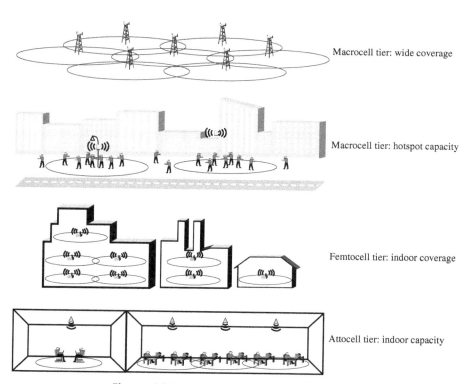

Macrocell tier: wide coverage

Macrocell tier: hotspot capacity

Femtocell tier: indoor coverage

Attocell tier: indoor capacity

Figure 16.1.   Overview of the attocell concept.

As such, attocells are targeted at static indoor UEs in locations such as homes, offices, and public areas such as airports, conference centers, and stadiums. They would be deployed in relatively large numbers, and consequently need to be easy to deploy. It is envisioned that attocells are as easy to install and as ubiquitous as light fixtures in a building, which are preferably deployed at ceilings. With the proposed ceiling mount, attocell BSs will provide coverage from ceiling to floor, that

TABLE 16.1   Typical Specifications of Different Elements in Heterogeneous Networks

Types of Nodes	Transmit Power	Coverage	Typical Backhaul	Spatial Reuse
Macrocell	46 dBm	Few km	Fiber, cable, wireless	Low
RRH	30 dBm	300 m	Fiber	None
Relay	30 dBm	300 m	Wireless	None
Metrocell	23–30 dBm	<300 m	Fiber, cable, wireless	Medium
Femtocell	<23 dBm	<50 m	Wireline IP	High
Attocell	<10 dBm	<10 m	Wireline IP	Very high

is, top-down, and thus BS-to-UE communications will be mostly line-of-sight (LOS) and channel degradation effects such as multi-path and inter-cell interference can be minimized. Because of their very low transmit power, attocell coverage areas can also be well isolated by walls.

Unlike other low-power radio access technologies that transmit on unlicensed radio bands such as wireless fidelity (Wi-Fi) and Bluetooth, attocells benefit from a better interoperability with the underlay cellular network as well as interference mitigation and cooperation. Attocells can be more easily integrated into the cellular operator network, and provide seamless mobility in conjunction with the umbrella macrocell, metrocells, or femtocells. Additionally, attocells, being part of the cellular network, benefit from quality of service provisioning and integrated security measures and services. They also avoid external sources of interference, common in unlicensed radio bands. Nevertheless, as a general concept, attocell Wi-Fi access points can also be implemented and be appealing to some service providers.

This vision of ubiquitous, very small cells can potentially provide many advantages, but it also entails many challenges from issues relating to architecture and hardware together with self-configuration, inter-cell interference, mobility management, and back-hauling. These are further discussed in the following.

16.2.1 Architecture

In an attocell environment, operators need to provide secure and scalable interfaces at reasonable costs to connect attocells to the core network. Depending on the scenario and the number of neighboring attocells deployed per household or in a building, different interfaces may be used.

Home scenarios are characterized by a low number of attocells per building (<5 attocells). In this case, attocells can adopt the same architecture as femtocells, and connect to a femtocell gateway through the public Internet using a high-speed broadband connection. The attocell would be seen as a femtocell by the femtocell gateway.

Small and medium enterprise scenarios are characterized by a larger number of deployed attocells per building (between 5 and 250 attocells). In this case, attocell traffic can be aggregated through a local gateway (i.e., the attocell gateway), and then routed to the core network through the operator's network or the public Internet using the enterprise broadband connection. The attocell gateway would be installed in intended buildings, and can help to quickly establish coordination among attocells in order to handle inter-cell interference coordination and other network functions without any core network supervision.

Large enterprise and shopping mall scenarios are characterized by a significant number of deployed attocells per building (>250 attocells). In this case, attocells may be operator-deployed and the attocell gateway should be connected to the core network through high-capacity wired or wireless interfaces (e.g., fiber, microwave). Proper back-haul dimensioning and tailored back-hauling solutions are necessary since the aggregated attocell traffic could be on the order of gigabits per second.

As can be derived from this discussion, in terms of architecture, the adopted solution depends on the particular scenario to be addressed as well as the level of agreement between the owner of the attocell network and the network operator. Attocell networks connected through an attocell gateway deployed by the network operator may benefit from a better connectivity to the core network. Another attocell network architecture, in line with the virtual radio access network concept, may allow attocell base-band processing and/or medium access control functionalities to be moved to an enhanced attocell gateway capable of performing centralized optimization such as clustering and joint scheduling. Evidently, this requires a more powerful attocell gateway and higher bandwidth connections to the attocells than for previous cases.

16.2.2 Hardware

The hardware of attocell BSs needs to be low-cost and physically small such that they can be integrated and deployed easily and in large numbers. With the continuing advancement of small cell BSs hardware, it is expected that the cost and size requirements of attocell BSs can be met.

A key hardware component of attocell BSs will be its antenna, due to its role played in inter-cell interference mitigation. A promising approach for inter-cell interference mitigation is the use of ceiling mounted directional antennas. In this way, attocells can be deployed in very high densities without causing excessive inter-cell interference, since they provide a much more localized coverage from up to down. The size of antennas of attocell BSs depends on aspects such as the frequencies used and the desired beamwidth. Generally, the higher the frequency and the wider the beamwidth, the smaller the antenna can be. For indoor attocell deployments, relatively small antenna arrays can be used. An example of the use of such directional antennas in attocell BS is given in Section 16.4.

16.3 CHALLENGES IN ATTOCELL DEPLOYMENTS AND OPERATION

From the wireless perspective, the deployment of attocells brings about opportunities but also challenges. Major technical challenges arise due to the coexistence of a large number of attocells with various types of other cells (e.g., macrocells, metrocells, femtocells), and are mainly related to self-organization, inter-cell interference, mobility management, energy efficiency as well as backhaul provisioning. These challenges need to be addressed to maximize the benefits from attocell networks. There is also a need for the industry to better understand the technical details and performance gains that attocell networks could bring. In this section, the major technical challenges that need to be addressed for successful attocell deployments are described.

16.3.1 Self-Organization

Since attocells will be deployed in very large numbers, human-based planning, configuration, and optimization of attocell networks is not cost-effective [2]. Therefore, attocells must be simple plug-and-play equipment with a large degree of self-organization, which should mainly work in two stages:

Self-Configuration: Attocells should automatically connect to the network, and then establish the necessary security contexts. Configuration files can be thereafter downloaded from the network, and provide the initial configuration for attocells. Once the latest software release has been downloaded, attocells should perform a self-test to check its installation and then start operation.

Self-Optimization: Since network load and UE traffic may quickly change depending on location and time (e.g., attocells switching on and off, UEs arriving, moving and departing) and because channel conditions can rapidly vary due to interference and fading, attocells should dynamically adapt their parameters to the fluctuating radio environment to maximize the overall network capacity.

The self-organization process for attocells can be implemented in a more flexible manner than that of femtocells if a local attocell gateway is available. It can be implemented through hybrid self-organizing approaches, where parts of the self-organizing schemes are executed at the attocell gateway, while others are executed at the attocells themselves. This hybrid approach can combine central and distributed intelligence. In this line, complex and computationally intensive optimization problems, such as the assignment of frequency carriers may be solved by a scheduler controlled by the attocell gateway, while power control and radio resource assignments may operate independently at each attocell much more frequently.

16.3.2 Inter-cell Interference

Similar to other small cell BS types, attocells can be deployed in open, closed or hybrid access modes as described in Chapter 4. However, open or hybrid access models are more appealing for the enterprise scenarios, as they provide more flexibility and avoid the potential inter-cell interference problem that could arise in the case of closed access. Due to the better inter-cell interference properties of open access, operators may also adopt co-channel deployment approaches where attocells share the spectrum with the underlying macrocells, metrocells, or femtocells. This shared carrier approach potentially results in enhanced spatial reuse in comparison with separate carrier and partially shared carrier approaches. However, because attocells are deployed in very high densities, the impact of inter-cell interference amongst neighboring attocells (inter-attocell interference) is very significant, particularly if they are placed within LOS of each other. As this can cause a very large degradation in performance in co-channel deployments, controlling inter-attocell interference effectively is a necessity to ensure that the potential gains of deploying attocells are not limited by excessive interference.

In order to mitigate inter-cell interference, LTE attocells benefit from their flexible orthogonal frequency division multiple access (OFDMA) physical layer and make use of standardized 3GPP Release 10 and Release 8 inter-cell interference coordination (ICIC) techniques [3]. Attocells may configure Release 10 almost blank subframes (ABSs) to avoid inter-cell interference toward pedestrian users connected to macrocells or metrocells. Attocells may additionally use Release 8 relative narrowband transmit power (RNTP) indicators, high-interference indicators (HIIs), and interference overload indicators (IOIs) over the attocell gateway for inter-cell interference coordination among attocells. Moreover, in-built sensing techniques that allow switching on and off attocells depending on whether active UEs are present in their coverage areas will play a key role in attocell networks to mitigate inter-cell interference as well as reduce unnecessary handovers and save energy.

However, since the configuration of ABS at attocells may result in significant attocell performance degradation due to the blanking of resources, attocells call for innovative power control and resource allocation algorithms (e.g., power-reduced subframes) to further mitigate inter-cell interference. These advanced techniques also introduce additional signaling to achieve coordination between cells and place additional delay constraints on the backhaul links. The added complexity also adds to the cost of deploying attocells in terms of additional hardware and software capability.

Apart from coordinated techniques, another way in which inter-cell interference can be managed is through the use of directional antennas. Placing attocell BSs on the ceiling with directional antennas pointing downward (90 degrees tilt) allows attocell signals to be constrained to a specific area, and therefore limit the inter-cell interference to neighboring attocells and other nodes. This can significantly decrease the amount of inter-cell interference between attocells, and help attocells overcome the interference from the underlay macrocell without introducing the complexities of coordinated interference management techniques. The impact of directional antennas is explored in detail in Section 16.4.

16.3.3 Mobility

Due to the reduced coverage of attocells, under current mobility management procedures, the presence of mobile and pedestrian users would cause an excessively large number of handovers in an attocell scenario, leading to high signaling overhead. This, in combination with the narrower cell boundaries, may lead to an increased call drop probability and performance instability if co-channel deployments are considered. These mobility management issues in multi-tier networks have already been identified as a problem with small cells in general [4], and it will be exacerbated with attocells. Therefore, new mechanisms are necessary so that UEs that are just moving past the attocell coverage do not initiate a handover to the attocell, while being able to maintain their connections with their serving cells.

Different mechanisms can be implemented to realize such mobility management depending on whether UE mobility state estimation (MSE) is available at the UEs. However, the reliability and accuracy of current MSE mechanisms based on UE measurements is still under investigation [5]. Alternatively, we may think to realize new MSE mechanisms, in which attocells themselves or the attocell gateway keeps track of the number of cell reselections and handovers requested per UE and unit of time to derive whether a UE is moving or static.

If MSE information is available, only semi-static UEs will request a handover to attocells. Otherwise, UEs may initiate a large handover time-to-trigger toward attocells once the triggering event conditions are met, so that UEs only request handovers if they have spent a long enough time under the attocell coverage. This indicates that these UEs are static UEs due to the reduced attocell coverage. Both non-MSE and MSE based methods require inter-cell interference coordination to avoid UEs declaring radio link failure while under the attocell coverage and connected to another cell due to skipped or delayed handovers. As mentioned in the previous section, in order to avoid radio link failure (or handover failure), LTE attocells may schedule ABSs so that macrocells and metrocells can schedule their UEs in subframes overlapping with the ABSs of attocells. Moreover, in order to handle mobility management, more advanced network architectures and techniques based on, for example, the splitting of the control and UE plane can also be adopted as suggested in [6]. The issue of mobility will be discussed in more detail in Section 16.5.

16.3.4 Energy Efficiency

With more legislation being introduced to protect the environment, operators are seeking for solutions to enhance their green credentials. Moreover, since electricity bills account for a considerable part of the overall network operational expense, there is a strong motivation for mobile operators to become more energy efficient. In addition to capacity and coverage benefits, small cells can be considered as an effective solution to reduce the overall network energy consumption. As shown in [7], with appropriate idle mode procedures in place, significant energy reduction gains (up to 46 fold) can be achieved when deploying small cells compared to a macrocell-only scenario. In addition to switching off small cells on a per-session basis, there are also proposals to introduce fast cell discontinuous transmission (DTX) in LTE, where the BS turns off the radio transmitted for orthogonal frequency division multiplexing (OFDM) symbols that do not carry any data or reference symbols [8]. Studies have shown that fast cell DTX reduces the energy consumption of heterogeneous small cell deployments by a further 31% to 37% [9].

For attocell BSs, the energy saving gains can potentially be even more significant if appropriate low-power sleep modes are efficiently used. In sleep mode, the BS switches off parts of its functionality in order to reduce its power consumption. By utilizing various sleep mode techniques as those described in [10], attocells can be made to autonomously go to sleep when not needed, and wake up when necessary.

Attocells are particularly well-suited to the implementation of sleep modes such as those described in Chapter 13, since their limited coverage area implies that there will be long periods of time when there will be no active UEs within its coverage. This fact allows attocells to enter sleep modes more often compared to cells with larger coverage areas.

If implemented well, sleep modes not only provide benefits in terms of increasing the overall energy efficiency, but they could also mitigate inter-cell interference as well as reduce the amount of unnecessary handovers that an attocell would cause if it would be powered on at all times.

16.3.5 Backhaul

Since the radio propagation conditions between the attocell and the UE tends to be favourable due to the small distance between them both, the achievable UE throughputs over the radio link are typically high as well. Taking, for example, a 10 MHz bandwidth LTE attocell, the maximum achievable downlink data throughput is on the order of 36.3 Mbps (assuming 50 resource blocks (RBs), 12 subcarriers per RB, 11 data OFDM symbols per subframe and 5.5 bit/symbol efficiency).

Given that attocells are expected to be deployed in very high densities, the local backhaul connecting the attocell to the local attocell gateway should be one that provides high capacity, but is also fairly low-cost. An example would be the use of wired Ethernet, which offers reliable, high-speed connections and is widely available in most commercial buildings, although some installation effort is required to pull the cables to the specific locations where attocells are located. In this case, attocells can also be powered by Power over Ethernet (PoE).

At the attocell gateway (assuming a local gateway at the premises is used), the aggregation of traffic from all attocells deployed within the premises may result in a significant link capacity required from the attocell gateway to the operator's core network, particularly if the number of attocells is large and the UE activity is high. Using features such as local breakout may reduce the amount of data traffic to be routed to the core network, but it does not reduce the overall backhaul capacity of the gateway to the external network. However, taking into account that attocell gateways may be deployed by network operators where attocell densities are high, and with the increasing availability and reducing costs of high-capacity backhaul connections such as fiber, it is expected that the required backhaul linking the attocell gateway to the core network or the public Internet would be sufficient in the near future.

16.4 SIMULATION OF AN ATTOCELL DEPLOYMENT

In this section, system-level simulations are used to examine the potential gains in terms of network throughput that can be achieved with an attocell deployment using ceiling mounted directional antennas.

16.4.1 Simulation Scenario

LTE attocells are deployed within a section of an office building, in the simulated area of 40 m by 25 m shown in Fig. 16.2. In this scenario, 29 attocells are deployed to provide coverage in the office cubicles and within enclosed rooms. Due to the importance of an accurate indoor radio propagation simulation to assess the performance of attocells, we have used the wireless system engineering (WiSE) tool [11], which is a comprehensive 3D ray tracing software capable of accurately capturing reflection and diffraction phenomena. The carrier frequency is 2.6 GHz and the ray-tracing is performed with the maximum number of reflections set to 5. The heat map in Fig. 16.2 shows an example of the received powers from one attocell using this ray-tracing tool.

The intended coverage area for each attocell is 5 m in diameter, with a macrocell tier underlay. Therefore, the attocells at the cubicles are placed roughly 5 m from each other, while one attocell is placed in each of the smaller rooms, and two are placed in larger rooms. For comparison purposes, the attocell BSs are deployed either on the ceiling of the building, using directional antennas with beams pointing directly downward onto the area of intended coverage, or at a height of 1.5 m, using omni-directional antennas.

The transmit powers of attocell BSs are set to between 0.5 mW and 160 mW. The simulations with high transmit powers are used for illustration purposes only, and those with omni-directional antennas are presented here to understand the effects of antenna directionality. In practice, the transmit power of attocells is expected to be low (10 mW or less) and their antennas should be directive. The macrocell tier underlay is comprised of 4 macrocell BSs that are located outside the office building, with the closest macrocell located approximately 250 m away from the building. The

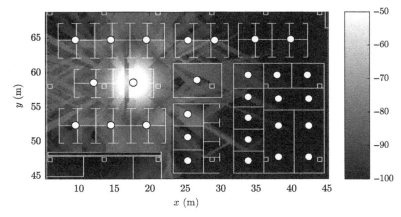

Figure 16.2. Illustration of simulated office scenario. Green dots are attocell BS positions, red lines are walls, and white lines are cubicle partitions. The heat map showing the RSCP of a narrowbeam antenna attocell is overlaid.

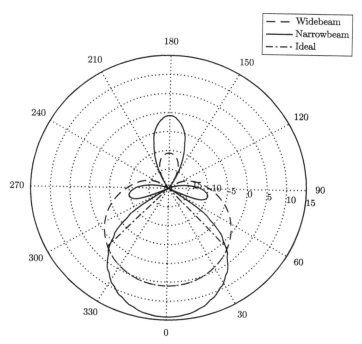

Figure 16.3. Directional attocell antenna gain patterns.

macrocell BSs provide good indoor coverage within the building, with an average macrocell reference signal received power (RSRP) of -70 dBm, and the RSRP never falling below -92 dBm within the simulated area.

Ideally, attocells should provide coverage only in the intended areas, and have minimal leakage outside the intended coverage in order to minimize inter-cell interference. Therefore, the attocell antenna gain patterns and directionality have a significant influence on the performance of the attocell network. In order to investigate this, three different directional attocell antennas are used in the simulations, whose antenna gain patterns are shown in Fig. 16.3. For the sake of brevity, we name them "widebeam," "narrowbeam," and "ideal" antennas. A fourth antenna used is an omnidirectional dipole antenna with a uniform gain along the horizontal axis of 2 dBi.

The widebeam antenna is derived from a patch antenna with some level of directionality, but with a relatively wide beamwidth. The narrowbeam antenna is derived from a multi-element array antenna with a narrower beamwidth than the widebeam antenna. The antenna array is composed of a 3×3 dipole antenna array, with a single element of the dipole array shown in Fig. 16.4. The physical dimension of the narrowbeam antenna array is approximately 180×180 mm. The ideal antenna is one with an artificially created antenna pattern that is derived from the widebeam antenna

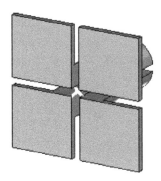

Figure 16.4. Diagram of the single-element dipole antenna model used for the narrow-beam antenna gain calculation.

pattern. It was created by using the same gains as the widebeam antenna between specific angles, and nulls placed at other angles. The angles are chosen such that when it is placed at the ceiling, it would provide a fairly uniform gain over a 5 × 5 m area directly below it and nulls outside that area. This contrived ideal antenna pattern is used here to illustrate the performance when an antenna that possesses the ideal characteristics described above is used.

A total of 165 static UEs with full buffer traffic models are placed randomly within the simulated area. 145 UEs are placed within the cubicles and rooms (and therefore within the attocell intended area of coverage), while the remaining 20 UEs are placed randomly in other areas, such as corridors. Subsequently, the UE's BS association is performed based on RSRP, and signal-to-interference-plus-noise ratios (SINRs) are calculated. UE SINRs are then translated to UE's downlink throughputs using an SINR to throughput mapping for LTE [12], with the assumption that the channel bandwidth is 10 MHz. A round robin scheduler is assumed, and no ICIC techniques are used. This process is performed 1000 times with different UE placements. The simulations were performed with the attocells, and to provide benchmark comparisons, without the attocells (i.e., with just macrocell coverage) and with four femtocells. The femtocells have a maximum transmit power of 100 mW, use an omni-directional antenna, and are deployed in locations that would provide the best coverage over the simulated area.

16.4.2 Simulation Results

Figures 16.5 and 16.6 show the mean UE SINR and the percentage of UEs offloaded to the attocells, respectively, in the simulations with different attocell antennas and transmit powers. With the widebeam and omni-directional antennas, the UE SINR is relatively low and saturates quickly, as the transmit powers are increased due to

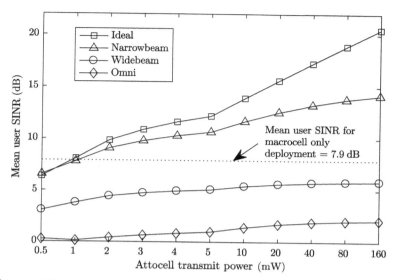

Figure 16.5. Mean UE SINR for different attocell antennas and transmit powers.

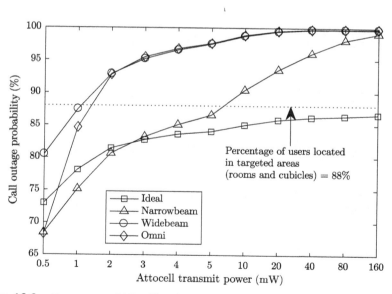

Figure 16.6. Percentage of UEs offloaded to attocells for different attocell antennas and transmit powers.

the high levels of inter-cell interference (interference limited scenario). With the narrowbeam and ideal antennas, the amount of inter-cell interference is reduced and the signal quality from the serving attocell is increased. As a result, the mean UE SINR is higher and improves with higher transmit powers, with the gains saturating at a much higher power level (transition toward a noise limited scenario).

For the ideal antenna, due to the absence of side-lobes and the sharp drop-off in gains at the precisely calculated angles, the leakage of attocell signals outside the intended area of coverage is largely eliminated. Therefore, the UE SINR increases steadily with higher transmit powers, and does not saturate within the simulated range of power levels (noise-limited scenario). This increase in SINR for the ideal antenna will start to level off eventually due to the effect of reflections, but this only starts to happen when the attocell transmit power is above 300 mW.

This antenna directionality also helps to constrain the coverage of the attocell within the targeted coverage areas, avoiding areas such as corridors and walkways. An optimum offloading in this scenario is therefore to offload the 88% of UEs located in the cubicles and rooms, and to avoid UEs outside these areas, which are the remaining 12% of UEs located in the corridors. The more predictable coverage area also allows to better optimize the offload toward the desired UEs, as shown in Fig. 16.6, and reduces the number of unnecessary handovers from users walking by.

Figure 16.7 shows the gains in throughput of the attocell deployments using directional antennas relative to the baseline of the attocell deployments using

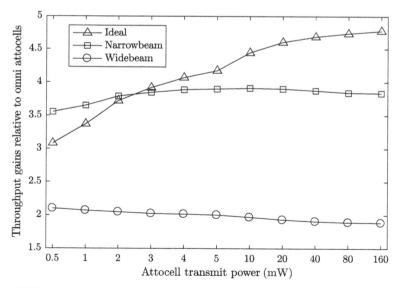

Figure 16.7. Gains in throughput achieved with attocells of different antennas and transmit powers, compared with a macrocell-only deployment.

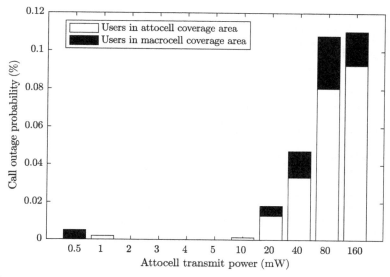

Figure 16.8. UE call outage probabilities for narrowbeam attocell deployment.

omni-directional antennas. The gains obtained due to the more targeted offloading and the overall improvement in SINR are significant, particularly with the narrowbeam and ideal antennas, where gains of more than four times are achieved.

While the deployment of attocells can provide an overall increase in system throughput, it also introduces higher interference for UEs located at the edge of the attocells. In order to gauge this effect, the call outage probability for the narrowbeam attocell is shown in Fig. 16.8, since the narrowbeam antenna is the most effective realistic one for use in attocells. A UE is defined as being in outage if the UE's SINR is less than −6.5 dB [13]. As a reference point, the call outage probability for the macrocell-only deployment is zero. In the latter case, the proportion of UEs experiencing outage is mostly located in the coverage area of an attocell since the majority of UEs are placed within cubicles and rooms.

When the attocell transmit power is below 2 mW, the attocell is not able to provide sufficient coverage over its targeted areas, resulting in low SINR for UEs located within those areas and consequently some outage occurring. As the transmit powers are increased, the coverage area of the attocells becomes more well-defined within the targeted areas, and the low SINR areas at the edge of the attocells start to disappear, since the antenna directionality keeps the received signals from the attocells high within its coverage area and low outside it. As a result, no outage is experienced between 2 mW and 5 mW. However, as the power is increased, the effect of interattocell interference becomes more significant due to the presence of the antenna sidelobes, and consequently call outages start to occur at 10 mW and increase as the power is increased.

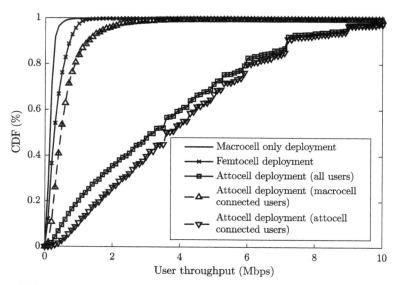

Figure 16.9. UE throughput CDF for macrocell-only, femtocell, and 5 mW narrowbeam attocell deployments.

The transmit powers of an attocell therefore should be set high enough to ensure that it is able to capture UEs within its intended area of coverage and achieve high SINR, but not too high as this will cause excessive inter-attocell interference. It should be noted, however, that the call outage probabilities do remain very low, with values of about 0.1% in the worst case.

Referring to the results shown in Fig. 16.8, it can be observed that a suitable atto-cell transmit power for the simulated scenario for the narrowbeam antenna is 5 mW. This configuration will be used as an exemplary case for the attocell deployment, to compare its UE throughput performance with two other baseline scenarios: the macrocell-only deployment and a deployment with four femtocells with the macro-cell underlay. In the femtocell deployment scenario, the femtocells have a maximum transmit power of 100 mW, use omni-directional antennas, and are deployed at a height of 1.5 m in locations that would provide the best coverage over the simulated area. The cumulative distribution function (CDF) of the UE throughputs for the three scenarios are shown in Fig. 16.9.

It can be seen that the throughputs achieved by UEs in the attocell deployment scenario is consistently higher than both the femtocell and macrocell-only deploy-ments, due to the increased number of cells and efficient spatial frequency reuse. One thing to note in particular is that the throughputs of the macrocell connected UEs are also higher than the macrocell only and femtocell deployments. This is mainly due to the offloading of UEs by the attocells, although the reduced interference experienced by the macrocell connected UEs due to the directionality of the attocell antennas is also a contributing factor. In contrast, the macrocell-connected UEs in the femtocell scenario does not benefit from this reduction in interference. This is reflected in the higher mean UE SINR for macrocell-connected UEs in the attocell scenario of 2.2 dB, compared with −0.5 dB for the femtocell scenario.

16.5 SIMULATIONS OF MOBILITY

One of the most obvious effects of deploying cells in such high densities is the impact of increased mobility. To examine this, mobility simulations of a UE moving in the office scenario are performed.

Figure 16.10 shows the path that the UE takes, moving at a speed of 4 km/h along a corridor between two rows of cubicles, and moving into and out of a cubicle before proceeding again down the corridor. The transmit powers of the attocells are set to 5 mW. The conditions for handover failure are modeled based on the assumptions in [14]. The handover time-to-trigger (TTT) was set to 40 ms, the hysteresis margin applied to the macrocell is 3 dB and to the attocells is 0 dB, and the Layer 3 filter coefficient K is set to 1. The attocells are all assumed to be actively transmitting on the downlink channels at full power, so the simulation captures a worst-case scenario from a handover failure point of view due to the increased interference at the cell edges.

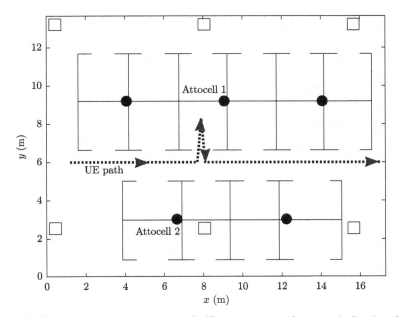

Figure 16.10. Map showing a portion of office scenario, with arrows indicating the UE path used in mobility simulations, and circles indicating attocell locations.

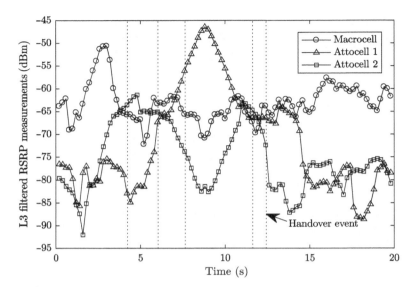

Figure 16.11. UE Layer 3 RSRP measurements moving along the path, with the dashed vertical lines indicating the times where handover occurs. Only measurements from BS the UE hands over to are shown.

Figure 16.11 shows the resulting RSRP measured by the UE, and the points at which handovers occur. Due to the attocell's directional antennas causing narrow cell borders with very sharp increases in signal strengths, the handover parameters are set to perform quick handover triggers to ensure no handover failures occurred. However, this has resulted in an excessively large number of handovers along the short distance that the UE has travelled, in this case, 6 handovers in 16 m (20 s). This increased number of handovers leads to high signaling overhead. These mobility management issues in multi-tier networks have already been identified as a problem with small cells in general [14], which is exacerbated with attocells. Therefore, new mechanisms are necessary so that UEs that are just moving past the attocell coverage do not initiate a handover to the attocell, while being able to maintain their connections with their serving cells.

16.6 FUTURE ENHANCEMENTS

The discussion of attocells has so far been done based on current LTE features and capabilities. While it was shown that significant gains in UE throughputs can be achieved with these, there are several future trends and developments that can further enhance the capability of attocells to provide even more capacity.

Owing to the small coverage areas of attocells, the distance between the UE and the attocell BS will be very small, and most likely within LOS. This makes attocells very suitable for the use with high-frequency carriers, including those in the millimetre wave region, where the amount of available bandwidth is on the order of hundreds of MHz or more. Such high frequencies have not been used in cellular networks in the past due to their poor propagation characteristics, which would not be an issue when applied to attocells. In fact, the poor propagation characteristics at such high frequencies would actually be desirable in attocell deployment scenarios, as this limits the amount of leakage of interfering signals to neighboring cells. Another way in which higher frequencies are beneficial to attocells is that they allow smaller antenna sizes to be used, which lowers the visual impact of the attocells and simplifies the installation process.

Another enhancement that can benefit the use of attocells is the use of higher-order modulations. With the short distances to UEs and the use of directional antennas, attocells are able to provide high levels of downlink SINR within its coverage area. Figure 16.12 shows the CDF of the SINR experienced by the UEs in the office simulation scenario described in the previous sections. As the curve shows, there is a significant portion of UEs that experiences SINR levels in excess of 20 dB, even with the assumption of a full buffer traffic model. With more realistic traffic models where not all UEs are active at the same time, high SINR values would be achieved even more often. However, these high levels of SINR are not fully exploited with 64-QAM. Figure 16.13 shows the spectral efficiency of different modulation schemes up to 64-QAM with different SINR conditions. The peak UE throughput stops increasing when the SINR reaches 20 dB, limited by the 64-QAM modulation.

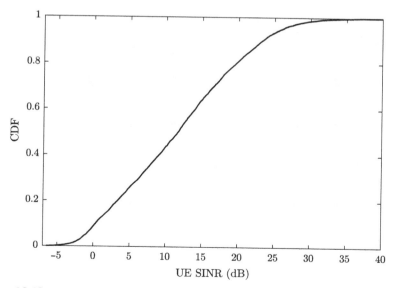

Figure 16.12. Cumulative distribution function of SINR experienced by UEs in office simulation scenario with 5 mW narrowbeam attocells.

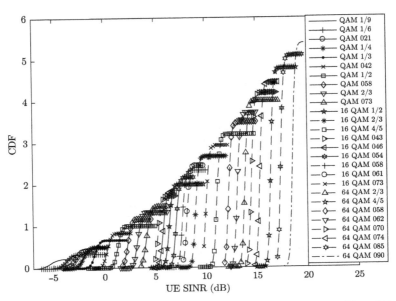

Figure 16.13.   Spectral efficiency of different modulation schemes to achieve 1% BLER.

It is clear that using higher-order modulations will allow small cells to exploit the high SINR they offer, and because of that, the usage of 256-QAM in LTE is appealing [15].

To show the impact of these enhancements, simulations of the previously presented office environment were performed. In the simulation, the attocells are assumed to use a 10 GHz carrier frequency with 100 MHz of bandwidth, while the macrocells are operating on the 2.6 GHz carrier frequency with 10 Hz of bandwidth, so the attocells are now operating on a dedicated carrier. The highest modulation of 256-QAM is also assumed to be available for both attocell and macrocell. UEs assumed to connect to the BS with the highest SINR. The overall effect of the higher bandwidth and modulation allows the peak UE throughput to increase to 488 Mbps, compared with 36 Mbps when 64-QAM modulation and 10 MHz bandwidth is used in the previous simulations. The use of a dedicated attocell carrier also eliminates interference between macrocell and attocells, which contributes to the better UE SINR. It also mitigates mobility problems.

Figure 16.14 shows the CDF of UE throughputs achieved with the use of the 10 GHz carrier with the narrowbeam and omni-directional antennas, as well as the 2 GHz carrier with narrowbeam antennas in the shared and dedicated carrier configuration. The advantages of the large available bandwidth with 10 GHz and higher-order modulation is very significant, with an approximately 10 times increase in the total system throughput compared with the 2 GHz dedicated carrier deployment. Figure 16.15 shows the UE SINR distribution for the 10 GHz attocell

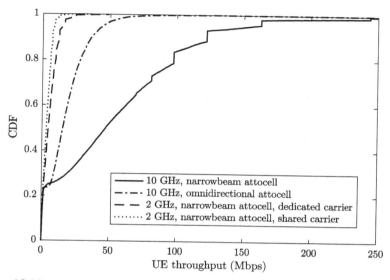

Figure 16.14. Cumulative distribution function of UE throughput for attocell deployments with up to 256-QAM available, for different carrier frequencies.

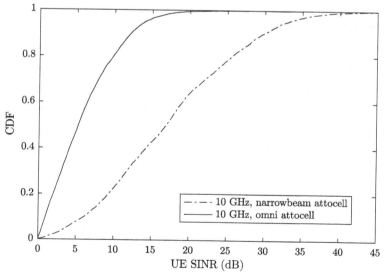

Figure 16.15. Cumulative distribution function of SINR experienced by UEs in office simulation scenario with 10 GHz attocells using narrowbeam and omni-directional antennas.

deployments with directional and omni-directional antennas. With directional antennas, over 30% of UEs experience an SINR higher than 20 dB, where the benefits of higher throughputs offered by 256-QAM modulation can be achieved. The advantage of this higher-order modulation is not present when omni-directional antennas are used as the number of UEs that achieve an SINR larger than 20 dB is less than 1% due to the high inter-attocell interference.

The combination of very dense attocell deployments with directional antennas enabling high spatial reuse of up to a single UE per cell, higher frequency carriers offering bandwidths on the order of hundreds of MHz, and increased spectral efficiencies of higher-order modulation is a way of achieving peak UE throughputs in the gigabit range for future 5G wireless networks.

16.7 SUMMARY AND CONCLUSIONS

In this chapter, the concept of deploying small cells in ultra-high densities, called attocells, to provide high capacity indoors was presented. In order to control the effect of inter-cell interference between attocells, the deployment of these attocell BSs at the ceiling with downward facing directional antennas was proposed, and a simulation of an office deployment scenario was performed. In the simulated scenario, attocells with directional antennas were able to provide gains four times higher than attocells with omni-directional antennas without outage conditions at the cell edges if the attocell transmit powers are set at an appropriate level. Simulation results also showed that when handover parameters were set to take into account the attocell's narrow cell edges, ping-pongs were very likely to happen when UEs moved past attocells. To overcome this, techniques such as UE MSE with ICIC, aggressive BS sleep modes and control and UE plane splitting can be used.

Although large capacity gains were achieved with current technology, the attocell concept is most ideally suited to exploit enhancements that would be introduced in the future such as high frequencies, carrier aggregation, and higher-order modulations.

There remain many different aspects to be addressed on the practicality and feasibility of the application of attocells, such as backhaul provisioning and energy efficiency, which will be the subject of further study. The extension of the use of ultra-dense attocells is a promising approach to provide the reliable, high-capacity connections required for future 5G services and applications.

REFERENCES

1. "Small cell enhancements for E-UTRA and E-UTRAN—physical layer aspects," 3GPP Technical Specification, Tech. Rep. TR 36.872 v 12.1.0, Dec. 2013.

2. G. Americas. (2011 Jul.). "Self-organizing networks: The benefits of SON in LTE," [Online]. Available: http://www.4gamericas.org/.

3. D. Lopez-Perez, I. Guvenc, G. de la Roche, M. Kountouris, T. Q. S. Quek, and J. Zhang, "Enhanced intercell interference coordination challenges in heterogeneous networks," *IEEE Wireless Commun.*, vol. 18, no. 3, pp. 22–30, Jun. 2011.

4. L. T. W. Ho and H. Claussen, "Effects of user-deployed, co-channel femtocells on the call drop probability in a residential scenario," in *2007 IEEE 18th Int. Symp. Pers., Indoor Mobile Radio Commun.*, Sep. 2007, pp. 1–5.

5. D. Lopez-Perez, I. Guvenc, and X. Chu, "Mobility management challenges in 3gpp heterogeneous networks," *IEEE Commun. Mag.*, vol. 50, no. 12, pp. 70–78, Dec. 2012.

6. S. Parkvall, E. Dahlman, G. Jongren, S. Landstrom, and L. Lindbom, "Heterogeneous network deployments in lte: The soft-cell approach," *Ericsson Technol. Rev.*, no. 2, Dec. 2011. Available: https://www.ericsson.com/res/thecompany/docs/publications/ericsson_review/2011/ER-hetnet-deployment.pdf

7. H. Claussen, L. Ho, and F. Pivit, "Leveraging advances in mobile broadband technology to improve environmental sustainability," *Telecommun. J. Australia*, vol. 59, no. 1, pp. 66–70, Dec. 2010.

8. P. Frenger, P. Moberg, J. Malmodin, Y. Jading, and I. Godor, "Reducing energy consumption in lte with cell dtx," in *2011 IEEE 73rd Veh. Technol. Conf. (VTC Spring)*, May 2011, pp. 1–5.

9. K. Hiltunen, "Utilizing enodeb sleep mode to improve the energy-efficiency of dense lte networks," in *2013 IEEE 24th Annu. Int. Symp. Pers., Indoor, Mobile Radio Commun. (PIMRC)*, Sep. 2013, pp. 3249–3253.

10. I. Ashraf, F. Boccardi, and L. Ho, "Sleep mode techniques for small cell deployments," *IEEE Commun. Mag.*, vol. 49, no. 8, pp. 72–79, Aug. 2011.

11. S. J. Fortune, D. M. Gay, B. W. Kernighan, O. Landron, R. A. Valenzuela, and M. H. Wright, "Wise design of indoor wireless systems: Practical computation and optimization," *IEEE Comput. Sci. Eng.*, vol. 2, no. 1, pp. 58–68, Spring 1995.

12. D. Lopez-Perez, X. Chu, A. V. Vasilakos, and H. Claussen, "On distributed and coordinated resource allocation for interference mitigation in self-organizing lte networks," *IEEE/ACM Trans. Netw.*, vol. 21, no. 4, pp. 1145–1158, Aug. 2013.

13. D. Lopez-Perez, A. Ladanyi, A. Jüttner, H. Rivano, and J. Zhang, "Optimization method for the joint allocation of modulation schemes, coding rates, resource blocks and power in self-organizing lte networks," in *2011 Proc. IEEE INFOCOM*, Apr. 2011, pp. 111–115.

14. D. Lopez-Perez, I. Guvenc, and X. Chu, "Mobility management challenges in 3gpp heterogeneous networks," *IEEE Commun. Mag.*, vol. 50, no. 12, pp. 70–78, Dec. 2012.

15. "Evolved Universal Terrestrial Radio Access (E-UTRA): User Equipment (UE) radio access capabilities," 3GPP Technical Specification, Tech. Rep. TS 36.306 v 12.10.0., Oct. 2016. Available: http://www.3gpp.org/ftp/Specs/archive/36_series/36.306/36306-ca0.zip

17

HETNET APPLICATIONS

17.1 INTRODUCTION

In the previous chapters, the benefits of small cells in providing improved coverage and capacity to end users were discussed. In addition to these benefits, owing to their special characteristics such as their short coverage range, small cells can be used as the technology enabler for a number of other applications. In fact, with residential, enterprise and outdoor small cell deployments rapidly growing across the globe, there is a major opportunity for operators to start offering unique new applications. This trend has already started and commercial offerings are now available in some countries. For example in Japan, mobile operators are offering commercial applications that send a short message service (SMS) to parents when children arrive home from school.

However, to be able to take advantage of this opportunity efficiently, mobile operators require simple yet effective means of managing the applications they offer. The management standards are central in achieving this capability. The Small Cell Forum, the independent industry and operator association that supports deployment of small cells worldwide, has already published an industry-wide agreed set of application

Small Cell Networks: Deployment, Management, and Optimization, First Edition. Holger Claussen,
David López-Pérez, Lester Ho, Rouzbeh Razavi, and Stepan Kucera.
© 2017 by The Institute of Electrical and Electronic Engineers, Inc. Published 2017 by John Wiley & Sons, Inc.

programming interface (API) specifications that enable advanced mobile applications based on small cells technology. The first applications have already been built based on these specifications by the Forum's vendor members for operator customers.

Additionally, the Small Cell Forum has worked with the Broadband Forum to update the standard for management of small cells to ensure that customer equipment can be remotely updated to support advanced applications. These specifications are for a network-based APIs, which will allow operators to drive the development of small cell applications. The API provides small cell awareness information so developers can incorporate enhanced presence, context and location-sensitive features into new and existing apps, and can also take advantage of the lower cost and faster data connections enabled by small cells [1].

To truly capitalize on the end-user interest in advanced services, communication service providers can work collaboratively with third-party application and content providers to harness the full power of small cells. Each cell provides unique end-user usage and positioning information such as whether the mobile device is active in a voice or data session. Access to this information enriches the small cell APIs. This can be combined with network capabilities, such as location, presence, quality of service (QoS) and trusted security to enable application development by in-house or external application and content providers.

In parallel to this, the Small Cell Forum has worked with the Broadband Forum to update its TR-196 Femto Access Point Service Data Model, a global femtocell management standard, to enable support for small cell applications. This will allow mobile operators to use their standard management systems to remotely provision and configure advanced small cell applications on their customers' equipment as well as issue repairs and updates. In terms of users' interest, a study [2] indicates that that almost 60% of consumers are interested in residential small cells. Of these, 68% found at least one of the advanced small cell services either very or extremely appealing. In the United States specifically, 72% of consumers who found residential small cells appealing were very interested in at least one advanced services and half of these were willing to pay $4.99/month for their single favorite service or $9.99/month for a bundle of their favorite three. This chapter covers a number of such advanced applications for residential, enterprise, and operators deployed small cells.

17.2 LOCALIZED AND PERSONALIZED PUSH SERVICES

Push technology, in which the user requests that specific data be automatically sent to his or her computer or mobile device, has been around for nearly two decades. The roots of push messaging can be traced back to the mid-1990s, and the advent of companies which offered a publish/subscribe model in which users could subscribe to data feeds that would be automatically pushed to them whenever new information arrived. Push was a game-changing departure from the traditional model of downloading or pulling information from online sources.

Content dissemination to mobile users has recently attracted particular attention. Examples of applications that rely on content delivery are notification services for weather or traffic reports, messaging systems for group discussions, or systems supporting the collaboration of mobile employees [3]. Location-based content delivery will be a premier feature in these systems [4]. From this perspective, small cells are very well positioned to provide location-based services. Compared to traditional macrocells, small cells are capable of providing much more accurate estimation of users' location and presence, due to their smaller coverage size. The following are just few examples of applications that are based on localized push services.

17.2.1 Localized Marketing

Hotspots such as shopping centers and malls as well as crowded shopping streets and areas with tourist attractions are considered as sweet spots for deploying small cells where they can boost the overall network capacity by offloading users. In such scenarios, local businesses are eager to engage with the crowd by real-time dissemination of information regarding their current or upcoming offerings, special deals, etc. In fact, local businesses have long been challenged to reach consumers near the point of purchase decision, particularly when those decisions are made impulsively like quick-service restaurants and certain other brick and mortar retailers. Similarly, it might be of great interest for consumers to receive such information. According to a research by BIA/Kelsey's Consumer Commerce Monitor [5], 35% of consumers who use smartphones when shopping locally regularly use apps to assist them with their shopping experience. Most are using more than one shopping application.

Currently, such marketing information is presented to the customers through traditional store signs and displays. Evidently, this is not the most efficient mechanism for information dissemination. To start with, the volume of the information that can be communicated to the customers is limited in this way. Moreover, such notes are either static, and effort and cost would be required to update them on a regular basis. Just as an example, consider a local restaurant, which has different menu options and deals for lunch and dinner. Applications enabled by small cells can easily address this gap. In this scenario, the locally deployed small cells can act as a proxy for pushing such notifications to the nearby customers. This can be considered as a major marketing channel for local small and medium-sized businessess (SMBs), which may not be interested in or able to afford other marketing channels suitable for larger geographic areas. Companies such as Yelp have been established to provide information and reviews about local businesses, as well as some additional services such as online reservation and delivery services. The company also trains small businesses on how to provide data about their businesses. According to Yelp, more than 64% of the search queries and 45% of consumer's reviews come from mobile devices.

Moreover, a common architecture behind push services is the publish/subscribe model. The service involves two types of entities: publishers and subscribers. Publishers are content sources that group and send data through channels. Subscribers are content destinations that subscribe and receive the corresponding data. Using small cells as proxy for push services, customers can define their preferences through the applications on their mobile devices. This includes the category of offers and deals that they might be interested in, or opting in for offers and information from particular brands. This will provide a way for delivering highly personalized and customized content to the consumers, which is shown to be a key factor for wide acceptance and success of push services [6]. Moreover, coupon reminders are shown to be a popular use of push notifications for local stores and retailers. According to [5], 43% of local purchases made by smartphone users were made through promotions or sales, such as discount deals, "daily deals," coupons, or similar types of discount offers.

17.2.2 Museums and Historic Sites

Museums and historic sites around the world today face the challenge of increasing and maintaining visitor numbers, especially with younger audiences. A fall in visitors is seen by most as a negative outcome, both financially and in terms of wider social and educational impact. This can happen due to a range of factors, but according to a study in [7], one of the most important is that museums can often find themselves competing with the products of the entertainment industry, which at its heart is in the business of telling a good story and engaging users.

In recent years, the integration of technology into cultural heritage centers, such as archaeological sites, historical buildings, art galleries, and museums has already resulted in an enhanced visitor experience. This is a direct consequence of providing rich and relevant information to the visitors during their visits. radio frequency identification (RFID) tags were traditionally used in such scenarios. In this setup, each tag is storing data on a corresponding object or section of the site. The visitor is equipped with an RFID reader and is then able to detect tag messages that provide the appropriate object description. This, however, requires the host venue to provide additional equipment and for visitors to carry such equipments with them at all times during their visit. In addition, the interactions are in most cases in one direction only, and the scope is limited to providing the visitors with textual information or audio description of objects. An alternative solution is to deploy small cells for localized dissemination of information in such venues. Instead of using specialized hardware, visitors can download mobile applications and use their own mobile devices to get relevant information as they walk around the venue. This also opens up new avenues of opportunity for having additional richer and more engaging services which may not have been possible if relying on traditional RFID technology.

One of the most successful real-world uses of interactive digital storytelling is augmented reality (AR), such as that employed to great effect at the British Museum. The technology transformed the museum experience for a child into a story puzzle using a dedicated tablet application. The application sets up a game, called "A Gift for Athena," which rewards the visitor for finding certain statues from their outlines by telling them more about the exhibit and directing them into the next stage of the game. The European Union-funded CHESS project (an acronym for Cultural Heritage Experiences through Socio-personal interactions and Storytelling) [8] takes digital storytelling much further and plans to make interactive content such as games and augmented reality available to the entire museum sector.

The CHESS project relies on visitor profiling, matching visitors to predetermined "personas," which are designed as a representative description of the various people that constitute a given museum's visitor base. These are created through data from surveys, visitor studies, and ethnographic observations. A given visitor is matched initially through a visitor survey to one of several representative personas, which in turn influences fundamentally the experience delivered by the CHESS system. Doing this makes the visitor experience non-linear. The system constantly adapts to a visitor's preferences. For example, if a visitor fails in a game or stays longer in front of certain artifacts, the system can adapt the storyline. It makes the experience more dynamic and relevant, so instead of sending the visitor to exhibit X, the system might instead choose to send her to exhibit Y, where she will get more information that is relevant to what she has shown an interest in. All of these are possible only if the location of the user can be accurately estimated. Owing to their limited coverage range, small cells can be envisioned as a suitable technology to enable such applications.

17.2.3 Bus/Train Stations and Airports

Similar to shopping malls, public transport stations such as bus and train stations as well as airports are considered as places with a high concentration of mobile users that are mostly stationary. This obviously makes such venues suitable candidates for deploying small cells to offload mobile users. Additionally, push applications can be developed to provide an enhanced experience for travelers. The application can provide information such as airport and train station information, mobile check-in, maps and directions, coupons for special offers at shops and food outlets, flights, trains, and gate alerts. The features perceived as most beneficial are real-time flight and train updates and alerts, flight check-in using the device, advice on best check-in and security queues based on wait times, and real-time directions in the airport and train stations. Just as an example, a new application has been recently developed by Vueling, which detects passengers' arrival at airport and provides access to the "Bring your flight forward" service. According to the company, one of the most interesting aspects of this new application is that it makes it possible to locate the passenger within the airport and to create the contextualized and personalized proposals that a passenger may need.

17.3 PROXIMITY AND PRESENCE DETECTION

One of the attractive applications of small cells is to detect proximity and presence of mobile users. For example, in Japan parents can setup their residential small cells to receive notifications when their children arrive or leave home. The notification can be in the form of either an SMS or an email. Such services have gained significant popularity in recent years.

Another example of applications where small cells can be used to detect the presence of a user with a mobile phone on premises such as government buildings and security-sensitive areas like prisons, where the use of mobile phones is prohibited. By setting up the small cells in open mode, a mobile user who is in proximity to the small cell would automatically handover to that small cell. Such handover events can then trigger a security alert. Moreover, the study in [9] proposes an infrastructure-based solution that provides spontaneous and transaction-oriented mobile device location authentication via small cell access points. In the proposed solution, a calling party can verify a (cooperating) called party's location by simply making a voice call while remotely monitoring the small cell's activity, even when the participants have no preexisting relationship. The study shows how such a traffic signature can be reliably detected even in the presence of heavy cross traffic introduced by other small cell users.

17.4 INDOOR LOCALIZATION

Indoor localization services have gained great attention recently while users have enjoyed their outdoor counterparts for years. Indoor navigation in shopping malls and tracking friends and family members in indoor public places are a few examples of how indoor location information can make services friendlier and more useful. More importantly, considering that 70% of emergency calls are placed from mobile phones with a large fraction being originated from indoors, accurate localization of indoor users is of great importance and value. In the United States, the Federal Communications Commission (FCC) adopted its wireless Enhanced 911 (E911) location accuracy rules in 1996, which seek to improve the effectiveness and reliability of wireless 911 emergency services by providing 911 dispatchers with additional information on wireless 911 calls including the location of callers.

Unfortunately, global positioning system (GPS) receivers do not work properly indoors due to several reasons. First, signals from the GPS satellites are inherently weak, and after traveling more than 21,000 km to reach the Earth's surface, the signal strength (-125 dBm) is barely enough to decode satellite information outdoors. Note that the thermal noise floor at GPS frequency is at about -111 dBm. Indoors, the signal strength is about 10 to 100 times weaker, and it is almost impossible for a typical GPS receiver to acquire any satellite. Second, even if signals from a satellite are detected indoors, the weaker signal strength combined with increased multi-path effects can cause the receiver to compute an inaccurate distance from the satellite, and yield an estimated location that is far away from the true location. Third, the location estimation in a typical GPS receiver requires at least four visible satellites, which is highly unlikely to be obtained in an indoor environment [10].

In the absence of GPS, small cells can be used as the enabling technology for radio frequency (RF)-based indoor positioning systems. These systems employ user devices which sense and measure radio signals transmitted by the small cells. The signal measured is converted into an estimated user position. Typical radio signal measurements include the received signal strength indicator (RSSI), the angle of arrival (AoA), the time of arrival (ToA), and the time difference of arrival (TDoA) [11]. Sensing the RSSI is considered to be most practical, since RSSI measurements require no additional hardware at the user's mobile device.

There are two main techniques for estimating a user's position from RSSI signal measurements, namely trilateration and signal pattern recognition.

The first technique is based on estimating the distance of the target mobile user to the transmitting devices from the radio signals measured, making use of triangle geometry. Positioning systems based on trilateration techniques are suitable for outdoor positioning systems. However, the complex propagation conditions in indoor environments reduce the accuracy of this technique [12].

The second positioning technique is performed via a two-stage process, one taking place offline and the other online. First, radio signals are measured during an offline calibration stage. These signals form a pattern which is unique to each scenario location. In the online phase, these patterns, called fingerprints, can be used to estimate the user location via pattern matching during the online phase [10].

The deployment of small cells has resulted in significant improvements in accuracy of positioning systems. This is because small cell deployments are usually dense, resulting in a larger number of visible cells, which itself results in enriching the fingerprint signatures. Moreover, as the receiver gets closer to the transmitter, the received RF signals not only becomes stronger but also more sensitive to the distance from the transmitter. As a direct consequence, the RSSI signatures become more distinctive when transmitters are in close proximity of the receiver. Finally, due to their short coverage range, the serving small cell already provides a good estimate of the user's location.

17.5 HOME AUTOMATION

The industry has witnessed rapid growth of home automation applications in recent years. Wireless home automation networks enable monitoring and control applications for home user comfort and efficient home management. A wireless home automation network typically comprises several types of severely constrained embedded devices, which may be battery-powered and are equipped with low-power RF transceivers. The use of RF communication allows flexible addition or removal of devices to or from the network and reduces installation costs since wired solutions require conduits or cable trays. The following are some of the home automation use cases [13]:

Light Control. A new light can be controlled from any switch, which reduces the need for new wired connections. Lights can also be activated in response to a command from a remote control. Furthermore, they can be turned on automatically when presence and luminance sensors detect that people are in a poorly illuminated room. The small cell can operate as a proxy between the mobile devices as controllers and light devices.

Remote Control. Infrared technology has been used for wireless communication between a remote control and devices such as televisions, high-fidelity (HiFi) equipment and heating, ventilating, and air conditioning (HVAC) systems. However, infrared requires line-of-sight (LOS) and short-distance communication. AGG RF technology overcomes these limitations.

Smart Energy. Window shades, HVAC, central heating, and so on may be controlled depending on the information collected by several types of sensors that monitor parameters such as temperature, humidity, light, and presence. The unnecessary waste of energy can thus be avoided. In addition, a number of studies have confirmed the effectiveness of smart meters to help customers reduce their consumption. The important role of usability and clarity of the feedback interface for smart meters is discussed in [14]. The study summarizes the results of 22 other studies between 1987 and 2006, and concludes that the most effective forms of energy consumption feedback to users involve electronic interaction with the households including mobile push messages. Similarly, the study in [14] finds that peak-time rebates reduced peak demand by 18–21%, and that adding an "Energy Orb," which reminds households of peak periods, increased this reduction by 23–27% [15]. Such applications can be effectively enabled by small cells, which can act as a proxy between the smart meters and household's mobile application.

Remote Care. Patients, disabled, and elderly citizens can benefit from at-home medical attention. Wearable wireless sensors can periodically report the levels of several body parameters (e.g., temperature, blood pressure, and insulin) for a precise diagnosis. If acceleration sensors suggest that a person has fallen, a push notification can be immediately sent to friends and family members.

Security and Safety. Advanced security systems can be based on several sensors (e.g., smoke detectors, glass-break sensors, and motion sensors) for detecting possible risk situations that trigger push messages to the home owners. In addition, small cells can be configured to record handover attempts from non-registered users. This by itself can be considered as an additional security measure, which does not require dedicated sensors.

17.6 LOCAL ACCESS CONTROL

The ever-increasing usage of mobile phones has introduced problems such as their potential use to invade privacy, disruption of silence in locations such as the hospitals, libraries, theatres, and churches, causing electromagnetic interference (e.g., in aircraft systems) as wells as the abuse of mobile phone communications on premises such as prison buildings. Selective restriction of mobile phone usage has been always a need in certain locations. The current known solution for this objective is to deploy mobile phone jammer devices. Jammer devices can overpower the cell phone by transmitting a signal on the same frequency and at a high enough power. However, numerous technical and legal issues are related to the use of jamming devices that has made their use illegal in most countries. For example, when using a jammer device, there is no mechanism to distinguish between users or services in use. A jammer makes all services unavailable for all users within its area of influence. A consequence of this is a health and safety concern when emergency call services become unavailable. Additionally, since mobile phones are capable of adapting their transmit power to overcome the interference (through power control mechanisms), for a jammer device to be effective, it needs to constantly transmit at a very high power. This may result in human exposure to some levels of electromagnetic radiation (EMR) that exceeds the health standards. For these reasons, the use of phone jammers has been banned in most countries. For example, the European Telecommunications Conformity Assessment and Market Surveillance Committee under European Directive 99/05/CE has determined that phone jammers should be banned, and that the local authorities must retire all devices present in the market and notify the European Commission of this fact.

One solution to realize local access control is to deploy small cells in open-access mode. Considering that in the 3G small cell architecture, the functionality of the legacy Node-B and radio network controller (RNC) are combined into the small cell base station (BS), once a user is camped to a small cell, that small cell will be in full control of the traffic flow. Therefore, if it is properly configured, an open-access small cell can provide the following functionalities.

- Particular services can be barred for all or a group of users. For example, a small cell installed in a library can allow for a wide range of data services but to block incoming and outgoing voice call requests.
- Selected destination numbers such as emergency services can be exceptionally allowed for all users.
- Certain users may be excluded from restrictions. In general, it is possible to implement a fully flexible service and user selectivity by defining access/privilege lists.
- Users that have been prevented from receiving incoming calls can be instantly notified of such events. In its simplest form, this can be achieved by sending a text message to the user.
- Once a user enters a premise equipped with such small cells, he can be immediately notified of services that are available and those that are blocked.
- For security reasons (e.g., in prison buildings) responsible personnel can be notified of forbidden access attempts.

The study in [16] describes a mechanism to optimize the coverage range of such small cell BS to avoid service disruption for users outside the intended area.

17.7 SUMMARY AND CONCLUSIONS

Originally developed to compliment the capabilities of macrocellular mobile networks, small cells are now envisioned as the technology enabler for a number of other applications. This is mainly due to some of their special characteristics, including their short coverage range. Some of the additional applications enabled by small cells have been briefly introduced in this chapter.

More specifically, the critical role of small cells in enabling localized and personalized push services was discussed. Push services have gained substantial popularity in mobile applications during recent years. However, such push notifications would only be of interest to users if they are personalized and relevant, which can be effectively realized when using small cells. As examples of such localized push services, mention can be made to proximity marketing services. In fact, according to a survey [5], more than half of the consumers are willing to share their current location to receive more relevant advertising. Other examples of location-based push services include interactions with visitors in museums and at sport events as well as passengers in bus/train stations and airports. These are only a few examples of scenarios where a group of co-located mobile users might be interested in receiving information that is specific to that particular location. Moreover, owing to their short coverage range, a small cell can be used as a proxy to determine presence or proximity of a mobile user. Sending a text notification to parents when their youngster arrived home is just one example of how this feature can be used in an application. Other examples include security applications where small cells are used to detect the presence of a mobile phone in a specific location. Indoor localization is another important class of applications that can be enabled or enhanced by small cells. An important application is locating indoor mobile users in case of emergency.

Home automation applications have gained attention and popularity in recent years. Residential small cells are a suitable technology to enable a number of such applications. Example use cases include the use of small cells for remote controlling of home appliances, remote monitoring of the house, and real-time notifications of energy consumption of the house. Since users can configure their residential small cells, it is possible to personalize such services. Finally, small cells can also be used as a means to control user's access and services available for different classes of users in specific premises. This includes buildings and venues where disruption of silence is not desirable (e.g., hospitals, libraries, churches etc.), or areas where the use of mobile phones are prohibited due to the security concerns (e.g., prison buildings). In this setup, the small cell is configured in an open-access mode allowing users to attach to it as the best serving cell, and then access regulation policies are applied.

Considering that small cell technology is still at its early years, it is anticipated that it will play even a more critical role moving forward. As discussed, delivery of localized and personalized services to mobile users is becoming the main focus for many application and service providers. Small cells can be envisioned as strategic technology enablers for a wide range of such applications and services. Furthermore, the development of management standards and API specifications that enable advanced mobile applications based on small cells technology will certainly facilitate the development of such applications in the future.

REFERENCES

1. "Femto forum addresses apps market with femtocell API specification and evolved management standard," Femto Forum press releases, 2011.

2. H. Wang, "Global consumer research on femtocells," Parks Associates, 2010.

3. I. Podnar, M. Hauswirth, and M. Jazayeri, "Mobile push: Delivering content to mobile users," in *Proc. Int. Conf. Distributed Comput. Syst.-Workshops*, IEEE, 2002, pp. 563–568.

4. A. Dornan, "Can e-commerce find a place in your network?" *Netw. Mag.*, vol. 16, no. 11, pp. 38–45, 2001.

5. BIA/Kelsey, "Small business research," *Market Research*, 2012.

6. A. Carzaniga, D. S. Rosenblum, and A. L. Wolf, "Design and evaluation of a wide-area event notification service," *ACM Trans. Comput. Syst. (TOCS)*, vol. 19, no. 3, pp. 332–383, 2001.

7. O. Bimber, L. M. Encarnaçao, and D. Schmalstieg, "The virtual showcase as a new platform for augmented reality digital storytelling," in *Proc. Workshop Virtual Environ. 2003*, ACM, 2003, pp. 87–95.

8. L. Pujol, M. Roussou, S. Poulou, O. Balet, M. Vayanou, and Y. Ioannidis, "Personalizing interactive digital storytelling in archaeological museums: the CHESS project," in *Proc. 40th Annu. Conf. Comput. Appl. Quant. Methods Archaeology*, Amsterdam University Press, 2012.

9. J. Brassil, S. Haber, P. Manadhata, R. A. Netravali and P. V. Rao, "Authenticating a user's location in a femtocell-based network," US Patent 9408025, Aug 2, 2016. Available: https://www.google.com/patents/US9408025

10. S. Nirjon, J. Liu, G. DeJean, B. Priyantha, Y. Jin, and T. Hart, "COIN-GPS: indoor localization from direct GPS receiving," in *Proc. 12th Annu. Int. Conf. Mobile Syst., Appl. Services*, ACM, 2014, pp. 301–314.

11. M. B. Kjærgaard, "A taxonomy for radio location fingerprinting," in *Location-and Context-Awareness*, Springer, 2007, pp. 139–156.

12. A. Narzullaev, Y. Park, K. Yoo, and J. Yu, "A fast and accurate calibration algorithm for real-time locating systems based on the received signal strength indication," *AEU-Int. J. Electron. Commun.*, vol. 65, no. 4, pp. 305–311, 2011.

13. C. Gomez and J. Paradells, "Wireless home automation networks: A survey of architectures and technologies," *IEEE Commun. Mag.*, vol. 48, no. 6, pp. 92–101, 2010.

14. C. Fischer, "Feedback on household electricity consumption: a tool for saving energy?" *Energy Efficiency*, vol. 1, no. 1, pp. 79–104, 2008.

15. J. Carroll, S. Lyons, and D. Eleanor, "Reducing electricity demand through smart metering: The role of improved household knowledge," Trinity College Dublin, Department of Economics, Tech. Rep., 2013. [Online]. Available: http://econpapers.repec.org/paper/tcdtcduee/tep0313.htm

16. R. Razavi and H. Claussen, "Controlling local service access in wireless cellular networks," in *Proc. Int. Symp. Pers. Indoor Mobile Radio Commun. (PIMRC)*, IEEE, 2012, pp. 1126–1130.

A

SIMULATING HETNETS

A.1 INTRODUCTION

In order to aid vendors and operators in the development and deployment of new heterogeneous cellular networks (HCNs), and the refinement of existing network procedures such as handover (HO) and radio resource management (RRM), network simulation, planning, and optimization tools that are able to evaluate the overall performance of complex cellular networks are required. In this context, system-level simulations have become a widely adopted methodology. In system-level simulations, the elements and operations of a cellular network are modeled by computer software. This approach is simpler and cheaper than a real implementation, and is more accurate and reliable than analytical modeling. The number of assumptions and simplifications made in system-level simulations depend on the computer software used, but they are usually much smaller than in analytical modeling. As a result, system-level simulations can model more complex cellular networks, but they usually require significant computing capabilities to obtain statistically representative results, mostly if the number of base stations (BSs) and user equipments (UEs) is large. In order to avoid a prohibitive computational costs, a trade-off between accuracy and complexity needs to be made. In this line, the Third-Generation Partnership Project (3GPP) provides useful guidelines on system-level simulations, and defines simulation procedures and parameters, which we follow in this book.

Small Cell Networks: Deployment, Management, and Optimization, First Edition. Holger Claussen, David López-Pérez, Lester Ho, Rouzbeh Razavi, and Stepan Kucera.
© 2017 by The Institute of Electrical and Electronic Engineers, Inc. Published 2017 by John Wiley & Sons, Inc.

In this chapter, important issues in system-level simulations for HCNs are discussed, and the simulation methodology used to obtain the simulation results of the different chapters of this book is described, including some enhancements over the existing 3GPP recommendations. In more detail, the typical HCN scenarios and the different models used to evaluate UE performance are presented. It is important to note that the total channel gain G between any transmitter and any receiver, the pillar of any system-level simulator, is modeled as the combination of individual channel gain and losses, that is, antenna gain G^a, path loss G^p, environment loss G^e, shadow fading G^s and multi-path fast fading G^{ff}, where G stands for gain/loss and the upper-index indicates the type of gain.

Based on this model, the rest of the chapter is organized as follows. In Section A.2, network layout models for both macrocell and small cell BSs are described. As indicated above, the total channel gain between a transmitter and a receiver is modeled as the composition of different gains and losses. In Sections A.3–A.7, the concepts and models for defining antenna gain G^a, path loss G^p, environment loss G^e, shadow fading G^s and multi-path fast fading G^{ff} are presented, respectively. In Section A.8, the models used to evaluate the received signal strength and quality at the receiver side are shown, providing details on maximal ratio combining (MRC) and exponential effective SINR mapping (EESM) techniques. In Section A.9, the models used to evaluate the throughput at the receiver side based on signal quality estimation are introduced. In Section A.10, UE traffic models, which have an important impact on the UE throughput distribution across the network, are presented. In Section A.11, UE mobility models that describe the UE movement across a scenario for mobility performance assessment are also discussed. Finally, in Section A.12, conclusions are drawn.

A.2 NETWORK LAYOUT MODELING

The network layout models the positions of the BSs within a geographical area. Macrocell and small cell BSs follow a different deployment nature, planned versus unplanned, and thus have different network layout models. In this section, widely used network layout models for both macrocell and small cell BSs are presented.

A.2.1 Macrocell Layout Modeling

Figure A.1 illustrates the homogeneous macrocell layout used in the 3GPP for analyzing macrocellular network performance in both suburban and urban environments [1]. This model is widely known as the hexagonal grid model since BSs are deployed following a hexagonal grid. It is fully characterized by the inter-site distance (ISD) D_{ISD}, which defines the regular distance between any two neighboring BSs, and the number of interfering tiers. In Fig. A.1, the number of interfering tiers is two and there are three sectors per BS; thus, the numbers of resulting BSs N_{BS} and sectors are

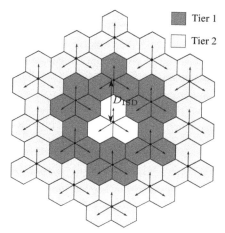

Figure A.1. Homogeneous macrocell layout following the hexagonal grid model.

19 and 57, respectively. The hexagonal grid model provides a well-defined macrocell scenario that represents the planned nature of macrocellular networks, facilitates the comparison of the results from distinct parties, and benefits from a low computational complexity. This is because multiple BS placements are not necessary to account for random BS deployments.

Since the central BS is surrounded by a larger number of neighboring cells than any first- or second-tier BS, it is obvious that the central BS will suffer from the highest inter-cell interference among all BSs in Fig. A.1. In order to analyze network performance without incurring border effects, it would be necessary to either collect data only from the central BS or consider a wrap-around model. The former approach benefits from a reduced computational complexity but decreases the statistical sampling, while the latter increases the statistical sampling at the expense of an increased computational complexity. Both models are equally used in the literature. For more details in the implementation of the wrap-around model, please refer to [2].

A.2.2 Small Cell Layout Modeling

Since small cell BSs have a more unplanned deployment nature than macrocell BSs—small cell BSs are deployed on demand to provide coverage or capacity where it is needed as it is needed—their layout usually follows a random fashion. As a result, small cell BS deployments are usually modeled using uniform or non-uniform distributions (the latter accounts for the existence of UE hotspots and thus depends on the UE distribution). Different small cell BS deployment flavors can be found in the literature.

- Small cell BSs could be uniformly or non-uniformly distributed within a given scenario according to a number of small cell BSs per unit area, or within a macrocell BS sector according to a number of small cell BSs per sector, as suggested in [3]. This scenario provides a realistic setup, but requires multiple small cell BS placements to obtain statistically reliable results.
- Small cell BSs may be deployed in a grid in order to assess the average performance of the network without needing multiple small cell BS placements, which are necessary when using random BS deployments. This is a more idealized deployment model than the one presented above, and usually results in a too optimistic performance assessment.
- In system-level simulations, small cell BSs may also be deployed in selected locations to analyze some particular aspect. For example, small cell BSs may be located at the boresight of the hosting macrocell BS sector to study mobility performance, as suggested in [4].

Moreover, it is important to note that some restrictions are usually imposed to small cell BS deployments. According to [3], the minimum distance between a macrocell BS and a small cell BS must be 75 m, while the minimum distance between two small cell BSs should be 40 m. This is because, from a deployment perspective, it does not make sense to deploy a small cell BS nearby a macrocell BS, since the macrocell should provide a good service in that area and the small cell coverage would be too small due to the large difference in transmit power between these two BS types. Other restrictions such as minimum inter-cluster or BS to UE distances also apply.

In order to account for more realistic scenarios, small cell BS deployments may also be coupled with UE locations, since small cell BSs are typically deployed in areas with high traffic demands to allow for macrocell off-loading and increase network capacity. In this case, the BS deployment should follow the UE distribution, and this aspect will be discussed in Section A.10. Moreover, each small cell BS can also be deployed within a given household structure, for example, residential home, apartment or building, which may follow the same deployment principles as outlined before. In this regard, the residential home, 5×5 apartment and dual strip models have attracted a lot of attention, where we would talk about deploying structures, for example, 5×5 apartments or dual strips, instead of just small cell BSs. For details in the implementation of these models, please refer to [2].

A.3 ANTENNA GAIN MODELING

As described in the introduction of this chapter, the total channel gain between a transmitter and a receiver is modeled as the composition of different gains and losses. In this section, the concept of antenna gain is described.

The antenna gain is a relative measure of the ability of an antenna to concentrate radio frequency energy in a particular direction. Using a more formal definition, the antenna gain can be defined as the ratio of the power produced by the antenna from a far-field source in a particular direction to the power produced by a hypothetical loss-less isotropic antenna in the same direction. Antenna gains are typically represented using antenna patterns and usually expressed in isotropic-decibels (dBi). Antenna patterns indicate the antenna gains toward every direction of space, and can be expressed as separate graphs in the horizontal plane and vertical plane, or as a three-dimensional (3D) graph. The former ones are often referred to as polar diagrams.

In the following, widely used antenna models for describing antenna patterns for both macrocell and small cell BS antennas are presented.

A.3.1 Macrocell Antenna Pattern

The coverage of a macrocell BS is typically divided into sectors in order to enhance spatial reuse, as illustrated in Fig. A.1, and the antenna of each macrocell BS sector is usually characterized by a vertical array of antenna elements, as the one shown in Fig. A.2a for a three-sector macrocell BS.

The horizontal and vertical antenna patterns of an usual macrocell BS sector can be expressed using the following model:

- Horizontal antenna pattern (see [1, 3])

$$G_H^a(\varphi)[\text{dB}] = -\min\left[12\left(\frac{\varphi}{\varphi_{3\text{dB}}}\right)^2, \kappa\right], \qquad (\text{A.1})$$

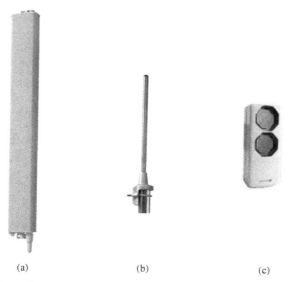

(a)   (b)   (c)

Figure A.2. Example of different types of antennas. (a) Macrocell antenna, (b) Small cell dipole array and (c) LightRadio.

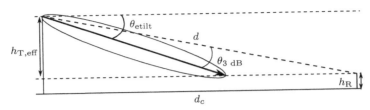

Figure A.3. Macrocell BS antenna downtilt.

- Vertical antenna pattern

$$G_V^a(\theta)[\text{dB}] = 20\log_{10}\left|\frac{\sin(K(\theta - \theta_{\text{etilt}}))}{K(\theta - \theta_{\text{etilt}})}\right|, \qquad (A.2)$$

where φ and θ are the angles of arrival, measured in radian, in the horizontal and vertical planes with respect to the main beam direction, respectively; $G_H^a(\varphi)$ and $G_V^a(\theta)$ are the attenuation offsets, measured in dB, introduced by the horizontal and vertical antenna patterns with respect to the maximum antenna gain in the direction of φ and θ, respectively; $\varphi_{3\text{dB}}$ and $\theta_{3\text{dB}}$ are the main beamwidths, measured in radian, of the horizontal and vertical antenna patterns within which half of the maximum transmit power is transmitted, respectively; κ is the front-to-back ratio, measured in dB, in the horizontal plane; θ_{etilt} is the angle, measured in radian, between the main beam direction and the horizon in the vertical plane (also known as downtilt); and $K = 164/\theta_{3\text{dB}}$. It is important to note that while the horizontal antenna pattern follows the 3GPP recommendation in [1, 3], the vertical antenna pattern has been enhanced to capture the effect of side lobes in our modeling (see Fig. A.4a).

According to [5], the downtilt θ_{etilt} can be calculated with respect to the effective transmitter antenna height $h_{\text{T,eff}}$ and the distance between the BS and the targeted cell boundary d_c (see Fig. A.3) as follows:

$$\theta_{\text{etilt}} = \arctan\left(\frac{h_{\text{T,eff}}}{d_c}\right) + z \cdot \theta_{3\text{dB}}, \qquad (A.3)$$

where z was empirically determined and set to 0.7 to achieve a good trade-off between the received power and inter-cell interference, and $h_{\text{T,eff}}$ and d_c are expressed in meter.

In order to consider the joint effect of the horizontal and vertical antenna patterns and create a 3-D like antenna pattern, the following model is used:

$$G^a(\varphi, \theta)[\text{dB}] = G_M^a[\text{dBi}] + G_H^a(\varphi)[\text{dB}] + G_V^a(\theta)[\text{dB}], \qquad (A.4)$$

where G_M^a is the maximum antenna gain in dBi.

For illustration purposes, Figs. A.4a and A.4b respectively show the antenna patterns and spatial antenna gains resulting when using the proposed models and the characteristic antenna parameters presented in Table A.1.

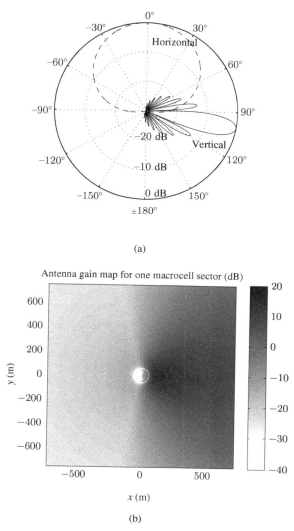

(a)

(b)

Figure A.4.   Macrocell BS antenna pattern and spatial antenna gains. (a) Macrocell BS horizontal and vertical antenna pattern and (b) Macrocell BS spatial antenna gains.

A.3.2 Small Cell Antenna Modeling

Due to the reduced form factor of small cell BSs, they use different antennas than macrocell BSs, and thus their antenna patterns are different. In the following, three-antenna pattern models for small cell BSs are presented, namely dipole (suitable for

TABLE A.1   Typical Parameters of a Macrocell Sector Antenna

Parameter	Value	
	Three-sector	Six-sector
G_M^a	15.5 dBi	19.8 dBi
φ_{3dB}	65 deg	33 deg
θ_{3dB}	11.5 deg	7.5 deg
κ	30 dB	30 dB

omni-directional femtocells), dipole array (suitable for omni-directional picocells), and LightRadio (suitable for directional small cells).

A.3.2.1 Dipole Model   The dipole antenna is the simplest practical antenna from a theoretical perspective, and is constructed based on two identical metal rods, oriented in parallel and displayed in line with respect to each other. The feeding current is applied in between the two rods. The most common form of dipole is the half-wave dipole, in which each of the two rods are approximately $1/4$ wavelength long. The half-wave dipole is typically used in femtocells for providing indoor residential coverage.

The horizontal and vertical antenna patterns of the half-wave dipole are modeled as follows:

- Horizontal antenna pattern

$$G_H^a(\varphi)[\text{dB}] = 0 \tag{A.5}$$

- Vertical antenna pattern

$$G_V^a(\theta)[\text{dB}] = 20\log_{10}\left(\frac{\cos\left(\frac{\pi}{2}\cos\left(\theta + \frac{\pi}{2}\right)\right)}{\sin\left(\theta + \frac{\pi}{2}\right)}\right) \tag{A.6}$$

and the combined gain is calculated as in (A.4), where the maximum antenna gain G_M^a equals 2.15 dBi

A.3.2.2 Dipole Array Model   The dipole array antenna model presented here is typically used in picocells for providing outdoor coverage and is characterized by four half-wave dipoles disposed in a vertical array and spaced by 0.6 λ_c, as illustrated in Fig. A.2b. As a result of arraying a number of half-wave dipoles, a vertical array factor gain $G_V^{a,array}(\theta)$ is obtained, which can be modeled as follows:

$$G_V^{a,array}(\theta) = \sum_{n=1}^{N_t} a(n) e^{(j(n-1) \times 2\pi d_\varepsilon(-\sin(\theta)) + \delta_{phase})}, \quad (A.7)$$

where N_t is the number of antenna elements in the vertical array, d_ε is the spacing between antenna elements in wavelength, $a(n)$ is the normalized voltage of antenna element n, and δ_{phase} is the phase increment within antenna elements in radian.

The combined gain of horizontal and vertical antenna patterns together with the vertical array factor gain is calculated as:

$$G^a(\varphi, \theta)[dB] = G_M^a[dBi] + G_H^a(\varphi)[dB] + G_V^a(\theta)[dB] + G_V^{a,array}(\theta)[dB], \quad (A.8)$$

where G_M^a, $G_H^a(\varphi)$, and $G_V^a(\theta)$ are the maximum antenna gain and the attenuation offsets on one element of the array, respectively.

In order to illustrate the performance of this dipole array, Figs. A.5a and A.5b respectively show the antenna patterns and spatial antenna gains resulting when using the proposed models and the characteristic antenna parameters presented in Table A.2. The obtained maximum antenna gain of this dipole array is 7.06 dBi.

A.3.2.3 LightRadio Model

The LightRadio antenna model presented here is used in outdoor metrocells, and is characterized by two directive antenna elements (referred to as lightRadio cubes) disposed in a vertical array and spaced by 0.5 λ_c, as illustrated in Fig. A.2c.

Figure A.6a shows the simulated horizontal and vertical antenna patterns of the lightRadio array resulting when using the characteristic antenna parameters presented in Table A.3.

The combined gain is calculated as indicated in (A.8), and Fig. A.6b shows the spatial antenna gains resulting from this configuration. The obtained maximum antenna gain of this lightRadio array is 11.19 dBi.

TABLE A.2 Typical Parameters of a
Dipole Array

Parameter	Value
N_t	4
G_M^a	2.15 dBi
d_ε	0.6 wavelength
$a(n)$	[0.97 1.077 1.077 0.86]
δ_{phase}	1.658 radian

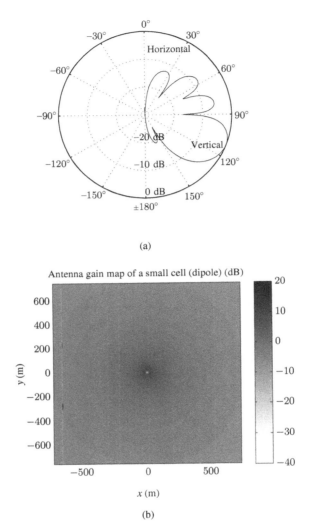

(a)

(b)

Figure A.5.   Dipole array horizontal and vertical antenna pattern (a) and spatial antenna gains (b).

A.4 PATH-LOSS MODELING

An important contributor to the total channel gain between a transmitter and a receiver is the path loss. The path-loss models the attenuation in signal strength that occurs when a radio signal travels a given distance from the antenna through space. A particular and important path-loss model is the free-space path-loss model, which indicates

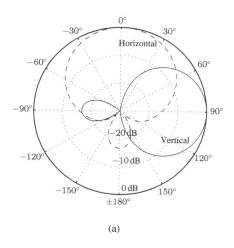

(a)

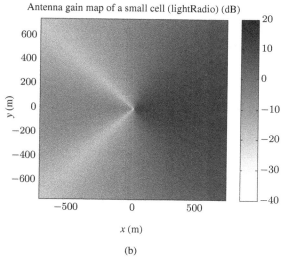

x (m)

(b)

Figure A.6. LightRadio array horizontal and vertical antenna pattern (a) and spatial antenna gains (b).

the attenuation in signal strength that occurs when a radio signal travels a given distance over a line-of-sight (LOS) path in free space, and can be modeled as:

$$G^{p,fs}[\mathrm{dB}] = -20\log_{10}(d) - 20\log_{10}(f_c) + 147.55, \qquad (A.9)$$

where d is the distance between the transmitter and the receiver in meter, and f_c is the frequency of the radio signal in GHz. However, the free-space path-loss model does

TABLE A.3   Typical Parameters of
a LightRadio Array

Parameter	Value
N_t	2
G_M^a	8.17 dBi
d_ϵ	0.5 wavelength
$a(n)$	[1 1]
δ_{phase}	0 radian

not hold for most environments due to its ideal conditions for example, the effects from the ground, building, vegetation, and others are not considered.

In order to enhance the accuracy of the free-space path-loss model and account for more realistic scenarios, deterministic or statistical (empirical) path-loss models are often adopted [2]. Deterministic path-loss models based on ray tracing, ray launching or finite-difference time-domain (FDTD) models can result in high accuracy at the expense of computation time and the need for detailed input data (e.g., city maps and material properties). In contrast, statistical models are appealing due to their lower complexity. These models are based on measured and averaged path losses in typical environments (e.g., dense urban, urban, rural, indoors), are created using curve-fitting approaches, and thus only require a few input parameters.

A widely used statistical path-loss model is the Okumura–Hata model [6], which was primarily built using measurements collected in the city of Tokyo, Japan, and enhanced later with different extensions. However, the Okumura–Hata model has some intrinsic drawbacks. For example, it neglects the terrain profile between the transmitter and the receiver, since it assumes that transmitters are located on hills, and only provides support up to 1.9 GHz.

In order to overcome such drawbacks and adapt to all types of conditions, a large number of statistical models have been developed ever since based on different measurement campaigns. In this book, the 3GPP path-loss models were adopted, which can be applied for different antenna heights and in the range of 2–6 GHz. A key advantage of the 3GPP modes is that they are widely used within industry and academia, and thus facilitate the comparison of the results from distinct parties.

A.4.1   Macrocell Path-Loss Modeling

The macrocell path-loss models considers LOS and non-line-of-sight (NLOS) propagation.

In this book, the urban macrocell path loss is modeled according to the urban macro (UMa) model in [3].

The LOS propagation is modeled as follows:

$$G_{\text{LOS}}^{\text{p,m}}[\text{dB}] = -22.0\log_{10}(d) - 20\log_{10}(f_c) - 28.0, \tag{A.10}$$

while the macrocell path loss in NLOS is modeled as follows:

$$\begin{aligned}
G_{\text{NLOS}}^{\text{p,m}}[\text{dB}] = &-161.04 + 7.1\log_{10}(W_s) - 7.5\log_{10}(h_b) \\
&+ \left(24.37 - 3.7\left(\frac{h}{h_T}\right)^2\right)\log_{10}(h_T) \\
&- (43.42 - 3.1\log_{10}(h_T))(\log_{10}(d) - 3) \\
&- 20\log_{10}(f_c) + (3.2(\log_{10}(11.75h_R))^2 - 4.97),
\end{aligned} \tag{A.11}$$

where f_c is the carrier frequency in GHz, d is the distance between transmitter and receiver (10 m $\leq d \leq$ 5000 m) h_b is the average building height, W_s is the average street width, h_T is the transmitter height and h_R is the receiver height, with all distances expressed in meter.

When using the following typical values for the presented parameters, $h_b = 20$ m, $W_s = 20$ m, $f_c = 2$ GHz, $h_T = 25$ m and $h_R = 1.5$ m, (A.10) and (A.11) respectively simplify to

$$G_{\text{LOS}}^{\text{p,m}}[\text{dB}] = 22.0\log_{10}(d) + 34.02, \tag{A.12}$$

$$G_{\text{NLOS}}^{\text{p,m}}[\text{dB}] = 39.01\log_{10}(d) + 21.56. \tag{A.13}$$

Having defined the path loss in LOS and NLOS conditions, it is now important to derive how often each type of propagation occurs. In the 3GPP, the LOS probability for the urban macro environment is modeled as follows:

$$\varrho_{\text{LOS}}^{\text{m}} = \min\left(\frac{18}{d_x}, 1\right) \cdot (1 - e^{\frac{-d_x}{63}}) + e^{\frac{-d_x}{63}}, \tag{A.14}$$

where d_x is the terrestrial distance between the transmitter and the receiver in meters (see d_c in Fig. A.3). During system-level simulations, in order to assess whether a LOS and NLOS transmission occurs, a uniform random variable between 0 and 1 is obtained and compared with the LOS probability. If the uniform random variable is smaller than $\varrho_{\text{LOS}}^{\text{m}}$, then LOS occurs; otherwise NLOS occurs.

In order to provide a smooth transition between LOS and NLOS propagation, the following modification is proposed over the presented 3GPP model, where LOS and NLOS are interpolated according to the LOS probability. In other words, the urban

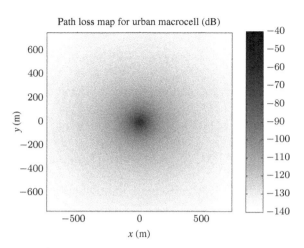

Figure A.7. Macrocell BS spatial path loss.

outdoor macrocell path loss $G^{p,m}$ is calculated as the expected value of the LOS and NLOS models as follows:

$$G^{p,m} = \varrho_{LOS}^m \cdot G_{LOS}^{p,m} + \left(1 - \varrho_{LOS}^m\right) \cdot G_{NLOS}^{p,m}. \tag{A.15}$$

Figure A.7 shows the spatial path losses resulting when using the presented urban macrocell path-loss model. The analysis of the proposed interpolation is provided for both macrocells and small cells later on in this section.

A.4.2 Small Cell Path-Loss Modeling

Small cell BSs are deployed at a much lower height than macrocell BSs, and therefore their path-loss profile is different. In the following, path-loss models for small cell BSs are presented, differentiating among outdoor and indoor models. Outdoor small cell models are suitable for metrocells and picocells, while indoor small cells models are suitable for femtocells.

A.4.2.1 Outdoor Small Cell Model   The urban outdoor small cell path loss in LOS is modeled according to the urban micro (UMi) model in [3] as follows:

$$G_{LOS}^{p,sc}[dB] = -22.0 \log_{10}(d) - 20 \log_{10}(f_c) - 28.0, \tag{A.16}$$

while the outdoor small cell path loss in NLOS is modeled as follows:

$$G_{NLOS}^{p,sc}[dB] = -36.7 \log_{10}(d) - 26 \log_{10}(f_c) - 22.7, \tag{A.17}$$

where $10 \text{ m} \le d \le 2000 \text{ m}$.

When using the following typical values in this model, $f_c = 2$ GHz, $f_c = 2$ GHz, $h_T = 10$ m and $h_R = [1-2.5]$ m, (A.16) and (A.17) respectively simplify to

$$G_{LOS}^{p,sc}[dB] = -22.0 \log_{10}(d) - 34.02, \qquad (A.18)$$

$$G_{NLOS}^{p,sc}[dB] = -36.7 \log_{10}(d) - 30.5. \qquad (A.19)$$

In this case, the outdoor small cell LOS probability for the urban micro environment is modeled in the 3GPP as follows:

$$\varrho_{LOS}^{sc} = \min\left(\frac{18}{d_x}, 1\right) \cdot (1 - e^{\frac{-d_x}{36}}) + e^{\frac{-d_x}{36}}. \qquad (A.20)$$

Then, the outdoor small cell path loss $G^{p,sc}$ is calculated using the proposed interpolation as follows:

$$G^{p,sc} = \varrho_{LOS}^{sc} \cdot G_{LOS}^{p,sc} + \left(1 - \varrho_{LOS}^{sc}\right) \cdot G_{NLOS}^{p,sc}. \qquad (A.21)$$

Figure A.8a presents the LOS probability when using the 3GPP models for both macrocells and outdoor small cells, that is, (A.14), (A.20) as well as the respective models when using the proposed smoothing interpolation. This figure shows how the proposed interpolation smoothes the transition from LOS to NLOS transition. Moreover, Fig. A.8b shows the path losses versus distance when using the presented urban macrocell and outdoor small cell path-loss models.

A.4.2.2 Indoor Small Cell Model   Similar to the outdoor small cell path loss, the indoor small cell path loss can be modeled using the indoor hotspot (InH) model in [3]. As a different approach, it could also be computed using just the LOS model (A.18) or any other one, and considering the transmission losses of the walls and other obstacles encountered by the radio signal based on the floor plan of the building, the materials, and the angle of incidence. This is the so-called environment loss, which is explained in the following section.

A.5 ENVIRONMENT LOSS MODELING

The environment loss (also referred to as wall penetration loss) models the attenuation that a radio signal suffers when traversing an obstacle, for example, buildings, walls, trees. The environment loss depends upon the frequency of the radio signal, as well as the thickness and properties of the obstructing materials, that is, relative permittivity, relative permeability and conductivity.

For evaluating the performance of a small cell, the effects of obstacles play a critical role, and can have both beneficial and detrimental effects. However, the previously

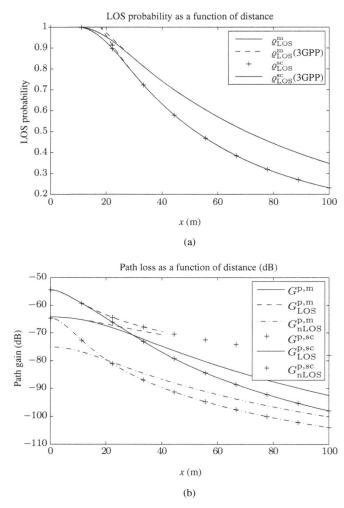

Figure A.8. Macrocell BS LOS probability (a) and path gain (b) as a function of distance.

presented statistical path-loss models only capture the average effect of obstacles for the measured scenarios, and does not accurately capture the effect of particular obstacles in specific scenarios.

One possibility to model these effects is using deterministic tools such as ray tracing, ray launching, or FDTD, which can take both transmission losses and reflections on obstacles into account. In this case, the environment loss of each obstacle can be precalculated according to the frequency of the radio signal, thickness of the obstacle, transmission and reflection properties, angle of arrival, etc. (if this values

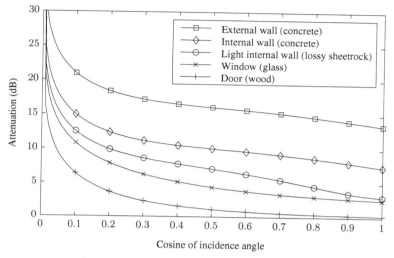

Figure A.9. City of Dublin environment map.

are known), and then can be applied to each radio signal when necessary. If the properties of the encountered objects are unknown, measurements can be taken and the properties of the obstacles can be calibrated, such that the predictions best fit the measurements [2]. The accuracy of ray tracing tools can be very good, with deviations after calibration typically on the order of less than 6 dB in path loss and 20 ns for the delay spread [7]. However, a disadvantage of ray tracing tools is their computational complexity, which can result in very long simulation times for large scenarios. One example of a ray tracing tool is WISE [8].

As a compromise between these full deterministic models and statistical models, the path loss can be modeled using the models presented in the previous section, and an environment loss G^e can be associated with each object (e.g., building, wall) and added to the path loss. This model has been widely used in the literature to analyze large scenarios, and is introduced in the following.

For indoor scenarios, the transmission losses of walls based on the floor plan of the building, the material and the angle of incidence can be taken into account. The environment loss G^e of the explicitly modeled walls can be calculated as the sum over all individual wall losses $G_w^e(m, \phi)$ between the transmitter and the receiver, each dependent on the material m and the angle of incidence ϕ, that is,

$$G^e = \sum_w G_w^e(m, \phi). \tag{A.22}$$

For the house models in this book, the transmission losses for the different materials used are shown in Fig. A.10, and are taken from WISE [8]. One example of the resulting path losses for a residential environment is shown in Fig. A.11.

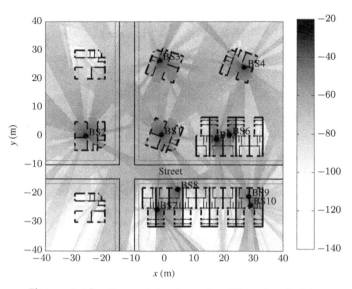

Figure A.10. Transmission losses for different materials.

For outdoor scenarios, because the floor plan of the outdoor area may not be available and because the transmitter is usually located at a much higher height than the receiver, an average environment loss G^e is considered. If the transmitter is indoors and the receiver is indoors in another building, the environment loss is $2 \times G^e$;

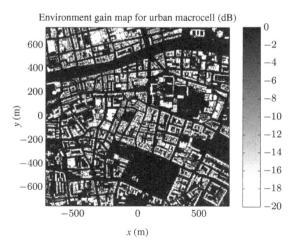

Figure A.11. Example of path gains using an explicit wall model based on transmission losses for a residential environment.

otherwise, the environment loss is 0 dB. Note that according to [3], the average environment loss is $G^e = -20$ dB. Figure A.9 illustrates environment losses in the city of Dublin, Ireland, when adopting the described model and considering an outdoor transmitter. However, G^e could also be object-specific, if the environment would also include information about the objects.

A.6 SHADOW FADING MODELING

Shadow fading models the random fluctuations of the received signal power at the receiver side created by the obstructions caused by the objects that a radio signal experiences in its propagation path. The locations, sizes, and dielectric properties of the obstructing objects, as well as the reflecting surfaces and scattering obstacles, are usually unknown. Due to such uncertainties, statistical models are generally used to model shadow fading. A widely used model is the log-normal shadowing model [9], which has been shown to be able to model shadow fading attenuations with good accuracy in both outdoor and indoor environments.

Accordingly, the shadow fading in the path between a transmitter and a receiver can be a priori modeled using a log-normal random variable, $G^s \sim \mathcal{N}(\mu_s, \sigma_s^2)$, where μ_s and σ_s are the mean and standard deviation of the log-normal random variable in dB, respectively. However, the modeling of shadow fading when considering multiple transmitters (e.g., BSs) and receivers (e.g., UEs) is more intricate due to the spatial auto-correlation and cross-correlation properties of shadow fading.

A.6.1 Auto-Correlated Shadow Fading

The shadow fading process is auto-correlated in space, that is, a moving UE may see similar shadow fading attenuations from a given BS at different but nearby locations, whereas these shadow fading attenuations may significantly differ between two distant positions.

A widely adopted auto-correlation model for shadowing is the Gudmundson model [10], which defines the auto-correlation coefficient as follows:

$$\rho_a(\Delta d) = \exp\left(-\frac{|\Delta d|}{d_{cor}} \cdot \ln 2\right), \tag{A.23}$$

where d_{cor} is the de-correlation distance, which is defined here as the distance at which the correlation coefficient $\rho_a(d_{cor})$ falls by $1/\exp$, and Δd is the distance between the two locations.

The auto-correlation of shadowing can be implemented as follows. If G_1^s is the log-normal shadowing in dB at position (x_1, y_1) and G_2^s is the auto-correlated log-normal shadowing in dB at position (x_2, y_2), which is Δd away from (x_1, y_1), then

G_2^s can be modeled as a normally distributed random variable in dB with mean $\mu_s' = \rho_a(\Delta d) \cdot G_s^1$ and standard deviation $\sigma_s' = \sqrt{(1 - \rho_a^2(\Delta d))} \cdot \sigma_s^2$ [11].

A.6.2 Cross-Correlated Shadow Fading

The shadowing process is cross-correlated in the network. Multiple links from different BSs to one UE may observe a highly cross-correlated shadow fading depending on the environment. For example, if two BSs are very close to each other, the two paths between the two BSs and a UE may traverse a similar environment, and thus the UE may observe a correlated shadow fading from the two BSs. Since the cross-correlation coefficient ρ_c among BSs and sectors is usually given in the simulation assumptions (e.g., $\rho_c = 1$ for collated sectors and $\rho_c = 0.5$ for neighboring BSs is assumed in the 3GPP modeling), cross-correlated shadowing can be modeled using Cholesky decomposition [11].

The cross-correlation of shadowing can be implemented as follows.

1. Define N_{BS} independent normally distributed random variables $\mathbf{G}^s = [G_1^s, \ldots, G_n^s, \ldots, G_{N_{SC}}^s]$ to represent the shadow fading in the links from the N different BSs to the studied UE, where $G_n^s \sim \mathcal{N}(\mu_s, \sigma_s^2)$ with both μ_s and σ_s in dB,

2. Define the cross-correlation coefficient matrix $\mathbf{B}$ as follows

$$
\mathbf{X} = \begin{pmatrix}
\rho_c^{1,1} & \rho_c^{1,2} & \rho_c^{1,n} & \cdots & \rho_c^{1,N_{SC}} \\
\rho_c^{2,1} & \rho_c^{2,2} & \rho_c^{2,n} & \cdots & \rho_c^{2,N_{SC}} \\
\vdots & \vdots & \vdots & \ddots & \vdots \\
\rho_c^{N_{SC},1} & \rho_c^{N_{SC},2} & \rho_c^{N_{SC},n} & \cdots & \rho_c^{N_{SC},N_{SC}}
\end{pmatrix}, \qquad (A.24)
$$

where $\rho_c^{1,2} = \rho_c^{2,1}$ is the cross-correlation coefficient of shadow fading between the links from BS 1 and BS 2 to the studied UE. Since $\mathbf{X}$ is a symmetric and positive definite matrix, it can be decomposed into a lower or upper triangular matrix by using Cholesky decomposition, for example, $\mathbf{X} = \mathbf{C}^T \mathbf{C}$, where $\mathbf{C}$ is an upper triangular matrix.

3. Then, the cross-correlated shadow fading attenuations for the N links can be modeled as $\hat{\mathbf{G}}^s = \mathbf{G}^s \mathbf{C} = [\hat{G}_1^s, \ldots, \hat{G}_n^s, \ldots, \hat{G}_{N_{SC}}^s]$.

A.6.3 Auto- and Cross-Correlated Shadow Fading Modeling

In order to consider both auto- and cross-correlated shadow fading, we can combine the auto-correlation and cross-correlation implementation procedures together as follows, and compute the resulting shadow fading on the fly. During the initialization, the cross-correlation procedure described above is performed for each UE, and $\hat{\mathbf{G}}^s$ is found. Then, when the UE locations are updated according to a mobility model, the cross-correlation procedure runs again. However, in step 1, the N independent normally distributed random variables $\mathbf{G}^s = [G_1^s, \dots, G_n^s, \dots, G_{N_{SC}}^s]$ are not new ones but the existing ones updated using the auto-correlation procedure described earlier. After running the cross-correlation procedure, a new $\hat{\mathbf{G}}'^s$ is found, which is auto- and cross-correlated with respect to $\hat{\mathbf{G}}^s$.

Due to complexity issues, in scenarios with a large number of BSs and UEs, computing both auto- and cross-correlated shadow fading attenuations in real time through Cholesky decomposition may significantly increase the running time and memory requirements of system-level simulations. In order to solve this issue, pre-computed shadowing maps obtained through low-complexity methods can be used. Similar to look-up tables (LUTs), shadowing maps can be imported in a system-level simulation, associated with BSs and queried during its execution to assess the shadow fading attenuations from each BS to each UE at a given location. Figure A.12a illustrates a shadowing map with inter-site correlation $\rho_c = 0.5$, correlation distance $d_{cor} = 20$ m, mean $\mu_s = 0$ dB and standard deviation $\sigma_s = 6$ dB (see Fig. A.12b for statistics). This map was computed using the low-complexity method described in [12], which also makes use of Cholesky decomposition.

A.7 MULTI-PATH FADING GAIN MODELING

The obstructing objects in the propagation path from the transmitter to the receiver may also produce reflected, diffracted, and scattered copies of the radio signal, resulting in multi-path components (MPCs). The MPCs may arrive at the receiver side attenuated in power, delayed in time and shifted in frequency (and/or phase) with respect to the first and strongest MPC, thus adding up constructively or destructively. As a consequence, the received power at the receiver side may vary significantly over very small distances on the order of a few wavelengths [9].

Since different MPCs travel over different paths of different lengths, a single impulse sent from a transmitter and suffering from multi-path will result in multiple copies of such single impulse being received at the receiver side at different times. As a result, the channel impulse response of a multi-path channel can be modeled using a tapped delay line as follows:

$$h(\tau, t) = \sum_{i=0}^{\nu-1} h_i(t) \cdot \delta(\tau - \tau_i), \qquad (A.25)$$

where ν is the number of taps or resolvable MPC, and $h_i(t)$ and τ_i are the complex channel gain and delay of tap i, respectively.

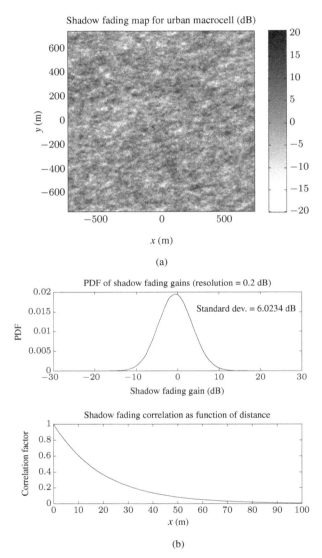

Figure A.12.   Auto- and cross-correlated shadow fading map (a) and its statistics (b).

Power-delay profiles are often used to model these tapped delay lines. A power-delay profile is characterized by the number of taps v, the time delay relative to the first tap, the average power relative to the strongest tap and the Doppler spectrum of each tap. The most frequently used power-delay profiles are those specified by International Telecommunication Union (ITU) and COST 207.

TABLE A.4   ITU Multi-path Channel Models

	Delay (ns)	Relative Power (dB)	Delay (ns)	Relative Power (dB)
		Pedestrian (≤3 km/h)		
Tap	Channel A (40%)		Channel B (55%)	
1	0	0	0	0
2	110	−9.7	200	−0.9
3	190	−19.2	800	−4.9
4	410	−22.8	1200	−8.0
5	-	-	2300	−7.8
6	-	-	3700	−23.9
		Vehicular (30, 60, 120 km/h)		
	Channel A (40%)		Channel B (55%)	
1	0	0	0	−2.5
2	310	−1	300	0
3	710	−9	890	−12.8
4	1090	−10	12,900	−10.0
5	1730	−15	17,100	−25.2
6	2510	−20	20,000	−16.0
		Indoor		
	Channel A (50%)		Channel B (45%)	
1	0	0	0	0
2	50	−3	100	−3.6
3	110	−10	200	−7.2
4	170	−18	300	−10.8
5	290	−26	500	−18.0
6	310	−32	700	−25.2

It is important to note that root mean square (RMS) delay spreads are usually relatively small, but occasionally and depending on the scenario, worst-case multi-path characteristics may result in large RMS delay spreads. In order to capture this delay spread variability, ITU has defined two power-delay profiles (i.e., channel A and channel B) for each of its three different test environments (i.e., vehicular, pedestrian, and indoor channels [13]). Within a test environment, channel A is the low-delay spread case and channel B is the median-delay spread case. Suburban and rural scenarios with low density of buildings have relatively low delay spreads (less than 1 μs), while dense urban scenarios with links of a few kilometres have higher delay spreads (less than 5 μs). The percentage of time for which Channel A and Channel B are expected to be encountered in each test environment is given in [13], and shown in Table A.4 together with the different ITU power-delay profiles.

TABLE A.5   TU Multi-path Channel Models

	Alternative 1		Alternative 2	
Tap	Delay (ns)	Relative Power (dB)	Delay (ns)	Relative Power (dB)
1	0.0	0.0	−4.0	−4.0
2	0.1	0.2	−3.0	−3.0
3	0.3	0.4	0.0	0.0
4	0.5	0.6	−2.6	−2.0
5	0.8	0.8	−3.0	−3.0
6	1.1	1.2	−5.0	−5.0
7	1.3	1.4	−7.0	−7.0
8	1.7	1.8	−5.0	−5.0
9	2.3	2.4	−6.5	−6.0
10	3.1	3.0	−8.6	−9.0
11	3.2	3.2	−11.0	−11.0
12	5.0	5.0	−10.0	−10.0

COST 207 has also specified power-delay profiles for typical rural areas (RAx), hilly terrain (HTx) and urban areas (TU) [14], where the TU model is widely used within the 3GPP (see Table A.5). The typical urban (TU) model is defined by 12 taps. However, due to complexity issues, it may not be always possible to simulate the complete model, and a reduced-complexity configuration with six taps was also defined. This reduced configuration may be used in particular for the multi-path simulation of interfering signals. Note that for each model, two equivalent alternative tap settings are proposed, denoted by Alternative 1 and Alternative 2 in Table A.5.

Calculating the complex multi-path fading G^{ff} using the presented multi-path channel models is usually time-consuming. For a state-of-the-art computer, it may take hours to generate an adequate number of multi-path channel realizations. In a system-level simulation, multi-path fading needs to be calculated for the carrier and all interfering signals in all frequency resources for each UE involved in the simulation, and thus the number of required multi-path fading predictions scales with the number of UEs, BSs, and frequency resources. Therefore, precomputed multi-path fading matrices are commonly used to reduce complexity and speed up system-level simulations. Multi-path fading matrices do not have a geographical meaning as previously presented maps. Each row and column is associated with a frequency and time resource, and the matrix contains the associated complex multi-path fading G^{ff}. Similar to the shadow fading maps, independent multi-path fading matrices can be imported in the system-level simulation, associated with different UEs, and queried during the simulation to assess the complex multi-path fading of each UE at a given frequency and time resource. Figure A.13 shows a multi-path fading matrix calculated using the TU model with 12 taps for a UE moving at 3 km/h. Results are shown in a map of 5 MHz by 100 ms, sampled every 45 kHz in the frequency domain and

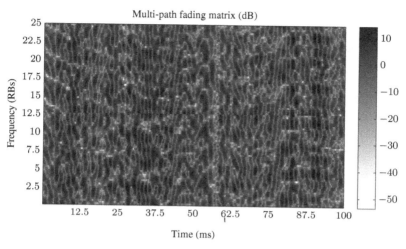

Figure A.13. Multi-path fading map.

every 0.25 ms in the time domain. The multi-path fading power gains $|G^{\mathrm{ff}}|^2$ are shown in dB.

It is important to note that for multi-path fading channel modeling, the 3GPP recommends that "if fast fading modeling is disabled in system-level simulations for relative evaluations, the impairment of frequency-selective fading channels shall be captured in the physical layer abstraction" [3]. This means that ordinary multi-path fading models like ITU and TU may be disabled, if their effects thereof are modeled somewhere else, for example, in the signal-to-interference-plus-noise ratio (SINR) to throughput mapping process. The effect of the multi-path fading has to be captured, but how to capture it is up to the implementation of the system-level simulator.

A.8 RECEIVED SIGNAL STRENGTH AND QUALITY MODELING

In this section, the received signal strength and quality perceived by a UE, which are key parameters to assess UE performance, are modeled as a function of the presented network layout and channel gains.

A.8.1 Received Signal Strength Modeling

Having defined the antenna gain G^{a}, the path loss G^{p}, the environment loss G_{e}, the shadow fading G_{s}, and the multi-path fading G_{ff} in the previous sections, the received signal strength $P^{\mathrm{rx}}_{t,r,k}$ of the signal transmitted by a transmitter T_t at a receiver R_r in a

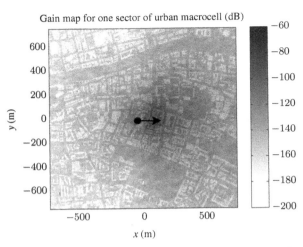

Figure A.14. City of Dublin received signal strength map of one urban macrocell BS sector.

frequency resource K_k can be calculated as:

$$P_{t,r,k}^{\mathrm{rx}} = P_{t,k}^{\mathrm{tx}} \cdot G_{t,r}^{\mathrm{a}} \cdot G_{t,r}^{\mathrm{p}} \cdot G_{t,r}^{\mathrm{e}} \cdot G_{t,r}^{\mathrm{s}} \cdot \left| G_{t,r,k}^{\mathrm{ff}} \right|^2, \qquad (A.26)$$

where $P_{t,k}^{\mathrm{tx}}$ is the transmit power applied by transmitter T_t in frequency resource K_k, $G_{t,r}^{\mathrm{a}}$ is the antenna gain from transmitter T_t to receiver R_r, $G_{t,r}^{\mathrm{p}}$ is the path loss from transmitter T_t to receiver R_r, $G_{t,r}^{\mathrm{e}}$ is the environment loss from transmitter T_t to receiver R_r, $G_{t,r}^{\mathrm{s}}$ is the shadow fading from transmitter T_t to receiver R_r and $|G_{t,r,k}^{\mathrm{ff}}|^2$ is the multi-path fading from transmitter T_t to receiver R_r in frequency resource K_k (note that $G_{t,r,k}^{\mathrm{ff}}$ was defined as a complex gain). All these variables are expressed in linear units.

Figure A.14 shows the received pilot signal strength map of one urban macrocell BS sector (indicated by an arrow) in the city of Dublin, Ireland, resulting from the combination of previously illustrated maps, as indicated by (A.26). The parameters used are presented in Table A.6. Note that multi-path fading is not considered in order to present an average received signal strength map, and that the transmit power $P_t^{\mathrm{tx,pilot}}$ applied by transmitter T_t to its pilot signal is set up in a way that this pilot signal is received indoors at the targeted cell radius d_c with a given signal-to-noise ratio (SNR) $\Gamma_{t,\mathrm{edge}(d_c),k}$, that is,

$$P_{t,k}^{\mathrm{tx,pilot}} = \frac{\Gamma_{t,\mathrm{edge}(d_c),k} \cdot \sigma^2}{G_{t,r}^{\mathrm{a}} \cdot G_{t,r}^{\mathrm{p}} \cdot G_{t,r}^{\mathrm{e}} \cdot G_{t,r}^{\mathrm{s}}}, \qquad (A.27)$$

TABLE A.6   System-Level Simulation Parameters

Parameters	Value
Scenario size	1500 m × 1500 m, around central macrocell
Scenario resolution	2 m
Carrier frequency	2000 MHz
Bandwidth	3.84 MHz (HSPA), 5 MHz (LTE)
Macro BS placement	7 macrocell BSs, 800 m inter-site distance
Small cell BS placement	4 small cell BSs per macrocell BS sector
Height	25 m (macro), 10 m (pico)
Total BS transmit power	21.6 W (macro), 1 W (picos)
SNR target at cell edge	5 dB
Macro BS antenna	See Section A.3.1
Small cell BS antenna	Dipole array, see Section A.3.2
Macro BS path loss	See Section A.4.1
Small cell BS path loos	Outdoor small cell, see Section A.4.2
Environment loss	$G_e = -20$ dB if indoor, 0 dB if outdoor
Shadow fading (SF)	6 dB std dev.
SF correlation	$R = e^{-1/20d}$, 50% inter-site
Noise density	-174 dBm/Hz
BS noise figure	4 dB
UE noise figure	7 dB

where $d_c = D_{ISD}/\sqrt{3}$ m, and σ^2 is the thermal noise power at the receiver side, which can be modeled as:

$$\sigma^2 = 1000 k_B T_b B_k \psi, \tag{A.28}$$

where k_B is the Boltzmann's constant in joules per Kelvin, T_b is the temperature in Kelvin, B_k is the bandwidth of frequency resource K_k in Hz, ψ is the noise figure, and $10\log_{10}(1000 k_B T) = -174$ dBm/Hz for $T = 300$ Kelvin.

It is important to note that the transmit power $P_{t,k}^{tx,pilot}$ applied to pilot signals represents a percentage of the total transmit power $P_{t,k}^{tx,total}$ available at the BS for frequency resource K_k, and thus $P_{t,k}^{tx,pilot}$ is bounded by $P_{t,k}^{tx,total}$ and this relationship. In high-speed packet access (HSPA), 10% of the total transmit power available at the BS is typically allocated to pilot signals, while in long-term evolution (LTE), the downlink (DL) reference symbol overhead is estimated to be around 9.52% for two transmitters and 14.29% for four transmitters [15].

A.8.2 Received Signal Quality Modeling

Having defined the received signal strength $P_{t,r,k}^{rx}$ of the signal transmitted by a transmitter T_t at a receiver R_r in a frequency resource K_k, the SINR $\gamma_{t,r,k}$ of this signal in a single input single output case can be modeled in linear units as:

$$\gamma_{t,r,k} = \frac{P_{t,r,k}^{\text{rx}}}{\displaystyle\sum_{\substack{t'=0 \\ t' \neq t}}^{T} P_{t',r,k}^{\text{rx}} + \sigma^2}. \tag{A.29}$$

For illustration, Fig. A.15 shows the SINR maps for a simulated hexagonal macrocell layout in the city of Dublin, Ireland, with and without small cells, where the maps have been calculated as indicated by (A.29), and the parameters used are presented in Table A.6. Note that here the SINR of the best server is plotted for each location, and multi-path fading is not considered to show an average SINR map, similar to Fig. A.14.

With regard to signal quality, it is important to note that the signal quality perceived at a receiver in each frequency resource varies due to the effects of multi-path fading. Moreover, it is also important to note that under the 3GPP system-level simulation assumptions, it is usual to find system models with two or more antennas at the receiver side, and that the signal quality perceived at each one of these antennas also varies due to the effects of multi-path fading.

MRC is often used to combine the signals received by different antennas [1]. Moreover, for estimating wideband SINRs, EESM, among others, is often used to compute the effective SINR across a given bandwidth based on the individual SINRs of the comprising frequency resources. In the following, we describe both MRC and EESM models, which are widely used in the 3GPP. However, it is important to mention that other types of combiners, for example, maximum likelihood, minimum mean square error, equal gain combining, generalized selected combining [16], and other types of effective SINR mappings, for example, capacity effective SINR mapping (CESM) and mutual information effective SINR mapping (MIESM) [17], could also be used.

A.8.2.1 Maximal Ratio Combining
Assuming perfect channel knowledge at the receiver side and independent channels observed at its different antennas, MRC is the optimum diversity combining technique over additive white Gaussian noise (AWGN) channels. Theoretically, when multiple independent copies of the same signal are combined through MRC, the instantaneous SNR at the output of the MRC combiner is maximized [16].

When having a receiver with A_r antennas, and under the assumptions of perfect channel knowledge and independent AWGN channels at both antennas, the SINR $\gamma_{t,r,k}^{\text{MRC}}$ at the output of the MRC combiner at receiver R_r when connected to transmitter T_t in frequency resource K_k can be calculated as:

$$\gamma_{t,r,k}^{\text{MRC}} = \sum_{a=1}^{A_r} \gamma_{t,r,k}^{a}, \tag{A.30}$$

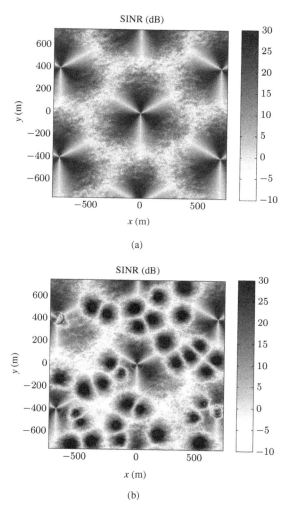

SINR (dB)

(a)

SINR (dB)

(b)

Figure A.15. City of Dublin SINR map of an hexagonal macrocell layout with $D_{ISD} = 800$ m and 1 tier of interferers with and without small cells. (a) Macrocell-only scenario and (b) Heterogeneous network (HetNet) scenario.

where $\gamma^a_{t,r,k}$ is the SINR of the received signal at antenna a of receiver R_r when connected to transmitter T_t in frequency resource K_k, and all values are given in linear units.

A.8.2.2 Exponential Effective SINR Mapping

In system-level simulations of orthogonal frequency division multiple access (OFDMA)-based networks, there is a need for calculating the effective SINR of each resource block (RB) based

TABLE A.7   EESM β Values

Modulation	Code Rate	β
QPSK	1/3	1.49
	1/2	1.57
	2/3	1.69
	3/4	1.69
	4/5	1.65
16QAM	1/3	3.36
	1/2	4.56
	2/3	6.42
	3/4	7.33
	4/5	7.68

on the SINRs of the subcarriers within the RB. There is also a need for calculating the effective SINR of the entire band (i.e., wideband SINR) based on the SINRs of the RBs within the band. The EESM mapping function serves this purpose and can be written as [18]:

$$\gamma_{t,r}^{\text{eff}} = -\beta \ln \left(\frac{1}{K} \sum_{k=1}^{K} \exp \left(-\frac{\gamma_{\text{t,r,k}}}{\beta} \right) \right), \tag{A.31}$$

where $\gamma_{\text{t,r}}$ is the set of SINRs to be mapped in linear units (e.g., SINRs obtained from MRC of all RBs in a band), which are typically different in a frequency selective channel, K is the cardinality of set $\gamma_{\text{t,r}}$, $\gamma_{t,r,k}$ is the kth element of set $\gamma_{\text{t,r}}$, and β is a modulation and coding scheme (MCS)-dependent parameter that compensates the difference between the exact block error rate (BLER) and the predicted BLER. Adequate values of β for different MCSs have been proposed in the literature. For example, Table A.7 presents β values for different MCSs suggested by the 3GPP [19], while Tables A.8 and A.9 present results based on ITU models [20].

TABLE A.8   ITU (Pedestrian A)-Based EESM β Values

Modulation	Code Rate	β
QPSK	1/2	1.59
	3/4	1.70
16QAM	1/2	5.33
	3/4	8.45
64QAM	1/2	16.10
	3/4	27.21

TABLE A.9   ITU (Pedestrian B)-Based EESM β Values

Modulation	Code Rate	β
QPSK	1/2	1.67
	3/4	1.75
16QAM	1/2	5.41
	3/4	7.94
64QAM	1/2	17.27
	3/4	32.33

A.9   THROUGHPUT MODELING

The Shannon–Hartley theorem states the maximum capacity of a wireless channel of a given bandwidth in the presence of noise (characterized by the SNR) [21], and it is modeled as:

$$C_{t,r,k} = B_k \cdot \log_2(1 + \gamma_{t,r,k}), \tag{A.32}$$

where the model can be extended to interfered systems under the assumption that the interference can be modeled as an AWGN. In this case, $C_{t,r,k}$ is the capacity in bps of the channel between transmitter T_t and receiver R_r in frequency resource K_k of bandwidth B_k in Hz when receiver R_r perceives a SINR $\gamma_{t,r,k}$ in linear units. A penalty of several dBs can be added to the received SINR $\gamma_{t,r,k}$ in order to account for physical layer imperfections, finite block lengths and realistic MCSs. Typical penalties are of 3 dB to 6 dB. A cap can also be added to account for the maximum MCS efficiency, as can be seen in Fig. A.16.

In order to assess the capacity of a wireless channel under the specifications of current cellular technologies, for example, HSPA, LTE, and account for mentioned imperfections, detailed link-level simulations are often used. Link-level and system-level simulations are independently executed due to their different time scales and computational costs, but they can interact through static interfaces called LUTs. Basically, LUTs are generated by link-level simulations, queried during system-level simulations, and characterize channel performance in a simplified manner. For example, LUTs may map SINR to throughput (usual in HSPA performance evaluation) or bits per data symbol (usual in LTE performance evaluation), as shown in Fig. A.16a and Fig. A.16b, respectively. This is a simple but efficient approach for including the effects of the physical layer and the fluctuations of the radio channel into system-level simulations.

Using these LUTs and knowing the receiver SINR $\gamma_{t,r,k}$, the performance of receiver R_r in frequency resource K_k can be assessed. In HSPA, since the bandwidth of

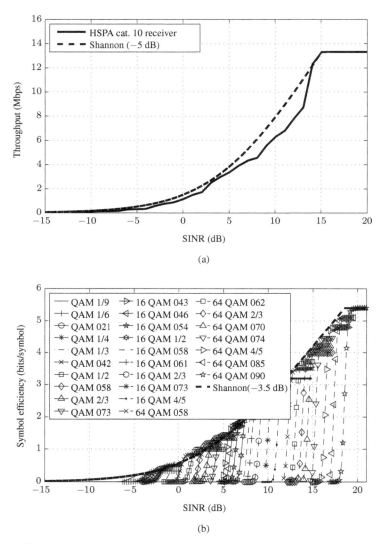

Figure A.16. LUT for channel performance assessment in HSPA and LTE systems. (a) Receiver SINR to throughput in HSPA and (b) Receiver SINR to bits per data symbol in LTE.

the carrier frequency is fixed, the receiver SINR can be directly mapped to throughput. However, in LTE, since the bandwidth of the carrier frequency can vary from network to network, and because the receiver may be allocated just a subset of the available frequency resources in a carrier frequency, the receiver SINR is mapped to bits per data symbol, and then the receiver throughput is proportional to the number of scheduled data symbols.

A.9.1 Scheduling

In LTE, the RB is the minimum resource unit that can be scheduled to a receiver, and the number of RBs allocated to a receiver is decided by the BS scheduler. There are a total of 12 subcarriers and 14 OFDMA symbols per RB (if normal cycle prefix (CP) is used), making a total of 168 symbols per RB. Moreover, within an RB, some of the symbols may be used as control symbols and the remaining as data symbols. The more symbols used for control, the less throughput for the UE. For details on the number of available data symbols per LTE RB, please refer to [15].

As one can derive from the above discussion, the BS scheduler has a significant impact on LTE performance since it decides how many and which specific resources are allocated to each receiver.

Sharing the resources fairly between receivers experiencing different channel qualities is a challenging task [22]. An opportunistic scheduler selects the receiver with the best channel quality at each time/frequency resource, aiming to solely maximize the overall throughput, whereas a round robin (RR) scheduler treats the receivers equally regardless of their channel quality, giving the same amount of time/frequency resources to all receivers. The former scheduler can increase system throughput remarkably compared to the latter, but it can be unfair since receivers with relatively bad channel quality may be never scheduled [23]. Different flavours of RR schedulers can be defined. For example, receivers are first listed according to some metric or rule, and then a selected receiver or a selected subset of receivers equally share the available spectrum during a subframe.

In contrast, proportional fair (PF) is an scheduling algorithm that exploits multiuser diversity using the receivers' channel quality indicators (CQIs), attempting to maximize the system throughput, while simultaneously forcing a degree of fairness in serving all the receivers. The CQIs of different receivers are simultaneously and continuously fed back to the serving BS, and it is worth noting that if the channel feedback delay is relatively small compared to the channel rate variation, the scheduler can have sufficient time to estimate the channel conditions of all receivers over all RBs within an appropriate time [22]. PF basically aims at weighting the receivers' potential instantaneous performance by its average performance, and this process consists of three stages. In the first stage, according to buffer information, the schedulable set of receivers is specified. In the second stage, in the time domain scheduler, R_{max} receivers are selected according to a time domain metric and passed to the frequency domain scheduler. In the frequency domain scheduler, RBs are allocated to receivers according to a frequency domain metric. The complexity of the frequency domain scheduler highly depends on the number of its input receivers, and thus the time domain scheduler has an impact on the complexity and performance of the frequency domain scheduler [24, 25].

A time domain PF metric as defined in [24] is:

$$M_{t,r}^{\mathrm{PF-TD}} = \frac{\hat{D}[r]}{R[r]}, \tag{A.33}$$

where $R[r]$ and $\hat{D}[r]$ are the past average throughput and potential instantaneous throughput of receiver R_r, respectively. The past average throughput of receiver R_r can be computed using a moving average as:

$$R_r(t+1) = \left(1 - \frac{1}{T_c}\right) R_r(t) + \frac{1}{T_c} \times c_{t,r}(t), \tag{A.34}$$

where T_c is the length of the moving average window and should be larger than the time elapsed between multiple schedules of the individual receiver, and $c_{t,r}$ is the potential instantaneous throughput between of receiver R_r.

A frequency domain PF metric as specified [24] is:

$$M_{t,r,k}^{\mathrm{PF-FD}} = \frac{\gamma_{t,r,k}}{\displaystyle\sum_{k=1}^{N_{\mathrm{RB}}} \gamma_{t,r,k}}, \tag{A.35}$$

where the numerator is the SINR of receiver R_r in RB K_k and the denominator is the sum of the receivers' SINRs over all RBs, which represents its average channel quality in such subframe.

The receivers will be then ranked in each RB according to the metric in (A.35), and the receiver with the maximum metric is selected to be served in such RB. For more information on PF schedulers, refer to [24]. It is important to note that other metrics may apply, and they are vendor-specific.

A.10 USER EQUIPMENT LOCATION AND TRAFFIC MODELING

The UE location and traffic models attempt to characterize UE behavior. The UE location models dictate the UEs locations within the simulation scenario, while the UE traffic models profile the UE data demands. In the following, the most common UE location and traffic models are described.

A.10.1 UE Location

Different UE location models can be found in the literature to model where the UEs are located within the simulation scenario. However, they can be mostly classified into two groups: uniform deployments and clustered deployments.

UEs can be uniformly distributed within a given scenario according to a number of UEs per unit of area, or within a macrocell BS sector or small cell coverage area according to a number of UEs per sector or small cell coverage area. These are simple and widely used models, due to their analytical tractability as well as easy reproducibility. However, there are concerns about their credibility, since it is known that UEs tend to cluster in given locations such as offices, coffee shops, and shopping malls

To this end, the 3GPP has proposed the following clustering model, which follows the next steps [3]:

1. Fix the total number N_{UEs}^m of UEs dropped within each macrocell BS coverage, where N_{UEs}^m is 30 or 60 in fading scenarios, and 60 in non-fading scenarios.
2. Randomly and uniformly drop the configured number N_{SC} of small cell BSs within each macrocell BS sector, where N_{SC} may take values from $\{1, 2, 4, 10\}$.
3. Randomly and uniformly drop N_{UEs}^{sc} hotspot UEs within a 40 m radius of each small cell BS, where $N_{UEs}^{sc} = \frac{\varphi_{hotspot} \cdot N_{UEs}^m}{N_{SC}}$, with $\varphi_{hotspot}$ defined in Table A.10 as the fraction of the number of hotspot UEs over the total number of UEs dropped within each macrocell BS coverage. Note that $N_{UEs}^{sc} = 0$ for non-clustered scenarios, that is, uniform UE distribution.
4. Randomly and uniformly drop the remaining $N_{UEs}^m - N_{UEs}^{sc} \cdot N$ UEs over the macrocell BS coverage area (including the small cell BS coverage areas).

This is just an example of a clustering model, but it clearly shows how a number of UEs are uniformly distributed around each small cell BS, while the remaining UEs are uniformly distributed within the entire simulation scenario. This is a common feature of most clustering models, and allows to account for more realistic scenarios, where small cell BS deployments are coupled with UE locations. Small cell BSs are

TABLE A.10   Configurations 4a and 4b for Clustered UE Deployment (Based on Table A.2.1.1.2-5 in [3])

Configuration	N_{UEs}^m	N	$\varphi_{hotspot}$
4a	30 or 60	1	1/15
		2	2/15
		4	4/15
		10	2/3
4b	30 or 60	1	2/3
		2	2/3
		4	2/3

typically deployed in areas with high traffic demands to allow macrocell off-loading and to increase network capacity.

Moreover, it is important to note that some restrictions are usually imposed to UE deployments. According to [3], the minimum distance between a macrocell BS and a UE should be 35 m, while the minimum distance between a small cell BS and a UE should be 10 m (if the small cell BS is outdoors) or 3 m (if the small cell BS is indoors).

Other clustering models with different constrains can be found in the literature. For example, in [26], multiple small cell BSs are deployed per hotspot area, and thus different parameters and constraints are used.

A.10.2 UE Traffic Modeling

Not only the UE location, but also the UE traffic will impact system-level simulation results. The packet rate and burstiness of UE traffic are the two major factors used to classify various traffic models. In [3], three basic traffic models are defined: full buffer model, file transfer protocol (FTP) model, and voice over IP (VoIP) model, which are widely used within industry and academia.

A.10.2.1 Full Buffer Model The full buffer model is a traffic model in which a UE tries to send and receive as many data as the air interface allows. This model is typically used to test the air interface in the sense of "worst-case scenario," and is useful to maximize the system load or throughput. The full buffer traffic model represents a theoretical maximum of the system, against which other traffic models may be compared for evaluating system performance. However, very few real applications exhibit the behavior of the full buffer traffic model.

A.10.2.2 FTP Model The single-file FTP model is a traffic model for best effort services, in which a UE downloads a single file of size on the order of 0.5 MB to 2 MB. One problem with the single-file FTP model is that many UEs may need to be created over the course of a simulation to have a constant ongoing traffic. Depending on how the system-level simulator is designed and implemented, this might lead to increased memory requirements. In [3], the single-file FTP model is described in detail and an alternative model is also presented to deal with this issue. In the alternative model, a UE downloads several files, where the time between downloads, the reading time, follows an exponential distribution with mean 5 s, In both models, the load of the network can also be adjusted by varying the number of UEs.

A.10.2.3 VoIP Model The VoIP model is a traffic model for real-time voice services, in which UEs take turns in sending data at a fixed rate, corresponding to the bit rate of a voice codec. A two-state Markov model can be used for modeling such VoIP service [27]. The length of each talk spurt is typically exponentially distributed, and the load of the system can be adjusted by varying the number of UEs, similar to the FTP model. The exact implementation of a typical VoIP model can be found in [15].

For VoIP services, the delays of packets decides whether a UE would be satisfied or incur outage. A UE is deemed satisfied if the radio interface delay is 50 ms or less for at least 98% of the packets. The voice capacity is typically defined as the number of UEs in the cell, when more than 95% of them are satisfied.

A.11 MOBILITY MODELING

The UE mobility models attempt to capture UE movement within the scenario for mobility performance assessment. In the following, the most common UE mobility models in system-level simulations are described, which can be classed as random walk and environment aware models.

A.11.1 Random Walk

In the random walk model, the environment is not considered and the UE freely moves within the scenario. In [4], two random walks models are proposed. Note that in both cases, 3, 30, 60, or 120 km/h are the selected UE velocities.

A.11.1.1 Hotspot simulations In this case, in order to analyze mobility performance when moving from a macrocell to a small cell and vice versa, and speed up simulations, UEs only move in a confined area around the selected small cell BS. This confined area is defined by a circle with a given radius around the small cell BS (e.g., 100 m). Note that the defined area has a larger radius than the UE hotspot dropping area, for example, 100 m versus 40 m. The UE should start its connection with the macrocell to simulate a macrocell to small cell HO, and finish it again with the macrocell to simulate a small cell to macrocell HO. In this model, first, UEs are placed randomly at the perimeter of the defined boundary. and then UEs move toward the small cell BS along a straight line in a randomly selected direction with a maximum deviation angle (e.g., 45 degrees) with respect to the imaginary line that joins the initial UE location and the small cell BS location. When the UE reaches the perimeter of the defined boundary, it bounces back toward the small cell BS in another randomly selected direction, again with the selected maximum deviation angle with respect to the imaginary line that joins the current UE location and the small cell BS location.

A.11.1.2 Large area simulations In this case, in order to analyze mobility performance in a more general heterogeneous network (HetNet), UEs are uniformly distributed over the entire simulation scenario, each with an independent random direction of motion. Then, UEs move along a straight line in its independent selected direction of motion, and each UE connects with its best server at each time.

When UEs hit the border of the simulation scenario, they bounce back inward to the simulation scenario and move along newly randomly selected directions. The border of the scenario is defined by a bouncing back ring with a 1.8 D_{ISD} radius.

A.11.2   Environment Aware Mobility Modeling

In scenarios where environmental obstructions such as walls, rivers, and lakes are modeled, mobility models that are aware of these obstructions are required. Here, two different approaches are used.

A.11.2.1 Obstacle Avoidance for Outdoor Simulations   In this case, the outdoor mobility model makes use of an environment map. The environment map is a two-dimensional matrix that describes the location of obstacles where UEs are not able to move through, according to a certain geographical resolution. The environment map shown in Fig. A.9 can be used here to describe the location of buildings, although additional information such as the location of rivers and lakes may be added. The resulting environment map matrix is therefore composed of elements with values indicating the locations of valid areas where UEs can move into and invalid areas where UEs cannot move into. The mobility model initializes the location of a UE first by randomly selecting one out of the four possible entry edges of the UE into the map, that is, from the north, south, west, or east. Once the direction of entry is chosen, the UE is randomly placed at a valid area at the chosen entry edge. The final step of the initialization is to set the initial direction of movement where the UE will move toward, which in this case is the opposite side of the entry edge. For example, if the entry edge is on the western side of the map, then the initial direction of movement will be set directly to the east.

The mobility model will then start determining the direction in which the UE will move in the next simulation time iteration. This is done by calculating the distances between the UE and the nearest obstacles that lie within a defined distance, and the angular range centered on the direction the UE was moving in the previous time iteration. In other words, the UE has a "field of vision" with a distance defining how far the field of vision reaches and an angle defining how wide the field of vision is. The UE will not change its direction if there are no obstacles directly in front of it, within its field of vision. If there is an obstacle detected in its current direction, then the UE will change its direction to a new angle with no obstacle within its range, or if that is not available, the angle that has the highest distance to an obstacle. If there are more than one angle that are candidates for a new direction, then the angle that is closest to the previous direction is chosen. The UE is then moved to its new location in the selected direction according to the set UE speed. This process of determining the best UE direction and moving the UE is repeated until the UE reaches an edge of the map, whereupon the UE is deemed to have left the simulated area and is removed from the simulation. Care should be taken when defining how wide the field of vision is, as defining one that is too narrow can result in UEs getting stuck in building corners and

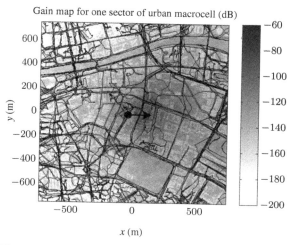

Gain map for one sector of urban macrocell (dB)

x (m)

Figure A.17. Generated routes over the city of Dublin.

one that is too wide can result in very erratic, uncorrelated movements. Figure A.17 shows an example of the routes generated over the Dublin city map presented in Fig. A.9. The UE routes shown were generated using a field of vision with a scan distance of 400 m and 120 degrees wide, with a scan resolution of 6 degrees.

A.11.2.2 Waypoint-Based Modeling In this case, the mobility model makes use of a set of predefined waypoints in the map. This approach is typically used for simulating indoor scenarios. Two types of waypoints are defined: waiting waypoints and routing waypoints. Waiting waypoints are locations where UEs will spend a certain amount of time. The amount of time that a UE spends at a waiting waypoint can be determined according to a distribution. Each waiting waypoint may be set to have a longer mean waiting time than others. For example, a waiting waypoint located in the bathroom of a house would have a much lower mean waiting time than one located in the bedroom. Routing waypoints are used to define the path the UEs take when moving from one waiting waypoint to another. Routing waypoints therefore need to be placed such that the UE will not walk through obstacles such as walls when moving between neighboring routing waypoints.

Once these waypoints have been defined, the locations of the UEs are initialized by placing them at waiting waypoints, either randomly or according to a certain distribution. The UE then remains static at the waiting waypoint for the set amount of time. Once the waiting time of the UE has expired, a new waiting waypoint is selected according to a defined probability. The route between the old waiting waypoint and the newly selected waiting waypoint is then calculated. This route is defined by the sequence of routing waypoints between the old and new waiting waypoint. The UE is

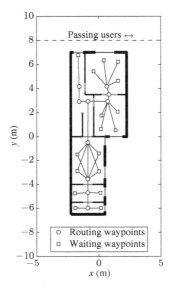

Figure A.18. Waypoints for mobility model in residential scenario.

then incrementally moved toward the first routing waypoint in a straight line, according to the UE speed and simulation time resolution. Once the UE has reached the first routing waypoint, it is then moved to subsequent routing waypoints until it arrives at the destination waiting waypoint. The UE then waits at the new waiting waypoint and the process is repeated.

Figure A.18 shows the waypoints used in a house for modeling a residential scenario, where waiting waypoints are placed around six rooms of the house, and routing waypoints connect the waiting waypoints. A UE inside the house will spend a certain amount of time at a waiting waypoint in a room, before moving to another waiting waypoint located in another room. The front of the house is assumed to directly face a public pavement, and therefore UEs pass by the house 1 m away from the front of the house. The UEs, both inside and outside the house, are assumed to move at 4 km/h. In this scenario, the time spent at a room is a normally distributed, with a mean of 1800 s for the lounge and study, 60 s for the corridor, 1200 s for the kitchen, 180 s for the utility room, and 300 s for the toilet. The standard deviation used for the distributions are 10% of the mean. Figure A.19 shows the probability of time spent at each room of the house for a UE in a simulation run over 2 weeks.

Figure A.20 also shows the waypoints used in an office building for modeling an enterprise scenario. Waiting waypoints are placed at various cubicles and offices in the building, and routing waypoints are defined to link the waiting waypoints together. A UE will spend a certain amount of time at a waiting waypoint before moving to another waiting waypoint. The time spent at a waiting waypoint is determined by

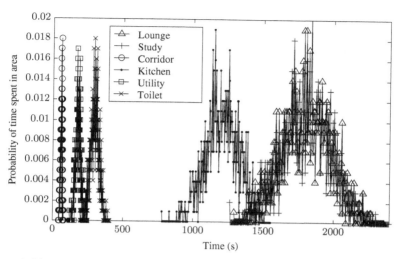

Figure A.19. Probability of the time a UE spends at an area in the residential scenario.

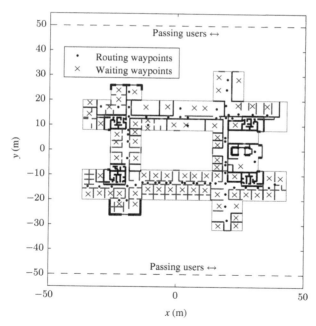

Figure A.20. Waypoints for mobility model in enterprise scenario.

using a normal distribution, with a mean of 1800 s and a standard deviation of 540 s. Two public pavements are assumed to exist 50 m to the north and the south of the center of the office building, and therefore UEs move past the building using these pavements. As with the residential scenario, all UEs are assumed to move at a speed of 4 km/h.

A.12 SUMMARY AND CONCLUSIONS

In this chapter, the basic building blocks of a typical system-level simulator have been reviewed. Models for computing the total channel gain between a transmitter and a receiver based on antenna gain, path loss, environment loss, shadow fading, and multi-path fading have been described. Moreover, models for assessing the received signal strength and quality at the receiver side based on such total gains have been presented. Models for calculating UE throughput under different technologies have also been shown, and different models for the deployment of BSs and UE, which may have a significant impact on the throughput distribution, have been described. UE traffic and mobility models have also been discussed.

REFERENCES

1. "Physical layer aspects for Evolved Universal Terrestrial Radio Access (E-UTRA)," 3GPP Technical Specification, Tech. Rep. TR 25.814 v 7.1.0.

2. X. Chu, D. López-Pérez, F. Gunnarsson, and Y. Yang, *Heterogeneous Cellular Networks: Theory, Simulation and Deployment*. Cambridge University Press, Jul. 2003.

3. "Further advancements for E-UTRA physical layer aspects," 3GPP Technical Specification, Evolved Universal Terrestrial Radio Access (E-UTRA), Tech. Rep. TR 36.814 v 9.0.0.

4. "Small cell enhancements for E-UTRA and E-UTRAN—physical layer aspects," 3GPP Technical Specification, Tech. Rep. TR 36.839 v 12.1.0.

5. G. Fischer, F. Pivit, and W. Wiesbeck, "EISL, the pendant to EIRP: A measure for the receive performance of base stations at the air interface," in *Proc. European Microwave Week*, Milan, Italy, Sep. 2002, pp. 1–4.

6. J. S. Seybold, *Introduction to RF Propagation*. John Wiley & Sons, Sep. 2005.

7. S. Y. Seidel and T. S. Rappaport, "A ray tracing technique to predict path loss and delay spread inside buildings," in *IEEE Globecom*, Dec. 1992, pp. 649–653.

8. S. J. Fortune, D. Gray, B. Keminghan, O. Landron, R. Valenzuela, and M. Wright, "WISE design of indoor wireless systems," *IEEE Comput. Sci. Eng.*, vol. 2, no. 1, pp. 58–68, 1995.

9. A. Goldsmith, *Wireless Communications*. Cambridge University Press, Sep. 2005.

10. M. Gudmundson, "Correlation model for shadow fading in mobile radio systems," *Electron. Lett.*, vol. 27, no. 23, pp. 2145–2146, Oct. 1991.

11. "Correlation models for shadow fading simulation," IEEE Tech. Rep. S802.16m-07/060, Mar. 2007.

12. H. Claussen, "Efficient modelling of channel maps with correlated shadow fading in mobile radio systems," in *Proc. IEEE Int. Symp. Pers., Indoor Mobile Radio Commun. (PIMRC)*, Sep. 2005, pp. 512–516.

13. "Guidelines for evaluation of radio transmission technologies for IMT-2000," ITU Tech. Rep. ITU-R M.1225 1, 1999.

14. "Technical specification group GSM/EDGE radio access network: Radio transmission and reception," 3GPP Technical Specification, Tech. Rep. TS 05.05 v 8.20.0.

15. "NGMN radio access performance evaluation methodology," NGMN Alliance, White Paper , Tech. Rep., Jan. 2008.

16. J. Proakis and M. Salehi, *Digital Communications*. McGraw Hill, Nov. 2007.

17. S. Ahmadi, *LTE-Advanced: A Practical Systems Approach to Understanding 3GPP LTE Releases 10 and 11 Radio Access Technologies*. Academic Press, Oct. 2013.

18. E. Tuomaala and H. Wang, "Effective SINR approach of link to system mapping in OFDM/multi-carrier mobile network," in *Proc. 2nd Int. Conf. Mobile Technol., Appl. Syst.*, Guangzhou, China, Nov. 2005.

19. "Feasibility study for OFDM for UTRAN enhancement," 3GPP Technical Specification, Tech. Rep. TR 25.892 v 6.0.0.

20. Fernando Andrés-Quiroga, "Link-to-system interfaces for system level simulations featuring hybrid ARQ," Master thesis, Technische Universitat Munchen, Munich, Germany, Nov. 2008.

21. C. E. Shannon, "Communication in the presence of noise," in *Proc. Inst. Radio Eng.*, vol. 37, no. 1, Jan. 1949, pp. 10–21.

22. T. Bu, L. Li, and R. Ramjee, "Generalized proportional fair scheduling in third generation wireless data networks," in *Proc. IEEE Int. Conf. Comput. Commun. (INFOCOM)*, Apr. 2006, pp. 1–12.

23. E. Liu and K. Leung, "Proportional fair scheduling: Analytical insight under rayleigh fading environment," in *Proc. IEEE Wireless Commun. Netw. Conf. (WCNC)*, Mar. 2008, pp. 1883–1888.

24. G. Mongha, K. Pedersen, I. Kovacs, and P. Mogensen, "QoS oriented time and frequency domain packet schedulers for the UTRAN long term evolution," in *Proc. IEEE Veh. Technol. Conf. (VTC)*, May 2008, pp. 2532–2536.

25. A. Jalali, R. Padovani, and R. Pankaj, "Data throughput of CDMA-HDR a high efficiency-high data rate personal communication wireless system," in *Proc. IEEE Veh. Technol. Conf. (VTC)*, vol. 3, Tokyo, Japan, May 2000, pp. 1854–1858.

26. "Further Advancements for E-UTRA—physical layer aspects," 3GPP Technical Specification, Evolved Universal Terrestrial Radio Access (E-UTRA), Tech. Rep. TR 36.872 v 9.0.0.

27. "Guidelines for evaluation of radio interface technologies for IMT-advanced," ITU Tech. Rep., ITU-R M.2135, Nov. 2008.

INDEX

Small Cell Networks: Deployment, Management, and Optimization, First Edition. Holger Claussen,
David López-Pérez, Lester Ho, Rouzbeh Razavi, and Stepan Kucera.
© 2017 by The Institute of Electrical and Electronic Engineers, Inc. Published 2017 by John Wiley & Sons, Inc.

IEEE Press Series on
Networks and Services Management

The goal of this series is to publish high quality technical reference books and textbooks on network and services management for communications and information technology professional societies, private sector and government organizations as well as research centers and universities around the world. This Series focuses on Fault, Configuration, Accounting, Performance, and Security (FCAPS) management in areas including, but not limited to, telecommunications network and services, technologies and implementations, IP networks and services, and wireless networks and services.

Series Editors:
Thomas Plevyak
Veli Sahin

---

1. *Telecommunications Network Management into the 21st Century*
   Edited by Thomas Plevyak and Salah Aidarous
2. *Telecommunications Network Management: Technologies and Implementations: Techniques, Standards, Technologies, and Applications*
   Edited by Salah Aidarous and Thomas Plevyak
3. *Fundamentals of Telecommunications Network Management*
   Lakshmi G. Raman
4. *Security for Telecommunications Network Management*
   Moshe Rozenblit
5. *Integrated Telecommunications Management Solutions*
   Graham Chen and Qinzheng Kong
6. *Managing IP Networks: Challenges and Opportunities*
   Thomas Plevyak and Salah Aidarous
7. *Next-Generation Telecommunications Networks, Services, and Management*
   Edited by Thomas Plevyak and Veli Sahin
8. *Introduction to IT Address Management*
   Timothy Rooney
9. *IP Address Management: Principles and Practices*
   Timothy Rooney
10. *Telecommunications System Reliability Engineering, Theory, and Practice*
    Mark L. Ayers
11. *IPv6 Deployment and Management*
    Michael Dooley and Timothy Rooney
12. *Security Management of Next Generation Telecommunications Networks and Services*
    Stuart Jacobs
13. *Cable Networks, Services, and Management*
    Mehmet Toy
14. *Cloud Services, Networking, and Management*
    Edited by Nelson L. S. da Fonseca and Raouf Boutaba